Illustrated Guide to the National Electrical Code®

6TH EDITION

Charles R. Miller

Illustrated Guide to the *National Electrical Code*®, 6E
Charles R. Miller

Senior Vice President, GM Skills & Product Planning: Dawn Gerrain

Product Team Manager: James DeVoe

Senior Director Development: Marah Bellegarde

Senior Product Development Manager: Larry Main

Senior Content Developer: John Fisher

Editorial Assistant: Andrew Ouimet

Vice President Marketing Services: Jennifer Baker

Market Manager: Linda Kuper

Senior Production Director: Wendy A. Troeger

Production Manager: Mark Bernard

Senior Content Project Manager: Glenn Castle

Senior Art Director: David Arsenault

Software Development Manager: Joe Pliss

Media Developer: Deborah Bordeaux

Cover Image: © iStock.com/Alex Yeung, © Joseph Villanova. Source taken from Simmons EGB, 4e.

For product information and technology assistance, contact us at
Cengage Learning Customer & Sales Support, 1-800-354-9706

For permission to use material from this text or product, submit all requests online at **www.cengage.com/permissions.**
Further permissions questions can be e-mailed to **permissionrequest@cengage.com**

Library of Congress Control Number: 2013954980

ISBN-13: 978-1-133-94862-9

ISBN-10: 1-133-94862-6

Delmar
5 Maxwell Drive
Clifton Park, NY 12065-2919
USA

Cengage Learning is a leading provider of customized learning solutions with office locations around the globe, including Singapore, the United Kingdom, Australia, Mexico, Brazil, and Japan. Locate your local office at: **international.cengage.com/region**

Cengage Learning products are represented in Canada by Nelson Education, Ltd.

To learn more about Delmar, visit **www.cengage.com/delmar**

Purchase any of our products at your local college store or at our preferred online store **www.cengagebrain.com**

Notice to the Reader

Publisher does not warrant or guarantee any of the products described herein or perform any independent analysis in connection with any of the product information contained herein. Publisher does not assume, and expressly disclaims, any obligation to obtain and include information other than that provided to it by the manufacturer. The reader is expressly warned to consider and adopt all safety precautions that might be indicated by the activities described herein and to avoid all potential hazards. By following the instructions contained herein, the reader willingly assumes all risks in connection with such instructions. The publisher makes no representations or warranties of any kind, including but not limited to, the warranties of fitness for particular purpose or merchantability, nor are any such representations implied with respect to the material set forth herein, and the publisher takes no responsibility with respect to such material. The publisher shall not be liable for any special, consequential, or exemplary damages resulting, in whole or part, from the readers' use of, or reliance upon, this material.

Printed in the United States of America
1 2 3 4 5 6 7 18 17 16 15 14

Contents

Illustrated Guide to the National Electrical Code offers an exciting new approach to understanding and applying the provisions of the *National Electrical Code*.* Unlike the *Code*, this text gathers and presents detailed information in a format, such as one-family or multifamily dwellings, based on "type of occupancy." *Code* specifications applicable to a given type of occupancy are logically organized in easy-to-read units and graphically enhanced by numerous technical illustrations. Going an extra step, the occupancy-specific material is subdivided into specific rooms and areas. Information relevant to more than one type of occupancy is organized into independent units for easier reference. For instance, items such as raceways and conductors are covered in Unit 5 but are related to every type of occupancy.

Students who wish to acquire a comprehensive grasp of all electrical codes will want to study this text section by section and unit by unit. Practicing electricians who have specialized in one type of occupancy and who wish to understand an unfamiliar segment may want to focus on those new areas. For example, an electrician who has been wiring commercial facilities for a number of years wants to wire a new house. Being unfamiliar with the codes concerning residential wiring, this individual can turn to Section 2, "One-Family Dwellings." Here, everything from receptacle placement to the placement of the service point is explained. Section 2 is made up of four units: Units 6 through 9. Unit 6, "General Provisions," contains general requirements for one-family dwellings, both interior and exterior. Unit 7, "Specific Provisions," addresses more complex issues, requiring additional provisions for specific areas such as kitchens, hallways, clothes closets, bathrooms, garages, basements, etc. Unit 8, "Load Calculations," simplifies the standard as well as optional load calculation methods for one-family dwellings. Unit 9, "Services and Electrical Equipment," is divided into five subheadings: Service-Entrance Wiring Methods; Service and Outside Wiring Clearances; Working Space around Equipment; Service Equipment and Panelboards; and Grounding.

The "what," "when," "where" adoption of the provisions of the *NEC* is under the discretionary control of state and local jurisdictions. State and local jurisdictions also have the liberty of appending additional codes, which in many cases may be more stringent than those outlined by the *NEC*. The *Code* may be adopted in whole or in part. For example, while some local codes do not allow the use of nonmetallic-sheathed cable for residential or commercial wiring, others allow its use in residential but not in commercial wiring applications. To ensure compliance, obtain a copy of any additional rules and regulations for your area.

This guide's objective is to provide the information needed to complete your project—without the necessity of learning the *NEC* from cover to cover. *Illustrated Guide to the National Electrical Code* will bring your project to life as quickly and as accurately as any text on the market today. In the electrical field, as in any career, the learning experience never ends. Whether you are an electrician's apprentice, a master electrician, or an electrical inspector, *Illustrated Guide to the National Electrical Code* has something for you. We believe you will find it to be a valuable addition to your reference library. In fact, you may want to include it in your toolbox or briefcase!

Take note that this guidebook was completed after all the normal steps in the National Fire Protection Association (NFPA) 70 review cycle—Proposals to Code-Making Panels, review by

---

National Electrical Code and *NEC*® are registered trademarks of the National Fire Protection Association, Inc., Quincy, MA 02169.

Technical Correlating Committee, Report on Proposals, Comments to Code-Making Panels, review by Technical Correlating Committee, Report on Comments, NFPA Annual Meeting, and ANSI Standards Council—and before the actual publication of the 2014 edition of the *NEC*. Every effort has been made to be technically correct, but there is always the possibility of typographical errors or appeals made to the NFPA Board of Directors after the normal review cycle that could change the appearance or substance of the *Code*.

If changes do occur after the printing of this book, they will be included in the Instructor's Guide and will be incorporated into the guidebook in its next printing.

Note also that the *Code* has a standard method to introduce changes between review cycles, called "Tentative Interim Amendment," or TIA. These TIAs and correction of typographical errors can be downloaded from the NFPA Web site, www.nfpa.org, to make your copy of the *Code* current.

SUPPLEMENTS:

An Instructor Resource CD is available for this text. It contains an Instructor's Guide, unit presentations in PowerPoint, and a computerized test bank. ISBN 978-1-133-95911-3

COURSEMATE

CourseMate complements the text and course content with study and practice materials. Cengage Learning's CourseMate brings course concepts to life with interactive learning, study, and exam preparation tools that support the printed textbook. Watch student comprehension soar as your class works with the printed textbook and the textbook-specific website.

CourseMate includes an integrated MindTap reader; interactive teaching and learning tools, including quizzes, flashcards, and Engagement Tracker, a first-of-its-kind tool that monitors student engagement in the course; and more. CourseMate goes beyond the book to deliver what you need!

To access additional course materials, including CourseMate, please visit www.cengagebrain .com. At the CengageBrain home page, search for the ISBN (from the back cover of the book), using the search box at the top of the page. This will take you to the product page where these resources can be found.

INSTRUCTOR SITE

An Instructor Companion website containing supplementary material is available. This site contains an Instructor Guide, an image gallery of text figures, unit presentations done in PowerPoint, and testing powered by Cognero.

Cengage Learning Testing Powered by Cognero is a flexible, online system that allows you to

- Author, edit, and manage test bank content from multiple Cengage Learning solutions.
- Create multiple test versions in an instant.
- Deliver tests from your LMS, your classroom or wherever you want.

Contact Cengage Learning or your local sales representative to obtain an instructor account.

Accessing an Instructor

Companion Website from
SSO Front Door

1. Go to http://login.cengage.com and log in, using the instructor e-mail address and password.
2. Enter author, title, or ISBN in the **Add a title to your bookshelf** search.
3. Click **Add to my bookshelf** to add instructor resources.
4. At the Product page, click the **Instructor Companion** site link.

ABOUT THE AUTHOR

For eighteen years, Charles R. Miller owned and operated a successful commercial electrical contracting company (Lighthouse Electric Co., Inc.) in Nashville, Tennessee. Throughout those years, he prided himself on solving problems abandoned by less-skilled or less-dedicated technicians. In 1988, he began operating a second company, dedicated to electrical-related training and known as Lighthouse Educational Services. Mr. Miller teaches custom-tailored classes and seminars covering various aspects of the *National Electrical Code* and NFPA 70E. Countless numbers of students have taken advantage of his extensive experience in electrical contracting; regulatory exams (current electrical codes); electrical-related business and law; and electrical safety–related work practices. Class and seminar attendees have included individuals employed by companies such as Ford, Textron, the Aerostructures Corporation, Aladdin Industries, Lorillard Tobacco Company, Smith & Wesson, and McKee Foods; by academic institutions such as Tennessee State University, Vanderbilt University, and Purdue University; and governmental agencies including the National Aeronautics and Space Administration (NASA).

In 1999, Charles started writing and illustrating the "Code In Focus" column in *Electrical Contractor* magazine. His attention-to-detail illustrations and writing style make this one of the top, if not the top, read columns in the monthly magazine. Charles Miller started writing for NFPA in 2003. Titles include *Pocket Guide to Residential Electrical Installations, Pocket Guide to Commercial and Industrial Electrical Installations, NFPA's Electrical References, NFPA's Pocket Electrical References, Electrician's Exam Prep Manual,* and *Ugly's Electrical Safety and NFPA 70E.* Besides teaching, writing, and illustrating, Charles cohosted a home improvement radio talk show in Nashville, Tennessee, for more than three years.

Charles Miller has dedicated over 5000 hours to making *Illustrated Guide to the National Electrical Code* a reality. His unsurpassed attention to detail is evident on every page. Since this book's inception, every day's waking hours have been consumed with careful planning and execution of content and design. His unwavering commitment to quality, from the first page in Unit 1 to the last page in Unit 19, has produced a technically superior, quintessentially user-friendly guide.

Acknowledgments

I would like to say "thank you" to my children, Christin and Adam, for being patient and understanding during the extremely long hours and endless days working on this text. My mother, Evelyn Miller, gets a special "thank you" and "I love you" for a lifetime of support and encouragement. She called every day to check on me and quite often sent encouraging greeting cards that always came at just the right time. "Thank you" to my wife, Linda, for all your love and support as I spend long hours writing and illustrating.

Thank you to the Senior Content Developer at Cengage Learning, John Fisher, for the privilege of writing for such a professional publishing company. I also would like to thank the entire Cengage Learning project team, comprising all those listed on the copyright page at the front of this book.

Last, but not least, the author and Cengage Learning would like to thank the following reviewers for their contributions:

Gary Daggett
Master Electrician

Kevin Weigman
Northeastern Wisconsin Technical College

Marvin Moak
Hinds Community College
Jackson, Missourri

Tom Collins
Gateway Community and Technical College
Florence, KY

Fred Copy
Northeast State Community College
Blountville, TN

Leland Floren
Ridgewater College
Wilmar, MN

Jim Richardson
Lee College
Baytown, TX

UNIT 1

Introduction to the National Electrical Code®

Objectives

After studying this unit, the student should:

▶ be able to give a brief account of electricity in its infancy.

▶ be able to identify the catalyst that brought about the *National Electric Code (NEC)*.

▶ understand how the *NEC* began and its purpose.

▶ understand how changes to the *Code* evolve.

▶ be familiar with the terminology, presentation, and format of the *NEC*.

▶ know what type of information is found in the *NEC* (its layout).

▶ understand the *NEC*'s concern with equipment and material standards.

▶ be able to recognize various trademark logos that denote listed and labeled products.

▶ comprehend the role of nationally recognized testing laboratories (NRTL) and the National Electrical Manufacturers Association (NEMA) as well as the expanded role of the National Fire Protection Association (NFPA).

▶ be familiar with this book's layout, text conventions, and illustration methods.

▶ be advised on how to study the *Illustrated Guide to the NEC*.

▶ be aware that electrical requirements in addition to the *NEC* may exist, and if so, that compliance is required.

THE *NATIONAL ELECTRICAL CODE*

Just as an extensive education is required for doctors to perform the duties of their chosen field, a working knowledge of the *NEC* is a necessity for anyone practicing a profession in the electrical industry. The *NEC* provides the standards by which all electrical installations are judged. Although other requirements, such as local ordinances and manufacturer instructions, must be applied, the *NEC* is the foundation on which successful installations are built. It is the most widely recognized and used compilation of technical rules for the installation and operation of electrical systems in the world today. Because of its widespread effect on the industry, it is important to understand the history of the *NEC*.

The Beginning

In 1882, New York City was home to the first central-station electric generating plant developed by Thomas A. Edison. The Pearl Street Station began operation at 3:00 P.M. on Monday, September 4. Fifty-nine customers had reluctantly consented to have their houses wired on the promise of three free months of electric light. They were given the option of discarding the service if it proved to be unsatisfactory. But this new way of lighting was more than satisfactory . . . it was a sensation. The number of customers tripled in only four months. And, as they say, the rest is history. The new industry swept the nation: New construction included the installation of electricity, and property owners demanded that existing structures be updated as well. New materials and equipment were developed and manufactured, and methods for installing and connecting these items to the electrical source were devised. For more than a decade, manufacturers, architects, engineers, inventors, electricians, and others worked independently to develop their contributions to the new technology. By 1895, there were as many as five different electric installation codes in use, and no single set of codes was accepted by all. To further complicate matters, there was an unexpected hazard darkening the prospects of this new industry.

Purpose and History of the *NEC*

Electrically caused fires were becoming commonplace and, by 1897, the problem was reaching epidemic proportions. A diverse group of knowledgeable, concerned individuals assembled to address this critical issue. The need for standardization was apparent. The consensus of more than 1200 individuals produced the first set of nationally adopted rules to govern electrical installations and operations—the *National Electrical Code*.

The *NEC* states its purpose as . . . *the practical safeguarding of persons and property from hazards arising from the use of electricity*. This objective has remained constant throughout the *NEC*'s existence, and the principles it contains continue to grow and change with the dynamic electrical industry.

Code Changes

The *NEC* is regularly revised to reflect the evolution of products, materials, and installation techniques. Since 1911, the National Fire Protection Association (NFPA) of Quincy, Massachusetts, has been responsible for the maintenance and publication of the *NEC*. The 2014 edition, which contains hundreds of reworded, as well as new, regulations, represents the diligent work of nineteen code-making panels and the technical correlating committee, composed strictly of volunteers from all professions within the electrical industry.

These panels are complemented by a host of private individuals who submit proposals or comment on proposals already submitted for changes to the *NEC*. Anyone who wishes to participate can contact the National Fire Protection Association, 1 Batterymarch Park, Quincy, MA 02169-7471, and request a free booklet, "The NFPA Standards-Making System." The current edition of the *NEC* provides information in the back of the book for submitting public inputs and public comments for changes to the next edition, a copy of which is reproduced on the next page for your reference.

Now let us examine what is inside the *NEC* and how we can go about understanding it.

The Public Input Closing Date for NFPA 70, National Electrical Code is:
October 3, 2014 for Paper Submittals
November 7, 2014 for Online Submittal (ePI)
For the most up-to-date schedule go to the NFPA website at www.nfpa.org/70next

NFPA has launched a comprehensive set of revisions to its current Regulations Governing Committee Projects, the Regulations Governing NFPA's Standards Development Process. These new regulations, which include changes to some familiar terms and add some new terms, will be in effect for standards reporting in the Fall 2013 Revision Cycle and all subsequent revision cycles. NFPA's intent is to take advantage of web-based technology and to make its standards development process more convenient, efficient, and easy for participants to use.

The development of new or revised NFPA Codes, Standards, Guides, or Recommended Practices (NFPA Standards) will continue to take place in two principal stages. Under the current regulations, those stages are known as the "Proposal Stage" and the "Comment Stage". Under the new regulations, the "Proposal Stage" has been renamed the "Input Stage"; the "Comment Stage" will operate much like the "Comment Stage" in the current regulations.

A revision cycle begins with a call for the public to submit proposed revisions ("Public Input", formerly called "Public Proposals").

Public Input Stage
NFPA accepts Public Input on documents via our online electronic submission system. To use the electronic submission system:

- To submit a Public Input to NFPA 70 go the NFPA website at www.nfpa.org/70next.
- Choose the link "The next edition of this standard is now open for Public Input". You will be asked to sign-in or create a free online account with NFPA before using this system.
- Follow the online instructions to submit your Public Input.
- Once a Public Input is saved or submitted in the system, it can be located on the "My Profile" page by selecting "My Public Input/Comments" in the left navigation bar.

Comment Stage
NFPA accepts Public Comments on NFPA documents via our online electronic submission system. To use the electronic submission system:

- To submit a Public Comment to NFPA 70 go to the NFPA website at www.nfpa.org/70next.
- Access the First Draft Report for use as background in the submission of comments.
- Choose the link "The next edition of this standard is now open for Public Comment". You will be asked to sign-in or create a free online account with NFPA before using this system.
- Follow the online instructions to submit your Public Comment.
- Once a Public Comment is saved or submitted in the system, it can be located on the "My Profile" page by selecting "My Public Input/Comments" in the left navigation bar.

For further instructions please go to www.nfpa.org/submitpipc

NEC Terminology, Presentation, and Format

Tables present a requirement's multiple application possibilities.

Table 210.21(B)(3) Receptacle Ratings for Various Size Circuits

Circuit Rating (Amperes)	Receptacle Rating (Amperes)
15	Not over 15
20	15 or 20
30	30
40	40 or 50
50	50

Diagrams, or figures, are used to further clarify *NEC* applications.

Receptacles   Caps

125 V, 20 A, 2-pole, 3-wire, grounding type

125 V, 15 A, 2-pole, 3-wire, grounding type

20 A, 125 V, 2-pole, 3-wire grounding type

30 A, 125 V, 2-pole, 3-wire, grounding type

50 A, 125/250 V, 3-pole, 4-wire, grounding type

Figure 551.46(C)(1) Configurations for grounding-type receptacles and attachment plug caps used for recreational vehicle supply cords and recreational vehicle lots.

© Cengage Learning 2012

Dictionary-style header—The left header shows the first section referenced, and the right header shows the last section referenced.

Informational Notes contain explanatory material such as references to other standards, references to related sections of the *Code*, or information related to a *Code* rule. These are informational only and do not require compliance ≫ *90.5(C)* ≪.

Exceptions appear in *italics* and explain when and where a specific rule does not apply.

Sections are numerical listings where the *Code* requirements are located.

Parts (subheadings) are used to break down articles into simpler topics. (Not all articles have subheadings.)

Not Shown

- **Bullets** (solid black circles) indicate areas where one or more complete paragraphs have been deleted since the last edition.

NFPA document number followed by a page number.

CAUTION

Be advised that the local authority having jurisdiction has the ability to amend the *Code* requirements. Consult the proper authority to obtain applicable guidelines.

Highlighted text within sections indicates changes, other than editorial, since the last *NEC* edition. Vertical lines are placed in outside margins to identify large blocks of changed or new text and for new tables and changed or new figures.

Normal black letters are used for basic *Code* definitions and explanations.

Permissive rules contain the phrases "shall be permitted" or "shall not be required." These phrases normally describe options or alternative methods. Compliance is discretionary ≫ *90.5(B)* ≪.

Mandatory rules use the terms "shall" or "shall not" and require compliance ≫ *90.5(A)* ≪.

110.24   **ARTICLE 110—REQUIREMENTS FOR ELECTRICAL INSTALLATIONS**

110.24 Available Fault Current.

(A) Field Marking. Service equipment in other than dwelling units shall be legibly marked in the field with the maximum available fault current. The field marking(s) shall include the date the fault-current calculation was performed and be of sufficient durability to withstand the environment involved.

Informational Note: The available fault-current marking(s) addressed in 110.24 is related to required short-circuit current ratings of equipment. *NFPA 70E*-2012, *Standard for Electrical Safety in the Workplace*, provides assistance in determining the severity of potential exposure, planning safe work practices, and selecting personal protective equipment.

(B) Modifications. When modifications to the electrical installation occur that affect the maximum available fault current at the service, the maximum available fault current shall be verified or recalculated as necessary to ensure the service equipment ratings are sufficient for the maximum available fault current at the line terminals of the equipment. The required field marking(s) in 110.24(A) shall be adjusted to reflect the new level of maximum available fault current.

Exception: The field marking requirements in 110.24(A) and 110.24(B) shall not be required in industrial installations where conditions of maintenance and supervision ensure that only qualified persons service the equipment.

110.25 Lockable Disconnecting Means. Where a disconnecting means is required to be lockable open elsewhere in this *Code*, it shall be capable of being locked in the open position. The provisions for locking shall remain in place with or without the lock installed.

Exception: Cord-and-plug connection locking provisions shall not be required to remain in place without the lock installed.

II. 600 Volts, Nominal, or Less

110.26 Spaces About Electrical Equipment. Access and working space shall be provided and maintained about all electrical equipment to permit ready and safe operation and maintenance of such equipment.

(A) Working Space. Working space for equipment operating at 600 volts, nominal, or less to ground and likely to require examination, adjustment, servicing, or maintenance while energized shall comply with the dimensions of 110.26(A)(1), (A)(2), and (A)(3) or as required or permitted elsewhere in this *Code*.

(1) Depth of Working Space. The depth of the working space in the direction of live parts shall not be less than that

70–40

Formal Interpretations

Section 90.6 states: *To promote uniformity of interpretation and application of the provisions of this Code, formal interpretation procedures have been established and are found in the NFPA Regulations Governing Committee Projects.* (The NFPA Regulations Governing Committee Projects are in the NFPA Directory. Contact NFPA for a copy of this annual publication.)

The *NEC* Layout

The table of contents in the *NEC* provides a breakdown of the information found in the book. Chapters 1 through 4 contain the most-often used articles in the *Code,* because they include general, or basic, provisions. Chapter 1, while relatively brief, includes definitions essential to the proper application of the *NEC.* It also includes an introduction and a variety of general requirements for electrical installations. More general requirements are found in Chapters 2, 3, and 4, addressing Wiring and Protection, Wiring Methods and Materials, and Equipment for General Use. Special issues are covered in Chapters 5 through 7. Chapter 5 contains information on Special Occupancies; Chapter 6, Special Equipment; and Chapter 7, Special Conditions. The contents of these chapters are applied in addition to the general rules given in earlier chapters. Chapter 8 covers Communications Systems and is basically independent of other chapters, except where cross-references are given. The final chapter, Chapter 9, contains Tables and Examples. Each chapter contains one or more articles, and each article contains sections. Sections may be further subdivided by the use of lettered or numbered paragraphs. The *Code* is completed by Annexes A through J, an index, and a proposal form.

WIRING SYSTEM PRODUCT STANDARDS

In addition to installation rules, the *NEC* is concerned with the type and quality of electrical wiring system materials. Two terms are synonymous with acceptability in this area: **labeled** and **listed.** Their definitions, found in Article 100, are very similar. Similarities within these definitions include: (1) an organization that is responsible for providing the listing or labeling, (2) that these organizations must be acceptable to the authority having jurisdiction, (3) that both are concerned with the evaluation of products, and (4) that both maintain periodic inspection of the production (or manufacturing) of the equipment or materials which have been listed or labeled. A manufacturer of labeled equipment (or material) must continue to comply with the appropriate standards (or performance) under which the labeling was granted. "Listed" also means that the equipment, materials, or services meet appropriate designated standards or have been tested and found suitable for a specified purpose. This information is compiled and published by the organization. The Informational Note under "Listed" states that each organization may have different means for identifying listed equipment. In fact, some do not recognize equipment as listed unless it is also labeled. Listed or labeled equipment must be installed and used as instructed ≫ *110.3(B)* ≪ .

The organizations described in the following directly affect the *Code* as it relates to equipment and material acceptability and play a role in developing and maintaining the standards set forth in the *NEC.*

Nationally Recognized Testing Laboratories

Prior to 1989, there were only two organizations perceived as capable of providing safety certification of products that would be used nationwide. Because there were only two, innovative technology was slow to be tested and approved. When Congress created the Occupational Safety and Health Administration (OSHA) in the early 1970s, OSHA was directed to establish safety regulations for the workplace and for the monitoring of those regulations. OSHA adopted an explanation from the *NEC* and included it in the *Code of Federal Regulations.* In part, it reads: "an installation or equipment is acceptable to the Assistant Secretary of Labor . . . if it is acceptable or certified, or listed, or labeled, or otherwise determined to be safe by a nationally recognized testing laboratory. . . ."

Testing by a nationally recognized testing laboratory (NRTL) was specified in the Code of Federal Regulations, but requirements for becoming an NRTL had not yet been identified. Although OSHA introduced "Accreditation of Testing Laboratories" in 1973, the process through which a laboratory would receive accreditation was still missing. Cooperative efforts produced the OSHA regulation finalized in 1988, and called "OSHA Recognition Process for Nationally Recognized Testing Laboratories."

OSHA's NRTL program greatly benefits manufacturers by providing a system that certifies that a product meets national safety standards. Just as important, the door was opened for a greater number of laboratories to provide certification, and manufacturers are now better able to meet the demands of today's highly competitive market.

The aim of NRTLs is to ensure that electrical products properly safeguard against reasonable, identifiable risks. An extensive network of field personnel conduct unannounced inspections at manufacturing facilities that use the laboratory's "seal of approval." Some of the better-known trademarks of testing laboratories are shown below:

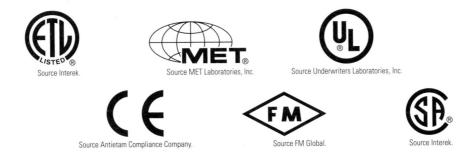

Some of the labels that appear on evaluated and certified electrical products, such as the ones that follow, carry the trademarks of the testing laboratory or the laboratory's standards being used for comparison.

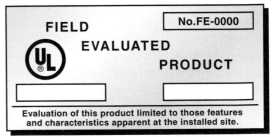

MET Laboratories

MET Laboratories, Inc., working with the Department of Labor as well as other agencies, served as a working example for the accreditation process for independent testing laboratories. In fact, MET became the first U.S. laboratory to successfully complete the process (1989), and thus became the first NRTL licensed by OSHA.

MET field inspectors interact with local electrical inspectors throughout the country to ensure product acceptance by all federal and state regulatory officials. The MET label is accepted by all fifty states, the federal government, and major retailers.

Underwriters Laboratories, Inc.

Prior to the formalization of NRTLs in 1989, electrical product standards were primarily written by Underwriters Laboratories, Inc. (UL), who also performed certification testing. Standards written by UL are still widely used. The appearance of the UL logo on a label indicates that the product complies with the UL standard. It does not mean, necessarily, that UL did the product testing. Although one of many NRTLs, Underwriters Laboratories is perhaps the most widely recognized and respected testing laboratory in operation today. Founded in 1894, UL is a not-for-profit corporation whose mission is to bring safer products to the marketplace and to serve the public through rigorous product safety testing. This organization offers a wide range of services, which include, but are not limited to, product listing, classification, component recognition, field certification, field engineering, facility registration, inspection, fact-finding, and research. As one can see from this list, UL plays a major role in guiding the safety of the electrical industry.

Intertek Testing Services

Select laboratories of Intertek Testing Services (ITS) have passed OSHA's stringent NRTL accreditation procedures and thereby have earned the right to issue product approvals and list products using the familiar ETL listed and CE marks. ITS has been conducting performance and reliability tests to nearly 200 safety standards applicable to workplace-related products since 1896. Intertek's comprehensive program includes testing, listing, labeling, and quarterly follow-up inspections. While recognized internationally by its many listed marks, the ETL listed mark is accepted throughout the United States, by all jurisdictions for electrical products, when denoting compliance with nationally recognized standards such as Wyle Laboratories (WL), International Electrotechnical Commission (IEC), UL, Canadian Standards Association (CSA), and FM Approvals (FM).

National Electrical Manufacturers Association

Founded in 1926, National Electrical Manufacturers Association (NEMA) comprises companies that manufacture equipment for all facets of electrical application, from generation through utilization. Its expansive objectives include product quality maintenance and improvement, safety standards for product manufacture and usage, and a variety of product standards, such as ratings and performance. NEMA contributes to the development of the *National Electrical Safety Code* as well as the *NEC*.

National Fire Protection Association

The NFPA, more than a century old, dedicates itself to safety standards, gathering statistical data, conducting research, providing crucial information on fire protection, prevention, and suppression methods, and much more. Boasting an internationally diverse membership of more than 75,000, this leading nonprofit organization publishes over 300 widely recognized consensus codes and standards, including the *NEC*. In addition, the NFPA is involved in training and education. Its primary pursuit is to protect lives and property from the often catastrophic hazards of fire.

THIS BOOK

The *Illustrated Guide to the NEC* is designed to teach through visualization. If a picture is truly worth a thousand words, this book should provide a more in-depth look at the *National Electrical Code* than can be found in any other single publication. Its highly detailed illustrations are complemented with concise, easy-to-understand written information. Not intended as a how-to book, the *Illustrated Guide to the NEC* instead strives to translate difficult material into simpler, straightforward principles. Once the reader understands how the *Code* translates in a specific area, the same techniques can be applied throughout.

Its Layout

Not only is the presentation of material in this text different from others on the market, but the organization of information also offers a new approach. After covering the fundamental provisions in the balance of Unit 1, this text proceeds to address code requirements by type of occupancy. Comprehensive information is given for one-family dwellings, multifamily dwellings, commercial locations, and special occupancies. To accomplish this task, information has been gathered logically from throughout the *Code* book and concentrated in one section, under the appropriate occupancy. Each occupancy type is broken down into its finite components, and each component is thoroughly discussed and illustrated (see table of contents).

Text Conventions

General text is grouped in small areas surrounding an illustration. **Notes** provide additional information considered relevant to the point being discussed. **Cautions** indicate that particular care is needed during application. **Warnings** indicate potential danger and are intended to prevent misunderstanding of a given rule.

Terms *Luminaire* and *Lighting Fixture*

The word *luminaire* is the international term for *lighting fixture*. As defined in Article 100, a luminaire is a complete lighting unit consisting of a light source such as a lamp or lamps, together with the parts designed to position the light source and to connect it to the power supply. It may also include parts to protect the light source, ballast, or distribute the light. A lampholder itself is not a luminaire. Starting with the 2002 edition, *luminaire* became the main term and *fixture* or *lighting fixture* followed in parentheses. In the 2008 edition, *fixture* and *lighting fixture* were removed and do not follow the term *luminaire*. Throughout this text, *fixture* and *lighting fixture* have also been omitted.

Studying This Text

As the title implies, frequent references are made to the *National Electrical Code*. Keep a copy of the latest edition of the *Code* close at hand. Any confusion about terminology not cleared up by the "Definitions" section of this text may be explained by consulting the *Code's Article 100—Definitions* section. Whenever direct references are made to the *Code*, benefits will be gained by taking the time to read the suggested article or section. The *Illustrated Guide to the NEC* is not intended, in any way, to replace the *Code*. Each unit's "Competency Test" requires a thorough understanding of related *NEC* subject matter. Use of this text alone is insufficient to successfully complete the test. It is, however, intended as an indispensable supplement to the *NEC*.

Note that when comparing calculations made by both the English and metric systems, slight differences will occur due to the conversion method used. These differences are not significant, and calculations for both systems are, therefore, valid.

ADDITIONAL ELECTRICAL REQUIREMENTS

Local Ordinances

The importance of local (state, city, etc.) electrical codes cannot be overemphasized. Local agencies can adopt the *NEC* exactly as written or can amend the *Code* by incorporating more or less stringent regulations. While the *NEC* represents the minimum standards for safety, some jurisdictions have additional restrictions. Obtain a copy of additional requirements (if any) for your area.

Engineers or architects who design electrical systems may also set requirements beyond the provisions of the *NEC*. For example, an engineer might require the installation of 20-ampere circuits in areas where the *NEC* allows 15-ampere circuits. Requirements from engineers or architects are found in additional documents, such as the following.

Plans and Specifications

If plans and specifications are provided for a project by knowledgeable engineers or architects, this information must be considered and, if need be, compared to the requirements set forth by the *NEC*. It is unlikely that the plans or specifications provided by competent professionals will conflict with or contradict the *Code*. Nonetheless, it is best to be diligent in applying the governing principals of the *NEC*.

Manufacturer Instructions

Equipment or material may include instructions from the manufacturer. In accordance with 110.3(B), these instructions must be followed. For example, baseboard heaters generally include installation instructions. The *NEC* does not prohibit the installation of receptacle outlets above baseboard heaters, but the manufacturer's instructions may prohibit the installation of its heater below receptacles.

CONCLUSION

While this unit briefly discusses the history of the *National Electrical Code*, it is not possible to do justice to the importance of the *Code* in a few short pages. With only a glimpse into its history and present-day supporting structure, this text moves on to the task of understanding the contents of the *Code*. The *Illustrated Guide to the NEC* presents visually stimulating information in an occupancy-organized, concise format. To begin the journey through the 2014 edition of the *National Electrical Code*, simply turn the page, read, look, and understand.

Definitions

Objectives

After studying this unit, the student should:

▶ understand the meaning of the term **accessible** (1) when applied to wiring methods and (2) when applied to equipment.

▶ be able to identify accessible equipment that is not readily accessible.

▶ be able to accurately evaluate a location as accessible, readily accessible, or not readily accessible.

▶ be able to identify equipment classified as appliances.

▶ be familiar with the four categories of branch circuits and be able to list their differences.

▶ be able to distinguish the difference between the terms **enclosed** and **guarded.**

▶ be able to determine whether a load is continuous or noncontinuous.

▶ know the difference between branch-circuit conductors and feeder conductors.

▶ understand the terminology associated with grounded and grounding.

▶ know the maximum distance permitted for **within sight** situations.

▶ be able to give examples of damp, dry, and wet locations.

▶ be able to determine which conductors are neutral conductors.

▶ comprehend the electrical vocabulary associated with the word **service.**

▶ be familiar with what constitutes a separately derived system.

▶ understand that the authority having jurisdiction (AHJ) could provide special permission, which is defined as **written consent.**

Introduction

What is the difference between accessible and readily accessible? Which is appropriate in a given application? When sizing a branch circuit or feeder, is the electrical load considered continuous or noncontinuous? What is the difference between a damp location and a wet location?

These and many other questions can be accurately answered only through a thorough understanding of *National Electrical Code (NEC)* terminology. Knowing the correct definition of words and phrases as found in Article 100 is crucial to installing a hazard-free electrical system. Article 100 does not include commonly defined general or technical terms from related codes and standards. Normally, only terms used in two or more articles are defined in Article 100. Other terms are defined within the article in which they appear but may be referenced in Article 100. Part 1 of Article 100 contains definitions to be applied wherever the terms are used throughout the *NEC*. Part 2 contains definitions applicable only to the parts of articles specifically covering installations and equipment operating at over 600 volts, nominal.

DEFINITIONS

Accessible (As Applied to Wiring Methods)

Wiring components are considered accessible when (1) access can be gained without damaging the structure or finish of the building or (2) they are not permanently closed in by the structure or finish of the building ≫ *Article 100* ≪.

Ⓐ Conductors in junction boxes behind luminaires are considered accessible if, by removing the luminaire, access to the conductors is available.

Ⓑ Conductors connected to switches and receptacles are accessible by removing the cover-plate and device.

Ⓒ Receptacles behind furniture are accessible because the furniture can be moved.

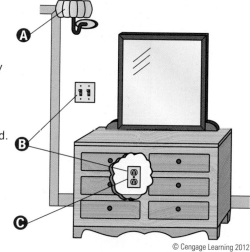

© Cengage Learning 2012

> **NOTE**
>
> Unlike readily accessible, wiring methods meet the definition of being accessible if access to the wiring method can be gained by using tools to remove covers or by climbing on ladders.

Accessible (As Applied to Equipment)

Ⓐ **Accessible** equipment is equipment not guarded by locked doors, elevators, or other effective means ≫ *Article 100* ≪. Equipment installed in locations requiring the use of portable means, such as a ladder, is considered accessible, but not *readily* accessible.

Ⓑ Overcurrent devices do not have to be **readily accessible** if located adjacent to the equipment where access is achieved by the use of portable means ≫ *240.24(A)(4)* ≪.

Accessible, Readily (Readily Accessible)

Readily accessible means capable of being reached quickly (for operation, renewal, or inspections) without having to take actions such as to use tools, to climb over or remove obstacles, or resort to portable ladders, etc. ≫ *Article 100* ≪.

Ⓐ The service disconnecting means must be readily accessible. It may be located either outside or inside, near the entry point of the service conductors ≫ *230.70(A)(1)* ≪.

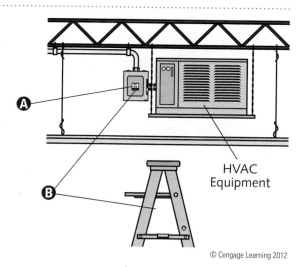

HVAC Equipment

© Cengage Learning 2012

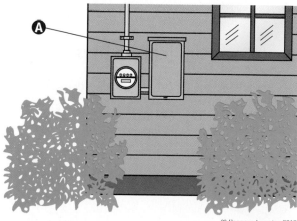

© Cengage Learning 2012

Accessible

Ⓐ Ready accessibility to wiring in luminaires is not required. In most cases, access can be gained through the use of a ladder, scaffolding, etc.

Ⓑ Conductors within junction boxes of recessed luminaires can be accessed by removing part of the luminaire, such as trims, lamps, internal shells, etc.

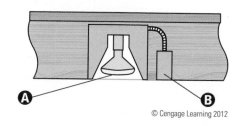

© Cengage Learning 2012

Appliance

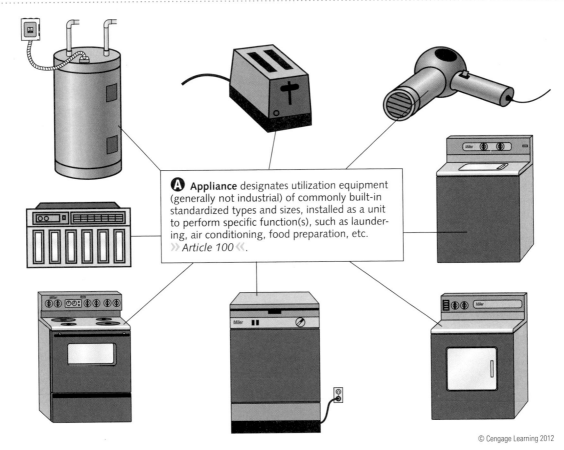

Ⓐ Appliance designates utilization equipment (generally not industrial) of commonly built-in standardized types and sizes, installed as a unit to perform specific function(s), such as laundering, air conditioning, food preparation, etc. ≫ *Article 100* ≪.

© Cengage Learning 2012

Not Readily Accessible

Ⓐ While a ceiling receptacle installed for a garage door opener is not readily accessible, it is accessible. Even though this receptacle is not readily accessible, it must have ground-fault circuit-interrupter (GFCI) protection for personnel ≫ *210.8(A)(2)* ≪.

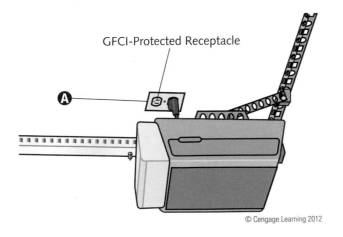

GFCI-Protected Receptacle

Ⓐ

© Cengage Learning 2012

Arc-Fault Circuit Interrupter (AFCI)

A device intended to provide protection from the effects of arc faults by recognizing characteristics unique to arcing and by functioning to de-energize the circuit when an arc fault is detected ≫ *Article 100* ≪.

© Cengage Learning 2012

Attachment Plug (Plug Cap) (Plug)

Ⓐ An attachment plug (plug cap) (plug) is a device that, when inserted into a receptacle, establishes connection between the conductors of the attached flexible cord and the conductors permanently connected to the receptacle ≫ *Article 100* ≪.

© Cengage Learning 2012

Bonding Conductor or Jumper; Supply-Side Bonding Jumper, and Main Bonding Jumper

Ⓐ When metal parts are required to be electrically connected, a reliable conductor (bonding jumper) is installed, thereby guaranteeing the required electrical conductivity ≫*Article 100*≪.

Ⓑ A supply-side bonding jumper is a conductor installed on the supply side of a service or within a service equipment enclosure(s) that ensures the required electrical conductivity between metal parts required to be electrically connected. A supply-side bonding jumper is also a conductor installed for a separately derived system that ensures the required electrical conductivity between metal parts required to be electrically connected.

Ⓒ The main bonding jumper is the connection at the service between the grounded circuit conductor and the equipment grounding conductor ≫*Article 100*≪.

> **N O T E**
>
> Main bonding jumpers must be made of copper or other corrosion-resistant material. A wire, bus, screw, or similar suitable conductor is acceptable as a main bonding jumper ≫*250.28(A)*≪.

Bonded (Bonding)

Bonded is connected to establish electrical continuity and conductivity ≫*Article 100*≪.

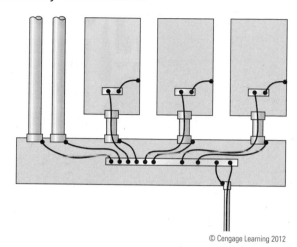

© Cengage Learning 2012

Bonding Jumper, Equipment

Ⓐ An equipment bonding jumper is the connection between two or more portions of the equipment grounding conductor ≫*Article 100*≪.

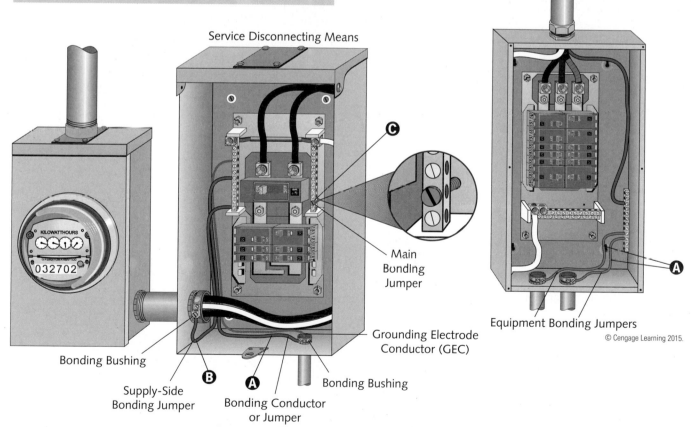

Service Disconnecting Means

Ⓒ — Main Bonding Jumper

Bonding Bushing

Ⓑ Supply-Side Bonding Jumper

Ⓐ Bonding Conductor or Jumper

Grounding Electrode Conductor (GEC)

Bonding Bushing

Equipment Bonding Jumpers — Ⓐ

© Cengage Learning 2015.

© Cengage Learning 2015.

Branch Circuit, Appliance

A An appliance branch circuit supplies energy to one or more outlets for the purpose of connecting appliance(s). These circuits exclude the connection of luminaires unless they are part of the appliance being connected ≫ *Article 100* ≪.

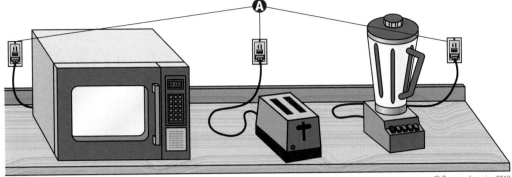

© Cengage Learning 2012

Branch Circuit

A The circuit conductors found between a circuit's final overcurrent protective device (such as the last fuse or circuit breaker) and the circuit's outlet(s) is called a branch circuit ≫ *Article 100* ≪. Branch circuits are divided into four categories: appliance, general purpose, individual, and multiwire.

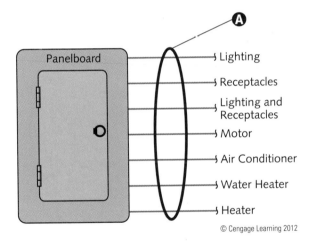

Panelboard

- Lighting
- Receptacles
- Lighting and Receptacles
- Motor
- Air Conditioner
- Water Heater
- Heater

© Cengage Learning 2012

Branch Circuit, Individual

A An individual branch circuit supplies only one piece of utilization equipment ≫ *Article 100* ≪.

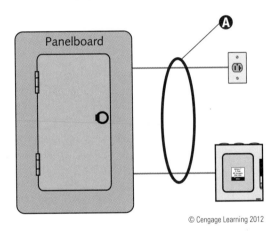

Panelboard

© Cengage Learning 2012

Branch Circuit, General Purpose

A A general purpose branch circuit supplies two or more receptacles or outlets for lighting and appliances ≫ *Article 100* ≪.

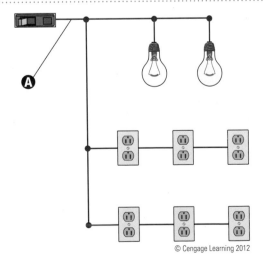

© Cengage Learning 2012

Branch Circuit, Multiwire

Ⓐ A voltmeter will not register a voltage (potential difference) when connected to the same ungrounded (hot) phase. Therefore, a multiwire circuit must consist of conductors connected to different phases.

Ⓑ A multiwire branch circuit consists of two or more ungrounded (hot) conductors that have a voltage between them 》*Article 100* 《.

Ⓒ The one grounded (neutral) conductor of a multiwire circuit must be connected to the neutral or grounded conductor of the system 》*Article 100* 《.

Ⓓ All conductors of a multiwire branch circuit must originate from the same panelboard or similar distribution equipment 》*210.4(A)* 《.

Ⓔ There must be only one grounded (neutral) conductor, and there must be an equal voltage between it and each ungrounded conductor of the circuit 》*Article 100* 《.

> **NOTE**
>
> See 210.4(A) through (D) for additional requirements for multiwire branch circuits.

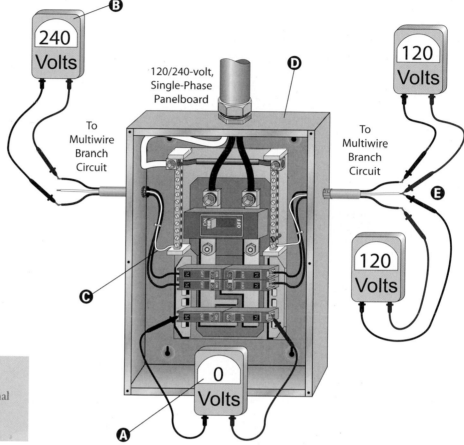

© Cengage Learning 2012

Concealed

Ⓐ Concealed means rendered inaccessible by the structure or finish of the building 》*Article 100* 《.

Ⓑ Conductors in concealed raceways, even though they may become accessible by withdrawing them, are still considered concealed 》*Article 100* 《.

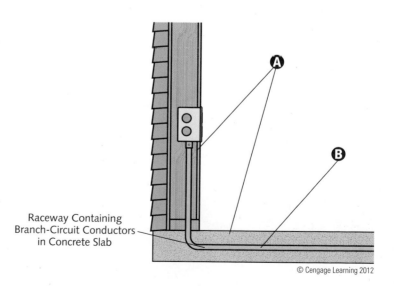

Raceway Containing Branch-Circuit Conductors in Concrete Slab

© Cengage Learning 2012

Receptacles on Multiwire Branch Circuits

A multiwire receptacle circuit consists of one (or more) duplex receptacles, one (or more) multiple receptacles, two (or more) single receptacles, or combinations thereof.

Ⓐ A duplex receptacle can be supplied by two branch circuits by removing the tab. This receptacle is fed from a multiwire branch circuit.

Ⓑ The tab has been removed to allow separate feed of each outlet.

Ⓒ Each multiwire branch circuit must be provided with a means that simultaneously disconnects all ungrounded (hot) conductors at the point where the branch circuit originates ⟫ *210.4(B)* ⟪. This is accomplished through the use of either one double-pole breaker or two single-pole breakers with identified handle ties.

Ⓓ Although the box contains a multiwire branch circuit, only one ungrounded (hot) conductor is feeding this duplex receptacle.

WARNING

In multiwire branch circuits, the continuity of a grounded conductor shall not be dependent on the device ⟫ *300.13(B)* ⟪. If breaking the grounded conductor at the receptacle breaks the circuit down the line, then the grounded conductors must not be connected to the receptacle. Simply splice the grounded conductors and install a jumper wire to the receptacle.

NOTE

Switch-controlled split-wire duplex receptacle(s) are sometimes installed in lieu of a lighting outlet ⟫*210.70(A)(1)* *Exception No. 1*⟪. One half of the duplex receptacle is controlled by a wall switch, while the other half is a typical receptacle. A split-wire receptacle receiving power from a single source (one breaker or one fuse) is not a multiwire receptacle.

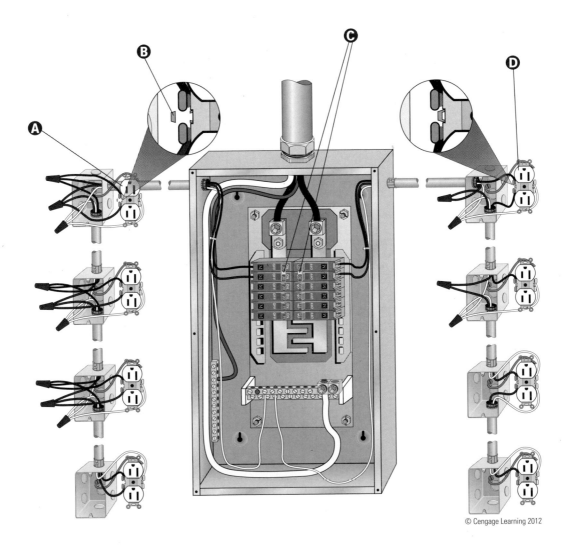

Conduit Body

A A conduit body is a separate portion of a conduit (or tubing) system providing access to the interior of the system through a removable cover(s) at a junction of multiple sections or at a terminal point of the system ≫ *Article 100* ≪ .

B FS, FD, and larger boxes (cast or sheet metal) are not classified as conduit bodies ≫ *Article 100* ≪ .

C A single conduit is not permitted as sole support for an FS-type or weatherproof junction box ≫ *314.23(E) and 314.23(F)* ≪ .

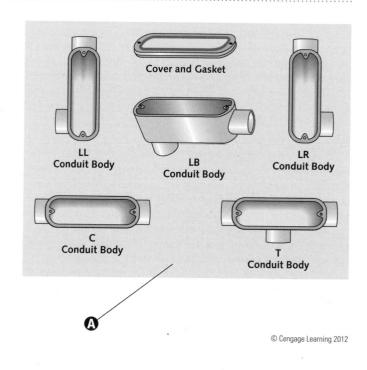

Cover and Gasket

LL Conduit Body

LB Conduit Body

LR Conduit Body

C Conduit Body

T Conduit Body

© Cengage Learning 2012

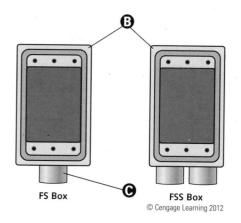

FS Box

FSS Box

© Cengage Learning 2012

Continuous Load

A load having the maximum level of current sustained for three hours or more is referred to as a continuous load ≫ *Article 100* ≪ . Office lighting is an example of a continuous load.

Device

A A unit of an electrical system, other than a conductor, that carries or controls electrical energy as its principal function is known as a device ≫ *Article 100* ≪ .

© Cengage Learning 2012

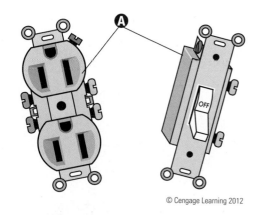

© Cengage Learning 2012

Effective Ground-Fault Current Path

Ⓐ An effective ground-fault current path is an intentionally constructed, low-impedance electrically conductive path designed and intended to carry current under ground-fault conditions from the point of a ground fault on a wiring system to the electrical supply source. The reason for this intentionally constructed, low-impedance electrically conductive path is to facilitate the operation of the overcurrent protective device or ground-fault detector during a ground-fault condition ≫ *Article 100* ≪.

Ⓑ A ground-fault current path is defined as an electrically conductive path from the point of a ground fault on a wiring system through normally non–current-carrying conductors, equipment, or the earth to the electrical supply source. Examples of ground-fault current paths are any combination of equipment grounding conductors, metallic raceways, metallic cable sheaths, electrical equipment, and any other electrically conductive material such as metal, water, and gas piping; steel framing members; stucco mesh; metal ducting; reinforcing steel; shields of communications cables; and the earth itself ≫ *Article 100* ≪.

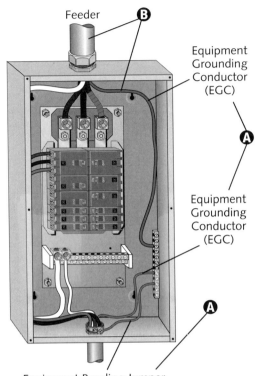

Feeder — **Ⓑ**

Equipment Grounding Conductor (EGC) — **Ⓐ**

Equipment Grounding Conductor (EGC)

Ⓐ

Equipment Bonding Jumper

© Cengage Learning 2015.

> **NOTE**
>
> Electrical equipment and wiring and other electrically conductive material likely to become energized shall be installed in a manner that creates a low-impedance circuit facilitating the operation of the overcurrent device or ground-fault detector. It shall be capable of safely carrying the maximum ground-fault current likely to be imposed on it from any point on the wiring system where a ground fault may occur to the electrical supply source. The earth shall not be considered as an effective ground-fault current path ≫ *250.4(A)(5)* ≪.

Enclosed

Equipment, conductors, etc., surrounded by a case, housing, fence, or walls that prevent persons from accidentally contacting energized parts are referred to as enclosed ≫ *Article 100* ≪.

Ⓐ Panelboards located within cabinets or cutout boxes are considered enclosed.

Ⓑ Electrical equipment installed within the perimeter of a fence, or similar area, qualifies as enclosed.

Ⓒ Conductors are also considered enclosed when installed in panels, wireways, raceways, etc.

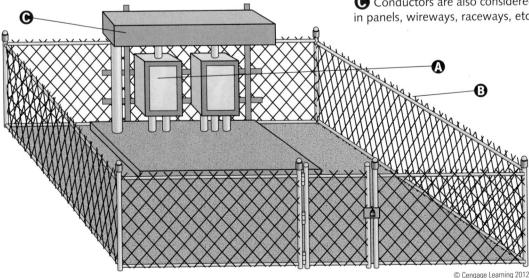

Ⓒ

Ⓐ

Ⓑ

© Cengage Learning 2012

Enclosure

A Any case, housing, apparatus, fence, or walls surrounding an installation, designed to prevent personnel from accidentally contacting energized parts or to protect the equipment from physical damage, serves as an enclosure »*Article 100*«.

Enclosure for a motor starter

© Cengage Learning 2012

Equipment

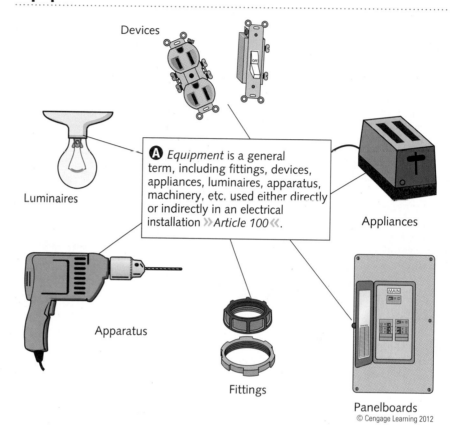

Devices

A *Equipment* is a general term, including fittings, devices, appliances, luminaires, apparatus, machinery, etc. used either directly or indirectly in an electrical installation »*Article 100*«.

Luminaires

Apparatus

Fittings

Appliances

Panelboards

© Cengage Learning 2012

Feeder

A A feeder consists of all circuit conductors located between the service equipment, the source of a separately derived system, or other power supply source and the final branch-circuit overcurrent device »*Article 100*«.

B Branch circuits (see definition).

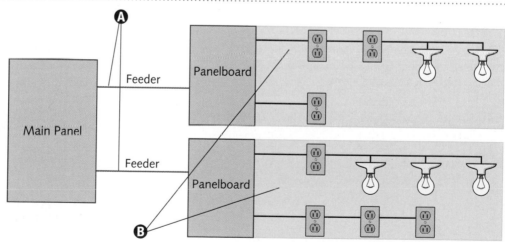

Main Panel

Feeder

Panelboard

Feeder

Panelboard

© Cengage Learning 2012

Festoon Lighting

Ⓐ Festoon lighting is a string of outdoor lights suspended between two points » *Article 100* « .

Ⓑ Overhead conductors for festoon lighting shall not be smaller than 12 AWG unless supported by messenger wires » *225.6(B)* « .

Ⓒ Messenger wire, together with strain insulators, is used to support conductors in all spans exceeding 40 ft (12 m) in length. Conductors shall not be attached to any fire escape, downspout, or plumbing equipment » *225.6(B)* « .

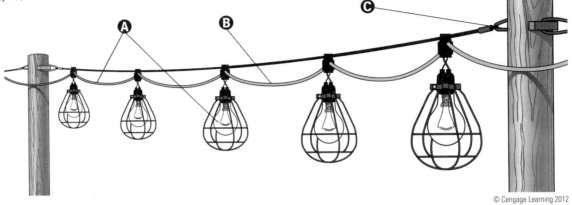

© Cengage Learning 2012

Fitting

Ⓐ A fitting is an accessory, such as a locknut or bushing, whose function is primarily mechanical, rather than electrical, in nature » *Article 100* « .

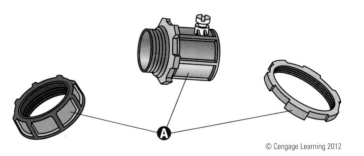

© Cengage Learning 2012

Grounded and Grounded Conductor

Ⓐ A grounded conductor is a system or circuit conductor that is intentionally grounded » *Article 100* « .

Ⓑ Grounded (grounding) is defined as being connected to ground or to a conductive body that extends the ground connection » *Article 100* « .

> **NOTE**
>
> Ground is defined as the earth
> » *Article 100* « .

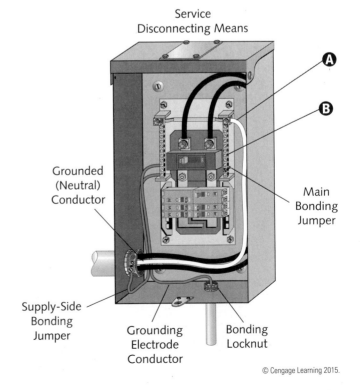

Service
Disconnecting Means

Grounded
(Neutral)
Conductor

Main
Bonding
Jumper

Supply-Side
Bonding
Jumper

Grounding
Electrode
Conductor

Bonding
Locknut

© Cengage Learning 2015.

Grounding Electrode; Grounding Electrode Conductor; and Equipment Grounding Conductor (EGC)

Ⓐ Main service equipment

Ⓑ The conductive path(s) that provides a ground-fault current path and connects normally non-current-carrying metal parts of equipment together and to the system grounded conductor, the grounding electrode conductor, or both is called an equipment grounding conductor » *Article 100* «.

Ⓒ System grounded (neutral) conductor

Ⓓ Supply-side bonding jumper

Ⓔ Bonding jumper or conductor

Ⓕ Main bonding jumper

Ⓖ The grounding electrode conductor is the conductor used to connect the system grounded conductor or the equipment to a grounding electrode or to a point on the grounding electrode system » *Article 100* «.

Ⓗ A grounding electrode is a conducting object that establishes a direct connection to earth » *Article 100* «.

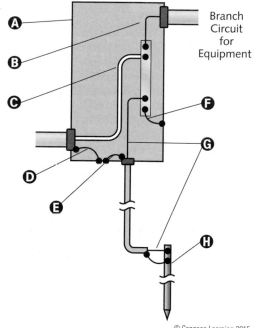

Branch Circuit for Equipment

© Cengage Learning 2015.

Guarded

Guarded is defined as covered, shielded, fenced, enclosed, or otherwise protected by means of suitable covers, casings, barriers, rails, screens, mats, or platforms effectively removing the likelihood of approach or contact by persons or objects » *Article 100* «.

Ⓐ Panelboards having doors or covers are considered guarded.

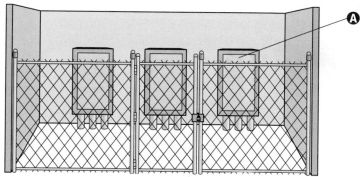

© Cengage Learning 2012

In Sight From (Within Sight From, Within Sight)

Ⓐ The *NEC* terms *in sight from, within sight from,* or *within sight,* etc., applied to equipment indicate that the specified items of equipment are visible and are not more than 50 ft (15 m) apart » *Article 100* «.

Ⓑ A motor disconnecting means must be located in sight from the motor location as required by 430.102(B).

Lighting Outlet

Ⓐ An outlet intended for the direct connection of a lampholder or luminaire is called a lighting outlet » *Article 100* «.

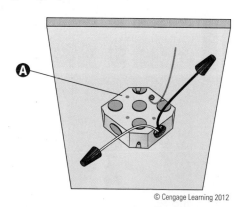

© Cengage Learning 2012

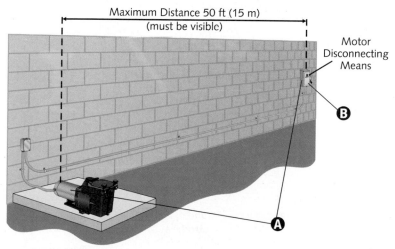

Maximum Distance 50 ft (15 m)
(must be visible)

Motor Disconnecting Means

© Cengage Learning 2012

Location, Damp

Ⓐ Damp locations are those subject to moderate degrees of moisture. Damp locations are those protected from weather and not subject to saturation with water or other liquids. Examples of such locations include partially protected locations under canopies, marquees, roofed (open) porches, and similar sites as well as interior locations subject to moderate degrees of moisture such as some basements, some barns, and some cold-storage warehouses ≫ *Article 100* and *Informational Note* ≪.

© Cengage Learning 2012

Location, Dry

Ⓐ Dry locations are those not normally subject to moisture, except on a temporary basis such as a building under construction ≫ *Article 100* ≪.

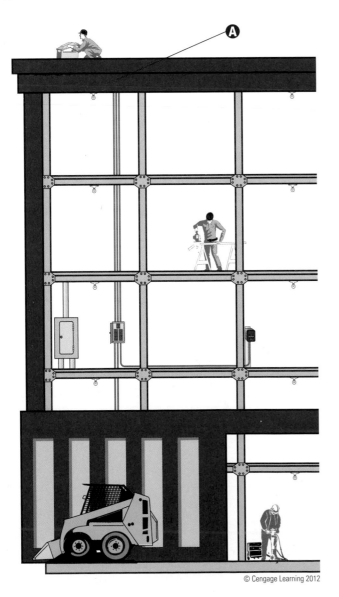

© Cengage Learning 2012

Location, Wet

Ⓐ Installations in any of the following categories are wet locations: underground, within concrete slabs, in masonry (directly contacting the earth), in areas subject to saturation (water or other liquids), and in locations unprotected from weather ≫ *Article 100* ≪.

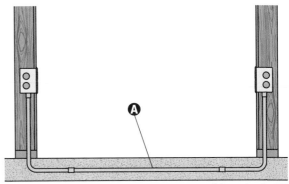

© Cengage Learning 2012

Luminaire

A luminaire is a complete lighting unit consisting of a light source such as a lamp or lamps, together with the parts designed to position the light source and to connect it to the power supply. It may also include parts to protect the light source, ballast, or distribute the light. A lampholder itself is not a luminaire.

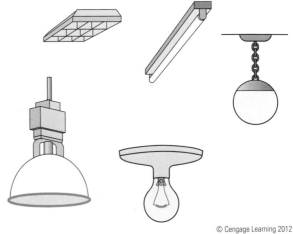

© Cengage Learning 2012

Multioutlet Assembly

Ⓐ A surface, flush, or freestanding raceway designed to hold conductors and receptacles (assembled in the field or at the factory) is called a multioutlet assembly »*Article 100*«.

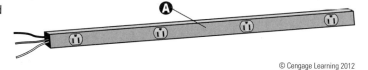

© Cengage Learning 2012

Neutral Conductor and Neutral Point

A neutral conductor is defined as the conductor connected to the neutral point of a system that is intended to carry current under normal conditions »*Article 100*«.

The neutral point is the common point on a wye-connection in a polyphase system or midpoint on a single-phase, 3-wire system, or midpoint of a single-phase portion of a 3-phase delta system, or a midpoint of a 3-wire, direct-current system »*Article 100*«.

Ⓐ The neutral point is the common point on a wye-connection in a polyphase system.

Ⓑ The neutral point is the midpoint on a single-phase, 3-wire system.

Ⓒ The neutral point is the midpoint of a single-phase portion of a 3-phase delta system.

Ⓓ The neutral point is the midpoint of a 3-wire direct-current system.

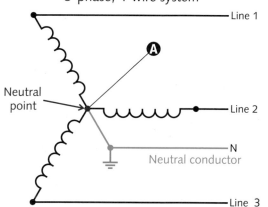

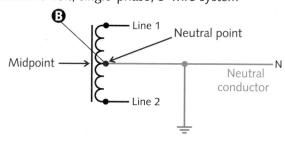

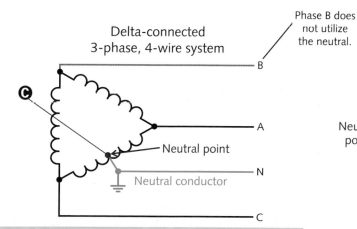

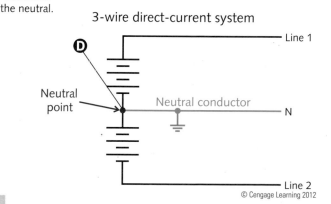

© Cengage Learning 2012

NOTE

At the neutral point of the system, the vectorial sum of the nominal voltages from all other phases within the system that utilize the neutral, with respect to the neutral point, is zero potential »*Informational Note to neutral point in Article 100*«.

WARNING

Neutral conductors must be identified in accordance with the requirements in 200.6.

Outlet

A A point in a wiring system from which current is taken to supply utilization equipment is known as an outlet >> *Article 100* <<.

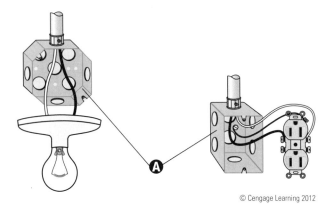

© Cengage Learning 2012

Overcurrent Protective Device, Branch-Circuit

A branch-circuit overcurrent protective device is a device capable of providing protection for service, feeder, and branch circuits and equipment over the full range of overcurrents between its rated current and its interrupting rating. Such devices are provided with interrupting ratings appropriate for the intended use but no less than 5000 amperes >> *Article 100* <<.

© Cengage Learning 2012

Plenum

A The space above a suspended ceiling used for environmental air-handling purposes is an example of **other space used for environmental air (plenum)** as described in 300.22(C).

B A compartment or chamber having one or more attached air ducts and forming part of the air distribution system is known as a plenum >> *Article 100* <<.

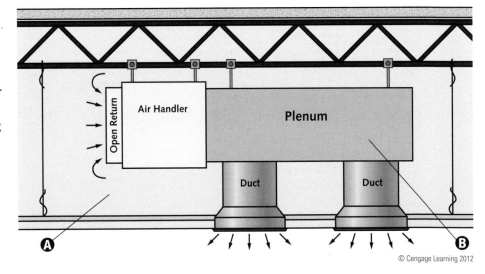

© Cengage Learning 2012

Receptacle

A A contact device installed at an outlet for the connection of an attachment plug is a receptacle >> *Article 100* <<.

B A single receptacle is a single contact device with no other contact device on the same yoke >> *Article 100* <<.

C A multiple receptacle is a single device consisting of two or more receptacles >> *Article 100* <<.

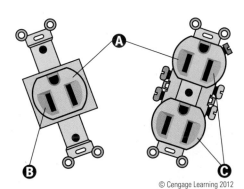

© Cengage Learning 2012

Separately Derived System

A Article 450 contains provisions for transformers.

B A separately derived system is an electrical source, other than a service, having no direct connection(s) to circuit conductors of any other electrical source other than those established by grounding and bonding connections 》*Article 100* 《.

> **N O T E**
>
> An alternate ac power source, such as an on-site generator, is not a separately derived system if the grounded conductor is solidly interconnected to a service-supplied system grounded conductor 》*250.30 Informational Note No. 1* 《.

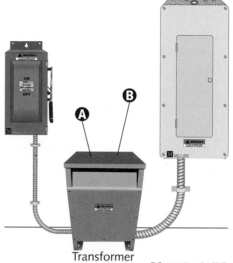

Transformer

© Cengage Learning 2015.

Service

A The conductors and equipment that deliver energy from the serving utility to the wiring system of the premises are called the service 》*Article 100* 《.

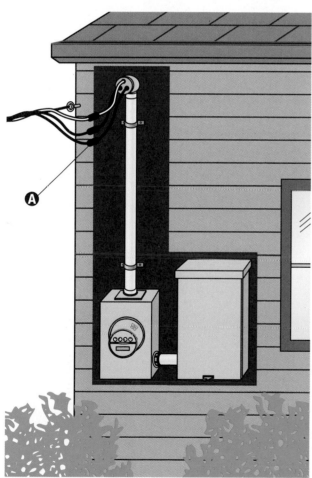

© Cengage Learning 2012

> **N O T E**
>
> Conductors and equipment are also defined as a service where receiving power underground.

Service Conductors

A The conductors from the service point to the service disconnecting means are known as service conductors 》*Article 100* 《.

© Cengage Learning 2012

> **N O T E**
>
> Overhead service conductors are the overhead conductors between the service point and the first point of connection to the service-entrance conductors at the building or other structure 》*Article 100* 《.

> **N O T E**
>
> Service conductors include: overhead service conductors, underground service conductors, overhead system service-entrance conductors, and underground system service-entrance conductors.

Service Drop

A The overhead conductors between the utility electric supply system and the service point »Article 100«.

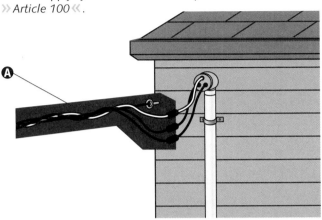

© Cengage Learning 2012

Service Equipment

A Service equipment is that equipment (usually circuit breaker[s], switch[es], fuse[s], and accessories) necessary to constitute the main control and cutoff that is connected to the load end of service conductors to a building or other structure, or an otherwise designated area »Article 100«.

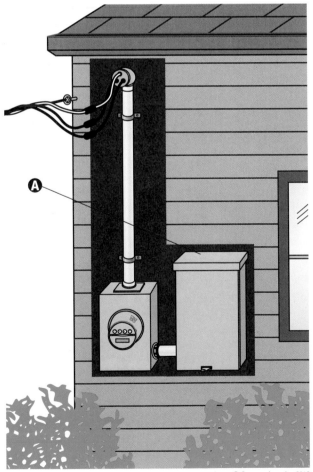

© Cengage Learning 2012

Service Lateral

A The service lateral is the underground conductors between the utility electric supply system and the service point »Article 100«.

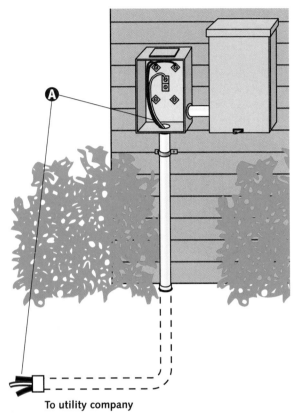

To utility company

© Cengage Learning 2012

NOTE

Underground service conductors are defined as the underground conductors between the service point and the first point of connection to the service-entrance conductors in a terminal box, meter, or other enclosure, inside or outside the building wall.
In accordance with the informational note under the definition of underground service conductors, where there is no terminal box, meter, or other enclosure, the point of connection is considered to be the point of entrance of the service conductors into the building »*Article 100*«.

NOTE

Underground system service-entrance conductors are the service conductors between the terminals of the service equipment and the point of connection to the service lateral or underground service conductors. As stated in the informational note, where service equipment is located outside the building walls, there may be no service-entrance conductors or they may be entirely outside the building »*Article 100*«.

Service Point

Ⓐ The service point is the point of the connection between the facilities of the serving utility and the premises wiring »*Article 100*«.

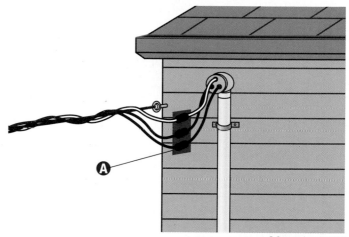

© Cengage Learning 2012

> **NOTE**
>
> The service point can be described as the point of demarcation between where the serving utility ends and the premises wiring begins. The serving utility generally specifies the location of the service point based on the conditions of service »*Article 100 Informational Note*«.

Special Permission and Authority Having Jurisdiction

Ⓐ Special permission is the written consent of the authority having jurisdiction (AHJ) »*Article 100*«.

Ⓑ The AHJ for enforcing the *NEC* may grant exception for the installation of conductors and equipment (not under the exclusive control of the electric utilities) used to connect the electric utility supply system to the service conductors of the premises served (provided such installations are outside a building or structure, or terminate inside at a readily accessible location nearest the point of entrance of the service conductors) »*90.2(C)*«.

Ⓒ AHJ is the organization, office, or individual responsible for enforcing the requirements of a code (or standard), or for approving equipment, materials, an installation, or a procedure »*Article 100*«.

© Cengage Learning 2012

Summary

- Conductors within junction boxes must be accessible without damaging the construction or finish of the building or structure.
- Certain equipment, such as the service disconnecting means, must be readily accessible.
- The term *appliance* denotes more than just kitchen equipment.
- Branch circuits are divided into four categories: appliance, general purpose, individual, and multiwire.
- General-purpose branch circuits may feed lights and receptacles or any combination thereof.
- An individual branch circuit feeds only one piece of equipment.
- The terms *bonded* and *grounded* are not interchangeable.

- A multiwire branch circuit must have a means to simultaneously disconnect all ungrounded (hot) conductors.
- A load having the maximum level of current sustained for three hours or more is a continuous load.
- *Equipment* is a general term encompassing a wide variety of items.
- A grounded conductor and a grounding conductor have different functions.
- A grounded conductor is not necessarily a neutral conductor.
- One duplex receptacle is not defined as a single receptacle.
- Special permission is the written consent of the AHJ.

Unit 2 Competency Test

NEC Reference	Answer

1. A(n) _____ branch circuit supplies two or more receptacles or outlets for lighting and appliances.

2. An electric circuit that controls another electric circuit through a relay is referred to as a(n) _____.

3. A(n) _____ branch circuit consists of two or more ungrounded conductors having a potential difference between them, and a grounded conductor having equal potential difference between it and each ungrounded conductor of the circuit and that is connected to the neutral or grounded conductor of the system.

4. An intermittent operation in which load conditions are regularly recurrent is the definition of _____.

5. The _____ is the connection between the grounded-circuit conductor and the equipment grounding conductor at the service.

6. A(n) _____ is used to connect the system grounded conductor or the equipment to a grounding electrode or to a point on the grounding electrode system.

7. Rainproof, raintight, or watertight equipment can fulfill the requirements for _____ where varying weather conditions other than wetness, such as snow, ice, dust, or temperature extremes, are not a factor.

8. A(n) _____ is an intentionally constructed, low-impedance electrically conductive path designed and intended to carry current under ground-fault conditions from the point of a ground fault on a wiring system to the electrical supply source and that facilitates the operation of the overcurrent protective device or ground-fault detectors.

9. A(n) _____ is a manually operated device used in conjunction with a transfer switch to provide a means of directly connecting load conductors to a power source and of disconnecting the transfer switch.

10. A type of surface, flush, or freestanding raceway, designed to hold conductors and receptacles, assembled in the field or at the factory is called a(n) _____.

11. A raceway encased in 4 in. (102 mm) of concrete on the ground floor (in direct contact with the earth) shall be considered a(n) _____ location.

12. A(n) _____ may consist of one or more sensing elements integral with the motor-compressor and an external control device.

13. An enclosure designed either for surface or flush mounting and provided with a frame, mat, or trim in which a swinging door or doors are or can be hung is called a(n) _____.

14. A(n) _____ is a shaftway, hatchway, well hole, or other vertical opening of space in which an elevator or dumbwaiter is designed to operate.

15. A multiwire branch circuit can supply:

    I. 120/240 volts to only one piece of utilization equipment

    II. 120/240 volts where all ungrounded conductors are opened simultaneously

    a) I only   b) II only   c) either I or II   d) neither I nor II

16. An overcurrent protective device with a circuit opening fusible part that is heated and severed by the passage of overcurrent through it is the definition of a(n)_____.

17. _____ is a string of outdoor lights suspended between two points.

NEC Reference	Answer

——————— ——————— 18. Name two items that must be present when defining an area as a bathroom.

——————— ——————— 19. A continuous load is where the _____ current is expected to continue for three hours or more.

      a) 80% b) average c) maximum d) 125%

——————— ——————— 20. Solidly grounded is defined as connected to ground without inserting any _____ or impedance device.

——————— ——————— 21. A(n) _____ is a point on the wiring system at which current is taken to supply utilization equipment.

——————— ——————— 22. A(n) _____ is defined as the circuit conductors between the final overcurrent device protecting the circuit and the outlet(s).

——————— ——————— 23. _____ enclosures are constructed or protected so that exposure to a beating rain will not result in the entrance of water under specified test conditions.

——————— ——————— 24. Continuous duty is an operation at a substantially constant load for _____.

      a) 1 hour or more   b) 1½ hours or more

      c) 3 hours or more   d) an indefinitely long time

——————— ——————— 25. Ampacity is defined as the current in amperes that a conductor can carry continuously under the conditions of use without exceeding _____.

——————— ——————— 26. When a disconnecting means must be located within sight from a motor, the disconnect must be visible and not more than _____ ft from the motor.

——————— ——————— 27. A building containing three dwelling units is called a(n) _____.

——————— ——————— 28. Ground is defined as _____.

——————— ——————— 29. At the neutral point of the system, the _____ of the nominal voltages from all other phases within the system that utilize the neutral, with respect to the neutral point, is zero potential.

UNIT 3

Boxes and Enclosures

Objectives

After studying this unit, the student should:

▶ be able to determine the cubic-inch capacity of boxes (metal and nonmetallic) when installing 6 American Wire Gauge (AWG) and smaller conductors.

▶ know which items are counted when calculating box fill and which are not.

▶ be aware that two identical switches, mounted side by side in a two-gang device box, could each have different cubic-inch volume allowances.

▶ be able to determine the minimum box size (including plaster ring, extension ring, etc.) that is needed if the number of conductors (6 AWG and smaller) is known.

▶ be familiar with box requirements when using nonmetallic-sheathed cable.

▶ know the minimum length of free conductor required to be left inside boxes.

▶ understand that boxes and conduit bodies must remain accessible after installation.

▶ be familiar with mounting and supporting provisions for boxes and conduit bodies.

▶ be able to determine the type of box needed for various applications.

▶ understand calculation procedures for junction boxes containing 4 AWG and larger conductors.

Introduction

Choosing the right type and size of box (or enclosure) is very important to installing a system essentially free from hazard. Article 314 covers a variety of provisions concerning boxes (outlet, device, pull, and junction), conduit bodies, and fittings. Box selection must be based on requirements for a given location (such as dry, damp, wet, or hazardous). Boxes have particular requirements concerning the maximum number of conductors. Boxes containing 6 AWG and smaller conductors are required to have a minimum cubic-inch capacity, which is determined by the size and number of conductors. Boxes containing 4 AWG and larger conductors are required to have a minimum height, width, and depth that is determined by the size and number of raceway entries. Article 314 contains provisions for installing, as well as supporting, boxes and conduit bodies. Boxes must be rigidly and securely fastened in place whether mounted *on* the surface, mounted to a framing member, or mounted *in* a finished surface. Under certain conditions, the only means of support a box needs is two threaded conduits. Access to conductors and devices located within boxes and conduit bodies must be available. Article 314 covers more information than the length of this book allows, such as manholes and boxes (pull and junction) for systems over 1000 volts, nominal.

BOX FILL CALCULATIONS

Metal Boxes

The maximum number of conductors (for sizes 18 AWG through 6 AWG) permitted in various standard size metal boxes is listed in Table 314.16(A). Minimum cubic-inch capacities are also shown.

Ⓐ Volume does not have to be marked on metal boxes listed in Table 314.16(A) ≫ *314.16(A)(2)* ≪ .

Ⓑ A 3- × 2-in. (75- × 50-mm) device box, 2½ in. (65 mm) deep, has a volume of 12.5 in.³ (205 cm³) ≫ *Table 314.16(A)* ≪ .

Ⓒ A 4-in. (100-mm) square box with a depth of 1½ in. (38 mm), has a volume of 21 in.³ (344 cm³) ≫ *Table 314.16(A)* ≪ .

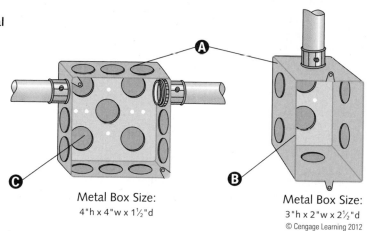

Metal Box Size:
4"h x 4"w x 1½"d

Metal Box Size:
3"h x 2"w x 2½"d

© Cengage Learning 2012

Volume Markings on Metal Boxes

It is permissible to use the volume marked on a metal box even if it is more than is shown in Table 314.16(A) for the same size box ≫ *314.16(A)(2)* ≪ .

Ⓐ The minimum volume for this size box in Table 314.16(A) is 21.5 in.³.

Ⓑ The volume marked on this box is 22.5 in.³

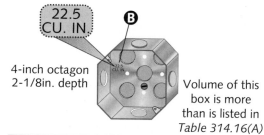

22.5 CU. IN. **Ⓑ**

4-inch octagon 2-1/8in. depth

Volume of this box is more than is listed in *Table 314.16(A)*

Table 314.16(A) Metal Boxes				
Box Trade Size			Minimum Volume	
mm	in.		cm³	in.³
100 x 54	4 x 2-1/8	Round/Octagonal	353	21.5

Ⓐ

© Cengage Learning 2012

Nonmetallic Boxes

Ⓐ Volume shall be marked on all nonmetallic boxes and boxes with a volume of 100 in.³ (1650 cm³) or less, except for boxes listed in Table 314.16(A) ≫ *314.16(A)(2)* ≪ .

18 CU. IN.

Ⓐ

© Cengage Learning 2012

Additional Capacity

Ⓐ Additional capacity can only be calculated when the plaster ring, extension ring, etc., is clearly marked with a volume or when it corresponds in size to a box listed in Table 314.16(A).

Ⓑ Volume can be increased by using plaster (mud) rings, domed covers, extension rings, or similar items ≫314.16(A)≪.

Ⓒ The combined volume of this box and plaster (mud) ring is 28.5 in.³ (467 cm³) ≫Table 314.16(A)≪.

Ⓓ The combined volume of the box and extension ring is 42 in.³ (688 cm³) ≫Table 314.16(A)≪.

Ⓔ A 4-in. (100-mm) square extension ring, having a depth of 1½ in. (38 mm), has a capacity of 21 in.³ (344 cm³) because it has the same dimensions as a box listed in Table 314.16(A).

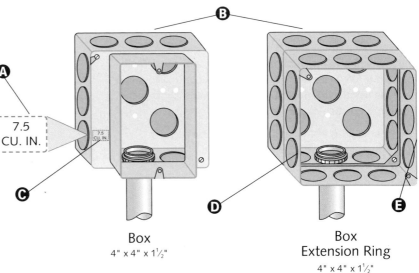

Ⓐ

7.5 CU. IN.

Box
4" x 4" x 1½"

Box
Extension Ring
4" x 4" x 1½"

© Cengage Learning 2012

ADDITIONAL MARKINGS

Ⓐ All nonmetallic boxes must be durably and legibly marked by the manufacturer with the volume of the box.

Ⓑ Some nonmetallic boxes are marked with the maximum number of certain sizes of conductors. This marking is not required.

Ⓒ The maximum number of conductors applies only if the box does not contain any fittings or devices.

Ⓓ The box fill calculation for this box is seven 12 AWG conductors. Because eight 12 AWG conductors are permitted, this installation is *Code* compliant.

Ⓔ The three equipment grounding conductors for this box count as one conductor.

Ⓕ Installing a device in this box is a violation of *314.16*.

Ⓖ Because the calculated number of 12 AWG conductors in this box is seven and a device counts as two 12 AWG conductors, the total number of 12 AWG conductors would be nine. In accordance with the marking on this box, only eight 12 AWG conductors are permitted.

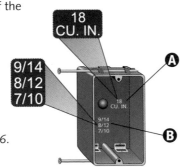

12-2 with ground
nonmetallic-sheathed
cable (typical)

9/14
8/12
7/10

14 AWG = 9
12 AWG = 8
10 AWG = 7

Blank cover

Wire connectors (wirenuts)
are not counted

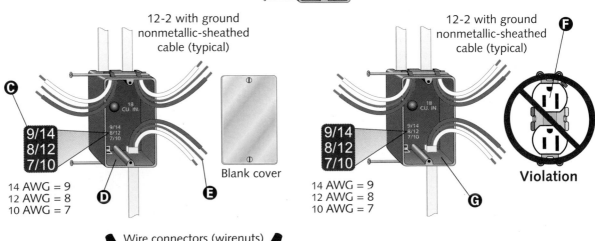

12-2 with ground
nonmetallic-sheathed
cable (typical)

9/14
8/12
7/10

14 AWG = 9
12 AWG = 8
10 AWG = 7

Violation

© Cengage Learning 2012

Equipment Grounding Conductor Fill

Ⓐ Pigtails are not counted.

Ⓑ Equipment grounding conductor(s) or equipment bonding jumpers entering a box count as one conductor ≫ 314.16(B)(5) ≪. Only the largest equipment grounding conductor shall be counted if multiple grounding conductors (of different sizes) enter the box.

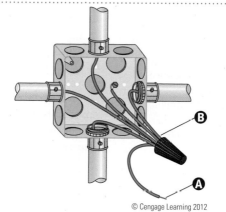

© Cengage Learning 2012

> **N O T E**
>
> One or more additional equipment grounding conductors (isolated equipment grounding conductors) as permitted in 250.146(D) shall be counted as one conductor ≫ **314.16(B)(5)** ≪. When multiple isolated equipment grounding conductors of different sizes enter the box, only the largest is counted.

Determining the Number of Conductors

Ⓐ Raceway fittings (connectors, hubs, etc.) are not counted.

Ⓑ Each conductor originating outside the box that is terminated or spliced inside the box counts as one conductor ≫ 314.16(B)(1) ≪. These red conductors are counted as two conductors.

Ⓒ This box, as pictured, contains five conductors. If all of the conductors are 12 AWG, the box could hold a maximum of nine conductors ≫ Table 314.16(A) ≪.

Ⓓ A conductor that does not leave the box, such as equipment bonding jumpers and pigtails, is not counted ≫ 314.16(B)(1) ≪.

Ⓔ A conductor that passes through the box without splice or termination (unbroken) counts as one conductor. Each loop or coil of unbroken conductor not less than twice the minimum length required for free conductors in 300.14 must be counted twice ≫ 314.16(B)(1) ≪. This black unbroken conductor is counted as one conductor.

Ⓕ Each conductor originating outside the box that is terminated or spliced inside the box counts as one conductor ≫ 314.16(B)(1) ≪. These white conductors count as two conductors.

Ⓖ Wire connectors are not counted.

Metal box
4" x 4" x 1½"

© Cengage Learning 2012

Cable Clamps and Connectors

Ⓐ External cable connectors are not counted.

Ⓑ Cable connector(s) with the clamping mechanism outside the box are not counted ≫ 314.16(B)(2) ≪.

Ⓒ Two internal cable clamps count as one conductor.

Ⓓ Internal cable clamp(s), whether factory or field supplied, shall be counted as one conductor ≫ 314.16(B)(2) ≪. Where more than one size conductor is present in the box, the clamp shall be counted as the largest conductor.

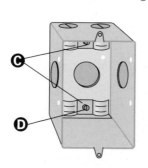
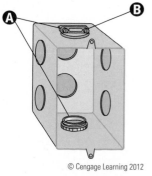

© Cengage Learning 2012

Luminaire Studs and Hickeys

Ⓐ Luminaire stud(s) shall be counted as one conductor ≫ 314.16(B)(3) ≪.

Ⓑ One or more hickeys shall be counted as one conductor ≫ 314.16(B)(3) ≪.

Ⓒ If conductors of different sizes are present in the box, each one shall be counted as the largest conductor ≫ 314.16(B)(3) ≪.

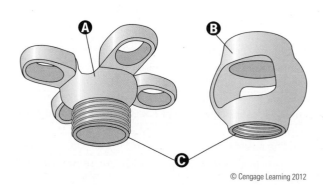

© Cengage Learning 2012

Devices or Equipment

A This duplex receptacle counts as two 12 AWG conductors.

B This single-pole switch counts as two 14 AWG conductors.

C Each mounting yoke or strap counts as two conductors. A mounting yoke or strap can contain one or more devices, such as a single receptacle, a duplex receptacle, a single switch, a double switch, a triple switch, or any combination. The size of the two conductors (when calculating box fill) shall be equal in size to the largest conductor connected to the device ≫ *314.16(B)(4)* ≪ .

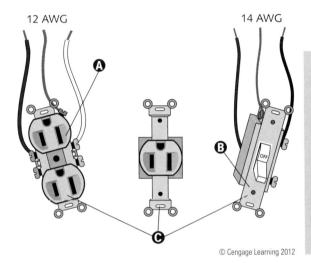

© Cengage Learning 2012

NOTE

A device or utilization equipment wider than a single 2-in. (50-mm) device box as described in Table 314.16(A) shall have double-volume allowances provided for each gang required for mounting ≫ *314.16(B)(4)* ≪ .

Volume per Conductor

Table 314.16(B) lists the cubic-inch volume for conductors, sizes 18 AWG through 6 AWG.

Boxes, enclosures, and conduit bodies containing conductors, size 4 AWG or larger, must also comply with 314.28 provisions ≫ *314.16* ≪ .

NOTE

Conductor insulation is not a factor when determining box fill calculations.

18 AWG = 1.50 in.³
16 AWG = 1.75 in.³
14 AWG = 2.00 in.³
12 AWG = 2.25 in.³
10 AWG = 2.5 in.³
8 AWG = 3.0 in.³
6 AWG = 5.0 in.³

© Cengage Learning 2012

Box Fill Calculation

Wire connectors, pigtails, locknuts, bushings, raceway connectors, grounding screws, and equipment bonding jumpers are not factors when calculating box fill.

A These three conductors, although spliced, count as *three* conductors.

B Two conductors that terminate in the box count as *two* conductors.

C One receptacle counts as *two* 12 AWG conductors.

D An unbroken conductor counts as *one* conductor unless the conductor is not less than twice the minimum length required for free conductors in 300.14.

E These three equipment grounding conductors that enter the box from the three raceways count as *one* conductor.

F These two conductors terminating in the box are counted as *two* conductors.

G A single unbroken conductor is counted as one conductor.

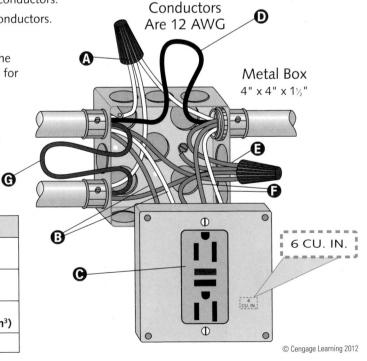

All Conductors Are 12 AWG

Metal Box 4" x 4" x 1½"

6 CU. IN.

© Cengage Learning 2012

Total 12 AWG conductors:	Twelve
Volume per 12 AWG	2.25 in.³ (each) (36.9 cm³ [each])
Minimum volume for conductors	12 × 2.25 in.³ = 27 in.³ (12 × 36.9 cm³ = 442.8 cm³)
Volume for box (including raised cover)	21 in.³ + 6 in.³ = 27 in.³ (344 cm³ + 98.4 cm³ = 442.4 cm³)
This installation complies with 314.16 provisions.	

Domed Covers and Canopies

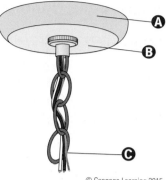

A Four or fewer luminaire wires (smaller than 14 AWG) and an equipment grounding conductor can be omitted from box fill calculations where they enter a box from a domed luminaire (or similar canopy) and terminate within that box »314.16(B)(1) Exception«.

B The size of the domed cover or canopy is not a factor.

C Two luminaire wires and one equipment ground are not counted.

Splices Inside Conduit Bodies

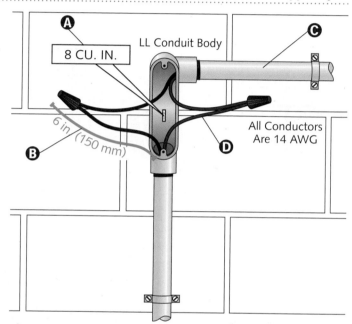

A Splices, taps, or devices are permitted inside conduit bodies, if the manufacturer durably and legibly marked the cubic-inch capacity on the conduit body »314.16(C)(2)«.

B At least 6 in. (150 mm) of free conductor (measured from the point in the conduit body where it emerges from its raceway or cable sheath) shall be left at each outlet, junction, and switch point »300.14«.

C A conduit body can be supported by rigid metal conduit, intermediate metal conduit, rigid nonmetallic conduit, or electrical metallic tubing »314.23(E) Exception«.

D The maximum number of conductors shall be calculated in accordance with 314.16(B). Four 14 AWG conductors with a volume of 2.0 in.³ (32.8 cm³) each require a total volume of 8 in.³ (131.2 cm³).

> **NOTE**
>
> The maximum number of conductors permitted shall be the maximum number permitted by Table 1 of Chapter 9 for the conduit to which it is attached »314.16(C)(1)«.

8 CU. IN.
LL Conduit Body
All Conductors Are 14 AWG
6 in. (150 mm)

GENERAL INSTALLATION

Securing Cables to Metal Boxes

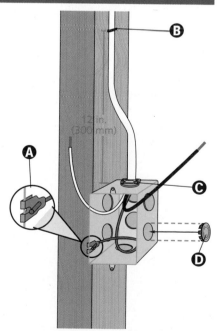

Conductors must be protected from abrasion where entering boxes, conduit bodies, or fittings »314.17«.

A A connection must be made between the equipment grounding conductor(s) and a metal box by means of a grounding screw that shall be used for no other purpose, equipment listed for grounding, or a listed grounding device »250.148(C)«. Grounding conductors must not be connected to enclosures with screws that engage less than two threads in the enclosure (or nut) such as sheet metal, drywall, or wood screws (see 250.8).

B Nonmetallic-sheathed cable shall be secured within 12 in. (300 mm) of every cabinet, box, or fitting. The cable can be secured by using staples, cable ties, straps, or similar fittings designed and installed so that the cable remains undamaged »334.30«.

C Cables must be securely fastened to any metal box or conduit body that they enter »314.17(B)«. The clamping mechanism can be internal or external to the box or conduit body.

D Unused cable or raceway openings in boxes and conduit bodies must be closed so that the protection provided is at least equal to that provided by the wall of the box or conduit body »110.12(A)«.

Single-Gang Nonmetallic Boxes

Ⓐ Nonmetallic-sheathed cable not fastened to the box must be secured within 8 in. (200 mm) of the box ≫ *314.17(C) Exception* ≪.

Ⓑ At least 6 in. (150 mm) of free conductor (measured from the point in the box where it emerges from its raceway or cable sheath) shall be left at each outlet, junction, and switch point ≫ *300.14* ≪.

Ⓒ The cable sheath shall extend into the box at least ¼ in. (6 mm) through the cable knockout or opening ≫ *314.17(C) Exception* ≪.

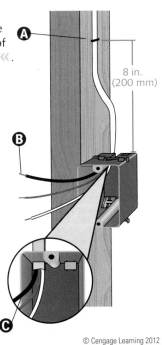

© Cengage Learning 2012

> **CAUTION**
>
> All permitted wiring methods must be secured to the box, unless the exception in 314.17(C) has been met.

Boxes in Combustible and Noncombustible Materials

Ⓐ Boxes employing a flush-type cover or faceplate in walls or ceilings constructed of noncombustible material (concrete, tile, etc.) shall be installed so that the front edge of the box is within ¼ in. (6 mm) of the finished surface ≫ *314.20* ≪.

Ⓑ Boxes employing a flush-type cover or faceplate in walls or ceilings constructed of combustible material, such as wood, shall be flush with or extend beyond the finished surface ≫ *314.20* ≪.

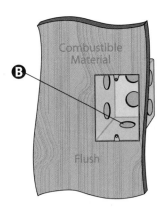

© Cengage Learning 2012

Securing Cables to Nonmetallic Boxes

Ⓐ Nonmetallic-sheathed cable shall be secured within 12 in. (300 mm) of every box ≫ *334.30* ≪.

Ⓑ Cables entering a nonmetallic box must be secured to the box, unless it is a single-gang nonmetallic box ≫ *314.17(C)* ≪.

Ⓒ The cable sheath shall extend into the box at least ¼ in. (6 mm) through the cable knockout or opening ≫ *314.17(C)* ≪.

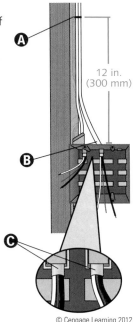

© Cengage Learning 2012

> **NOTE**
>
> The exception pertaining to nonmetallic-sheathed cable entering a box which has no means of securing the cable to the box applies only to *single-gang* nonmetallic boxes ≫*314.17(C) Exception*≪.

Back-to-Back Boxes in Fire-Resistant-Rated Wall

Ⓐ Qualified testing laboratories publish electrical construction material directories listing installation restrictions that apply to maintaining fire-resistive ratings of assemblies involving penetrations, or openings. (An example is the minimum 24-in. [600-mm] horizontal separation usually required between boxes on opposite sides of the wall.) These fire-resistance directories, product listings, and building codes offer assistance in 300.21 compliance ≫ *300.21 (Informational Note)* ≪.

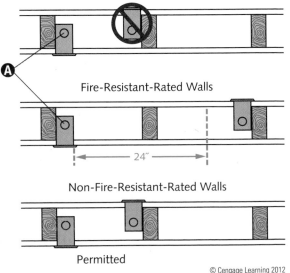

Fire-Resistant-Rated Walls

Fire-Resistant-Rated Walls

24″

Non-Fire-Resistant-Rated Walls

Permitted

© Cengage Learning 2012

Gaps or Open Spaces

A Damaged or incomplete plaster, drywall, or plasterboard surfaces around boxes employing a flush-type cover or faceplate must be repaired so that no gap or open space greater than $\frac{1}{8}$ in. (3 mm) surrounds the box or fitting ≫ *314.21* ≪.

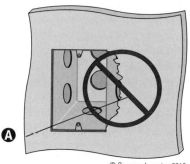

© Cengage Learning 2012

Surface Extensions

Equipment grounding, where required, must be in accordance with Part VI of Article 250.

A Surface extensions must have an extension ring mounted and mechanically secured to the box ≫ *314.22* ≪.

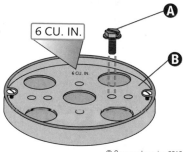

Extension Ring

Flush-Mounted Box

© Cengage Learning 2012

Minimum Internal Depth

A Except as permitted in 250.112(I), all metal boxes shall be grounded and bonded in accordance with Parts I, IV, V, VI, VII, and X of Article 250, as applicable ≫ *314.4* ≪.

B Outlet boxes that do not enclose devices or utilization equipment must have an internal depth of at least $\frac{1}{2}$ in. (12.7 mm) ≫ *314.24(A)* ≪.

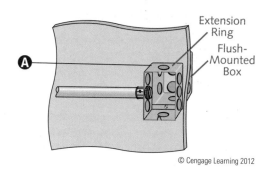

6 CU. IN.

© Cengage Learning 2012

Surface Extensions Made from Covers

A The cover of a flush-mounted box can provide a surface extension where the cover is designed so that it is unlikely to fall off or be removed if its securing means becomes loose. The wiring method shall be flexible for an approved length that permits removal of the cover and provides access to the box interior and shall be arranged so that any grounding continuity is independent of the connection between the box and the cover ≫ *314.22 Exception* ≪.

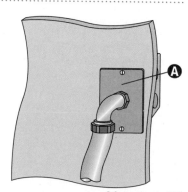

© Cengage Learning 2012

Metal Faceplates Covering Nonmetallic Boxes

Ⓐ Snap switches (including dimmer switches) must be connected to an equipment grounding conductor and must also provide a means to connect metal faceplates to the equipment grounding conductor, even if a metal faceplate is not installed ≫ *404.9(B)* ≪ .

Ⓑ Both metal and nonmetallic covers and plates shall be permitted. Metal covers or plates, when used, must comply with 250.110 grounding requirements ≫ *314.25(A)* ≪ .

Ⓒ Should the snap switch enclosure, or the wiring method used, not have an equipment ground, a snap switch without a grounding connection can be used for replacement purposes only. A snap switch wired under the provisions of this exception and located within 8 ft (2.5 m) vertically, or 5 ft (1.5 m) horizontally, of ground or exposed grounded metal objects shall be provided with a faceplate of nonconducting noncombustible material with nonmetallic attachment screws, unless the switch mounting strap or yoke is nonmetallic or the circuit is protected by a ground-fault circuit interrupter ≫ *404.9(B) Exception No. 1 to (B)* ≪ .

Ⓓ There are two acceptable methods for grounding snap switches effectively: (1) The switch is mounted with metal screws to a metal box or metal cover that is connected to an equipment grounding conductor or to a nonmetallic box equipped with integral means for connecting to an equipment grounding conductor. (2) An equipment grounding conductor, or equipment bonding jumper, is connected to the equipment grounding termination on the snap switch ≫ *404.9(B)* ≪ .

CAUTION

Isolated ground receptacles, in nonmetallic boxes, must be covered with either a nonmetallic faceplate or with an effectively grounded metal faceplate ≫ *406.3(D)(2)* ≪ .

NOTE

Metal receptacle faceplates (cover plates) shall be grounded ≫ *406.6(B)* ≪ .

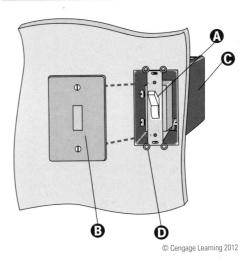

© Cengage Learning 2012

NOTE

Listed kits or listed assemblies shall not be required to be connected to an equipment grounding conductor if all of the following conditions are met:
(1) The device is provided with a nonmetallic faceplate that cannot be installed on any other type of device,
(2) The device does not have mounting means to accept other configurations of faceplates,
(3) The device is equipped with a nonmetallic yoke, and
(4) All parts of the device that are accessible after installation of the faceplate are manufactured of nonmetallic materials ≫ *404.9(B) Exception No. 2 to (B)* ≪ .

Covers and Canopies

Ⓐ To complete the installation, each box must have a cover, faceplate, lampholder, or luminaire canopy, except where the installation complies with 410.24(B) ≫ *314.25* ≪ .

Ⓑ Screws used for the purpose of attaching covers, or other equipment, to the box shall be either machine screws matching the thread gauge or size that is integral to the box or shall be in accordance with the manufacturer's instructions ≫ *314.25* ≪ .

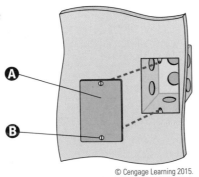

© Cengage Learning 2015.

Receptacles Mounted on Covers

Ⓐ Where receptacles are mounted and supported by a cover, they must be secured by more than one screw, unless the box cover or device assembly is listed and identified as single-screw mounting ≫ *406.5(C)* ≪ .

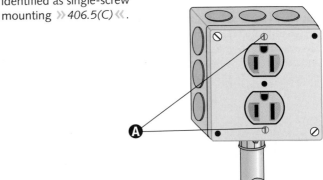

© Cengage Learning 2012

Floor Boxes

A Boxes containing receptacles, located in the floor, must be listed for the specific application ≫*314.27(B)*≪.

> **N O T E**
>
> Boxes located in elevated floors (such as show windows) do not have to be listed specifically as floor boxes if the authority having jurisdiction (AHJ) judges them free from likely exposure to physical damage, moisture, and dirt. Receptacles and covers shall be listed as an assembly for this type of location ≫*314.27(B) Exception*≪.

> **CAUTION**
>
> Only a limited number of boxes are listed specifically for wood floor construction.

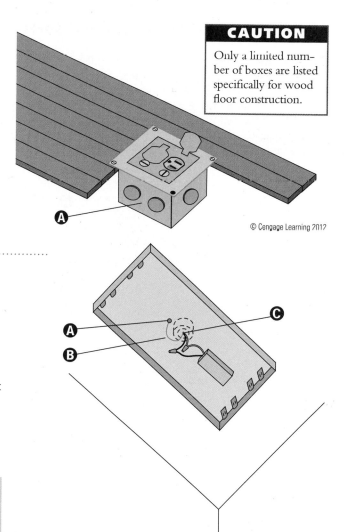

© Cengage Learning 2012

Access to Outlet Boxes

A Unless meeting one of the three exceptions, luminaires and equipment shall be mechanically connected to an equipment grounding conductor as specified in 250.118 and sized in accordance with 250.122 ≫*410.44*≪.

B A means for connecting an equipment grounding conductor must be provided for luminaires with exposed metal parts ≫*410.46*≪.

C Electric-discharge luminaires, such as fluorescent luminaires, that are surface mounted over concealed boxes (outlet, pull, or junction) and are designed not to be supported solely by the outlet box shall have suitable openings in back of the luminaire providing access to the wiring in the box ≫*410.24(B)*≪. This provision also applies to LED luminaires.

> **N O T E**
>
> Supplementary overcurrent protective devices, when used within luminaires, do not have to be readily accessible ≫*240.10*≪.

© Cengage Learning 2012

BOX AND LUMINAIRE SUPPORT

Box and Enclosure Supports

A Nails, where used, can attach brackets on the outside of the enclosure or extend through the interior within ¼ in. (6 mm) of the back or ends of the enclosure ≫*314.23(B)(1)*≪.

B An enclosure supported from a structural member or from grade shall be rigidly supported either directly or by use of a brace (metal, polymeric, or wood) ≫*314.23(B)*≪.

C Wood braces must have a cross section of at least 1 in. by 2 in. (25 mm by 50 mm) ≫*314.23(B)(2)*≪.

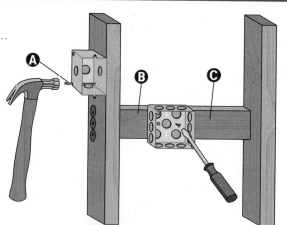

© Cengage Learning 2012

> **N O T E**
>
> An enclosure mounted on a building or other surface shall be rigidly and securely fastened in place. If the mounting surface does not provide rigid and secure support, additional support must be provided in accordance with 314.23 provisions ≫*314.23(A)*≪.

> **CAUTION**
>
> Screws passing through the box (with exposed threads in the box) are not permitted, unless they are protected using approved means. Exposed threads could be abrasive to conductor insulation ≫*314.23(B)(1)*≪.

Mounting Enclosures in Finished Surfaces

A No support is required where the cable is fished between access points, concealed in the finished surface, and where such supporting of the cable is impractical. Provisions are found in 320.30(D)(1) for armored cable support, in 330.30(D)(1) for metal-clad cable support, and in 334.30(B)(1) for nonmetallic-sheathed cable support.

B An enclosure mounted in a finished surface must be rigidly secured to the surface by clamps, anchors, or fittings identified for the application ≫314.23(C)≪.

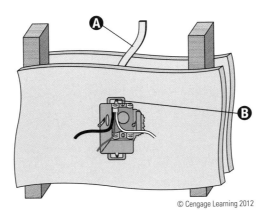

© Cengage Learning 2012

> **NOTE**
>
> No support is required for unbroken lengths (without coupling) of electrical metallic tubing (EMT) that is fished between access points, concealed in the finished surface, and where such securing of the raceway is impractical ≫*358.30(A) Exception No. 2*≪.

Enclosures in Suspended Ceilings

A Raceways shall not be supported by ceiling grid support wires. Raceways can be secured to independent (additional) support wires that are secured at both ends ≫300.11(A)≪.

B An enclosure mounted in a suspended ceiling system shall be fastened to framing members by mechanical means (bolts, screws, rivets, clips, etc.) identified for use with the enclosure(s) and ceiling framing member(s) employed. The framing members shall be supported in an approved manner and securely fastened to each other as well as to the building structure ≫314.23(D)(1)≪.

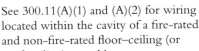

> **CAUTION**
>
> See 300.11(A)(1) and (A)(2) for wiring located within the cavity of a fire-rated and non-fire-rated floor–ceiling (or roof–ceiling) assembly.

> **NOTE**
>
> Boxes can be secured to independent (additional) support wires that are attached at both ends ≫*300.11(A)*≪.

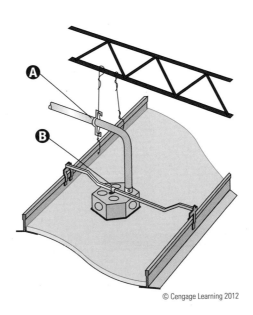

© Cengage Learning 2012

Enclosures (without Device, Luminaires, or Lampholders) Supported by Raceways

Ⓐ An enclosure that contains no devices **or** supports no luminaires can be supported by entering raceways when *all* of the following conditions are met: (1) the enclosure does not exceed 100 in.³ (1650 cm³) size; (2) the enclosure has threaded entries or identified hubs; (3) the enclosure is supported by two or more conduits threaded wrenchtight into the enclosure or hubs; and (4) each conduit is secured within 36 in. (900 mm) of the enclosure, unless all entries are on the same side ≫ *314.23(E)* ≪.

Ⓑ Only rigid or intermediate metal conduit with threaded ends are permitted in 314.23(E).

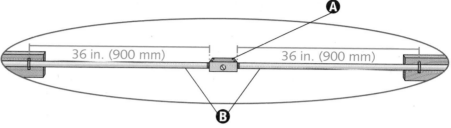

© Cengage Learning 2012

Enclosures (with Devices, Luminaires, or Lampholders) Supported by Raceways

Ⓐ An enclosure that contains devices (other than splicing devices) or supports luminaires, lampholders, or other equipment can be supported by entering raceways when *all* of the following conditions are met: (1) the enclosure does not exceed 100 in.³ (1650 cm³) in size; (2) the enclosure has threaded entries or identified hubs; (3) the enclosure is supported by two or more conduits threaded wrenchtight into the enclosure or hubs; and (4) each conduit is secured within 18 in. (450 mm) of the enclosure ≫ *314.23(F)* ≪.

Ⓑ An outlet box can support a luminaire weighing 50 pounds (23 kg) or less, unless the outlet box is listed and marked on the interior of the box to indicate the maximum weight the box shall be permitted to support ≫ *314.27(A)(2)* ≪.

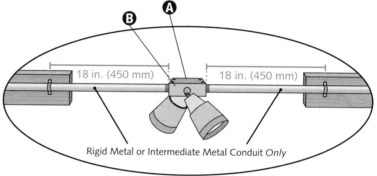

Rigid Metal or Intermediate Metal Conduit *Only*

© Cengage Learning 2012

Enclosures with Conduit Entries on One Side

Ⓐ An enclosure can be supported by conduits entering on the same side when *all* of the following conditions are met: (1) the enclosure does not exceed 100 in.³ (1650 cm³) in size; (2) the enclosure has threaded entries or identified hubs; (3) the enclosure is supported by two or more conduits threaded wrenchtight into the enclosure or hubs; and (4) each conduit is secured within 18 in. (450 mm) of the enclosure ≫ *314.23(E)* ≪.

Ⓑ Unused cable or raceway openings, other than those intended for the operation of equipment, those intended for mounting purposes, or those permitted as part of the design for listed equipment, must be closed so that the protection provided is at least equal to that provided by the wall of the equipment ≫ *110.12(A)* ≪.

Ⓒ Boxes must be supported within 18 in. (450 mm), whether or not they contain devices or support fixtures, if all the conduits enter on the same side, unless the requirements of 314.23(F) Exception No. 2 are met.

> **NOTE**
> Boxes shall not be supported by one conduit unless the requirements of **314.23(F) Exception No. 2** are met.

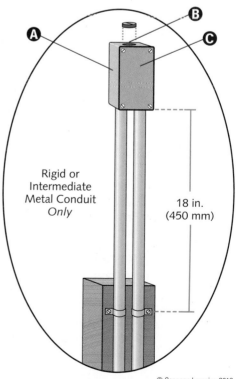

Rigid or Intermediate Metal Conduit *Only*

18 in. (450 mm)

© Cengage Learning 2012

Support for Conduit Bodies

A Any size conduit body not containing a device(s), luminaire(s), lampholder(s), or other equipment can be supported by rigid metal conduit, intermediate metal conduit, rigid polyvinyl chloride conduit, reinforced thermosetting resin conduit or electrical metallic tubing, provided the conduit body's trade size is no larger than the largest trade size of the supporting raceway ≫ *314.23(E) Exception* ≪.

B Conduit body support must be rigid and secure ≫ *314.16(C)(2)* ≪.

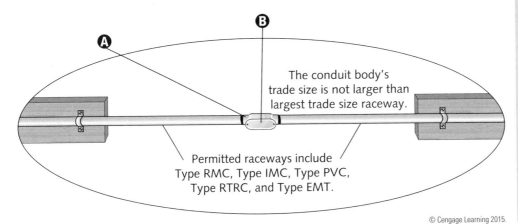

The conduit body's trade size is not larger than largest trade size raceway.

Permitted raceways include Type RMC, Type IMC, Type PVC, Type RTRC, and Type EMT.

© Cengage Learning 2015.

Supporting Luminaires Using Lengths of Conduit Longer than 18 in.

A Section 314.23(F) Exception No. 2 permits unbroken length(s) of rigid or intermediate metal conduit to support a box used for luminaire or lampholder support or to support a wiring enclosure within a luminaire where all of the following conditions are met:

B The length of conduit extending beyond the last point of securely fastened support does not exceed 3 ft (900 mm).

C A luminaire supported by a single conduit does not exceed 12 in. (300 mm), in any direction, from the point of conduit entry.

D The weight supported by any single conduit shall not exceed 20 pounds (9 kg).

E At the luminaire end, each conduit (if more than one) is threaded wrenchtight into the box or wiring enclosure or into identified hubs.

F Where accessible to unqualified persons, the luminaire's lowest point is at least 8 ft (2.5 m) abovegrade (or standing area) and at least 3 ft (900 mm) [measured horizontally to the 8 ft (2.5 m) elevation] from windows, doors, porches, fire escapes, or similar locations.

G The unbroken conduit before the last point of support is 12 in. (300 mm) or greater and that portion of the conduit is securely fastened not less than 12 in. (300 mm) from its last point of support.

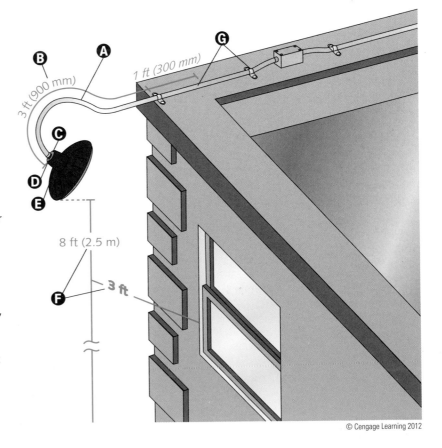

© Cengage Learning 2012

Strain Relief for Flexible Cords

A Multiconductor cord or cable used to support a box must be protected in an approved manner so that the conductors are not subjected to strain. A strain-relief connector threaded into a box with a hub would be acceptable »*314.23(H)(1)*«.

© Cengage Learning 2012

Luminaire Hanger

A Section 314.27(A)(2) requires an independent support, such as a luminaire hanger, for luminaires that weigh more than 50 pounds (23 kg) unless the outlet box is listed and marked on the interior of the box to indicate the maximum weight the box shall be permitted to support.

B When raceway fittings are used to support luminaire(s), they must be capable of supporting the combined weight of the luminaire assembly and lamp(s) »*410.36(E)*«.

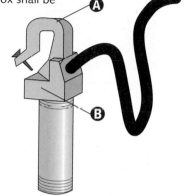

© Cengage Learning 2012

Conduit Stems Supporting Luminaires

A An outlet box used exclusively for lighting shall be designed or installed so that a luminaire or lampholder may be attached. Boxes must be able to support a luminaire weighing at least 50 pounds (23 kg). A luminaire that weighs more than 50 pounds (23 kg) shall be supported independently of the outlet box, unless the outlet box is listed and marked on the interior of the box to indicate the maximum weight the box shall be permitted to support »*314.27(A)(2)*«.

B Stems longer than 18 in. (450 mm) shall be connected to the wiring system with flexible fittings suitable for the location. At the luminaire end, the conduit(s) shall be threaded wrenchtight into the box or wiring enclosure or into identified hubs »*314.23(H)(2)*«.

C A box supporting lampholders, luminaires, or wiring enclosures within luminaires used in lieu of boxes in compliance with 300.15(B) must be supported by rigid or intermediate metal conduit stems »*314.23(H)(2)*«.

D A luminaire supported by a single conduit shall not exceed 12 in. (300 mm) in any horizontal direction from the point of conduit entry »*314.23(H)(2)*«.

CAUTION

Any point of a luminaire supported by a single conduit must be at least 8 ft (2.5 m) abovegrade (or standing area) and at least 3 ft (900 mm) (measured horizontally to the 8 ft [2.5 m] elevation) from windows, doors, porches, fire escapes, or similar locations, unless an effective means to prevent the threaded joint from loosening (such as a set screw) is used »*314.23(H)(2)*«.

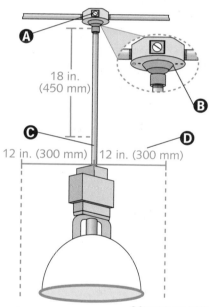

18 in. (450 mm)

12 in. (300 mm) 12 in. (300 mm)

© Cengage Learning 2012

Device Boxes Supporting Luminaires or Lampholders

A A vertically mounted luminaire or lampholder weighing no more than 6 pounds (3 kg) can be supported by boxes (such as device boxes) not specifically designed to support luminaires or lampholders, provided the luminaire or its supporting yoke, or the lampholder, is secured to the box with at least two No. 6 or larger screws. Plaster rings, secured to other boxes, are also acceptable »*314.27(A) Exception*«.

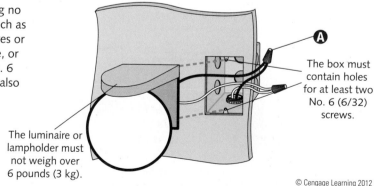

The box must contain holes for at least two No. 6 (6/32) screws.

The luminaire or lampholder must not weigh over 6 pounds (3 kg).

© Cengage Learning 2012

Mounting Nonmetallic Boxes

Ⓐ Supporting screws for nonmetallic boxes must be mounted outside of the box, unless the box is constructed in a manner that prevents contact between the conductors in the box and the supporting screws 》314.43《.

> **N O T E**
>
> Approved drainage openings not larger than 1/4 in. (6 mm) shall be permitted to be installed in the field in boxes or conduit bodies listed for use in damp or wet locations. For installation of listed drain fittings, larger openings are permitted to be installed in the field in accordance with manufacturer's instructions 》*314.15*《.

Luminaire Outlet Boxes

Ⓐ Boxes used at luminaire or lampholder outlets in a ceiling must be designed for the purpose and shall be required to support a luminaire weighing a minimum of 50 pounds (23 kg). Every box used exclusively for lighting must be designed and installed so that a luminaire may be attached 》*314.27(A)(2)*《.

Ⓑ An outlet box can support a luminaire weighing no more than 50 pounds (23 kg), unless the outlet box is listed and marked on the interior of the box to indicate the maximum weight the box shall be permitted to support 》*314.27(A)(2)*《.

Ⓒ Inspection of the connections between luminaire conductors and circuit conductors must be possible without having to disconnect any part of the wiring (unless the luminaires are connected by attachment plugs and receptacles) 》*410.8*《.

> **N O T E**
>
> Typically, two No. 8 (8/32) screws are used to attach a luminaire or its supporting yoke to an outlet box.

> **N O T E**
>
> Boxes used at luminaire or lampholder outlets in or on a vertical surface shall be identified and marked on the interior of the box to indicate the maximum weight of the luminaire that is permitted to be supported by the box if other than 50 lb (23 kg) 》*314.27(A)(1)*《.

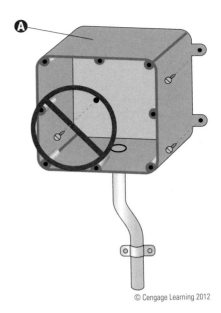

© Cengage Learning 2012

Ceiling-Suspended (Paddle) Fan Boxes

Ⓐ An outlet box or outlet box system shall not be used as the sole means of support for a ceiling-suspended (paddle) fan unless the box is (1) listed, (2) marked by the manufacturer as suitable for the purpose, and (3) supporting a ceiling-suspended (paddle) fan weighing 70 pounds (32 kg) or less. For outlet boxes or outlet box systems designed to support ceiling-suspended (paddle) fans weighing more than 35 pounds (16 kg), the required marking must include the maximum weight that can be supported 》*314.27(C)*《.

Ⓑ Ceiling-suspended (paddle) fans must be supported independently of an outlet box, unless the listed outlet box or outlet box system is identified for the use and installed in accordance with 314.27(C) 》*422.18*《.

> **N O T E**
>
> Where spare, separately switched, ungrounded conductors are provided to a ceiling-mounted outlet box, in a location acceptable for a ceiling-suspended (paddle) fan in single-family, two-family, or multi-family dwellings, the outlet box or outlet box system shall be listed for sole support of a ceiling-suspended (paddle) fan 》*314.27(C)*《.

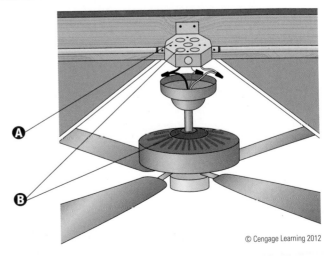

© Cengage Learning 2012

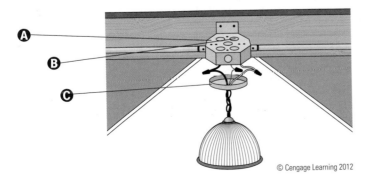

© Cengage Learning 2012

JUNCTION AND PULL BOX SIZING

Straight Pull—Two Raceways

Section 314.16 is used to determine box size requirements for 6 AWG and smaller conductors. Calculations are based on the sizes and numbers of *conductors*. Section 314.28 is used to determine the box size requirements for 4 AWG and larger conductors (under 600 volts). Calculations here are based on the sizes and numbers of *raceways*.

A Box calculations for 4 AWG and larger conductors (under 600 volts) are performed based on the size and numbers of raceways ⟫⟫314.28⟪⟪.

B Boxes or conduit bodies containing straight pulls are sized according to the largest raceway entering the box. The length must be at least eight times the trade size (metric designator) of the largest raceway ⟫⟫314.28(A)(1)⟪⟪.

N O T E

Use the trade dimension that is applicable to the installation. For example, if millimeters or centimeters are needed to size the junction or pull box, use the metric designator in millimeters instead of the trade size in inches. A junction box is needed for a straight pull with two metric designator 53 raceways. (A 2-in. trade size raceway has a metric designator of 53 mm.) Calculate the minimum length by multiplying the metric designation by eight (53 mm × 8 = 424 mm = 42.4 cm). The minimum size pull box required is 424 mm or 42.4 cm.

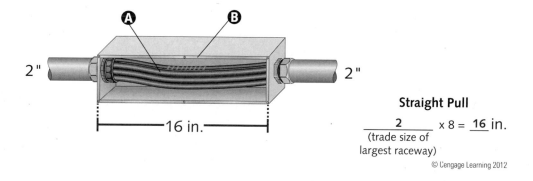

2" 2"

16 in.

Straight Pull

$$\frac{2}{\text{(trade size of largest raceway)}} \times 8 = \underline{16} \text{ in.}$$

© Cengage Learning 2012

WIDTH AND DEPTH OF THE BOX

A The box width must be large enough to provide proper installation of the raceway or cable, including locknuts and bushings.

B The box depth must be large enough to provide proper installation of the raceway or cable, including locknuts and bushings.

C No requirement specifies the depth of the box, unless a raceway enters the back of the box.

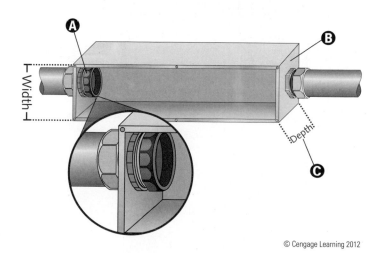

Width

Depth

© Cengage Learning 2012

Straight Pull—Multiple Raceways

Ⓐ The length must be at least eight times the trade size (metric designator) of the largest single raceway. No extra space is required for additional raceways when calculating the minimum length of straight pulls. However, additional space is needed for the width of additional raceways, including locknuts and bushings ⟫*314.28(A)(1)*⟪.

Ⓑ Conduit bodies and boxes (junction, pull, and outlet) must be installed so that the wiring they contain can be made accessible without removing any part of the building or structure ⟫*314.29*⟪.

> ### CAUTION
> Except as permitted in 250.112(l), all metal boxes shall be grounded and bonded in accordance with Parts I, IV, V, VI, VII, and X of Article 250 as applicable ⟫*314.4*⟪.

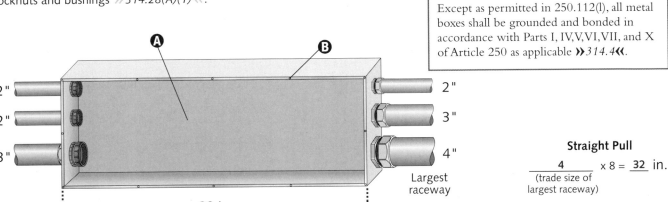

Straight Pull

$$\underset{\substack{\text{(trade size of} \\ \text{largest raceway)}}}{\underline{\quad 4 \quad}} \times 8 = \underline{\textbf{32}} \text{ in.}$$

© Cengage Learning 2012

Angle or U Pulls (or Splices)—Multiple Raceways

Ⓐ Where splices or where angle or U pulls are made, the distance between each raceway entry into the box and the opposite wall must be at least six times the trade size (metric designator) of the largest raceway in a row. This distance is increased for additional raceway entries (in the same row on the same wall of the box) by the sum of the diameters of all other raceway entries ⟫*314.28(A)(2)*⟪.

Ⓑ To calculate the dimension of a box with angle pulls, start with one wall where the raceways enter the box, and find the distance to the opposite wall of the box. The path of the conductors is irrelevant to this calculation.

Ⓒ Pick one wall and multiply the largest raceway (trade diameter) by 6. Add to that number the trade diameter of all other raceway(s) in the same row, on the same side of the box.

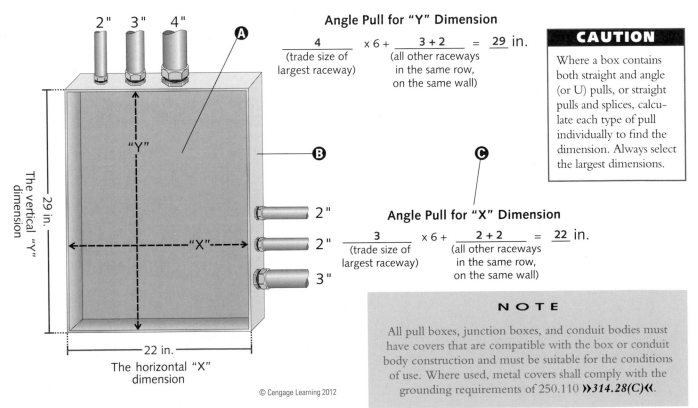

Angle Pull for "Y" Dimension

$$\underset{\substack{\text{(trade size of} \\ \text{largest raceway)}}}{\underline{\quad 4 \quad}} \times 6 + \underset{\substack{\text{(all other raceways} \\ \text{in the same row,} \\ \text{on the same wall)}}}{\underline{\quad 3 + 2 \quad}} = \underline{\textbf{29}} \text{ in.}$$

> ### CAUTION
> Where a box contains both straight and angle (or U) pulls, or straight pulls and splices, calculate each type of pull individually to find the dimension. Always select the largest dimensions.

Angle Pull for "X" Dimension

$$\underset{\substack{\text{(trade size of} \\ \text{largest raceway)}}}{\underline{\quad 3 \quad}} \times 6 + \underset{\substack{\text{(all other raceways} \\ \text{in the same row,} \\ \text{on the same wall)}}}{\underline{\quad 2 + 2 \quad}} = \underline{\textbf{22}} \text{ in.}$$

> ### NOTE
> All pull boxes, junction boxes, and conduit bodies must have covers that are compatible with the box or conduit body construction and must be suitable for the conditions of use. Where used, metal covers shall comply with the grounding requirements of 250.110 ⟫*314.28(C)*⟪.

© Cengage Learning 2012

Straight Pulls with Splices—Multiple Raceways

Where a junction box contains 4 AWG or larger conductors that are spliced and the box contains only straight pulls, compliance with 314.28(A)(1) and (A)(2) is required. Perform both the straight pull calculation and the angle pull calculation and then compare the results. The minimum dimension will be the larger of the two calculations.

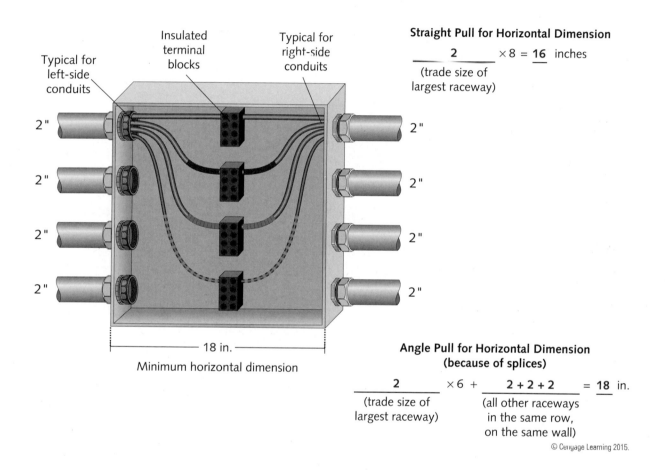

Straight Pull for Horizontal Dimension

$$\frac{2}{\text{(trade size of largest raceway)}} \times 8 = \underline{\mathbf{16}} \text{ inches}$$

Angle Pull for Horizontal Dimension (because of splices)

$$\frac{2}{\substack{\text{(trade size of} \\ \text{largest raceway)}}} \times 6 + \frac{2 + 2 + 2}{\substack{\text{(all other raceways} \\ \text{in the same row,} \\ \text{on the same wall)}}} = \underline{\mathbf{18}} \text{ in.}$$

© Cengage Learning 2015.

Labels in figure:
Typical for left-side conduits
Insulated terminal blocks
Typical for right-side conduits
2″ (×4 left), 2″ (×4 right)
18 in.
Minimum horizontal dimension

Raceways Enclosing the Same Conductors

The distance between raceway entries enclosing the same conductor shall not be less than six times the trade size (metric designator) of the largest raceway 》*314.28(A)(2)*《. This provision is applicable even if the raceway entries are on different walls.

Ⓐ Because no other raceways enter on the same wall of the box, no additional raceway diameters are added.

Ⓑ The minimum dimension required for this box is 12 in. × 12 in. The distance between the raceway entries must not be less than six times the trade size of the larger raceway. Because each raceway is located in the center of the wall of the box, the distance between the raceway entries is less than 12 in. Therefore, this installation is not permitted.

Ⓒ This 12 in. × 12 in. junction (or pull) box is permitted if the raceways can be installed so the distance between the raceway entries is at least 12 in. Because the distance between the raceway entries is 12 in., this installation is permitted.

Angle Pull for "X" Dimension

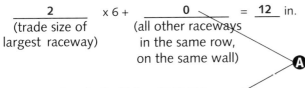

$$\underbrace{2}_{\substack{\text{(trade size of} \\ \text{largest raceway)}}} \times 6 + \underbrace{0}_{\substack{\text{(all other raceways} \\ \text{in the same row,} \\ \text{on the same wall)}}} = \underline{12} \text{ in.}$$

Angle Pull for "Y" Dimension

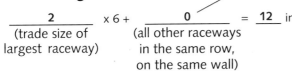

$$\underbrace{2}_{\substack{\text{(trade size of} \\ \text{largest raceway)}}} \times 6 + \underbrace{0}_{\substack{\text{(all other raceways} \\ \text{in the same row,} \\ \text{on the same wall)}}} = \underline{12} \text{ in.}$$

Raceways Enclosing the Same Conductors

$$\underbrace{2}_{\substack{\text{(trade size of} \\ \text{larger raceway)}}} \times 6 = \underline{12} \text{ in.}$$

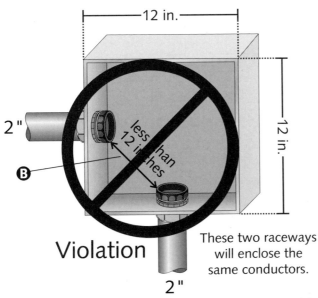

Violation

These two raceways will enclose the same conductors.

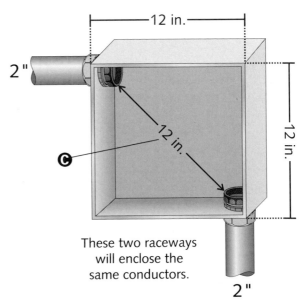

These two raceways will enclose the same conductors.

Conductors in these boxes will be larger than 6 AWG.

Angle Pull—Multiple Rows

A Where multiple rows of raceways enter a box, calculate each row separately. Use the single row that provides the maximum distance ❯❯*314.28(A)(2)*❮❮.

B Create an imaginary divider line separating each row.

C Perform each row calculation as if in a separate box.

Angle Pull for "Y" Dimension (Back Row)

$$\underset{\substack{\text{(trade size of} \\ \text{largest raceway)}}}{4} \times 6 + \underset{\substack{\text{(all other raceways,} \\ \text{in the same row,} \\ \text{on the same wall)}}}{3\frac{1}{2} + 3\frac{1}{2}} = \underline{31} \text{ in.}$$

Angle Pull for "Y" Dimension (Front Row)

$$\underset{\substack{\text{(trade size of} \\ \text{largest raceway)}}}{3\frac{1}{2}} \times 6 + \underset{\substack{\text{(all other raceways,} \\ \text{in the same row,} \\ \text{on the same wall)}}}{3 + 3 + 2\frac{1}{2} + 2\frac{1}{2}} = \underline{32} \text{ in.}$$

Angle Pull for "X" Dimension (Back Row)

$$\underset{\substack{\text{(trade size of} \\ \text{largest raceway)}}}{4} \times 6 + \underset{\substack{\text{(all other raceways,} \\ \text{in the same row,} \\ \text{on the same wall)}}}{4 + 4} = \underline{32} \text{ in.}$$

Angle Pull for "X" Dimension (Front Row)

$$\underset{\substack{\text{(trade size of} \\ \text{largest raceway)}}}{3\frac{1}{2}} \times 6 + \underset{\substack{\text{(all other raceways,} \\ \text{in the same row,} \\ \text{on the same wall)}}}{3 + 3 + 2 + 2} = \underline{32} \text{ in.}$$

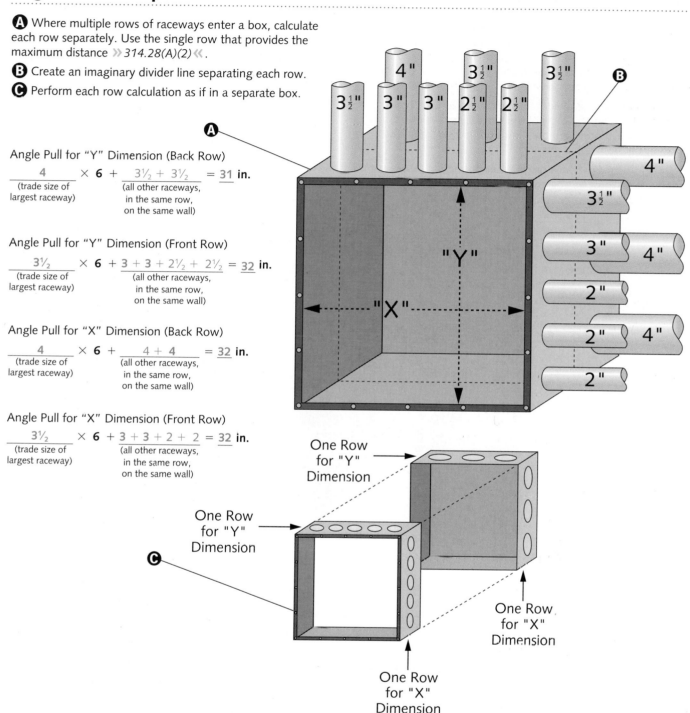

© Cengage Learning 2012

NOTE

In pull or junction boxes having any dimension over 6 ft (1.8 m), all conductors shall be cabled (or racked) in an approved manner ❯❯*314.28(B)*❮❮.

U Pull—Two Raceways

Ⓐ Use the angle pull method to calculate U pulls. A box with conduit entries only on one wall has a minimum distance to the opposite wall. Multiply the largest raceway by six and add the sum of the trade size (metric designator) of the other raceway(s) entering the same wall ≫*314.28(A)(2)*≪.

Ⓑ The minimum box width must include 12 in. between raceways plus the thickness of the two raceways (including enough area to provide proper installation of locknuts and bushings).

Ⓒ The distance between raceways enclosing the same conductor(s) must be at least six times the trade size (metric designator) of the largest raceway ≫*314.28(A)(2)*≪.

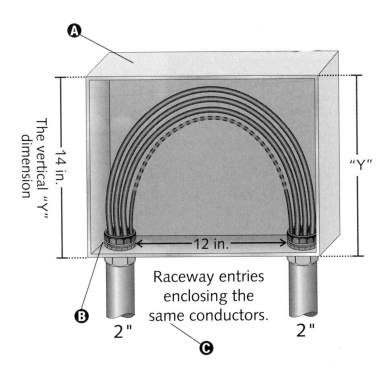

Raceway entries enclosing the same conductors.

U Pull for "Y" Dimension

$$\underset{\substack{\text{(trade size of} \\ \text{largest raceway)}}}{\underline{2}} \times 6 + \underset{\substack{\text{(all other raceways} \\ \text{in the same row,} \\ \text{on the same wall)}}}{\underline{2}} = \underline{14}\ \text{in.}$$

Raceways Enclosing the Same Conductors

$$\underset{\substack{\text{(trade size of} \\ \text{larger raceway)}}}{\underline{2}} \times 6 = \underline{12}\ \text{in.}$$

Raceways Entering Opposite Removable Covers

Ⓐ Where angle pulls are made, the distance between each raceway entry into the box and the opposite wall of the box must be at least six times the trade size (metric designator) of the largest raceway »*314.28(A)(2)*«. Where no other raceways enter the same wall of the box, no additional raceway diameters are added.

Ⓑ Where a raceway or cable enters the wall of a box (or conduit body) opposite a removable cover, the distance from the entry wall to the cover can be determined by the distance requirements for one wire per terminal found in Table 312.6(A) »*314.28(A)(2) Exception*«.

Ⓒ The minimum distance between raceways enclosing the same conductor(s) is six times the trade size (metric designator) of the largest raceway »*314.28(A)(2)*«.

Raceways or Cables Entering Boxes Opposite from Removable Covers		
Wire Size (AWG or kcmil)	Minimum Distance from Wall to Cover	
	in.	mm
4–3	2	50.8
2	2½	63.5
1	3	76.2
1/0–2/0	3½	88.9
3/0–4/0	4	102
250	4½	114
300–350	5	127
400–500	6	152
600–700	8	203
750–900	8	203
1000–1250	10	254
1500–2000	12	305

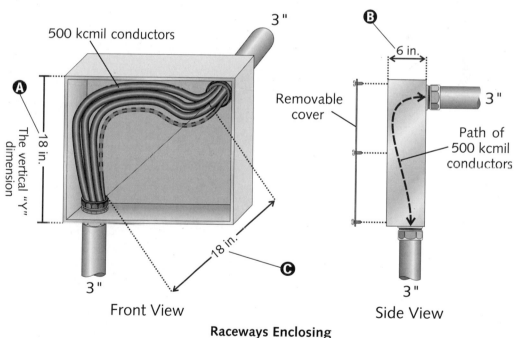

Angle Pull for "Y" Dimension

$$\underset{\substack{\text{(trade size of} \\ \text{largest raceway)}}}{\underline{\quad 3 \quad}} \times 6 + \underset{\substack{\text{(all other raceways} \\ \text{in the same row,} \\ \text{on the same wall)}}}{\underline{\quad 0 \quad}} = \underline{\ \mathbf{18}\ } \text{ in.}$$

Raceways Enclosing the Same Conductors

$$\underset{\substack{\text{(trade size of} \\ \text{larger raceway)}}}{\underline{\quad 3 \quad}} \times 6 = \underline{\ \mathbf{18}\ } \text{ in.}$$

Boxes and Conduit Bodies Not Meeting 314.28 and Chapter 9

A Listed boxes or listed conduit bodies of dimensions less than those required in 314.28(A)(1) and (A)(2) shall be permitted for installations of combinations of conductors that are less than the maximum conduit or tubing fill (of conduits or tubing being used) permitted by Table 1 of Chapter 9 *314.28(A)(3)*.

B Listed conduit bodies of dimensions less than those required in 314.28(A)(2), and having a radius of the curve to the center-line not less than that indicated in Table 2 of Chapter 9 for one-shot and full-shoe benders, shall be permitted for installations of combinations of conductors permitted by Table 1 of Chapter 9. These conduit bodies shall be marked to show they have been specifically evaluated in accordance with this provision.

Where the permitted combinations of conductors for which the box or conduit body has been listed are less than the maximum conduit or tubing fill permitted by Table 1 of Chapter 9, the box or conduit body shall be permanently marked with the maximum number and maximum size of conductors permitted.

C Three 500 kcmil conductors are the maximum number and size permitted in this conduit body.

D By comparison, the maximum number of 500 kcmil THHN conductors permitted in 4-inch rigid metal conduit is seven.

E By comparison, the maximum number of 300 kcmil THHN conductors permitted in a 4-inch rigid metal conduit is eleven.

F Three 300 kcmil conductors are the maximum number and size permitted in this conduit body.

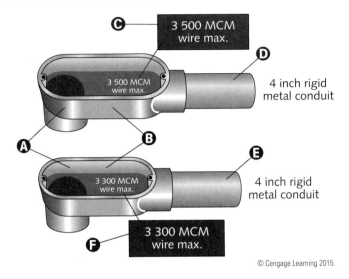

3 500 MCM wire max.

4 inch rigid metal conduit

3 300 MCM wire max.

4 inch rigid metal conduit

© Cengage Learning 2015.

> **WARNING**
>
> Conduit bodies do not have to be large enough to contain the same size and number of conductors permitted in the raceway that enters the conduit body.

> **CAUTION**
>
> Listed or labeled equipment shall be installed and used in accordance with any instructions included in the listing or labeling *110.3(B)*.

> **NOTE**
>
> Smaller-size conduit bodies can generally hold the same number and sizes of conductors (6 AWG and smaller) as the raceway entering the conduit body.

> **NOTE**
>
> If a conduit body is durably and legibly marked by the manufacturer with the volume, it can contain splices, taps, or devices *314.16(C)(2)*.

Summary

- Limitations apply to the size and number of conductors permitted in boxes (junction, pull, etc.) and conduit bodies.
- One strap (yoke) counts as two of the largest conductors connected to the device.
- Nonmetallic boxes may list the maximum number of conductors permitted in the box, but that number must be reduced if switches, receptacles, etc. are installed in the box.
- Plaster rings, raised covers, extension rings, etc. can provide additional cubic-inch capacity to boxes.
- Conduit bodies can contain splices, taps, or devices if the conduit body is durably and legibly marked by the manufacturer with the volume.
- The minimum length of free conductor (6 in. [150 mm]) is measured from the point in the box where it emerges from its raceway or cable sheath.

- Except as permitted in 250.112(l), all metal boxes shall be grounded and bonded in accordance with Parts I, IV, V, VI, VII, and X of Article 250 as applicable.
- Boxes and conduit bodies must remain accessible.
- In suspended ceilings, boxes can be secured by support wires that are installed in addition to the ceiling grid support wires.
- Where boxes enclose 4 AWG and larger conductors that are required to be insulated, different calculation methods apply to straight pulls than to angle pulls.
- There is a minimum distance required between raceways enclosing the same conductors (4 AWG and larger) within a pull or junction box.
- Conduit bodies may not allow the same size and number of conductors as is permitted in the conduit (or tubing) that enters the conduit body.

Unit 3 Competency Test

NEC **Reference**   **Answer**

314.28(B)   6 Ft.

342.15   2

314   6

1. In pull boxes or junction boxes having any dimension over _____ in., all conductors shall be cabled or racked up in an approved manner.

    a) 6   b) 12   c) 36   d) 72

2. A nonmetallic extension can be run in any direction from an existing outlet, but shall not be run on the floor or within _____ in. from the floor.

3. What are the minimum dimensions for "X" and "Y" in the drawing below?

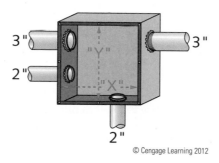

© Cengage Learning 2012

316.   1/2

B/5   6

312   1/4

370   .0625

1926, 405   bushings

312   D

4. Outlet boxes that do not enclose devices or utilization equipment shall have a minimum internal depth of _____ in.

5. A 4¹¹⁄₁₆-in. square box that is 2⅛ in. deep already contains two receptacles, two internal cable clamps, four 12 AWG THHN copper conductors (two black and two white), two grounding conductors, three wire nuts, two pigtails, and one equipment bonding jumper. (A flat plaster ring was used to secure the receptacles to the box.) How many more 12 AWG THHN copper conductors can be added to this box?

6. In damp or wet locations, surface-type enclosures (within the scope of Article 312) shall be so placed or equipped as to prevent moisture or water from entering and accumulating within the cabinet or cutout box and shall be mounted so there is at least _____ in. airspace between the enclosure and the wall or other supporting surface.

7. The minimum thickness in inches for a steel box measuring 6 in. × 4 in. × 3½ in. is _____ in. thick.

8. Covers of outlet boxes and conduit bodies having holes through which flexible cord pendants may pass shall be provided with approved _____ or shall have smooth, well-rounded surfaces on which the cord may bear.

9. A weatherproof metal junction box measuring 4 in. × 4 in. × 6 in. has been installed to support a luminaire. Two rigid metal conduits have been threaded wrenchtight into the threaded entries of the enclosure. The box has not yet been secured to the brick wall, but both conduits have been strapped to the wall 16 in. from the box. What additional support (from the following list) is required for this box?

    I. Two metal screws with plastic anchors

    II. Two toggle bolts

    III. Two drive pins

        a) none   b) I or II only   c) I or III only   d) I, II, or III

NEC Reference	Answer
3 70	Box
316	4
317	EGC
314	1/4 in
314	current carrying
315	2
314	6
362	8
370	4
314	Bushings
314	8
314	3 times
314	1/4

10. Where permanent barriers are installed in a junction box, each section shall be considered as a(n) _____.

11. What is the maximum number of 12 AWG THW conductors permitted in a 4-in. octagon box that is 1½ in. deep? (The box contains a fixture stud and a hickey.)

12. A means must be provided in each metal box for the connection of a(n) _____.

13. In walls or ceiling of noncombustible material, boxes shall be installed so that the front edge of the box will not be set back of the finished surface more than _____ in.

14. Nonmetallic boxes shall be suitable for the lowest _____ conductor entering the box.

15. An electrician needs to install a surface-mounted, nonmetallic, weatherproof cabinet on the outside of a concrete block building. Since the cabinet must be mounted in a wet location, it must be mounted so there is at least _____ in. of airspace between the cabinet and the wall.

16. How many 14 AWG THW conductors are permitted in a 3 in. × 2 in. × 2½ in. device box?

17. Nonmetallic surface extensions must be secured in place by approved means at intervals not exceeding _____ in.

18. A 2½-in. rigid metal conduit enters the back of a pull box with a removable cover on the opposite side. The raceway encloses four 4/0 AWG THHN copper conductors and one 4 AWG equipment grounding conductor. The minimum depth (distance from the entry wall to the cover) for this pull box is _____ in.

19. Where nonmetallic boxes are used with open wiring or concealed knob-and-tube wiring, the conductors must enter the box through _____.

20. Luminaires weighing no more than _____ pounds can be supported by boxes (such as device boxes) not specifically designed to support luminaires, provided the luminaire, or its supporting yoke, is secured to the box with at least two No. 6 or larger screws.

21. For straight pulls, the length of the box shall not be less than _____ the outside diameter, over sheath, of the largest shielded conductor or cable entering the box on systems over 600 volts, nominal.

22. Approved drainage openings not larger than _____ shall be permitted to be installed in the field in boxes or conduit bodies listed for use in damp or wet locations.

UNIT 4

Cables

Objectives

After studying this unit, the student should:

▶ know that cables must be installed at least 1¼ in. (32 mm) from the nearest edge of wood framing members, unless a steel plate (or bushing) has been installed.

▶ understand that nonmetallic-sheathed cable passing through metal framing must be protected by bushings (or grommets) covering all metal edges.

▶ be aware that openings around electrical penetrations through fire-resistant-rated construction must be sealed using approved methods to maintain the fire-resistant rating.

▶ be able to determine what cables are permitted in spaces used for environmental air-handling purposes.

▶ know the support requirements for MC, AC, and nonmetallic-sheathed cable.

▶ be aware that a minimum bending radius must be maintained with cables.

▶ understand that cables must be protected from physical damage.

▶ be familiar with both general and specific installation provisions for MC, AC, and nonmetallic-sheathed cable.

▶ understand conductor identification and the permissible reidentification of certain conductors.

▶ have a good grasp of underground installation provisions.

▶ be introduced to flat conductor, integrated gas-spacer, mineral-insulated, and medium-voltage cables.

Introduction

Unit 4 contains regulations for cable systems that are used often (such as metal-clad cable and nonmetallic-sheathed cable), as well as cable systems that are used rarely (such as integrated gas-spacer cable and flat conductor cable). A key provision found in most of the cable articles (and some of the raceway articles) states that the cable (or raceway) must not be installed where subject to physical damage. Cable protection can be achieved by several methods, including installing the cable a minimum distance from the outside edge of a framing member; installing a steel plate on the wood framing member, over the cable; or installing the cable in an approved raceway system. Cables installed in attics have different provisions depending on the type of attic entrance. Minimum support distances (from box or enclosure to the first support, or between supports) are included in this unit. Conductor identification (and permissible reidentification) for cables is also presented in this unit.

The cable articles, found in Chapter 3 of the *NEC* have sections titled "Uses Permitted" and "Uses Not Permitted." These sections (*XXX.10* and *XXX.12*) give locations (some specific) where the cable system can (or cannot) be installed. Since most of these provisions are clearly understandable, this book does not repeat much of this information. Also, be aware that state and local jurisdictions restrict the usage of certain cables in some (if not all) types of occupancies.

GENERAL INSTALLATION

Setback for Bored Holes

Ⓐ Holes must be bored so that the edge of the hole is at least 1¼ in. (32 mm) from the nearest edge of the wood member 》*300.4(A)(1)* 《.

Ⓑ The 1¼-in. (32-mm) setback applies to all locations (concealed and exposed).

Ⓒ The 1¼-in. (32-mm) setback applies to studs, joists, rafters, etc.

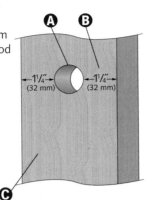

© Cengage Learning 2012

> ### NOTE
>
> The largest hole that can be bored in a 3½-in. wood stud without using a steel plate (clip or sleeve) is 1 in. (provided the drill bit cuts straight through the center of the stud). The 1¼-in. (32-mm) setback must be maintained from both edges (front and back); 3½ in. minus 2½ in. (1¼ setback, two edges) leaves a remainder of 1 in.

Metal Framing Members

Ⓐ Nonmetallic-sheathed cable passing through holes (or slots) in metal framing must be protected by listed bushings (or grommets) covering all metal edges and securely fastened in the opening prior to cable installation 》*300.4(B)(1)* 《.

Ⓑ Section 300.4 applies to all locations (concealed and exposed).

Ⓒ Openings (slots or holes) can be punched (either factory or field), cut, or drilled into the metal members 》*300.4(B)(1)* 《.

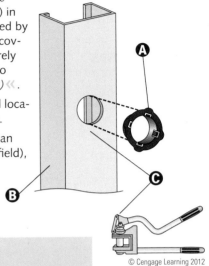

© Cengage Learning 2012

> ### NOTE
>
> Where nails (or screws) are likely to penetrate nonmetallic-sheathed cable (or electrical nonmetallic tubing), a steel plate (clip or sleeve) at least 1/16 in. (1.6 mm) thick shall be used to protect the cable or tubing 》***300.4(B)(2)*** 《.

Steel Plates

Ⓐ Where the bored hole is less than 1¼ in. (32 mm) from the edge, the cable (or raceway-type wiring method) shall be protected by a steel plate(s) or bushing(s) from penetration by screws or nails 》*300.4(A)(1)* 《.

Ⓑ Cables, installed in a groove, covered by wallboard, paneling, carpeting, etc., shall be protected by 1/16-in. (1.6-mm) thick steel plate, sleeve, or equivalent or by at least 1¼ in. (32 mm) free space for the full length of the groove in which the cable is installed 》*300.4(F)* 《.

Ⓒ Where there is no potential of weakening building structure, cables (or raceways) can be laid within notches in wood studs, joists, rafters, etc. 》*300.4(A)(2)* 《.

Ⓓ The steel plate(s) or bushing(s) must be at least 1/16 in. (1.6 mm) thick and of appropriate length and width to adequately cover the wiring 》*300.4(A)(1)* 《.

Ⓔ The steel plate must be at least 1/16 in. (1.6 mm) thick and of appropriate length and width installed to cover the area of the wiring. The steel plate must be installed before the building finish is installed 》*300.4(A)(2)* 《.

Ⓕ Requirements for notched wood members apply to all locations (concealed and exposed).

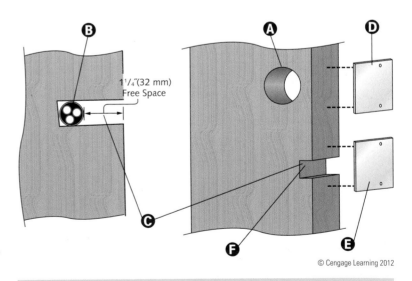

1¼" (32 mm) Free Space

© Cengage Learning 2012

> ### NOTE
>
> A listed and marked steel plate that provides equal or better protection against nail or screw penetration shall be permitted, even if it is less than 1/16 in. (1.6 mm) thick 》***300.4(A)(2) Exception No. 2*** 《.

Parallel to Framing Members and Furring Strips

(A) Cables installed parallel to framing members (studs, joists, etc.) or installed parallel to furring strips shall be installed and supported so that the nearest outside surface of the cable is at least 1¼ in. (32 mm) from the nearest edge of the framing member or furring strip where nails (or screws) are likely to penetrate ≫*300.4(D)*≪.

(B) If a 1¼-in. (32-mm) setback is not possible, the cable must be protected from penetration by nails (or screws) by a minimum ⅛₆-in. (1.6-mm) thick steel plate (sleeve, etc.) ≫*300.4(D)*≪. A listed and marked steel plate less than ⅛₆ in. (1.6 mm) thick that provides equal or better protection against nail or screw penetration can be installed ≫*300.4(D) Exception No. 3*≪.

N O T E

For concealed work in finished buildings or finished panels for prefabricated buildings, where such support is impracticable, cables can be fished between access points ≫***300.4(D) Exception No. 2***≪.

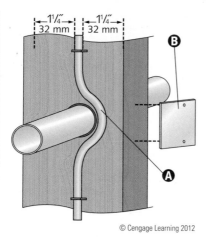

© Cengage Learning 2012

Maintaining the Integrity of Fire-Resistant-Rated Construction

(A) Electrical installation openings that penetrate into or through fire-resistant-rated structures (walls, partitions, floors, ceilings, etc.) must be firestopped using approved methods that maintain the fire-resistance rating ≫*300.21*≪.

CAUTION

Some state and local jurisdictions may require that all penetrations (fire-rated and non-fire-rated) be sealed.

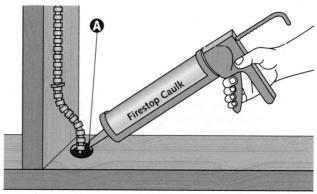

© Cengage Learning 2012

N O T E

In hollow spaces, vertical shafts, and ventilation (or air-handling) ducts, electrical installations must not substantially increase the possible spread of fire or products of combustion ≫***300.21***≪.

Wiring in Ducts, Plenums, and Other Air-Handling Spaces

(A) Only wiring methods consisting of Type MI cable without an overall nonmetallic covering, Type MC cable employing a smooth or corrugated impervious metal sheath without an overall nonmetallic covering, electrical metallic tubing, flexible metallic tubing, intermediate metal conduit, or rigid metal conduit without an overall nonmetallic covering shall be installed in ducts specifically fabricated to transport environmental air. Flexible metal conduit shall be permitted, in lengths not to exceed 4 ft (1.2 m), to connect physically adjustable equipment and devices permitted to be in these fabricated ducts ≫*300.22(B)*≪.

(B) Section 300.22(C) applies to spaces not specifically fabricated for environmental air-handling purposes but used for air-handling purposes as a plenum. This section shall not apply to habitable rooms or areas of buildings, the prime purpose of which is not air handling. Type MI cable without an overall nonmetallic covering, Type MC cable without an overall nonmetallic covering, Type AC cable, or other factory-assembled multiconductor cables (control or power) listed specifically for the use within an air-handling space, or listed prefabricated cable assemblies of metallic manufactured wiring systems without nonmetallic sheath shall be permitted to be installed in these spaces. All other types of cables and conductors must be installed in one of the raceways (or wireways) named in 300.22(C)(1).

(C) The space over a suspended ceiling used for environmental air-handling purposes is an example of the type of other space to which 300.22(C) applies ≫*300.22(C) Informational Note No. 1*≪.

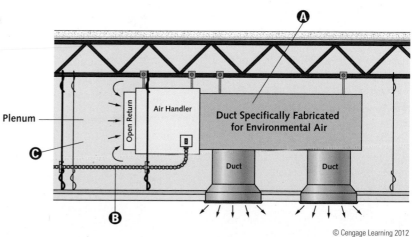

© Cengage Learning 2012

Wiring in Ducts, Plenums, and Other Air-Handling Spaces *(continued)*

> **CAUTION**
>
> Ducts used to transport dust, loose stock, or flammable vapors shall not contain wiring systems of any type. No wiring system (of any type) shall be installed within a duct or shaft containing only such ducts used for vapor removal or for ventilation of commercial-grade cooking equipment ≫*300.22(A)*≪.

> **NOTE**
>
> Nonmetallic cable ties and other nonmetallic cable accessories used to secure and support cables shall be listed as having low smoke and heat release properties ≫*300.22(C)(1)*≪.

> **NOTE**
>
> Metallic manufactured wiring systems (without nonmetallic sheath) having listed prefabricated cable assemblies are permitted in this type of installation ≫*300.22(C)(1)*≪.

Wiring within Air-Handling Spaces in Dwelling Units

Ⓐ Section 300.22 does not apply to **dwelling unit** joist or stud spaces where wiring passes perpendicular to the long dimension of such spaces ≫*300.22(C) Exception*≪.

Ⓑ Conductors must remain at least 1¼ in. (32 mm) from the nearest edge of a wood member ≫*300.4(A)(1)*≪.

Ⓒ Joist or stud spaces, used for environmental air-handling purposes, shall not contain nonmetallic-sheathed cable that runs along the long dimension ≫*300.22(C) Exception*≪.

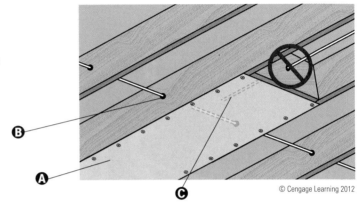

© Cengage Learning 2012

Securing Cables

Generally, all cables must be secured to the box, cabinet, etc. ≫*314.17(B) and (C)*≪.

Staples, cable ties, straps, etc. must be designed to secure cables and shall be installed so that cables are not damaged.

Ⓐ Type MC cable must be supported and secured at intervals not exceeding 6 ft (1.8 m) ≫*330.30*≪.

Ⓑ Type AC cable must be supported and secured at intervals not exceeding 4½ ft (1.4 m) ≫*320.30*≪.

Ⓒ Nonmetallic-sheathed cable must be supported and secured at intervals not exceeding 4½ ft (1.4 m) ≫*334.30*≪.

Ⓓ Type MC cables containing no more than four conductors, size 10 AWG or smaller, must be secured within 12 in. (300 mm) of every box, cabinet, fitting, etc. ≫*330.30*≪.

Ⓔ Type AC cable must be secured within 12 in. (300 mm) of every box, cabinet, or fitting ≫*320.30*≪.

Ⓕ Nonmetallic-sheathed cable (secured to the box) must be secured within 12 in. (300 mm) of every box, cabinet, or fitting ≫*334.30*≪.

Ⓖ Nonmetallic-sheathed cable *not* secured to a single-gang box must be secured within 8 in. (200 mm) ≫*314.17(C) Exception*≪.

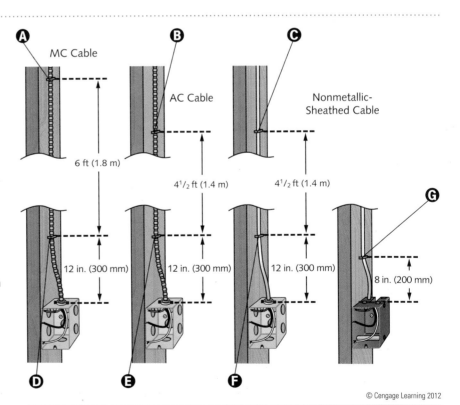

MC Cable
AC Cable
Nonmetallic-Sheathed Cable
6 ft (1.8 m)
4½ ft (1.4 m)
4½ ft (1.4 m)
12 in. (300 mm)
12 in. (300 mm)
12 in. (300 mm)
8 in. (200 mm)

© Cengage Learning 2012

> **NOTE**
>
> Cables fished between access points, where concealed in finished buildings (or structures) and where support is impracticable, can remain unsupported and unsecured ≫*320.30(D)(1), 330.30(D)(1), and 334.30(B)(1)*≪.

Cables Passing through Framing Members

Ⓐ Cables running horizontally (or diagonally) are considered supported and secured when passing through a framing member (wood, metal, etc.), unless the support intervals exceed those listed for the specified cable. Cables installed in notches and protected by a ¹⁄₁₆-in. (1.6-mm) steel plate are also considered secured ≫*320.30(C), 330.30(C), and 334.30(A)*≪ .

Ⓑ Cables must be secured within 12 in. (300 mm) of every box, cabinet, fitting, etc.

Ⓒ This cable is not "secured." The support interval exceeds the maximum distance permitted.

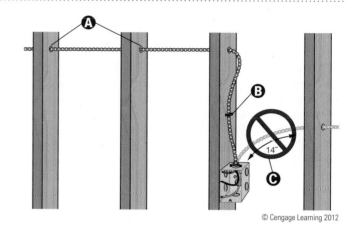

© Cengage Learning 2012

Attics without Permanent Stairs or Ladders

Ⓐ Attics and roof spaces not accessible by permanent stairs or ladders require protection only within 6 ft (1.8 m) of the nearest edge of the scuttle hole or attic entrance ≫*320.23(A), 330.23, and 334.23*≪ .

Ⓑ Protection is required for cables located within 6 ft (1.8 m) (measured both vertically and horizontally) of the attic entrance.

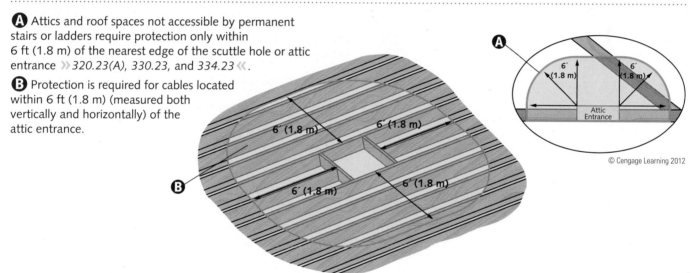

© Cengage Learning 2012

Bending Radius

Type AC cable must have a bending radius at least five times the diameter of the cable ≫*320.24*≪ .

Type MC cable (interlocked-type or corrugated sheath) has a bending radius of seven times the external diameter of the metallic sheath ≫*330.24(B)*≪ .

Nonmetallic-sheathed cable has a minimum bending radius of five times the cable diameter ≫*334.24*≪ .

Ⓐ A curve radius is measured from the inner edge of the bend.

Example: If the outside diameter of Type MC cable (interlocked-type) is ¹⁄₂ (0.5) in. (13 mm), then the radius will need a measurement of 3¹⁄₂ in. (89 mm) (0.5 in. × 7 = 3.5 in.).

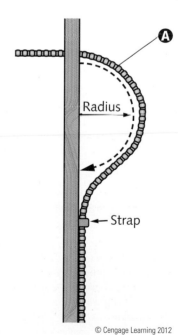

CAUTION
All bends shall be made so that the cable will not be damaged.

© Cengage Learning 2012

Attics with Permanent Stairs or Ladders

Ⓐ Cables within 7 ft (2.1 m) of the floor (or floor joist) can run through the studs if meeting the requirements of 300.4(D). Cables more than 7 ft (2.1 m) above the floor (or floor joist) can run across the face of framing members.

Ⓑ If cables are installed across the face of rafters or studding and they are located within 7 ft (2.1 m) of the floor (or floor joist) in attics or roof spaces that are accessible, the cable must be protected by guard strips that are at least as high as the cable ≫ 320.23(A) ≪.

Ⓒ Cables running across the top of floor joists must be protected by guard strips that are of sufficient height to adequatel protect the cable ≫ 320.23(A) ≪.

Ⓓ Cable installed parallel to the sides of rafters, studs, or floor joists does not require guard strips or running boards; however the installation must also comply with 300.4(D) ≫ 320.23(B) ≪

Ⓔ Cables not installed on the face (or surface) do not require guard strips or running boards.

Ⓕ Cables installed in attics (and roof spaces) accessible by permanent stairs (or ladders) must meet 320.23 provisions.

> **NOTE**
>
> Type MC cable and nonmetallic-sheathed cable must also comply with 320.23 ≫**330.23 and 334.23**≪.

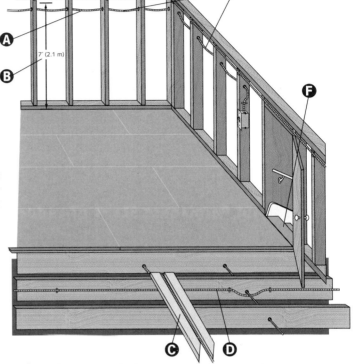

© Cengage Learning 2012

Insulating (Antishort) Bushings

Ⓐ All armor termination points of AC cable shall have a fitting to protect the wires from abrasion, unless the outlet boxes (or fittings) are designed to afford equivalent protection. In addition, an insulating bushing (or its equivalent protection) shall be provided between the conductors and the armor ≫ 320.40 ≪.

Ⓑ Type AC cable shall have a flexible metal tape armor ≫ 320.100 ≪.

Ⓒ The connector or clamp that fastens Type AC cable to boxes or cabinets must be designed so that the insulating bushing (or its equivalent) is visible for inspections ≫ 320.40 ≪.

Ⓓ Type AC cables shall have an internal bonding strip (copper or aluminum) in intimate contact with the armor for its entire length ≫ 320.100 ≪.

Ⓔ Some insulating (antishort) bushings are manufactured so that part of the bushing extends past the connector or clamp (once installed), effectively acting as a flag. This flag increases the visibility of the bushing after installation.

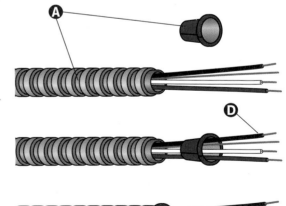
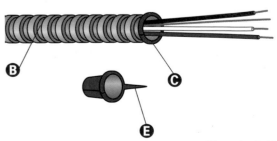

> **NOTE**
>
> Insulating bushings in Type MC cable are optional rather than required. Although Article 330 does not require an insulating bushing between the conductors and the armor, it is a good practice to use one.

> **NOTE**
>
> Fittings used for connecting Type MC cable to boxes, cabinets, or other equipment shall be listed and identified for such use ≫**330.40**≪.

© Cengage Learning 2012

Exposed Runs of Type AC and MC Cables under Joists

A Exposed runs of Type AC cable installed on the underside of joists must be supported (and secured) at *every* joist and must be located so that physical damage is avoided ⟫*320.15*⟪.

B Type MC cable must be supported (and secured) at intervals of 6 ft (1.8 m) or less ⟫*330.30(B)* and *(C)*⟪.

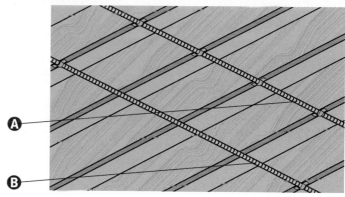

© Cengage Learning 2012

Exposed Nonmetallic-Sheathed Cable in Unfinished Basements and Crawl Spaces

A Exposed nonmetallic-sheathed cable (smaller than 8 AWG if three-conductor or 6 AWG if two-conductor) running at angles with joists in unfinished basements and crawl spaces must either be run through bored holes in the joists or on running boards ⟫*334.15(C)*⟪.

B The nearest outside surface of installed and supported cable must be at least 1¼ in. (32 mm) from the nearest edge of the framing member where nails or screws are likely to penetrate ⟫*300.4(A)(1)*⟪.

C Three-conductor 8 AWG, or two-conductor 6 AWG, or larger, can be secured directly to the lower edges of the joists ⟫*334.15(C)*⟪.

D Where current-carrying conductors in multiconductor cables are bundled or stacked longer than 24 in. (600 mm) without maintaining spacing, reduce the allowable ampacity of each conductor as shown in Table 310.15(B)(3)(a) ⟫*310.15(B)(3)(a)*⟪.

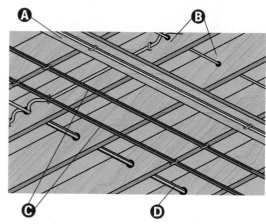

© Cengage Learning 2012

Exposed Nonmetallic-Sheathed Cable Passing through a Floor

Cables entering or exiting conduit (or tubing) used for support or protection against physical damage require fittings on the conduit (or tubing) ends to prevent cable abrasion ⟫*300.15(C)*⟪.

A Nonmetallic-sheathed cable passing through a floor must be enclosed in rigid (or intermediate) metal conduit, electrical metallic tubing, Schedule 80 PVC conduit, Type RTRC marked with the suffix -XW, or other approved means that extends at least 6 in. (150 mm) above the floor ⟫*334.15(B)*⟪.

6″ (150 mm)

Nonmetallic-Sheathed Cable Enclosed in EMT

Nonmetallic-Sheathed Cable Exposed

© Cengage Learning 2012

NOTE

Raceway (or cable) openings in a fire-resistant-rated floor must be firestopped using approved methods to maintain the fire-resistance rating ⟫*300.21*⟪.

If subject to physical damage, conductors, raceways, and cables must be protected ⟫*300.4*⟪. Nonmetallic-sheathed cable can be protected by rigid (or intermediate) metal conduit, electrical metallic tubing, Schedule 80 PVC conduit, Type RTRC marked with the suffix –XW, or other approved means ⟫*334.15(B)*⟪.

CONDUCTOR IDENTIFICATION

General Conductor Identification Provisions

A Generally, a conductor with a continuous white or gray covering shall be used only as a grounded-circuit conductor »200.7«.

B Equipment grounding conductors can be bare, covered, or insulated. Individually covered (or insulated) equipment grounding conductors must have a continuous outer finish that is either green or green with yellow stripe(s) »250.119«.

C Conductors used as ungrounded (hot) conductors, whether single conductors or in multiconductor cables, must be finished in a way that clearly distinguishes them from grounded and grounding conductors »310.110(C)«. Ungrounded (hot) conductors (except for a **high-leg** conductor) can be any color other than white, gray, green, or green with yellow stripe(s).

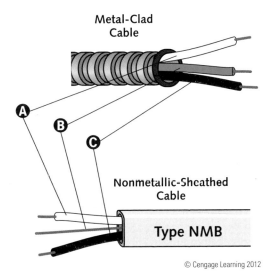

Metal-Clad Cable

Nonmetallic-Sheathed Cable

Type NMB

© Cengage Learning 2012

Three-Phase Conductor Identification

A Generally, ungrounded (hot) conductors can be any color except white, gray, green, or green with yellow stripe(s). A widely accepted practice is to identify 3-phase ungrounded (hot) conductors as black, red, and blue in 208/120 volt, 4-wire, wye-connected systems; and brown, orange, and yellow in 480/277 volt, 4-wire, wye-connected systems.

B Interlocking-armor Type MC cable requires an equipment grounding conductor.

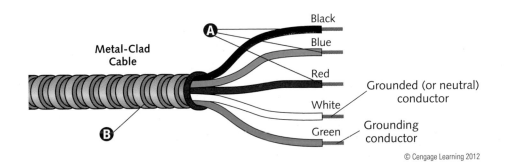

Metal-Clad Cable

Black
Blue
Red
White — Grounded (or neutral) conductor
Green — Grounding conductor

© Cengage Learning 2012

White Conductor Used as Ungrounded Switch Leg (Loop)

A A cable assembly's white or gray conductor can be used as a switch loop (leg) in single-pole, three-way, and four-way switch installations, even though it is an ungrounded (hot) conductor »200.7(C)(1)«.

B The white or gray conductor's new use as an ungrounded (hot) conductor must be permanently reidentified by marking tape, painting, or other effective means at its termination and at each location where the conductor is visible and accessible. Identification shall encircle the insulation and shall be a color other than white, gray, or green »200.7(C)(1)«.

C Red marking tape has been wrapped around the white conductor, reidentifying it as an ungrounded (hot) conductor.

D The white or gray conductor can be used to feed a switch or can be a traveler in three-way (or four-way) switch installations. The white or gray conductor shall not be used as a return conductor from the switch to the outlet »200.7(C)(1)«.

E A grounded circuit conductor is not required at the switch location where the switch controls a receptacle load »404.2(C)(7)«.

CAUTION
The grounded circuit conductor for the controlled lighting circuit shall be provided at the location where switches control lighting loads that are supplied by a grounded general-purpose branch circuit for other than the locations stated in 404.2(C)(1) through (7) »404.2(C)«.

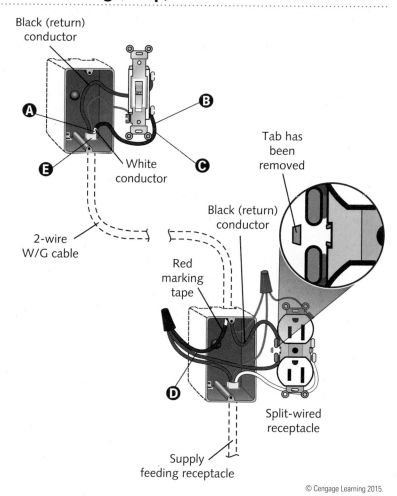

Black (return) conductor

B

Tab has been removed

White conductor

C

Black (return) conductor

2-wire W/G cable

Red marking tape

Split-wired receptacle

Supply feeding receptacle

© Cengage Learning 2015.

White Conductor Used as an Ungrounded (Hot) Conductor

A The white or gray conductor can be used as an ungrounded (hot) conductor if it is part of a cable assembly. The insulation must be permanently reidentified as an ungrounded conductor by marking tape, painting, or other effective means at its termination and at each location where the conductor is visible and accessible »200.7(C)(1)«.

WARNING
A white or gray conductor in a conduit (or raceway) shall not be used (or reidentified) as an ungrounded conductor.

NOTE
Identification must encircle the insulation and shall be a color other than white, gray, or green »**200.7(C)(1)**«.

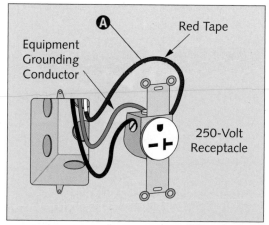

A Red Tape

Equipment Grounding Conductor

250-Volt Receptacle

© Cengage Learning 2012

GROUNDED CONDUCTORS PROVIDED AT SWITCH LOCATIONS

Ⓐ A grounded circuit conductor for the controlled lighting circuit is not required where multiple switch locations control the same lighting load such that the entire floor area of the room or space is visible from the single or combined switch locations ≫404.2(C)(5)≪.

Ⓑ Switching in three-way and four-way configurations is done only in the ungrounded (hot) circuit conductor ≫404.2(A)≪.

Ⓒ In this installation, the supply is feeding one of the three-way switches. Therefore, the grounded circuit conductor is already in each switch location.

WARNING

A grounding conductor (green, green with one or more yellow stripes, or bare) cannot be used (or reidentified) as an underground (hot) or a grounded (white) conductor.

NOTE

The grounded circuit conductor for the controlled lighting circuit shall be provided at the location where switches control lighting loads that are supplied by a grounded general-purpose branch circuit for other than the following:
(1) Where conductors enter the box enclosing the switch through a raceway, provided that the raceway is large enough for all contained conductors, including a grounded conductor
(2) Where the box enclosing the switch is accessible for the installation of an additional or replacement cable without removing finish materials
(3) Where snap switches with integral enclosures comply with 300.15(E)
(4) Where a switch does not serve a habitable room or bathroom
(5) Where multiple switch locations control the same lighting load such that the entire floor area of the room or space is visible from the single or combined switch locations
(6) Where lighting in the area is controlled by automatic means
(7) Where a switch controls a receptacle load ≫404.2(C)≪.

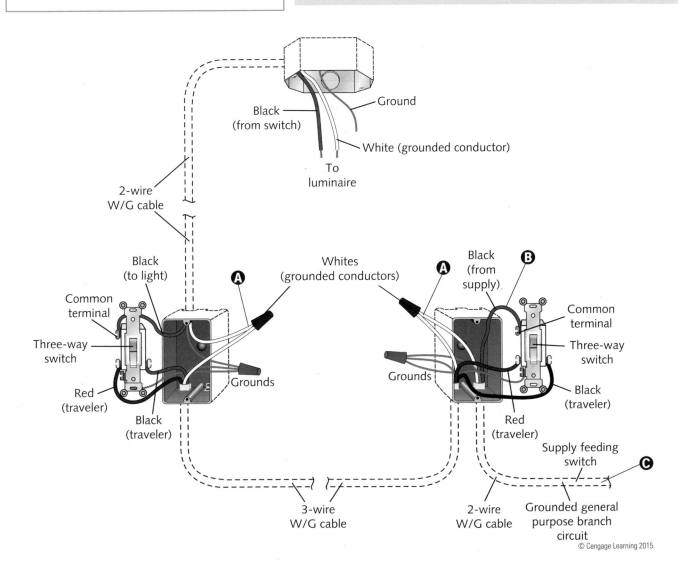

© Cengage Learning 2015.

GROUNDED CONDUCTOR PROVIDED FOR THE FUTURE

A This cable installation does not meet any of the seven exemptions in 404.2(C).

B Although the grounded conductor will not be utilized in this current design, it must still be provided at the switch location unless meeting one of the exemptions in 404.2(C)(1) through (7) ≫ *404.2(C)* ≪.

C The provision for a (future) grounded conductor is to complete a circuit path for electronic lighting control devices ≫ *404.2(C) Informational Note* ≪.

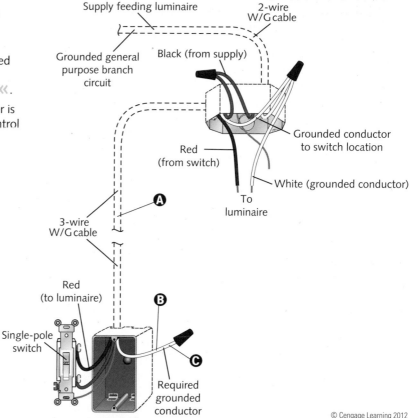

© Cengage Learning 2012

UNDERGROUND INSTALLATIONS

Cover for Direct-Burial Cables and Conductors

Cover measurement commences at the top surface of any direct-burial wiring method (conductor, cable, conduit, or other raceway) and ends at the closest point of the finished grade's top surface (including concrete, etc.) ≫ *Table 300.5* ≪.

CAUTION

Raceways must enclose underground cables and conductors installed under buildings unless the wiring method is Type MC or Type MI cable that is identified for direct burial ≫ *300.5(C)* and *Table 300.5* ≪.

NOTE

If the cover depth requirements of Table 300.5 cannot be met due to the presence of solid rock, the wiring must be installed in a raceway (metal or nonmetallic) approved for direct burial. The raceways must then be covered by a minimum of 2 in. (50 mm) of concrete that extends down to the solid rock surface ≫ *Table 300.5, Note 5* ≪.

All locations **NOT** specified in Table 300.5

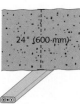

24" (600 mm)

Streets, highways, roads, alleys, driveways, and parking lots

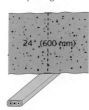

24" (600 mm)

Under minimum of 4-in. (100-mm) thick concrete exterior slab with no vehicular traffic (the slab must extend at least 6 in. [150 mm] beyond the underground installation)

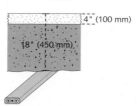

4" (100 mm)
18" (450 mm)

In trench below 2-in. (50-mm) thick concrete or equivalent

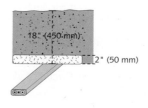

18" (450 mm)
2" (50 mm)

One- and two-family dwelling driveways (and outdoor parking areas) used only for dwelling-related purposes **AND** airport runways, including adjacent areas where trespassing is prohibited

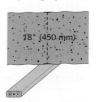

18" (450 mm)

© Cengage Learning 2012

Protection of Conductors and Cables

Underground cables and conductors installed under a building shall be in a raceway unless the wiring method is Type MC or Type MI cable that is identified for direct burial ≫ *300.5(C)* ≪.

A Protection must extend at least 8 ft (2.5 m) above finished grade ≫ *300.5(D)(1)* ≪.

B Direct-buried conductors and cables emerging from the ground must be protected by enclosures or raceways ≫ *300.5(D)(1)* ≪.

C Conductors entering a building shall be protected to the point of entry ≫ *300.5(D)(2)* ≪.

D Raceways and boxes shall be fastened in place securely ≫ *300.11(A)* ≪.

E Protection must extend below grade to a depth matching Table 300.5 cover requirements, up to a maximum depth of 18 in. (450 mm) ≫ *300.5(D)(1)* ≪.

F Direct-buried conductors, raceways, or cables subject to movement by settlement (or frost) must be arranged to prevent damage to both the enclosed conductors and the equipment connected to the raceways ≫ *300.5(J)* ≪.

G Excavation backfill shall not contain large rock, paving materials, cinders, large or sharp substances, or corrosive materials where damage to raceways or cables may occur. Provide protection in the form of granular or selected material, suitable running boards, suitable sleeves, or other approved means to prevent physical damage to the raceway or cable ≫ *300.5(F)* ≪.

H Where the conductors or cables emerge as a direct-burial wiring method, install a bushing or fitting to protect conductors from abrasion on the end of the conduit (or tubing) that terminates underground. A seal incorporating the same protective characteristics can be used in lieu of a bushing ≫ *300.5(H) and 300.15(C)* ≪.

I Section 300.5 recognizes "**S**" loops in underground direct burial to raceway transitions, expansion joints in raceway risers to fixed equipment, and the provision of flexible connections to equipment subject to settlement or frost heaves ≫ *300.5(J) Informational Note* ≪.

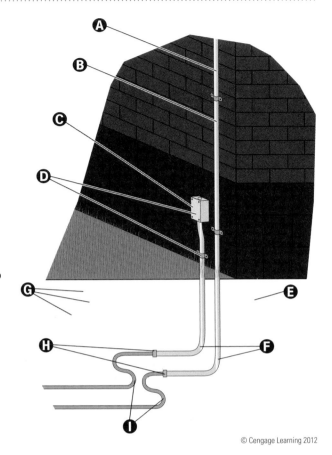

© Cengage Learning 2012

Cables Permitted Underground

Cable Types	Article
IGS Integrated Gas-Spacer Cable	326
MV Medium Voltage Cable	328
MI Mineral-Insulated Metal-Sheathed Cable	332
MC Metal-Clad Cable (where identified for such use)	330
USE Underground Service-Entrance Cable	338
UF Underground Feeder (and Branch-Circuit) Cable	340

© Cengage Learning 2015.

Cover for Residential Branch Circuits

Residential branch circuits rated 120 volts (or less) with ground-fault circuit-interrupter (GFCI) protection and maximum overcurrent protection of 20 amperes ❯❯*Table 300.5* ❮❮.

CAUTION

Raceways must enclose underground cables and conductors installed under buildings unless the wiring method is Type MC or Type MI cable that is identified for direct burial ❯❯*300.5(C)*❮❮.

N O T E

The minimum cover for residential branch circuits, in or under airport runways (including adjacent areas where trespassing is prohibited), is 18 in. (450 mm) ❯❯*Table 300.5*❮❮. Lesser depths are permitted where cables and conductors rise for terminations (or splices) or where access is otherwise required ❯❯*Table 300.5, Note 3*❮❮.

All locations **NOT** specified in Table 300.5

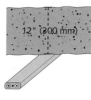

12" (300 mm)

One- and two-family dwelling driveways (and outdoor parking areas) used only for dwelling-related purposes

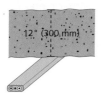

12" (300 mm)

Under minimum of 4-in. (100-mm) thick concrete exterior slab with no vehicular traffic (the slab must extend at least 6 in. [150 mm] beyond the underground installation)

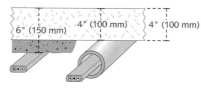

6" (150 mm) 4" (100 mm) 4" (100 mm)

In trench below 2-in. (50-mm) thick concrete or equivalent

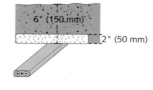

6" (150 mm) 2" (50 mm)

Streets, highways, roads, alleys, driveways, and parking lots

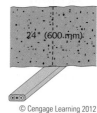

24" (600 mm)

© Cengage Learning 2012

Cover for Low-Voltage Circuits

Circuits for control of irrigation and landscape lighting limited to not more than 30 volts and installed with Type UF or in other identified cable (or raceway) ❯❯*Table 300.5* ❮❮.

N O T E

Where a wiring method listed in Table 300.5, columns 1–3, is used for a circuit type in columns 4 and 5, the shallower depth of burial is allowed ❯❯*Table 300.5, Note 3*❮❮.

All locations **NOT** specified in Table 300.5

6" (150 mm)

Under minimum of 4-in. (100-mm) thick concrete exterior slab with no vehicular traffic (the slab must extend at least 6 in. [150 mm] beyond the underground installation)

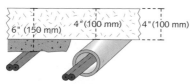

6" (150 mm) 4" (100 mm) 4" (100 mm)

In trench below 2-in. (50-mm) thick concrete or equivalent

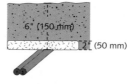

6" (150 mm) 2" (50 mm)

Streets, highways, roads, alleys, driveways, and parking lots

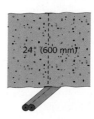

24" (600 mm)

One- and two-family dwelling driveways (and outdoor parking areas) used only for dwelling-related purposes **AND** airport runways, including adjacent areas where trespassing is prohibited

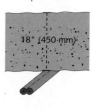

18" (450 mm)

© Cengage Learning 2012

SPECIAL APPLICATION CABLES

Flat Conductor Cable (for Installation Under Carpet Squares)

A Power feed, grounding connection, and shield system connection between the flat conductor cable (FCC) system and other wiring systems shall be accomplished through a transition assembly identified for this use ≫ *324.40(D)* ≪.

B Carpet squares are required to cover floor-mounted Type FCC cable, cable connectors, and insulating ends. They shall not be larger than 39.37 in. × 39.37 in. (1.0 m × 1.0 m). Release-type adhesives must be used when adhering carpet squares to the floor ≫ *324.41* ≪.

C All receptacles, receptacle housings, and self-contained devices used with the FCC system shall be identified for this use and shall be connected to the Type FCC cable as well as the metal shields. Connection from any grounding conductor of the Type FCC cable to the shield system must be made at *each* receptacle ≫ *324.42(A)* ≪.

D All FCC system components must be securely anchored to the floor (or wall) using an adhesive or mechanical system identified for this use. Floors must be prepared to ensure attachment of the FCC system to the floor until the carpet squares are placed ≫ *324.30* ≪.

E Type FCC cable can be installed on hard, sound, smooth, continuous surfaces made of concrete, ceramic, composition flooring, wood, or similar materials ≫ *324.10(C)* ≪.

F A metal top shield shall be installed over all floor-mounted Type FCC cable, connectors, and insulating ends. The top shield must completely cover all cable runs, corners, connectors, and ends ≫ *324.40(C)(1)* ≪.

G Type FCC cable consists of three or more flat copper conductors placed edge to edge and separated by an enclosing insulation assembly ≫ *324.100(A)* ≪. While the maximum current rating for general-purpose and appliance branch circuits is 20 amperes, individual branch circuits can have a current rating up to 30 amperes ≫ *324.10(B)(2)* ≪.

H All Type FCC cable, connectors, and insulating ends require a bottom shield, either metallic or nonmetallic ≫ *324.40(C)(2)* and *324.100(B)(1)* ≪.

I FCC systems are not permitted on wall surfaces unless enclosed in surface metal raceways ≫ *324.10(D)* ≪.

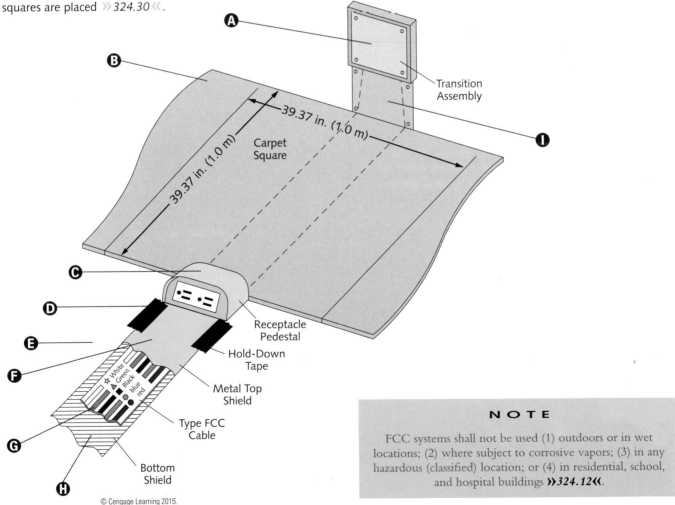

Transition Assembly

39.37 in. (1.0 m)

39.37 in. (1.0 m)

Carpet Square

Receptacle Pedestal

Hold-Down Tape

Metal Top Shield

Type FCC Cable

Bottom Shield

White
Green
Black
blue
red

© Cengage Learning 2015.

NOTE

FCC systems shall not be used (1) outdoors or in wet locations; (2) where subject to corrosive vapors; (3) in any hazardous (classified) location; or (4) in residential, school, and hospital buildings ≫ *324.12* ≪.

Integrated Gas-Spacer Cable (Type IGS)

The maximum bending radius is found in Table 326.24.

A run of Type IGS cable between pull boxes (or terminations) shall not contain more than the equivalent of four quarter bends (360 degrees total), including those bends located immediately adjacent to the pull box (or terminations) 》*326.26*《.

Terminations and splices must be identified as suitable for maintaining the gas pressure within the conduit. A valve and cap must be provided for each length of the cable and conduit for monitoring and maintenance of gas pressure in the conduit 》*326.40*《.

Each conductor contains ½-In. (12.7-mm) diameter solid aluminum rod(s). A conductor assembly contains from one to nineteen of these rods, which are laid parallel. Conductor size ranges from a 250-kcmil minimum to a 4750-kcmil maximum 》*326.104*《.

A Type IGS cable is a manufactured assembly of conductor(s), each individually insulated and enclosed in a loose-fit nonmetallic flexible conduit as an integrated gas-spacer cable rated 0 to 600 volts 》*326.2*《.

> **NOTE**
>
> Type IGS cable shall not be (1) used as interior wiring, (2) exposed in contact with buildings, or (3) used above ground 》***326.10 and 12***《.

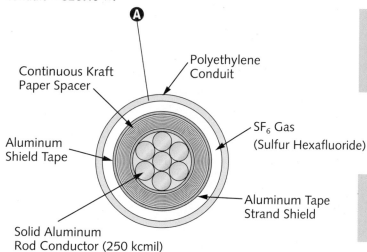

Continuous Kraft Paper Spacer

Polyethylene Conduit

Aluminum Shield Tape

SF_6 Gas (Sulfur Hexafluoride)

Aluminum Tape Strand Shield

Solid Aluminum Rod Conductor (250 kcmil)

> **NOTE**
>
> Table 326.80 provides the maximum ampacity for Type IGS cable.

© Cengage Learning 2012

Medium-Voltage Cable (Type MV)

A Type MV is a solid dielectric insulated cable (single or multiconductor) rated 2001 volts or higher 》*328.2*《.

B Type MV cable shall not be used where exposed to direct sunlight, unless identified for the use 》*328.12*《.

C Type MV cables must have copper, aluminum, or copper-clad aluminum conductors and shall comply with Table 310.104(C) and Table 310.104(D) or Table 310.104(E) 》*328.100*《.

D Conductor ampacity is found in 310.60, unless the cable is installed in a cable tray, in which case the ampacity shall be determined in accordance with 392.80(B) 》*328.80*《.

> **WARNING**
>
> Type MV cable shall be installed, terminated, and tested by qualified persons 》*328.14*《.

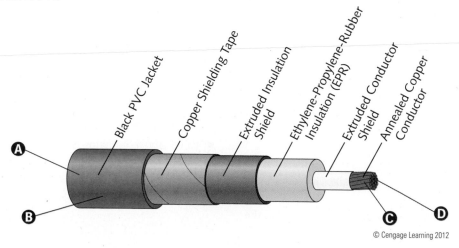

Black PVC Jacket

Copper Shielding Tape

Extruded Insulation Shield

Ethylene-Propylene-Rubber Insulation (EPR)

Extruded Conductor Shield

Annealed Copper Conductor

© Cengage Learning 2012

Mineral-Insulated Metal-Sheathed Cable (Type MI)

Type MI cable must be securely supported at intervals no greater than 6 ft (1.8 m) except where fished »332.30«.

A Type MI (mineral-insulated metal-sheathed) cable is a manufactured assembly of conductor(s) insulated with a highly compressed refractory mineral insulation and enclosed in a liquidtight (and gastight) continuous copper (or alloy steel) sheath »332.2«.

B The cable must not be damaged when bent. The radius of the inner edge of any bend shall not be less than five times the external diameter of the metallic sheath for cable that is not more than ¾ in. (19 mm) in external diameter. Cable diameters greater than ¾ in. (19 mm) but not more than 1 in. (25 mm) in external diameter must have a bend radius not less than ten times the external diameter »332.24«.

C When using single-conductor cables, all phase conductors and the neutral conductor (if any) must be grouped together to minimize induced voltages on the sheath »332.31 and 300.20(A)«.

D Only fittings identified for this use can connect Type MI cable to boxes, cabinets, or other equipment »332.40(A)«.

E If these single-conductor cables had entered a ferrous enclosure, compliance with 300.20(B) would be required to prevent heating from induction.

F Where Type MI cable terminates, a seal shall be provided *immediately* after stripping to prevent the entrance of moisture into the insulation. Each conductor extending beyond the sheath must be provided with an insulating material »332.40(B)«.

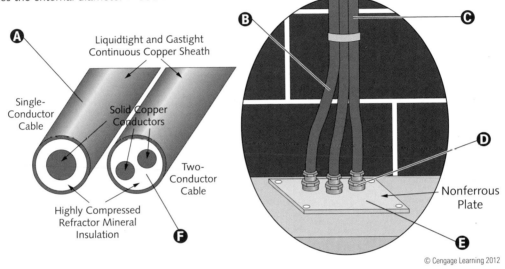

Liquidtight and Gastight Continuous Copper Sheath

Single-Conductor Cable

Solid Copper Conductors

Two-Conductor Cable

Highly Compressed Refractor Mineral Insulation

Nonferrous Plate

© Cengage Learning 2012

Summary

- When installed, cables must be protected from the possibility of physical damage.
- Conductors must be installed at least 1¼ in. (32 mm) from the nearest edge of the framing member, unless a steel plate (at least 1⁄16 in. [1.6 mm] thick) has been installed to protect the conductor(s).
- Conductors passing through metal studs (or framing members) may require protection in the form of bushings or steel plates.
- Fire-resistant-rated construction must remain intact once the electrical system has been installed.
- In dwellings, wiring can pass through the short dimension of a joist (or stud) space being used for environmental air-handling purposes.
- Cables must be secured within a certain distance from boxes (junction, device, etc.), unless the cable has been fished between access points.
- The maximum distance between supports depends on the type of cable.

- The minimum bending radius must be observed when installing a cable.
- Unless meeting one of the exemptions in 404.2(C), a grounded circuit conductor for the controlled lighting circuit shall be provided at the location where switches control lighting loads that are supplied by a grounded general-purpose branch circuit.
- Certain sizes of nonmetallic-sheathed cable can be attached directly to the underside of exposed joists in unfinished basements.
- A cable assembly's white or gray conductor can be used as a switch loop (leg) in single-pole, three-way, and four-way switch installations after it has been reidentified.
- Table 300.5 contains minimum cover requirements for direct-buried cables and conductors.
- Chapter 3 of the *NEC* contains provisions for certain specialized cables.

Unit 4 Competency Test

NEC Reference	Answer
334	12
320	39-39
320	4 1/2
326	35
330	MC cable
330	moisture
320	B
330	MI cable
342	D
328	transition
314	B
314	GCE
330	12
300	1 1/4

1. Nonmetallic-sheathed cable must be secured within _____ in. from a two-gang plastic device box mounted in a wall.

2. Floor-mounted Type FCC cable, cable connectors, and insulating ends shall be covered with carpet squares no larger than _____ in. square.

3. Type AC cable shall be secured by staples, cable ties, straps, hangers, or similar fittings at intervals not exceeding _____ ft.

4. The minimum bending radius for 3-in. diameter integrated gas-spacer cable is _____ in.

5. _____ is a factory assembly of one or more insulated circuit conductors with or without optical-fiber members enclosed in an armor of interlocking metal tape or a smooth or corrugated metallic sheath.

6. Where Type MI cable terminates, a seal shall be provided immediately after stripping to prevent the entrance of _____ into the insulation.

7. Where nonmetallic-sheathed cable is run at angles with joists in unfinished basements, it shall be permissible to secure cables not smaller than three _____ AWG conductors directly to the lower edges of the joists.

   a) 12   b) 10   c) 8   d) 6

8. _____ is a single or multiconductor solid dielectric insulated cable rated 2001 volts or higher.

9. The overall covering of Type UF cable shall be _____.

   I. flame retardant

   II. suitable for direct burial in the earth

   III. moisture-, fungus-, and corrosion-resistant

   a) II only   b) I and II only   c) II and III only   d) I, II, and III

10. Power feed, grounding connection, and shield system connection between the FCC system and other wiring systems shall be accomplished in a(n) _____ identified for this use.

11. Type AC cable shall be permitted to be unsupported where the cable is not more than _____ in. in length at terminals where flexibility is necessary.

    a) 2   b) 6   c) 24   d) 72

12. The _____ for the controlled lighting circuit shall be provided at the location where switches control lighting loads that are supplied by a grounded general-purpose branch circuit unless meeting one of the seven exemptions.

13. The minimum bending radius for metal-clad (corrugated sheath) cable is _____ times the external diameter of the metallic sheath.

14. In both exposed and concealed locations, where a cable or raceway-type wiring method is installed through bored holes in joists, rafters, or wood members, holes shall be bored so that the edge of the hole is not less than _____ in. from the nearest edge of the wood member.

NEC Reference	Answer
320	DA
314	B
330	MI
314	A
314	D
314	C
326	continuous
324	1/16
350	20
330	4½
326	6

15. Where run across the top of floor joists or within _____ ft of the floor or floor joists across the face of rafters or studding, in an attic that has permanent stairs, nonmetallic-sheathed cable must be protected by substantial guard strips that are at least as high as the cable.

16. Type FCC cable shall be clearly and durably marked on both sides at intervals of not more than 24 in. with the information required by 310.120(A) and with what additional information?

    I. Material of conductors and ampacity

    II. Rated frequency and number of phases

    III. Maximum temperature rating

        a) I only   b) I and II only   c) I and III only   d) I, II, and III

17. _____ is a factory assembly of one or more conductors insulated with a highly compressed refractory mineral insulation and enclosed in a liquidtight and gastight continuous copper or alloy steel sheath.

18. Armored cable installed in thermal insulation shall have conductors rated at _____. The ampacity of the cable installed in these applications shall not exceed that of a 60°C-rated conductor.

    a) 60°C   b) 75°C   c) 90°C   d) 90°F

19. Type MC cable must be supported (and secured) at intervals not exceeding _____ in.

    a) 4½   b) 6   c) 54   d) 72

20. The maximum ampacity of a 500-kcmil, integrated gas-spacer cable (Type IGS) is _____ amperes.

21. Type UF cable, buried underground, fed from a 15-ampere, ground-fault circuit interrupter at a one-family dwelling must have a minimum cover of _____ in.

    a) 6   b) 12   c) 18   d) 24

22. Splices in knob-and-tube wiring shall be _____ unless approved splicing devices are used.

23. A nonmetallic-sheathed cable, passing through a bored hole less than 1¼ in. from the edge of a wood framing member, must be protected by a steel plate (or bushing) at least _____ in. thick.

24. In FCC systems, general-purpose and appliance branch circuits have a maximum current rating of _____ amperes, while individual branch circuits have a maximum current rating of _____ amperes.

25. Type MI cable shall be supported securely at intervals not exceeding _____ ft by straps, staples, hangers, or similar fittings designed and installed so as not to damage the cable.

26. Type UF cable, buried under a one-family dwelling driveway, must have a minimum cover of _____ in.

UNIT 5

Raceways and Conductors

Objectives

After studying this unit, the student should:

▸ know the maximum number of bends for raceways installed between pull points.

▸ have a good understanding of rigid and intermediate metal-conduit provisions.

▸ be able to determine the type of raceway best suited for each installation.

▸ be familiar with intermediate metal conduit (IMC), rigid metal conduit (RMC), and electrical metallic tubing (EMT) support requirements.

▸ understand the unique characteristics of the four types of rigid nonmetallic conduits.

▸ thoroughly understand rigid nonmetallic conduit and electrical nonmetallic tubing provisions, both general and specific.

▸ be familiar with the different types of flexible conduit, and be able to locate each related article.

▸ be aware of raceway types other than conduit and tubing.

▸ successfully calculate the electrical trade-size conduit required for each project.

▸ have a basic grasp of conductor properties.

▸ be familiar with conductor temperature limitations.

▸ know and understand the provisions pertaining to conductors connected in parallel.

▸ understand ampacity correction factors, including continuous loads, and how to apply them.

Introduction

Raceways are the conduits, or similar housing, whose dual function is to facilitate the installation (or removal) of electrical conductors and to provide protection from mechanical damage. Article 100 defines raceway as an enclosed channel of metallic (or nonmetallic) materials designed expressly for holding wires, cables, or busbars, with additional functions as permitted in the *National Electrical Code*. The term **raceway** applies to more than just tubular conduit.

This unit contains regulations for raceway systems that are frequently used (such as electrical metallic tubing and wireways), as well as raceway systems that are rarely used (such as cellular concrete floor raceways). In addition to defining different raceways, other important information is covered, such as minimum and maximum size, installation provisions, fittings, and support requirements. Instructions are presented on how to perform raceway fill calculations, a prerequisite in determining the size of raceway needed for particular circuits. A brief explanation of conductors, including ampacity correction factors and conductor temperature limitations, can be found near the conclusion of this unit. Understanding this unit will make finding the best raceway system for a particular project much easier.

Most raceway articles have sections entitled "Uses Permitted" and "Uses Not Permitted" (*NEC* Chapter 3). These sections give locations (some specific) where the raceway system can (or cannot) be installed. Because most of these provisions are clearly understandable, this book does not repeat much of this information. Although not addressed in this unit, another important area is underground raceway installation (300.5). Underground installations are covered in Unit 4, which explains minimum cover requirements for underground cables and conductors. Use Table 300.5 (Minimum Cover Requirements) to find the minimum burial depth for different types of raceways installed in a variety of locations.

GENERAL PROVISIONS

Suspended (Drop) Ceilings

Ⓐ Raceways can be supported by independent (additional) support wires, secured at both ends. Support wires and associated fittings must provide adequate support ≫ *300.11(A)* ≪.

Ⓑ Boxes can be secured by attaching independent (additional) support wires at both ends. Secure support must be provided by the support wires and associated fittings ≫ *300.11(A)* and *314.23(D)(2)* ≪.

Ⓒ Ceiling grid support wires shall not be used to support raceways.

Ⓓ Certain boxes can also be mounted to suspended ceiling framing members ≫ *314.23(D)(1)* ≪.

Ⓔ EMT conduit must be fastened securely within 3 ft (900 mm) of all boxes (outlet, junction, device, etc.) ≫ *358.30(A)* ≪.

Ⓕ Suspended ceiling system framing members used to support luminaires must be securely fastened to one another as well as to the building structure at appropriate intervals. Luminaires must be securely fastened to the ceiling framing member by mechanical means (bolts, screws, or rivets, etc.). Listed clips, identified for use with the type of ceiling framing member(s) and luminaire(s), are also permitted ≫ *410.36(B)* ≪.

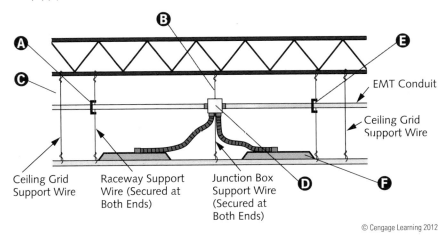

Ceiling Grid Support Wire

Raceway Support Wire (Secured at Both Ends)

Junction Box Support Wire (Secured at Both Ends)

EMT Conduit

Ceiling Grid Support Wire

© Cengage Learning 2012

Maximum Bends in One Run

Ⓐ The equivalent of four quarter bends (360° total) is the maximum allowed between pull points, for example, conduit bodies and boxes. Because the total bends in this conduit run is 340°, this installation falls within *NEC* specifications.

Ⓑ The bend maximum of 360° applies to the following raceways: IGS (326.26), IMC (342.26), RMC (344.26), FMC (348.26), LFMC (350.26), PVC (352.26), HDPE (353.26), NUCC (354.26), RTRC (355.26), LFNC (356.26), EMT (358.26), and ENT (362.26).

Ⓒ Generally, raceway installation must be complete between outlet, junction, or splicing points prior to the installation of conductors ≫ *300.18(A)* ≪.

Ⓓ All bends are counted, even those located immediately adjacent to the pull box (or termination). A box offset with two 10° bends counts as 20°. .

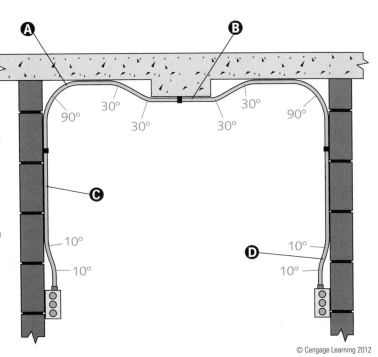

© Cengage Learning 2012

> ### NOTE
> Metal raceways must not be supported, terminated, or connected by welding unless specifically permitted by design or *Code* specifications ≫ ***300.18(B)*** ≪.

Insulated Fittings

Ⓐ An insulated fitting is not required if a smoothly rounded (or flared) entry for conductors is provided by threaded hubs (or bosses) that are an integral part of a cabinet, box enclosure, or raceway ≫ *300.4(G) Exception* ≪.

Ⓑ Where raceways containing 4 AWG or larger insulated circuit conductors enter a cabinet (box, enclosure, or raceway), the conductors must be protected by an identified fitting that provides a smoothly rounded insulating surface, unless the conductors are separated from the fitting (or raceway) by other securely attached identified insulating material ≫ *300.4(G)* ≪.

Ⓒ Conduit bushings constructed completely of insulating material shall not be used to secure a fitting or raceway ≫ *300.4(G)* ≪. An insulating (plastic, thermoplastic, etc.) bushing shall not replace a locknut.

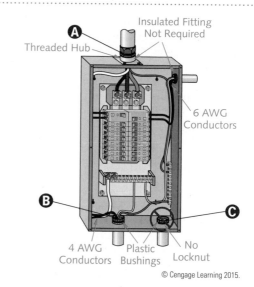

© Cengage Learning 2015.

Raceways Supporting Raceways

Ⓐ Conduits (not used as raceways) may support raceways if fastened securely in place with approved fittings ≫ *300.11(B)* ≪.

Ⓑ Raceways *shall not* be used to support other raceways, cables, conductors, or nonelectric equipment unless the raceway or means of support is identified as a means of support ≫ *300.11(B)(1)* ≪.

Ⓒ This conduit contains power supply conductors for HVAC unit.

Ⓓ Raceways containing power supply conductors for electrically controlled equipment can support Class 2 circuit conductors (or cables) used solely for connection to the equipment control circuits ≫ *300.11(B)(2)* ≪.

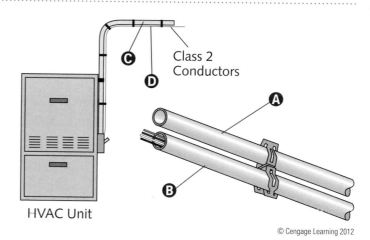

© Cengage Learning 2012

Conduit Thickness

Ⓐ Internal diameters for each electrical trade size conduit and tubing are listed in Table 4 of Chapter 9.

Ⓑ Ten 10 AWG THHN conductors are permitted in ¾-in. EMT.

Ⓒ Eleven 10 AWG THHN conductors are permitted in ¾-in. IMC.

Ⓓ Ten 10 AWG THHN conductors are permitted in ¾-in. RMC.

Ⓔ Annex C (preceding the index in the back of the *Code* book) can be used to find the maximum number of conductors permitted in each electrical trade size conduit or tubing. The conductors must be the same size (total cross-sectional area including insulation) when using Annex C.

Ⓕ Nine 10 AWG THHN conductors are permitted in ¾-in. schedule 40 PVC.

Ⓖ Seven 10 AWG THHN conductors are permitted in ¾-in. schedule 80 PVC.

Ⓗ The internal diameters of different types of conduit vary. Therefore, a group of conductors that fit into one conduit may not fit into another type, even though it is the same electrical trade size.

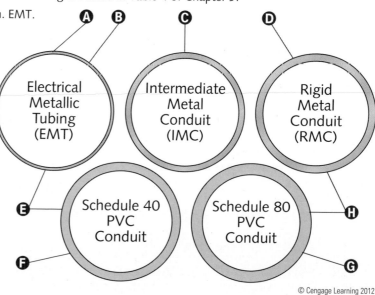

© Cengage Learning 2012

Bending Radius

A Bends must be made so that the conduit or tubing remains undamaged with its internal diameter basically undiminished. For any field bend, the radius of the curve to the centerline of the conduit must not be less than indicated in Table 2 of Chapter 9. Raceways permitted to use the "One Shot and Full Shoe Benders" column include IMC ≫ 342.24 ≪, RMC ≫ 344.24 ≪, and EMT ≫ 358.24 ≪.

B The bending radius of certain other raceways must not be less than shown in the column titled "Other Bends." They include
FMC ≫ 348.24 ≪,
LFMC ≫ 350.24 ≪,
PVC ≫ 352.24 ≪,
LFNC ≫ 356.24 ≪, and
ENT ≫ 362.24 ≪.

© Cengage Learning 2012

Table 2 (Chapter 9) Radius of Conduit and Tubing Bends

Conduit Size	One Shot and Full Shoe Benders	Other Bends
Trade Size	Bending Radius (in in.)	Bending Radius (in in.)
½	4	4
¾	4½	5
1	5¾	6
1¼	7¼	8
1½	8¼	10
2	9½	12
2½	10½	15
3	13	18
3½	15	21
4	16	24
5	24	30
6	30	36

© Cengage Learning 2012.

NONFLEXIBLE CONDUIT (AND TUBING)

Rigid Metal Conduit: Type RMC

A Rigid metal conduit (RMC) is a threadable raceway of circular cross section designed for the physical protection and routing of conductors and cables and for use as an equipment grounding conductor when installed with its integral or associated coupling and appropriate fittings ≫ 344.2 ≪.

B RMC is the heaviest (thickest-walled) classification of metal conduit.

C Threadless couplings and connectors used with conduit must be made tight. Those buried in masonry (or concrete) must be the concrete-tight type. If installed in wet locations, they must comply with 314.15 ≫ 344.42(A) ≪.

D The minimum approved electrical trade size for RMC is ½ in. ≫ 344.20(A) ≪. The maximum approved electrical trade size for RMC is 6 in. ≫ 344.20(B) ≪.

E RMC shall be permitted as an equipment grounding conductor ≫ 344.60 ≪.

F Galvanized steel and stainless steel RMC can be used under all atmospheric conditions, within all types of occupancies ≫ 344.10(A)(1) ≪.

G Contact of dissimilar metals (except for the combination of aluminum and steel) should be avoided wherever possible to lessen the potential for galvanic action ≫ 344.14 ≪.

H RMC usually ships in standard lengths of 10 ft (3.05 m), including the coupling. Normally one coupling is furnished with each length ≫ 344.130 ≪.

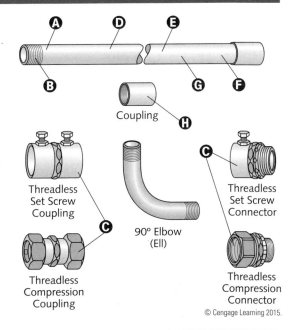

Coupling

Threadless Set Screw Coupling

Threadless Compression Coupling

90° Elbow (Ell)

Threadless Set Screw Connector

Threadless Compression Connector

© Cengage Learning 2015.

N O T E

RMC shall be made of one of the following: (1) Steel (ferrous), with or without protective coatings; (2) Aluminum (nonferrous); (3) Red brass; or (4) Stainless steel ≫344.100≪.

N O T E

Galvanized steel, stainless steel, and red brass RMC, elbows, couplings, and fittings can be installed in concrete, in direct contact with the earth, or in areas subject to severe corrosive influences where corrosion protection is provided and where judged suitable for the condition ≫344.10(B)(1)≪.

Intermediate Metal Conduit: Type IMC

(A) The minimum approved electrical trade size for IMC is ½ in. »*342.20(A)*«.

(B) The maximum approved electrical trade size for IMC is 4 in. »*342.20(B)*«.

(C) The definition for intermediate metal conduit (IMC) and rigid metal conduit (RMC) is essentially the same »*342.2*«. However, as the name implies, IMC is lighter in weight and is constructed with thinner walls than RMC. Unlike RMC, IMC is only manufactured of steel that provides protective strength equivalent to thicker-walled conduits.

(D) IMC is a steel threadable raceway of circular cross section designed for the physical protection and routing of conductors and cables and for use as an equipment grounding conductor when installed with its integral or associated coupling and appropriate fittings »*342.2 and 342.60*«.

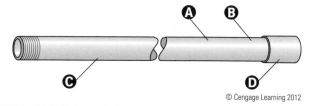

© Cengage Learning 2012

> ### N O T E
> For other similarities between IMC and RMC, read Article 342.

RMC, IMC, and EMT Support Requirements

(A) The maximum support interval for RMC, IMC, and EMT is 10 ft (3 m) »*342.30(B)(1), 344.30(B)(1), and 358.30(A)*«.

(B) In addition, each conduit (or tube) must be securely fastened within 3 ft (900 mm) of each conduit termination (outlet box, junction box, device box, cabinet, conduit body, etc.) »*342.30(A)(1), 344.30(A)(1), and 358.30(A)*«.

(C) If structural members are not available within 3 ft (900 mm), a distance of 5 ft (1.5 m) is acceptable for IMC and RMC »*342.30(A)(2) and 344.30(A)(2)*«.

(D) Unbroken lengths of EMT, i.e., without coupling, can be fastened within 5 ft (1.5 m) where structural members do not readily permit fastening within 3 ft (900 mm) »*358.30(A) Exception No. 1*«.

> ### N O T E
> RMC, IMC, and EMT must be installed as a complete system before installing conductors »**342.30, 344.30, and 358.30**«. It is not required (where approved) that RMC and IMC be securely fastened within 3 ft (900 mm) of the service head for above-the-roof mast termination »**342.30(A)(3) and 344.30(A)(3)**«.

Cutting, Reaming, and Threading

(A) Electrical metallic tubing must not be threaded. Factory threaded integral couplings can be used »*358.28(B)*«.

(B) All EMT cut ends must be reamed (or otherwise finished) to remove rough edges »*358.28(A)*«.

(C) All cut ends of IMC and RMC must be reamed (or otherwise finished) to remove rough edges »*342.28 and 344.28*«.

(D) Running threads shall not be used on conduit for coupling connections »*342.42(B) and 344.42(B)*«.

(E) Threading conduit in the field requires a standard cutting die with a ¾-in. taper per ft (1 in 16) »*342.28 and 344.28*«.

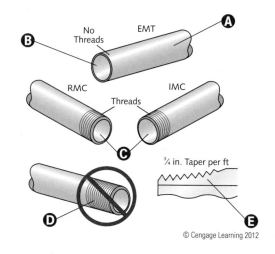

© Cengage Learning 2012

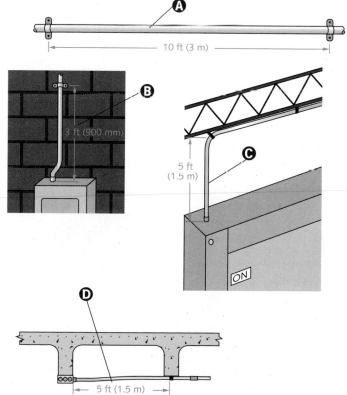

© Cengage Learning 2012

RMC and IMC with Threaded Couplings

(A) Straight runs of RMC and IMC made up with threaded couplings can be supported in accordance with Table 344.30(B)(2), provided such supports prevent transmission of stresses to termination where conduit is deflected between supports ⟫ *344.30(B)(2)* ⟪.

(B) The distance between supports increases as the conduit size increases.

(C) The distance between supports can be increased to 20 ft (6 m) for exposed vertical risers from industrial machinery or fixed equipment, provided (1) the conduit is made up with threaded couplings; (2) the conduit is supported and securely fastened at the top and bottom of the riser; and (3) no other means of intermediate support is readily available ⟫ *344.30(B)(3)* ⟪.

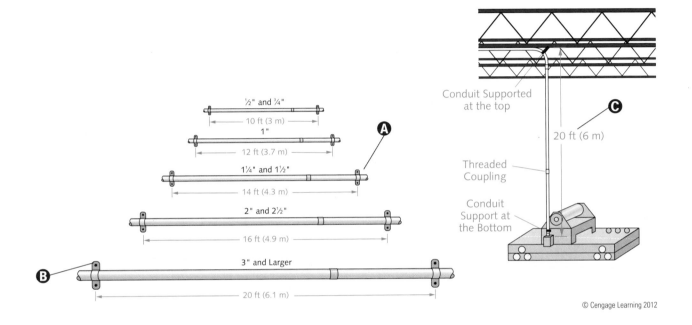

½" and ¾"
10 ft (3 m)
1"
12 ft (3.7 m)
1¼" and 1½"
14 ft (4.3 m)
2" and 2½"
16 ft (4.9 m)
3" and Larger
20 ft (6.1 m)

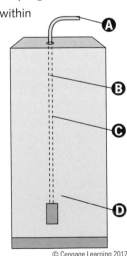

Conduit Supported at the top

20 ft (6 m)

Threaded Coupling

Conduit Support at the Bottom

© Cengage Learning 2012

Horizontal Runs through Framing Members

(A) Horizontal runs of RMC, IMC, and EMT that are supported by openings through framing members (at intervals not greater than 10 ft [3 m]) and are fastened securely within 3 ft (900 mm) of termination points, are permitted ⟫ *342.30(B)(4), 344.30(B)(4), and 358.30(B)* ⟪.

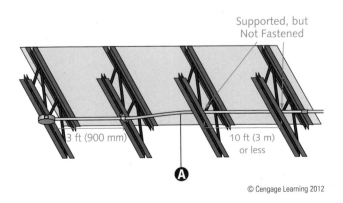

Supported, but Not Fastened

3 ft (900 mm)
10 ft (3 m) or less

© Cengage Learning 2012

EMT Fished in Walls

(A) For concealed work in finished buildings (or prefinished wall panels) where standard securing is impracticable, unbroken lengths (without coupling) of electrical metallic tubing can be fished ⟫ *358.30(A) Exception No. 2* ⟪.

(B) To comply with this exception, the fished portion of tubing in the wall must not have any couplings.

(C) Normally, EMT must be fastened within 3 ft (900 mm) of device boxes.

(D) Finished wall provides no access to secure tubing.

© Cengage Learning 2012

EMT Used as an Equipment Grounding Conductor

Ⓐ EMT can serve as an equipment grounding conductor 》*250.118(4)* and *358.60* 《.

Ⓑ A raceway used as the equipment grounding conductor, as provided in 250.118 and 250.134(A), must comply with 250.4(A)(5) or 250.4(B)(4) 》*250.122(A)* 《.

Ⓒ EMT must be supported according to 358.30 provisions.

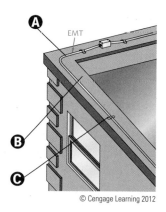

© Cengage Learning 2012

Electrical Metallic Tubing: Type EMT

Ⓐ Electrical metallic tubing (EMT) is an unthreaded thin-wall raceway of circular cross section designed for the physical protection and routing of conductors and cables and for use as an equipment grounding conductor when appropriate fittings are installed 》*358.2* 《. EMT is also referred to as thin wall.

Ⓑ EMT, the thinnest-walled classification of metal nonflexible raceways, provides protection from all but severe physical damage. Read 358.10 and 12 for additional information on uses (permitted and not permitted).

Ⓒ The minimum approved electrical trade size for EMT is ½ in. 》*358.20(A)* 《.

Ⓓ The maximum approved electrical trade size for EMT is 4 in. 》*358.20(B)* 《.

Ⓔ All EMT cut ends must be reamed (or otherwise finished) to remove rough edges 》*358.28(A)* 《.

Ⓕ Couplings and connectors used with tubing must be made tight. Buried in masonry (or concrete), they must be the concrete-tight type. Installed in wet locations, they must comply with 314.15 》*358.42* 《.

NOTE

The number of conductors permitted in EMT must not exceed the percentage fill specified in Table 1, Chapter 9 》*358.22* 《.

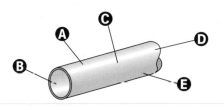

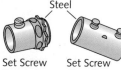

Die Cast   Steel

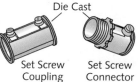

Set Screw Coupling   Set Screw Connector   Set Screw Connector   Set Screw Coupling

Compression Coupling   Compression Connector

© Cengage Learning 2012

Rigid Polyvinyl Chloride Conduit: Type PVC

NEC identifies five types of rigid PVC conduit:

1. Schedule 80 PVC is an extra-heavy-walled raceway with a wall thickness conforming to Schedule 80-Iron Pipe Size (IPS) dimensions.

2. Schedule 40 PVC is a heavy-walled raceway with a wall thickness conforming to Schedule 40-IPS dimensions.

3. Type A PVC is a thin-walled raceway with wall thickness conforming to Schedule A-IPS dimensions. Limited to underground installations, Type A PVC conduit must be laid with its entire length in concrete.

4. Type EB PVC is a thin-walled raceway with wall thickness designed to achieve a minimum pipe stiffness of 20 lbs./in./in. Type EB PVC conduit, limited to underground installations, must be laid with its entire length in concrete in outdoor trenches.

5. HDPE Schedule 40 is a high-density polyethylene raceway with a wall thickness conforming to Schedule 40-IPS dimensions. HDPE Schedule 40 conduit, also limited to underground installations, can be direct buried with or without being encased in concrete.

A Rigid polyvinyl chloride (PVC) conduit is a rigid nonmetallic raceway of circular cross section, with integral or associated couplings, connectors, and fittings for the installation of electrical conductors and cables 》 *352.2* 《.

B All joints between lengths of conduit, and between conduit and couplings, fittings, and boxes, must be made by an approved method 》 *352.48* 《.

C All cut ends must be trimmed inside and out to remove rough edges 》 *352.28* 《.

D Expansion fittings must be provided for PVC conduit to compensate for thermal expansion and contraction where the length change is expected to be 0.25 in. (6 mm) or greater, in accordance with Table 352.44, in a straight run between securely mounted items such as boxes, cabinets, elbows, or other conduit terminations 》 *352.44* and *300.7(B)* 《.

E The minimum approved electrical trade size for PVC conduit is 1/2 in. 》 *352.20(A)* 《.

F The maximum approved electrical trade size for PVC conduit is 6 in. 》 *352.20(B)* 《.

G Expansion and contraction problems generally do not arise in underground PVC conduit applications.

H Only listed factory elbows and associated fittings shall be used with PVC conduit 》 *352.6* 《.

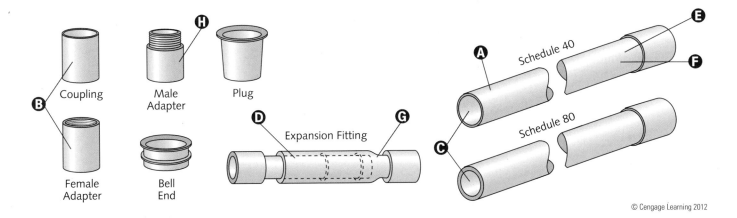

Coupling Male Adapter Plug Female Adapter Bell End Expansion Fitting Schedule 40 Schedule 80

© Cengage Learning 2012

NOTE

Where a raceway crosses a structural joint intended for expansion, contraction, or deflection, used in buildings, bridges, parking garages, or other structures, a listed expansion/deflection fitting or other approved means shall be used 》*300.4(H)*《.

CAUTION

Where conductors or cables are rated at a temperature higher than the listed temperature rating of PVC conduit, it shall be permissible to install those conductors or cables in PVC conduit, but those conductors or cables must not be operated at a temperature higher than the listed temperature rating of the PVC conduit 》*352.10(I)*《.

NOTE

Read 352.10 and 12 for a description of uses (permitted and not permitted) of rigid PVC conduit.

Securing Rigid PVC Conduit

(A) Each conduit must be fastened securely within 3 ft (900 mm) of all termination points 》*352.30(A)*《.

(B) Table 352.30 support provisions must be followed for PVC conduit 》*352.30(B)*《.

(C) PVC must be fastened so that movement from thermal expansion or contraction is permitted 》*352.30*《.

> **NOTE**
>
> Horizontal runs of PVC that are supported by openings through framing members (at intervals not greater than those in Table 352.30) and are securely fastened within 3 ft (900 mm) of termination points are permitted 》*352.30(B)*《.
> PVC conduit, listed for securing at other than 3 ft (900 mm), can be installed in accordance with the listing 》*352.30(A)*《.

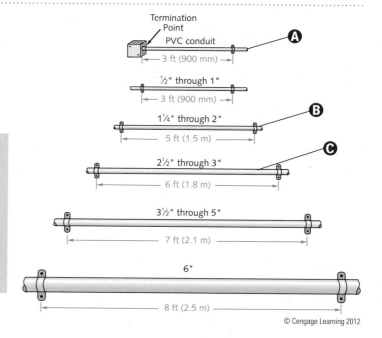

© Cengage Learning 2012

Bending Rigid PVC Conduit

(A) PVC bends must be made so that the conduit remains undamaged and the internal diameter of the conduit is not reduced 》*352.24*《.

(B) The conduit must not be damaged.

(C) Field bends must be made only with identified bending equipment. The radius of the curve to the centerline of such bends must not be less than shown in Table 2 of Chapter 9 》*352.24*《.

(D) Installing PVC plugs in the ends of larger conduits (2 to 6 in.) can prevent conduit collapses or other deformities.

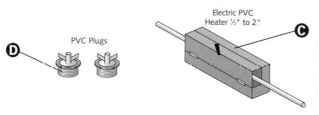

CAUTION

Field bends shall be made with identified bending equipment.

© Cengage Learning 2012

Securing Electrical Nonmetallic Tubing (ENT)

(A) The equivalent of four quarter bends (360° total) is the maximum between pull points, for example, conduit bodies and boxes 》*362.26*《.

(B) Electrical nonmetallic tubing must be secured every 3 ft (900 mm) or less 》*362.30(A)*《.

(C) Bends must be made so that the tubing is undamaged and the tubing's internal diameter is not reduced. Although bends can be made manually (without auxiliary equipment), the radius of the curve of the inner edge of each bend must not be less than shown in Table 2 of Chapter 9 and using the column titled "Other Bends" 》*362.24*《.

(D) ENT must be fastened securely within 3 ft (900 mm) of each outlet box, device box, junction box, cabinet, or other termination point 》*362.30(A)*《.

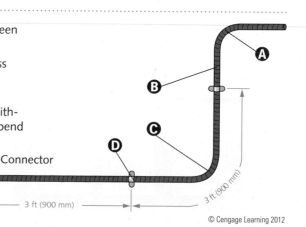

© Cengage Learning 2012

Electrical Nonmetallic Tubing: Type ENT

For the complete listing of ENT uses (permitted and not permitted), read 362.10 and 12.

A Electrical nonmetallic tubing (ENT) is a pliable corrugated raceway of circular cross section with integral (or associated) couplings, connectors, and fittings listed for the installation of electrical conductors. It is composed of a material that is flame-retardant as well as resistant to moisture and chemical atmospheres ≫362.2≪.

B All cut ends must be trimmed inside and out to remove rough edges ≫362.28≪.

C The minimum approved electrical trade size for ENT is ½ in. ≫362.20(A)≪.

D The maximum approved electrical trade size for ENT is 2 in. ≫362.20(B)≪.

E A pliable raceway can be bent by hand with reasonable force, but without other assistance ≫362.2≪.

F Outside diameters are such that standard rigid PVC conduit couplings and connectors can be used on ENT of PVC construction. ENT installation instructions outline the procedure to follow when installing PVC conduit fittings that are cemented in place. Specific cement requirements and application methods are provided.

G An approved method must be used for all joints between lengths of tubing and between tubing and couplings, fittings, and boxes ≫362.48≪.

CAUTION

ENT must not be stored or installed where exposed to direct sunlight, unless identified as "sunlight resistant" ≫362.12(8)≪.

Snap-in Connector

Male Adapter

Coupling

Schedule 40 PVC Male Adapter

© Cengage Learning 2012

NOTE

ENT must comply with 300.4 protection requirements.

ENT through Framing Members

A If the bored hole is less than 1¼ in. (32 mm) from the edge of the framing member, the raceway must be protected from penetration (by screws, nails, etc.) by a steel plate(s) or bushing(s). The steel plate must be at least 1/16 in. (1.6 mm) thick and of appropriate length and width to adequately cover the wiring ≫300.4(A)(1)≪.

B Holes must be bored so that the edge of the hole is at least 1¼ in. (32 mm) from the nearest edge of the wood member ≫300.4(A)(1)≪.

C ENT must be securely fastened in place within 3 ft (900 mm) of all termination points ≫362.30(A)≪.

D Horizontal runs of ENT can be supported by openings through framing members at intervals not greater than 3 ft (900 mm) ≫362.30(B)≪.

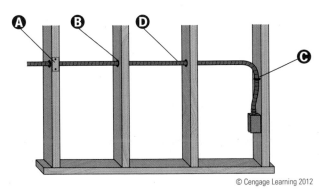

© Cengage Learning 2012

ENT Luminaire Whips

A ENT must be secured within 3 ft (900 mm) of all termination points ≫362.30(A)≪.

B Securing is not required for ENT in lengths not more than 6 ft (1.8 m) from a luminaire terminal connection (for tap connections to luminaires) ≫362.30(A) Exception No. 1≪.

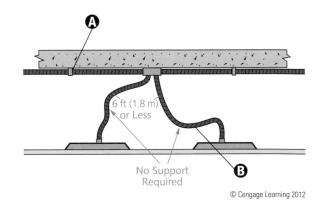

6 ft (1.8 m) or Less

No Support Required

© Cengage Learning 2012

NOTE

Securing is not required for ENT in lengths not exceeding 6 ft (1.8 m) from the last point of support for connections within an accessible ceiling to luminaire(s) or other equipment ≫362.30(A) Exception No. 2≪.

FLEXIBLE CONDUIT

Flexible Metal Conduit: Type FMC

(A) Flexible metal conduit (FMC) is a raceway of circular cross section made of helically wound, formed, and interlocked metal strip ≫*348.2*≪. FMC must be listed and can be used in both exposed and concealed locations ≫*348.6* and *348.10*≪.

(B) The minimum electrical trade size for FMC is ½ in., unless a 348.20(A) provision is met.

(C) Flexible metal conduit is often referred to as *Greenfield,* or simply *flex.*

(D) Fittings used with FMC must be listed ≫*348.6*≪.

(E) The maximum approved electrical trade size for FMC is 4 in. ≫*348.20(B)*≪.

(F) The equivalent of four quarter bends (360° total) is the maximum between pull points, for example, conduit bodies and boxes ≫*348.26*≪.

(G) Bends must be made so that the conduit is not damaged and the conduit's internal diameter is not effectively reduced. The radius of the curve of the inner edge of any field bend must not be less than shown in Table 2 of Chapter 9 using the column titled "Other Bends" ≫*348.24*≪.

(H) All cut ends must be trimmed (or otherwise finished) to remove rough edges, except where fittings are used that thread into the convolutions ≫*348.28*≪.

(I) Angle connectors shall not be used for concealed raceway installations ≫*348.42*≪.

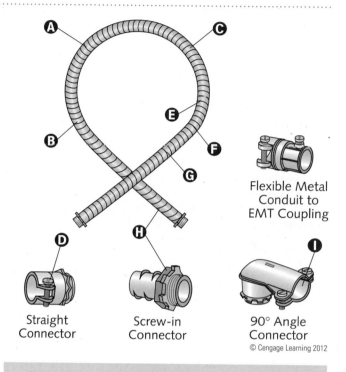

Flexible Metal Conduit to EMT Coupling

Straight Connector

Screw-in Connector

90° Angle Connector

© Cengage Learning 2012

NOTE

For areas where FMC is not allowed, see 348.12.

FMC Support

Support is not required for fished FMC ≫*348.30(A) Exception No. 1*≪.

(A) FMC must be fastened securely in place (by an approved means) at intervals of 4½ ft (1.4 m) or less ≫*348.30(A)*≪.

(B) FMC must be secured within 12 in. (300 mm) of each box, cabinet, conduit body, or other conduit termination ≫*348.30(A)*≪.

(C) Refer to Table 348.22 for the maximum number of conductors permitted in ⅜ in. flex.

(D) If the length of FMC from a luminaire terminal for tap connections, as permitted in 410.117(C), to luminaires is 6 ft (1.8 m), or less, supporting is not required ≫*348.30(A) Exception No. 3*≪.

(E) FMC of ⅜ in. electrical trade size can be used (1) for enclosing the leads of motors as permitted in 430.245(B); (2) in lengths not in excess of 6 ft (1.8 m) as part of a listed assembly, for tap connections to luminaires as permitted in 410.117(C), or for utilization equipment; (3) for manufactured wiring systems as permitted in 604.6(A); (4) in hoistways, as permitted in 620.21(A)(1); or (5) as part of a listed assembly to connect wired luminaire sections as permitted in 410.137(C) ≫*348.20(A)*≪.

NOTE

Horizontal runs of FMC can be supported by openings through framing members (at intervals not greater than 4½ ft [1.4 m]), if securely fastened within 12 in. (300 mm) of each termination point ≫***348.30(B)***≪.

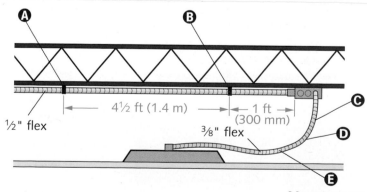

½" flex

4½ ft (1.4 m)

1 ft (300 mm)

⅜" flex

© Cengage Learning 2012

FMC Installation

Ⓐ Equipment grounding conductors are required for circuits over 20 amperes ⟫*250.118(5)*⟪

Ⓑ For lengths of 3 ft (900 mm) or less requiring flexibility, the 12-in. (300-mm) securing distance can be waived ⟫*348.30(A) Exception No. 2*⟪.

Ⓒ If an equipment bonding jumper is required around FMC, it must be installed in accordance with 250.102 ⟫*348.60*⟪.

Ⓓ Listed FMC can serve as a grounding means if (1) the conduit is terminated in listed fittings; (2) the circuit conductors contained therein are protected by overcurrent devices rated at 20 amperes or less; (3) the total length in any ground-fault current path is 6 ft (1.8 m) or less; and (4) the conduit is not installed where flexibility is necessary to minimize the transmission of vibration from equipment or to provide flexibility for equipment that requires movement ⟫*250.118(5)*⟪.

Ⓔ FMC must be secured within 12 in. (300 mm) of each box, cabinet, conduit body, or other conduit termination ⟫*348.30(A)*⟪.

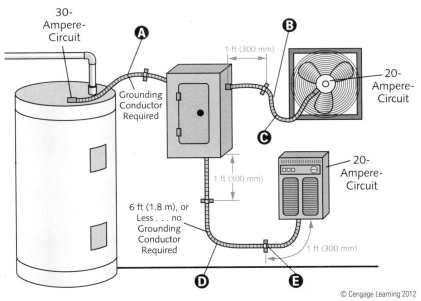

© Cengage Learning 2012

CAUTION

An equipment grounding conductor must be installed in FMC used to connect equipment where flexibility is necessary to minimize the transmission of vibration from equipment or to provide flexibility for equipment that requires movement after installation ⟫*348.60*⟪.

Liquidtight Flexible Metal Conduit: Type LFMC

The equivalent of four quarter bends (360° total) is the maximum allowed between pull points, for example, conduit bodies and boxes ⟫*350.26*⟪.

Ⓐ Liquidtight flexible metal conduit (LFMC) is a raceway of circular cross section having an outer liquidtight, nonmetallic, sunlight-resistant jacket over an inner flexible metal core with associated couplings, connectors, and fittings and approved for the installation ⟫*350.2*⟪.

Ⓑ The minimum electrical trade size for LFMC is ½ in., except as provided in 348.20(A) ⟫*350.20(A)*⟪. The number of conductors allowed in ⅜ in. conduit must not exceed the limit set in Table 348.22, "Fittings Outside Conduit" columns ⟫*350.22(B)*⟪.

Ⓒ LFMC and associated fittings must be listed ⟫*350.6*⟪.

Ⓓ Angle connectors shall not be concealed ⟫*350.42*⟪.

Ⓔ The maximum approved electrical trade size for LFMC is 4 in. ⟫*350.20(B)*⟪.

Ⓕ LFMC is often referred to as *Sealtite* (a registered trademark).

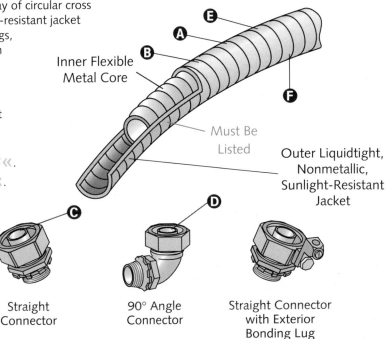

© Cengage Learning 2015.

NOTE

Securing and support requirements for LFMC are listed in 350.30.

Grounding Liquidtight Flexible Metal Conduit

Ⓐ Listed LFMC, ³⁄₈ in. through 1¹⁄₄ in., can be used as a grounding means if (1) the total length of flexible conduit in any ground-fault current path is not more than 6 ft (1.8 m); (2) the conduit is terminated in listed fittings; and (3) the circuit conductors contained therein are protected by overcurrent devices rated at 20 amperes or less for ³⁄₈-in. and ¹⁄₂-in. electrical trade sizes and 60 amperes or less for ³⁄₄-in. through 1¹⁄₄-in. electrical trade sizes ≫ *250.118(6)* ≪.

Ⓑ LFMC can serve as a grounding means as covered in 250.118(6). If an equipment bonding jumper is required around LFMC, it must be installed in accordance with 250.102 ≫ *350.60* ≪.

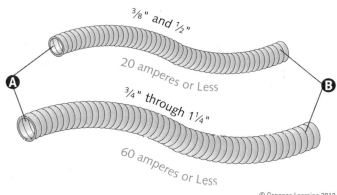

© Cengage Learning 2012

CAUTION

LFMC used to connect equipment, where flexibility is necessary to minimize the transmission of vibration from equipment or to provide flexibility for equipment that requires movement after installation, must have an equipment grounding conductor ≫ *350.60* and *250.118(6)e* ≪.

NOTE

An equipment bonding jumper can be installed either inside or outside of a raceway. If installed on the raceway's exterior, the jumper shall not exceed 6 ft (1.8 m) in length and must be routed with the raceway ≫ *250.102(E)* and *(E)(2)* ≪.

Liquidtight Flexible Nonmetallic Conduit: Type LFNC

The equivalent of four quarter bends (360° total) is the maximum allowed between pull points, for example, conduit bodies and boxes ≫ *356.26* ≪.

Ⓐ Liquidtight flexible nonmetallic conduit (LFNC) is a raceway of circular cross section of various types, including FNMC-A, FNMC-B, and FNMC-C ≫ *356.2* ≪.

Ⓑ Type LFNC-A has a smooth seamless inner core surrounded by reinforcement layer(s) and bonded to a smooth seamless cover ≫ *356.2(1)* ≪.

Ⓒ Type LFNC-B has a smooth inner surface having integral reinforcement within the raceway wall ≫ *356.2(2)* ≪. (Outside appearance is similar to LFMC.)

Ⓓ Type LFNC-C has a corrugated internal and external surface having no integral reinforcement within the raceway wall ≫ *356.2(3)* ≪. (Appearance is similar to ENT, except more flexible.)

Ⓔ LFNC and associated fittings must be listed ≫ *356.6* ≪.

Ⓕ Angle connectors shall not be used for concealed raceway installations ≫ *356.42* ≪.

Ⓖ The maximum approved electrical trade size for LFNC is 4 in. ≫ *356.20(B)* ≪.

Ⓗ The minimum electrical trade size for LFNC is ¹⁄₂ in., unless a 356.20(A) provision applies.

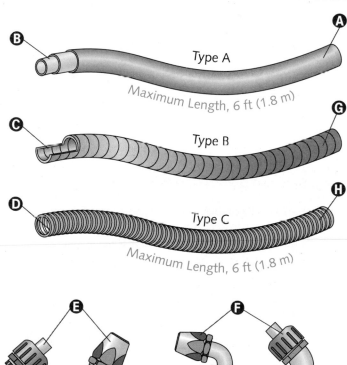

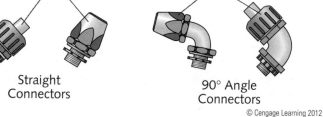

Straight Connectors

90° Angle Connectors

© Cengage Learning 2012

Liquidtight Flexible Nonmetallic Conduit Installation

Strapping is not required for LFNC lengths of 6 ft (1.8 m) or less from a luminaire terminal connection for tap conductors to luminaires as permitted in 410.117(C) ≫*356.30(2)*≪.

Support is not required for fished LFNC ≫*356.30(2)*≪.

Horizontal runs of LFNC can be supported by openings through framing members [at intervals not greater than 3 ft (900 mm)], if securely fastened within 12 in. (300 mm) of each termination point ≫*356.30(3)*≪.

Securing and supporting is not required for LFNC-B where installed in lengths not exceeding 6 ft (1.8 m) from the last point of support for connections within an accessible ceiling to luminaire(s) or other equipment. For the purpose of 356.30, listed liquidtight flexible nonmetallic conduit fittings shall be permitted as a means of support ≫*356.30(4)*≪.

A LFNC must be securely fastened at intervals not greater than 3 ft (900 mm) ≫*356.30(1)*≪.

B LFNC must be secured within 12 in. (300 mm) of each connection to every outlet box, junction box, cabinet, or fitting ≫*356.30(1)*≪.

C Where flexibility is necessary, LFNC can be secured within 3 ft (900 mm) of the termination ≫*356.30(2)*≪.

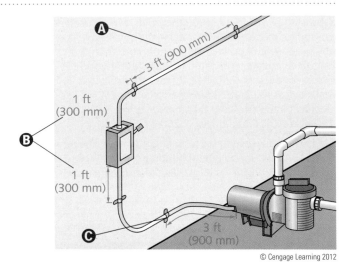

© Cengage Learning 2012

N O T E

If an equipment grounding conductor is required for the circuits installed in LFNC, it can be installed either inside or outside of the conduit. Where installed on the outside, the length of the equipment grounding conductor must not exceed 6 ft (1.8 m) and must be routed with the raceway ≫*250.102(E)* **and** *(E)(2)*≪.

OTHER RACEWAYS

Surface Nonmetallic Raceways

Where surface nonmetallic raceways are used in combination for both signaling and for lighting and power circuits, the different wiring systems must be run in separate compartments identified by stamping, imprinting, or color coding of the interior finish ≫*388.70*≪.

A Surface nonmetallic raceway construction must be visibly distinguishable from other raceways. Surface nonmetallic raceways and their fittings must be designed so that sections can be mechanically coupled together and installed without subjecting the wires to abrasion ≫*388.100*≪.

B The size and number of conductors installed in any raceway must not exceed the raceway's design limitations ≫*388.21* and *388.22*≪.

C Splices and taps are permitted within surface nonmetallic raceway having a removable cover that remains accessible once installed. The conductors, at the point of splices and taps, must not fill the raceway to more than 75% of its area. Splices and taps in surface nonmetallic raceways without removable covers must be made only in boxes. All splices and taps must be made using approved methods ≫*388.56*≪.

D Surface nonmetallic raceways can be installed in dry locations ≫*388.10(1)*≪. NEC 388.12 lists the locations where these raceways shall not be used.

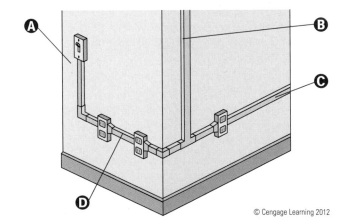

© Cengage Learning 2012

N O T E

Unbroken lengths of surface nonmetallic raceways can pass transversely through dry walls, partitions, and floors. Conductors must be accessible on both sides of the wall, partition, or floor ≫*388.10(2)*≪.
Where equipment grounding is required, a separate equipment grounding conductor must be installed within the raceway ≫*388.60*≪.

Surface Metal Raceways

A Surface metal raceway construction must be distinguishable from other raceways. Surface metal raceways and their fittings must be designed so that the sections can be electrically and mechanically coupled together and installed without subjecting the wires to abrasion. Nonmetallic covers and accessories can be used on surface metal raceways only if identified for such use ≫ *386.100* ≪.

B Uses (permitted and not permitted) are located in 386.10 and 386.12.

C The number and size of conductors installed in any raceway must not exceed the raceway's design limitations ≫ *386.21* and *386.22* ≪.

D The adjustment factors of 310.15(B)(3)(a) do not apply to conductors installed in surface metal raceways where all of the following conditions are met: (1) the cross-sectional area of the raceway exceeds 4 in.^{2} (2500 mm^2); (2) the current-carrying conductors do not exceed 30 in number; and (3) the sum of the cross-sectional areas of all contained conductors does not exceed 20% of the interior cross-sectional area of the surface metal raceway ≫ *386.22* ≪.

E Unbroken lengths of surface metal raceway can pass transversely through dry walls, dry partitions, and dry floors. Access to the conductors must be maintained on both sides of the wall, partition, or floor ≫ *386.10(4)* ≪.

F Where surface metal raceways are used in combination for both signaling and for lighting and power circuits, the different wiring systems must be run in separate compartments identified by stamping, imprinting, or color coding of the interior finish ≫ *386.70* ≪.

G Splices and taps are permitted within surface metal raceway having a removable cover that remains accessible after installation. At the point of splices and taps, the conductors must not fill the raceway to more than 75% of its area. Splices and taps in surface metal raceways without removable covers can be made only in boxes. Use only approved methods for splices and taps ≫ *386.56* ≪.

H Multioutlet assemblies are covered in Article 380.

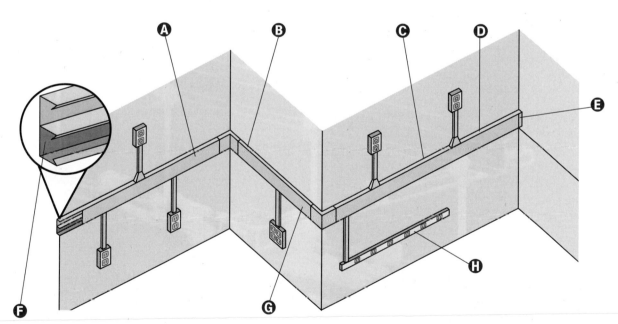

© Cengage Learning 2012

NOTE

If a surface metal raceway enclosure provides a transition from another wiring method, it must have a means to connect an equipment grounding conductor ≫ *386.60* ≪.

Strut-Type Channel Raceways

Use Table 384.22 to determine the maximum number of conductors permitted in strut-type channel raceway. Apply the appropriate Chapter 9 Tables for the cross-sectional area of the type and size of wire used ≫ *384.22* ≪ .

The adjustment factors of 310.15(B)(3)(a) do not apply to conductors installed in strut-type channel raceways where all of the following conditions are met: (1) the cross-sectional area of the raceway exceeds 4 in.^{2} (2500 mm^2); (2) the current-carrying conductors do not exceed 30 in number; and (3) the sum of the cross-sectional areas of all contained conductors does not exceed 20% of the interior cross-sectional area of the strut-type channel raceway ≫ *384.22* ≪ .

Conductors larger than that for which the strut-type channel raceway is listed must not be used ≫ *384.21* ≪ .

A surface mount strut-type channel raceway must be secured to the mounting surface with external retention straps at intervals not exceeding 10 ft (3 m), and within 3 ft (900 mm) of each raceway termination (outlet box, cabinet, junction box, etc.) ≫ *384.30(A)* ≪ .

Ⓐ Strut-type channel raceways, closure strips, and accessories must be listed and identified for such use ≫ *384.6* ≪ .

Ⓑ Splices and taps are permitted in raceways that provide access after installation via a removable cover. At any point of splices or taps, the conductors must not fill the raceway to more than 75%. All splices and taps must be made by approved methods ≫ *384.56* ≪ .

Ⓒ Uses (permitted and not permitted) are covered in 384.10 and 384.12.

Ⓓ Strut-type channel raceway enclosures providing a transition to (or from) other wiring methods must accommodate the connection of an equipment grounding conductor. A strut-type channel raceway can be used as an equipment grounding conductor in accordance with 250.118(13) ≫ *384.60* ≪ .

Ⓔ Strut-type channel raceways shall be permitted to be suspension mounted in air with identified methods at intervals not to exceed 10 ft (3 m) and within 3 ft (900 mm) of channel raceway terminations and ends ≫ *384.30(B)* ≪ .

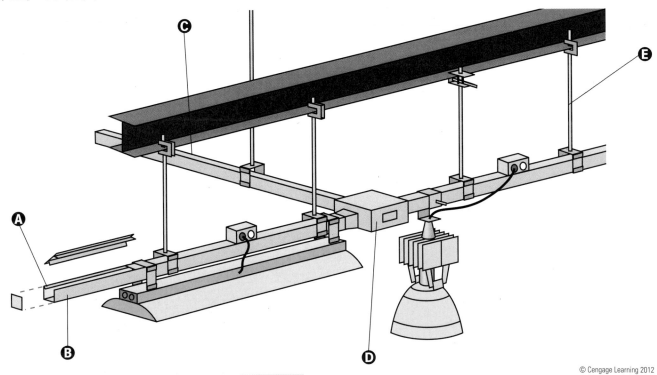

© Cengage Learning 2012

N O T E

Unbroken lengths of strut–type channel raceway can extend through walls, partitions, and floors provided closure strips are removable from either side and the portion within the wall, partition, or floor remains covered ≫*384.10(7)*≪.

Underfloor Raceways

An underfloor raceway is defined as a raceway and associated components designed and intended for installation beneath or flush with the surface of a floor for the installation of cables and electrical conductors ≫ *390.2* ≪.

A Connections between raceways and distribution centers and wall outlets must be made by approved fittings or by any of the wiring methods in Chapter 3, where installed according to the provisions of the respective articles ≫ *390.15* ≪.

B Underfloor raceways can be installed beneath a concrete (or other flooring material) surface. In office occupancies, installations flush with the concrete floor and covered with linoleum (or equivalent covering) are acceptable ≫ *390.3(A)* ≪.

C The combined cross-sectional area of all conductors or cables must not exceed 40% of the interior cross-sectional area of the raceway ≫ *390.6* ≪.

D The size of conductors installed must not be larger than that for which the underfloor raceway is designed ≫ *390.5* ≪.

E Inserts must be leveled and sealed to prevent the entrance of concrete. Metal raceway inserts must also be metal and must be electrically continuous with the raceway ≫ *390.14* ≪.

F Seal and level junction boxes to the floor grade to prevent entrance of water or concrete. Underfloor metal raceway junction boxes must also be metal and must be electrically continuous with the raceway ≫ *390.13* ≪.

G Splices and taps are acceptable only within junction boxes. Continuous, unbroken conductor connecting the individual outlets (so-called loop wiring) is not considered a splice or tap ≫ *390.7* ≪.

H Underfloor raceways must be laid so that a straight line from the center of one junction box to the center of the next junction box coincides with the centerline of the raceway system. Raceways must be held firmly in place to prevent misalignment during construction ≫ *390.9* ≪.

I Raceway dead ends must be closed ≫ *390.11* ≪.

J Install a suitable marker at, or near, each end of straight raceway runs to locate the last insert ≫ *390.10* ≪.

CAUTION

When an outlet is abandoned, discontinued, or removed, the supplying circuit conductors must be removed from the raceway. Splices or reinsulated conductors, such as would occur in the case of abandoned outlets on loop wiring, are not allowed in raceways ≫*390.8*≪.

NOTE

390.3(B) lists areas where underfloor raceways are not permitted.

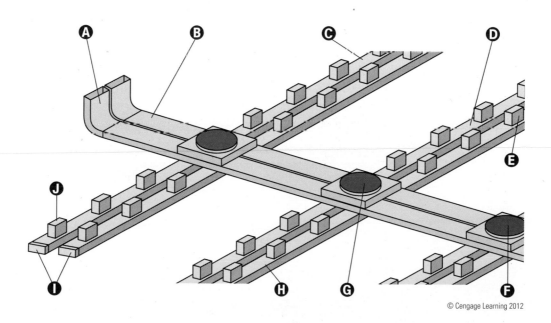

© Cengage Learning 2012

Underfloor Raceway Covering

A Half-round and flat-top raceways not over 4 in. (100 mm) in width must be covered by ¾ in. (20 mm) or more of concrete (or wood) 》 *390.4(A)* 《.

B Flat-top raceways greater than 4 in. (100 mm) but not more than 8 in. (200 mm) wide, with a minimum of 1-in. (25-mm) spacing between the raceways, must be covered with at least 1 in. (25 mm) of concrete 》 *390.4(B)* 《.

C Trench-type flush raceways having removable covers can be laid flush with the floor's surface. Such approved raceways shall be designed so that the cover plates provide adequate mechanical protection and rigidity equivalent to junction box covers 》 *390.4(C)* 《.

D In office occupancies, approved flat-top metal raceways (of 4 in. [100 mm] or less in width) can be flush with the concrete floor surface, provided they are covered with a minimum ¹⁄₁₆-in. (1.6-mm) floor covering, such as linoleum. Where more than one (but fewer than four) single raceways are installed flush with the concrete, they must be contiguous with one another and be joined forming a rigid assembly 》 *390.4(D)* 《.

E Flat-top raceways greater than 4 in. (100 mm) but not more than 8 in. (200 mm) wide that are spaced less than 1 in. (25 mm) apart must be covered with at least 1½ in. (38 mm) of concrete 》 *390.4(B)* 《.

F Splices and taps are permitted in trench-type flush raceways whose removable covers are accessible after installation. At the point of any splices or taps, the conductors must not fill more than 75% of the raceway's area 》 *390.7 Exception* 《.

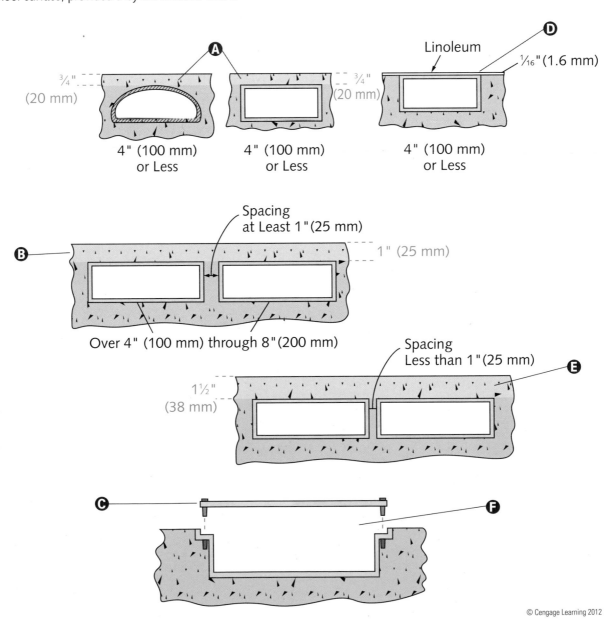

© Cengage Learning 2012

Cellular Metal Floor Raceways

A The combined cross-sectional area of all conductors or cables must not exceed 40% of the cell or header's interior cross-sectional area ≫374.5≪.

B Junction boxes must be leveled to the floor grade and sealed against the entrance of water or concrete. Junction boxes used with these raceways must be metal and must be electrically continuous with the raceway ≫374.9≪.

C A cellular metal floor raceway is defined in Article 370 as the hollow spaces of cellular metal floors, together with suitable fittings, that may be approved as enclosed channel for electric conductors ≫374.2≪.

D Conductors larger than 1/0 AWG can be installed only by special permission ≫374.4≪.

E Inserts must be leveled to the floor grade and sealed against the entrance of concrete. Only metal inserts can be used with metal raceways, and each must be electrically continuous with the raceway ≫374.10≪.

F Connections between raceways and distribution centers and wall outlets shall be made by means of liquidtight flexible metal conduit, flexible metal conduit where not installed in concrete, rigid metal conduit, intermediate metal conduit, electrical metallic tubing, or approved fittings. Where there are provisions for the termination of an equipment grounding conductor, rigid polyvinyl chloride conduit, reinforced thermosetting resin conduit, electrical nonmetallic tubing, or liquidtight flexible nonmetallic conduit shall be permitted. Where installed in concrete, liquidtight flexible metal conduit and liquidtight flexible nonmetallic conduit shall be listed and marked for direct burial ≫374.11≪.

G Splices and taps are made only in header access units or junction boxes. Continuous, unbroken conductor connecting the individual outlets (so-called loop wiring) is not considered a splice or tap ≫374.6≪.

H A "header" is a transverse raceway for electric conductors, providing access to predetermined cells of a cellular metal floor, accommodating the installation of electric conductors from a distribution center to the cells ≫374.2≪.

I A "cell" is a single, enclosed tubular space in a cellular metal floor member, whose axis is parallel to the axis of the metal floor member ≫374.2≪.

NOTE

Section 374.3 lists areas where cellular metal floor raceways shall not be used.

CAUTION

When an outlet is removed, abandoned, or discontinued, remove the section of circuit conductors supplying the outlet from the raceway. Raceways must not contain splices or reinsulated conductors as would occur with abandoned outlet loop wiring ≫374.7≪.

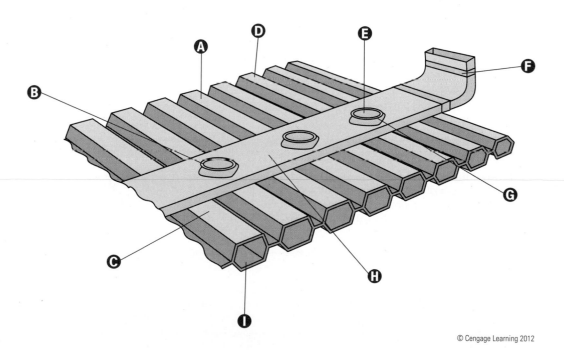

Cellular Concrete Floor Raceways

A Splices and taps are allowed only in header access units or junction boxes. A continuous unbroken conductor connecting the individual outlets is not a splice or tap ››372.12‹‹.

B A "header" is a transverse metal raceway for electric conductors that provides access to predetermined cells of a precast cellular concrete floor and permits the installation of conductors from a distribution center to the floor cells ››372.2‹‹.

C Inserts must be leveled and sealed against the entrance of concrete. Inserts must be metal and must be fitted with grounded-type receptacles. A grounding conductor must connect the insert receptacles to a positive ground connection on the header ››372.9‹‹.

D The header must be installed in a straight line and at right angles to the cells ››372.5‹‹.

E Mechanically secure the header to the top of the precast cellular concrete floor. The header must be electrically continuous throughout its entire length and must be electrically bonded to the distribution center enclosure ››372.5‹‹.

F Conductors larger than 1/0 AWG can be installed only by special permission ››372.10‹‹.

> **CAUTION**
>
> When an outlet is removed, abandoned, or discontinued, remove the section of circuit conductors supplying the outlet from the raceway. Raceways must not contain splices or reinsulated conductors, such as loop wiring of abandoned outlets ››372.13‹‹.

G The combined cross-sectional area of all conductors or cables must not exceed 40% of the cell or header's cross-sectional area ››372.11‹‹.

H A "cell" is a single, enclosed tubular space in a floor made of precast cellular concrete slabs, whose direction is parallel to the direction of the floor member ››372.2‹‹.

I Cellular concrete floor raceways are hollow spaces in floors constructed of precast cellular concrete slabs (together with suitable metal fittings) designed to provide access to the floor cells ››372.1‹‹.

> **NOTE**
>
> Section 372.4 lists areas where cellular concrete floor raceways shall not be used.

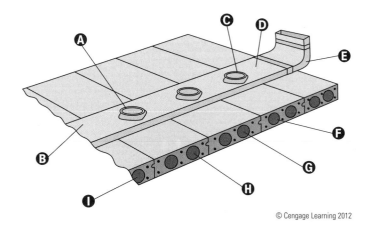

© Cengage Learning 2012

Wireways (Metal and Nonmetallic)

A Wireways (sheet metal and flame-retardant, nonmetallic) are troughs with hinged or removable covers for housing and protecting electric wires (and cables). The complete raceway system is installed before conductors are laid in place ››376.2 and 378.2‹‹.

B The sum of the cross-sectional area of all contained conductors must not exceed 20% of the wireway's interior cross-sectional area ››376.22(A) and 378.22‹‹.

C No conductor larger than that for which the wireway is designed shall be installed ››376.21 and 378.21‹‹.

D Within **metal wireways,** the derating factors in 310.15(B)(3)(a) are only applicable where the number of current-carrying conductors, including neutral conductors classified as current-carrying under the provisions of 310.15(B)(5), exceeds 30 at any cross section of the wireway ››376.22(B)‹‹.

E Derating factors specified in 310.15(B)(3)(a) are applicable to **nonmetallic wireways** containing more than three current-carrying conductors ››378.22‹‹.

> **CAUTION**
>
> Wireways containing 4 AWG or larger conductors and used as junction or pull boxes must be sized according to 314.28(A) or 314.71.

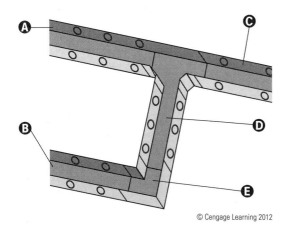

© Cengage Learning 2012

> **NOTE**
>
> Conductors for signaling circuits or controller conductors between a motor and its starter (used only for starting duty) are not considered as current-carrying conductors ››376.22(B) and 378.22‹‹. The manufacturer's name (or trademark) on all wireways must remain visible after installation. In addition, the interior cross-sectional area (in square inches) must be marked on all nonmetallic wireways ››376.120 and 378.120‹‹.

Wireway Installation

Listed **nonmetallic** wireways are permitted (1) only for exposed work, except as permitted in 378.10(4); (2) where subject to corrosive environments where identified for the use; and (3) in wet locations where listed for the purpose ≫ *378.10* ≪.

Nonmetallic wireways shall not be used (1) where subject to physical damage; (2) in any hazardous (classified) location, except as permitted by other articles in the *Code;* (3) where exposed to sunlight unless listed and marked as suitable for the purpose; (4) where subject to ambient temperatures other than those for which nonmetallic wireway is listed; and (5) for conductors whose insulation temperature limitations would exceed those for which the raceway is listed ≫ *378.12* ≪.

Ⓐ Accessible splices and taps are permitted within a wireway. At the point of any splice or tap, the conductors must not fill the wireway to more than 75% of its area ≫ *376.56* and *378.56* ≪.

Ⓑ Close all dead ends with listed fittings ≫ *376.58* and *378.58* ≪.

Ⓒ Where insulated conductors are deflected within a wireway, either at the ends or where conduits, fittings, or other raceways (or cables) enter (or leave) the wireway, or where the direction of the wireway is deflected greater than 30°, 312.6(A) dimensions corresponding to one wire per terminal apply ≫ *376.23(A)* and *378.23(A)* ≪.

Ⓓ Where 4 AWG or larger insulated conductors enter a wireway, the distance between raceway and cable entries (enclosing the same conductor) must not be less than that required in 314.28(A)(1) for straight pulls and 314.28(A)(2) for angle pulls ≫ *376.23(B)* and *378.23(B)* ≪.

Ⓔ Unbroken wireway lengths can pass transversely through walls. Access to the conductors must be maintained on both sides of the wall ≫ *376.10(4)* and *378.10(4)* ≪.

Ⓕ **Metal** wireways can be used (1) for exposed work except as permitted in 376.10(4), or (2) in any hazardous (classified) locations as permitted by other articles in the *NEC* ≫ *376.10* ≪.

Ⓖ **Metal** wireways are not permitted where subject to severe physical damage or severe corrosive environments ≫ *376.12* ≪.

Ⓗ Extensions from wireways can be made with cord pendants (installed in accordance with 400.10) or with any Chapter 3 wiring method that includes an equipment grounding means. Where a separate equipment grounding conductor is employed, connection of the wiring method's equipment grounding conductors to the wireway must comply with 250.8 and 250.12 ≫ *376.70* and *378.70* ≪.

CAUTION

Extreme cold may cause nonmetallic wireways to become brittle and, therefore, more susceptible to damage from physical contact ≫ *378.10(3) Informational Note* ≪.

NOTE

Wireways installed in wet locations must be listed for the purpose ≫ *376.10(3)* **and *378.10(3)* ≪.**

NOTE

When transposing cable size into raceway size, the minimum trade size (metric designator) raceway required for the number and size of conductors in the cable shall be used ≫ *376.23(B)* **and *378.23(B)* ≪.**

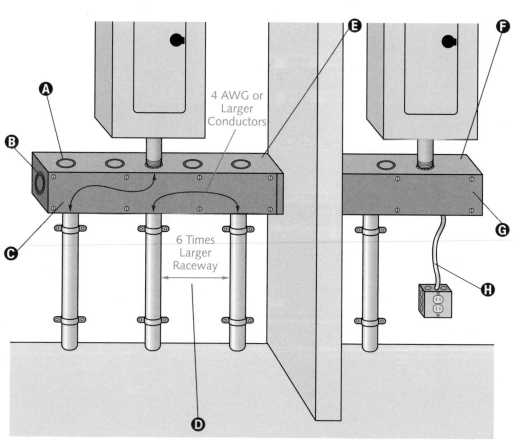

4 AWG or Larger Conductors

6 Times Larger Raceway

Wireway Supports

A Horizontally run wireways must be supported at each end ≫376.30(A)≪.

B The maximum distance between supports for individual lengths of metal wireway must not exceed 10 ft (3 m) ≫376.30(A)≪.

C Nonmetallic wireway support intervals must not exceed 3 ft (900 mm) ≫378.30(A)≪.

D Horizontally run individual nonmetallic wireway lengths of more than 3 ft (900 mm) must be supported at each end or joint ≫378.30(A)≪.

E Individual nonmetallic wireways must be supported as listed, and in no case shall the distance between supports exceed 10 ft (3 m) ≫378.30(A)≪.

F Metal wireway support intervals must not exceed 5 ft (1.5 m) ≫376.30(A)≪.

G Individual metal wireway lengths (longer than 5 ft [1.5 m]), run horizontally, must be supported at each end or joint, unless listed otherwise ≫376.30(A)≪.

H Horizontally run wireways must be supported at each end ≫378.30(A)≪.

I Nonmetallic wireway expansion fittings must be provided to compensate for thermal expansion and contraction where the change in length is expected to be 0.25 in. (6 mm) or more in a straight run ≫378.44≪. Expansion characteristics of PVC nonmetallic wireway (which is the same as PVC rigid nonmetallic conduit) are listed in Table 352.44.

> **NOTE**
>
> Nonmetallic wireway vertical runs must be securely supported at intervals of 4 ft (1.2 m) or less (except as otherwise listed) and can have only one joint between supports. Adjoining nonmetallic wireway sections must be securely fastened together to form a rigid joint ≫*378.30(B)*≪.
>
> Metal wireway vertical runs must be securely supported at intervals not exceeding 15 ft (4.5 m) and are limited to one joint between supports. Adjacent wireway sections must be securely fastened together to form a rigid joint ≫*376.30(B)*≪.

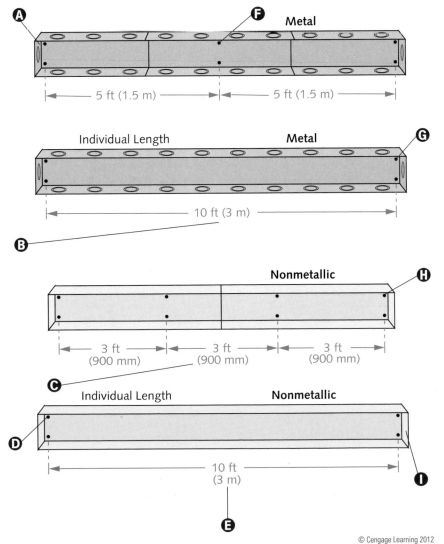

© Cengage Learning 2012

RACEWAY FILL

Same Size Conductors

Ⓐ Informative Annex C (located in the back of the *Code* book) can be used to find the maximum number of conductors permitted in a particular raceway (conduit or tubing). The conductor maximum is based on all conductors (in the raceway) being the same size (total cross-sectional area including insulation) ≫ *Chapter 9, Note 1* ≪. Compact conductors are also listed in Informative Annex C.

Ⓑ Tables C1 through C12 (in Informative Annex C) are based on the 40% fill for three or more conductors as permitted in Table 1 of Chapter 9.

Ⓒ Informative Annex C also lists the maximum number of fixture wires permitted in each type of conduit (or tubing).

© Cengage Learning 2012

Different Size Conductors

Ⓐ Tables 4, 5, and 5A (of Chapter 9) are used when a combination of different size conductors is installed in a single raceway ≫ *Chapter 9, Note 6* ≪. If the actual values of the conductor diameter and area are known, they shall be permitted to be used ≫ *Chapter 9, Note 10* ≪.

Ⓑ Use the 40% column of Table 4 when there are three or more conductors.

Ⓒ Use Table 5 for dimensions of insulated conductors and Table 5A for dimensions of insulated compact conductors.

Ⓓ Include equipment grounding or bonding conductors, if any, when calculating raceway fill ≫ *Chapter 9, Note 3* ≪.

Ⓔ Table 4 contains trade size information for 12 different types of conduits and tubing. Columns include internal diameters and total area, also referred to as the cross-sectional area. Table 4 also shows the maximum percent fill for tubing or conduits containing different numbers of conductors.

> **NOTE**
>
> The approximate area for bare conductors is found in Table 8 of Chapter 9.

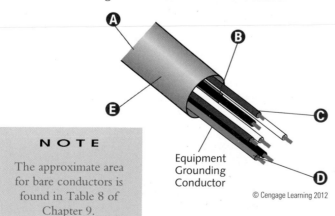

Equipment
Grounding
Conductor

© Cengage Learning 2012

Raceways Containing Multiconductor Cables

Ⓐ A multiconductor cable, optical fiber cable, or flexible cord of two or more conductors is treated as a single conductor for percentage conduit fill area calculations. To calculate the cross-sectional area of cables with elliptical cross sections, use the major diameter of the ellipse as a circle diameter ≫ *Chapter 9, Note 9* ≪.

> **NOTE**
>
> Although Table 1 (in Chapter 9) applies only to complete conduit or tubing systems, it does not apply to sections of conduit used to protect exposed wiring from physical damage ≫ *Chapter 9, Note 2* ≪.

One
Conductor 53%

© Cengage Learning 2012

Nipple Fill

Ⓐ A length of conduit (or tubing) measuring 24 in. (600 mm) or less is considered a nipple ≫ *310.15(B)(3)(a) and Chapter 9, Note 4* ≪.

Ⓑ For nipples, disregard 310.15(B)(3)(a) adjustment factors ≫ *Chapter 9, Note 4* ≪.

Ⓒ Installed between boxes, cabinets, and similar enclosures, conduit or tubing nipples can be filled to 60% of their total cross-sectional area ≫ *Chapter 9, Note 4* ≪.

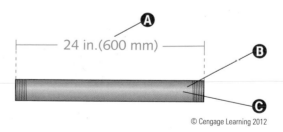

24 in.(600 mm)

© Cengage Learning 2012

> **NOTE**
>
> When calculating the maximum number of same size (total cross-sectional area including insulation) conductors or cables permitted in a conduit or tubing, the next-higher whole number is used when the calculation results in a decimal of 0.8 or larger ≫ *Chapter 9, Note 7* ≪.

Raceway Fill Percentage

A A conduit having a single conductor or cable can be filled to 53% of the conduit's cross-sectional area.

B A conduit containing exactly two conductors and/or cables can only be filled to 31% of the cross-sectional area of the conduit or tubing.

C A conduit containing three or more conductors and/or cables can be filled to 40% of its cross-sectional area.

D When pulling three conductors or cables into a raceway, if the ratio of the raceway (inside diameter) to the conductor or cable (outside diameter) is between 2.8 and 3.2, jamming can occur. Jamming is less likely to occur when pulling four or more conductors or cables into a raceway ≫ *Chapter 9, Table 1, Informational Note No. 2* ≪.

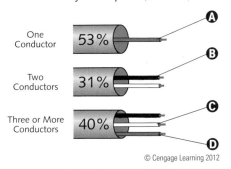

One Conductor **53%** **A**

Two Conductors **31%** **B**

Three or More Conductors **40%** **C** **D**

© Cengage Learning 2012

NOTE

Tables 1 and 4 (in Chapter 9) list maximum fill percentages for conduits and tubing.

NOTE

To perform the calculation ratio of the raceway to the conductor or cable, divide the total (100%) cross-sectional area of the conduit or tubing by the approximate area of the three conductors and/or cables.

For example, three 3 AWG THHN conductors will be installed in a 1-inch EMT raceway. Find the ratio of the raceway to the conductors.

The total cross-sectional area of 1-inch EMT is 0.864 in.^{2}. The approximate area for three 3 AWG THHN conductors is 0.2919 in.^{2} ($0.0973 \times 3 = 0.2919$). The ratio of the raceway to the conductors is 2.96 ($0.864 \div 0.2919 = 2.9599 = 2.96$). Because 2.96 is between 2.8 and 3.2, there is a higher probability that jamming will occur.

CONDUCTORS

Conductors in Parallel

A Conductors of one phase, polarity, neutral, grounded-circuit conductors, or equipment grounding conductors can have different physical characteristics from those of another phase, neutral, grounded-circuit conductors, or equipment grounding conductors and still achieve balance. For example, neutral conductors do not have to be the same length as Phase A conductors, and Phase A conductors do not have to be the same length as Phase B conductors, etc. ≫*310.10(H)(2)*≪.

B As a general rule, 1/0 AWG and larger size conductors can be connected in parallel (electrically joined at both ends to form a single conductor) ≫*310.10(H)(1)*≪.

C The paralleled conductors in each phase, polarity, neutral, grounded-circuit conductor, equipment grounding conductors, or equipment bonding jumper must have the same characteristics ≫*310.10(H)(2)*≪. For example, all paralleled, Phase C conductors must:

   (1) be the same length,

   (2) consist of the same conductor material,

   (3) be the same size in circular mil area,

   (4) have the same insulation type, and

   (5) be terminated in the same manner.

D Conductors carrying alternating current, installed in ferrous metal enclosures or ferrous metal raceways, must be so arranged as to avoid heating the surrounding metal by induction ≫*300.20(A)*≪.

E If run in separate raceways (or cables), the raceways (or cables) must have the same physical characteristics. Where conductors are in separate raceways or cables, the same number of conductors must be used in each raceway or cable ≫*310.10(H)(3)*≪.

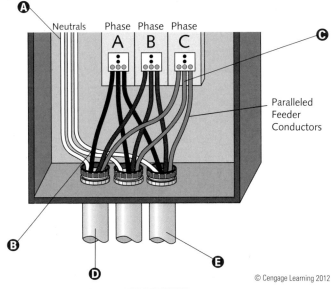

A
Neutrals Phase Phase Phase
A B C
C

Paralleled Feeder Conductors

B **D** **E**

© Cengage Learning 2012

NOTE

Equipment grounding conductors used with conductors in parallel must comply with 310.10(H) requirements, applying the sizing requirements of 250.122.

Maximum Ampacities

Ⓐ Ampacities for conductors rated 0 to 2000 volts are specified in the Allowable Ampacity Table 310.15(B)(16) through Table 310.15(B)(19), and Ampacity Table 310.15(B)(20) and Table 310.15(B)(21) as modified by 310.15(B)(1) through (B)(7) ⟫*310.15(B)*⟪.

Ⓑ Table 310.104(A) contains specific conductor information, such as trade name, type letter, maximum operating temperature, application provisions, insulation, size, and outer covering.

Ⓒ Tables 310.15(B)(16) through 310.15(B)(21) list aluminum (or copper-clad aluminum) as well as copper conductors.

Ⓓ Maximum ampacities are listed in Tables 310.15(B)(16) through 310.15(B)(21). Other factors must be considered before using these ampacities: temperature limitations ⟫*110.14(C)*⟪, continuous loads ⟫*210.19(A)* and *215.2(A)*⟪, ambient temperature ⟫*Tables 310.15(B)(2)(a) and (b)*⟪, and the number of current-carrying conductors ⟫*310.15(B)(3)(a)*⟪, to name a few.

60°C (140°F)	75°C (167°F)	90°C (194°F)
70 amperes	85 amperes	95 amperes
TW	THW	THHN
4 AWG Copper	4 AWG Copper	4 AWG Copper

© Cengage Learning 2012

Conductor Temperature Limitations—100 Amperes or Less

Ⓐ Equipment termination provisions for circuits rated 100 amperes or less or marked for 14 AWG through 1 AWG conductors are used only for conductors rated 60°C (140°F) ⟫*110.14(C)(1)(a)*⟪. Because the lowest temperature (weakest link) is 60°C (140°F), the ampacity of this conductor shall not exceed 70 amperes.

Ⓑ A 4 AWG copper conductor with a temperature rating of 60°C (140°F) has an ampacity of 70 amperes ⟫*Table 310.15(B)(16)*⟪.

Ⓒ A temperature rating of 60/75°C means that the termination is suitable for 60°C (140°F) or 75°C (167°F) conductors.

Ⓓ Conductor temperature limitations can be compared to the strength of a chain. A chain is only as strong as its weakest link. One potential weak link for conductors is the termination (connection) point. The conductor ampacity shall not be higher than the lowest temperature rating of any connected termination, conductor, or device ⟫*110.14(C)*⟪.

Ⓔ The ampacity of a 4 AWG THHN copper conductor is 95 amperes ⟫*Table 310.15(B)(16)*⟪.

Ⓕ A 4 AWG copper conductor with a temperature rating of 75°C (167°F) has an ampacity of 85 amperes ⟫*Table 310.15(B)(16)*⟪.

Ⓖ Equipment termination provisions for circuits rated 100 amperes or less or marked for 14 AWG through 1 AWG conductors can be used for conductors up to their maximum ampacities if the equipment is listed and identified for use with such conductors ⟫*110.14(C)(1)(a)(3)*⟪. The ampacity of this conductor now has a rating of 85 amperes because the lowest temperature (weakest link) is 75°C (167°F).

NOTE

When using conductors with temperature ratings higher than those specified for terminations, the higher ampacity can be used for ampacity adjustment, correction, or both ⟫*110.14(C)*⟪. For example, an adjustment factor of 80% must be applied to 4 AWG THHN copper conductors because there will be four current-carrying conductors in the raceway. The connected load requires an overcurrent protection and conductor ampacity rating of at least 70 amperes, but the lowest temperature rated termination is 60°C. A 70-ampere overcurrent protective device can be installed with 4 AWG THHN (90°C) copper conductors because the allowable ampacity (after the adjustment factor has been applied) is 76 amperes (95 × 80% = 76 amperes). Had derating not been possible (using the higher ampacity), the maximum conductor ampacity would be 56 amperes (70 × 80% = 56 amperes).

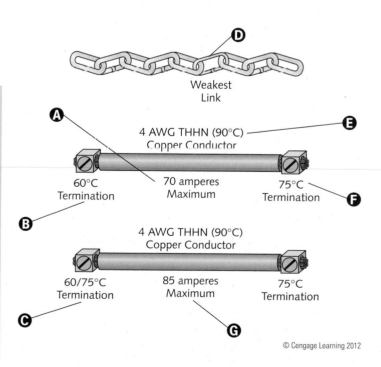

© Cengage Learning 2012

Conductor Properties

Ⓐ A 6 AWG or smaller insulated grounded conductor must be identified by a continuous white or gray outer finish or by three continuous white stripes on other than green insulation along the conductor's entire length ≫ 200.6(A) ≪ .

Ⓑ Table 310.104(A) lists conductor insulations and applications.

Ⓒ An insulated grounded conductor larger than 6 AWG shall be identified by either a continuous white or gray outer finish or by three continuous white or gray stripes on other than green insulation along the conductor's entire length or, at the time of installation, by a distinctive white marking at its terminations. This marking must encircle the conductor or insulation ≫ 200.6(B) ≪ .

Ⓓ Ampacities of insulated conductors rated 0 through 2000 volts, 60°C through 90°C (140°F through 194°F), with three or fewer current-carrying conductors installed in raceway, cable, or earth (directly buried) are found in Table 310.15(B)(16). The ampacities listed there are based on an ambient temperature of 30°C (86°F). A variety of insulations are listed for both copper and aluminum (or copper-clad aluminum) conductors.

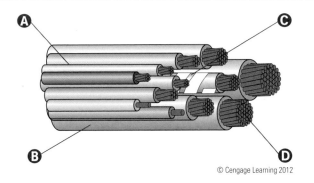

© Cengage Learning 2012

> **NOTE**
>
> Where installed in raceways, conductors 8 AWG and larger, not specifically permitted or required elsewhere in the *NEC* to be solid, shall be stranded ≫ 310.106(C) ≪ .

Conductor Temperature Limitations—Over 100 Amperes

Equipment termination provisions for circuits rated over 100 amperes or marked for conductors larger than 1 AWG are used only for conductors with higher temperature ratings provided the conductor ampacity is based on the 75°C (167°F) ampacity of the conductor size used ≫ *110.14(C)(1)(b)* ≪ .

> **NOTE**
>
> In supervised industrial installations where the overcurrent device is rated over 800 amperes, the ampacity of the conductors it protects shall be equal to or greater than 95 percent of the rating of the overcurrent device specified in 240.6 in accordance with: (1) the conductors are protected within recognized time vs. current limits for short-circuit currents and (2) all equipment in which the conductors terminate is listed and marked for the application ≫ *240.91(B)* ≪ .

> **CAUTION**
>
> Three paralleled sets of 500 kcmil THHN copper conductors do not meet *NEC* provisions if the overcurrent device is 1200 amperes, unless the equipment is listed and identified for use with 90°C (194°F) conductors. Because of 110.14(C)(1)(b), the maximum ampacity (at 75°C [167°F]) for these conductors is only 380 amperes. The total combined ampacity for each phase is, therefore, only 1140 amperes; Section 240.4(C) requires a minimum conductor ampacity of 1200 amperes.

> **NOTE**
>
> The minimum size conductors (aluminum, copper-clad aluminum, or copper) that can be connected in parallel (electrically joined at both ends to form a single conductor) is size 1/0 AWG, unless meeting an exception ≫ *310.10(H)(1)* ≪ .

Where the overcurrent device is rated over 800 amperes, the ampacity of the conductors it protects must be equal to or greater than the overcurrent device rating ≫ *240.4(C)* ≪ .

Ⓐ Switchboard and panelboard provisions are found in Article 408.

Ⓑ When running *three* sets of paralleled copper conductors (for a 1200-ampere overcurrent device), the minimum size is 600 kcmil, unless the equipment is listed and identified for use with 90°C (194°F) conductors, or unless meeting the provision in 240.91(B).

Ⓒ When running *four* sets of paralleled copper conductors (for a 1200-ampere overcurrent device), the minimum size is 350 kcmil, unless the equipment is listed and identified for use with 90°C (194°F) conductors.

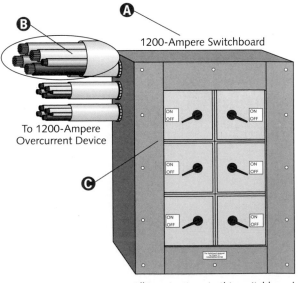

1200-Ampere Switchboard

To 1200-Ampere Overcurrent Device

All terminations in this switchboard are rated 75°C.

© Cengage Learning 2012

Ampacity Correction and Adjustment Factors

Where conductors of different systems, as provided in 300.3, are installed in a common raceway or cable, the adjustment factors shown in Table 310.15(B)(3)(a) shall apply only to the number of power and lighting conductors (Articles 210, 215, 220, and 230) ≫ *310.15(B)(3)(a)* ≪.

The following **are not** counted as current-carrying conductors when using Table 310.15(B)(2)(a):

- Conductors in cable trays, unless required by 392.80 ≫ *310.15(B)(2)(a)(1)* ≪.

- Conductors in a raceway having a length not exceeding 24 in. (600 mm) ≫ *310.15(B)(2)(a)(2)* ≪.

- Adjustment factors shall not apply to underground conductors entering or leaving an outdoor trench if those conductors have physical protection in the form of rigid metal conduit (RMC), intermediate metal conduit (IMC), rigid polyvinyl chloride conduit (PVC), or reinforced thermosetting resin conduit (RTRC) having a length not exceeding 10 ft (3.05 m), and if the number of conductors does not exceed four ≫ *310.15(B)(3)(a)(3)* ≪.

- Neutral conductors of normally balanced circuits containing three or more conductors ≫ *310.15(B)(5)(a)* ≪.

- Grounding and bonding conductors ≫ *310.15(B)(6)* ≪.

The following **shall** be counted as current-carrying conductors when using Table 310.15(B)(3)(a):

- The neutral conductor in a 3-wire circuit consisting of two phase (hot) wires and the neutral that is fed from a 4-wire, 3-phase wye-connected system ≫ *310.15(B)(5)(b)* ≪.

- The neutral conductor of a 4-wire, 3-phase wye circuit where the major portion of the load consists of **nonlinear loads,** because harmonic currents are present in the neutral conductor ≫ *310.15(B)(5)(c)* ≪. Examples: Electronic equipment (such as computers), electronic/electric-discharge lighting (such as fluorescent lights and lights with ballast), adjustable-speed drive systems, and similar equipment ≫ *Article 100—Definitions* ≪.

- A grounded conductor of any 2-wire circuit.

Ⓐ When there are more than three current-carrying conductors in a raceway or cable, the allowable ampacity of each conductor must be reduced (derated) as shown in Table 310.15(B)(3)(a) ≫ *310.15(B)(3)(a)* ≪.

Ⓑ Ambient temperature is the temperature of the conductor's surrounding environment. It is the temperature of the environment in which the conductor, raceway, or cable is installed. This environment may include air, water, earth, or a combination of those elements.

Ⓒ A grounding (or bonding) conductor does not count as a current-carrying conductor ≫ *310.15(B)(6)* ≪.

Ⓓ When the ambient temperature exceeds 30°C (86°F), the table ampacity has to be multiplied by the ambient temperature correction factor in either Table 310.15(B)(2)(a) or Table 310.15(B)(2)(b).

Ⓔ Ampacity correction factors for ambient temperatures are found in Tables 310.15(B)(2)(a) or (b). These tables are used to correct the ampacity of conductors by multiplying the ampacity of the conductor by the appropriate factor.

> ### NOTE
>
> The maximum ampacities given in Table 310.15(B)(16) are based on a maximum of three current-carrying conductors and a maximum ambient temperature of 30°C (86°F). When the ambient temperature is something other than 30°C (86°F), the ampacity of the conductor shall be corrected. While ampacities in Tables 310.15(B)(16) and (17) are based on a maximum ambient temperature of 30°C (86°F), ampacities in Tables 310.15(B)(18) through (21) are based on a maximum ambient temperature of 40°C (104°F).

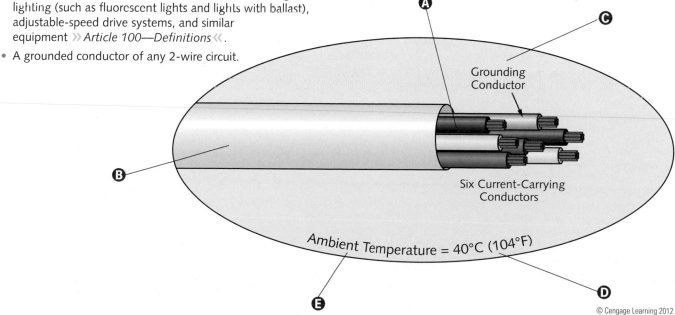

Grounding Conductor

Six Current-Carrying Conductors

Ambient Temperature = 40°C (104°F)

Adjustment Factors for Type AC Cable and Type MC Cable

(A) The adjustment factors in 310.15(B)(3)(a) shall not apply to Type AC cable or to Type MC cable under the following conditions:

a. The cables do not have an overall outer jacket.

b. Each cable has not more than three current-carrying conductors.

c. The conductors are 12 AWG copper.

d. Not more than 20 current-carrying conductors are installed without maintaining spacing, are stacked, or are supported on "bridle rings" ≫ *310.15(B)(2)(a)(4)* ≪.

(B) Normally, Table 310.15(B)(3)(a) adjustment factors apply to cables bundled together, without maintaining spacing, for a continuous length longer than 24 inches. But, it will **not** be necessary to apply the adjustment factors to Type AC and Type MC cables meeting the conditions in 310.15(B)(3)(a)(4).

(C) An adjustment factor of 60 percent shall be applied to Type AC cable or Type MC cable under the following conditions:

a. The cables do not have an overall outer jacket.

b. The number of current-carrying conductors exceeds 20.

c. The cables are stacked or bundled longer than 24 in. (600 mm) without spacing being maintained ≫ *310.15(B)(2)(a)(5)* ≪.

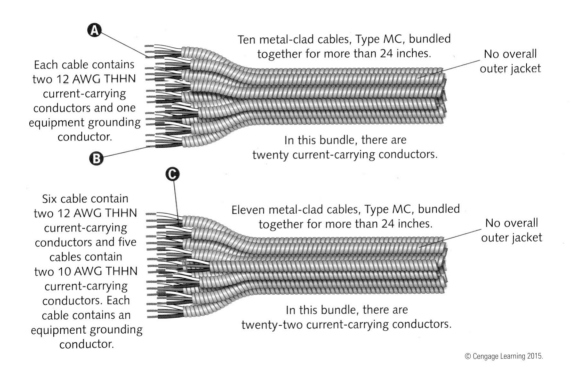

(A) Each cable contains two 12 AWG THHN current-carrying conductors and one equipment grounding conductor.

Ten metal-clad cables, Type MC, bundled together for more than 24 inches.

No overall outer jacket

In this bundle, there are twenty current-carrying conductors.

(B)

(C) Six cable contain two 12 AWG THHN current-carrying conductors and five cables contain two 10 AWG THHN current-carrying conductors. Each cable contains an equipment grounding conductor.

Eleven metal-clad cables, Type MC, bundled together for more than 24 inches.

No overall outer jacket

In this bundle, there are twenty-two current-carrying conductors.

Raceways and Cables Exposed to Sunlight on Rooftops

© Cengage Learning 2015.

A Where raceways or cables are exposed to direct sunlight on or above rooftops, the adjustments shown in Table 310.15(B)(3)(c) shall be added to the outdoor temperature to determine the applicable ambient temperature for application of the correction factors in Table 310.15(B)(2)(a) or Table 310.15(B)(2)(b) ≫310.15(B)(3)(c)≪.

B Weather-resistant type, ground-fault circuit-interrupter (GFCI) receptacle with a, outlet box hood listed and identified as "extra duty." [See 210.8(B)(3); 210.8(B)(3) Exception No. 1; and 406.9(A)(1)]

> **NOTE**
>
> One source for the ambient temperatures in various locations is the ASHRAE Handbook — Fundamentals ≫*310.15(B)(3)(c) Informational Note*≪.

> **NOTE**
>
> Informational Note to Table 310.15(B)(3)(c): The temperature adders in Table 310.15(B)(3)(c) are based on the measured temperature rise above the local climatic ambient temperatures due to sunlight heating ≫*310.15(B)(3)(c) Informational Note to Table 310.15(B)(3)(c)*≪.

Example 1: A raceway containing 2 AWG THWN-2 copper conductors is exposed to sunlight on a rooftop. The bottom of the raceway is 4 inches above the roof. The ambient temperature will be 110°F.

The Table 310.15(B)(16) ampacity (in the 90°C column) is 130 amps.

In accordance with Table 310.15(B)(3)(c), the temperature adder for a raceway 4 inches above the roof is 30°F.

Because of the temperature adder, the ambient temperature is 140°F (110 + 30 = 140).

The Table 310.15(B)(2)(a) correction factor for 140°F (in the 90°C column) is 0.71.

After correcting the ampacity, the maximum ampacity for these 2 AWG THWN-2 conductors is 92 amps (130 × 0.71 = 92.3 = 92).

> **NOTE**
>
> Type XHHW-2 insulated conductors shall not be subject to this ampacity adjustment for raceways and cables exposed to sunlight on rooftops ≫*310.15(B)(3)(c) Exception*≪.

Example 2: A raceway containing 2 AWG XHHW-2 copper conductors is exposed to sunlight on a rooftop. The bottom of the raceway is 4 inches above the roof. The ambient temperature will be 110°F.

The Table 310.15(B)(16) ampacity (in the 90°C column) is 130 amps.

In accordance with the exception to 310.15(B)(3)(c), Type XHHW-2 insulated conductors shall not be subject to this ampacity adjustment for raceways and cables exposed to sunlight on rooftops. Therefore, there is no temperature adder.

The ambient temperature is 110°F.

The Table 310.15(B)(2)(a) correction factor for 110°F (in the 90°C column) is 0.87.

After correcting the ampacity, the maximum ampacity for these 2 AWG XHHW-2 conductors is 113 amps (130 × 0.87 = 113.1 = 113).

Continuous Loads

Ungrounded (hot) conductors must independently meet requirements for (1) termination and (2) correction and adjustment factors throughout the raceway. (See 210.19(A)(1) for branch-circuit conductors, 215.2(A)(1) for feeder conductors, and 230.42(A) for service conductors.)

Because of the overcurrent device termination [110.14(C)], apply the continuous load correction factor in 210.19(A)(1), using 75°C ampacity column in Table 310.15(B)(16):

$$25 \div 125\% = 20 \text{ amperes}$$

In a separate calculation, apply Table 310.15(B)(3)(a) adjustment factors and Table 310.15(B)(2)(a) correction factors, using 90°C ampacity column in Table 310.15(B)(16):

$$30 \times 0.80 \times 0.87 = 20.88 = 21 \text{ amperes}$$

Although these conductors have an allowable ampacity of 21 amperes after adjustment and correction factors, the overcurrent device limits the maximum ampacity to 16 amperes.

In accordance with 210.20(A), the maximum on a 20-ampere circuit breaker is 16 amperes:

$$20 \div 125\% = 16 \text{ amperes}$$

Ⓐ The maximum continuous load current permitted on a 20-ampere overcurrent device is 16 amperes (20 ÷ 125% = 16).

Ⓑ A continuous load is expected to maintain maximum current for three hours or more ≫ *Article 100—Definitions* ≪. Fluorescent office lighting represents a continuous load.

Ⓒ Branch-circuit conductors shall have an ampacity not less than the maximum load to be served. Conductors shall be sized to carry not less than the larger of 210.19(A)(1)(a) or (b).

(a) Where a branch circuit supplies continuous loads or any combination of continuous and noncontinuous loads, the minimum branch-circuit conductor size shall have an allowable ampacity not less than the noncontinuous load plus 125 percent of the continuous load.

(b) The minimum branch-circuit conductor size shall have an allowable ampacity not less than the maximum load to be served after the application of any adjustment or correction factors ≫ *210.19(A)(1) and 215.2(A)(1)* ≪.

Ⓓ Where a branch circuit (or feeder) supplies any combination of continuous and noncontinuous loads, the rating of the overcurrent device shall not be less than the total of the noncontinuous load plus 125% of the continuous load ≫ *210.20(A) and 215.3* ≪.

Ⓔ After conductor adjustment and correction factors have been applied, overcurrent protection can be determined. Where the adjusted ampacity does not correspond to a standard size fuse or circuit breaker, the next higher standard size can be used only (1) if it does not exceed 800 amperes; (2) the conductors being protected are not part of a branch circuit supplying more than one receptacle for cord-and plug-connected portable loads; and (3) the conductor's ampacity does not correspond with the standard ampere rating of a fuse or circuit breaker without overload trip adjustments above its rating (but that shall be permitted to have other trip or rating adjustments) ≫ *240.4(B)* ≪.

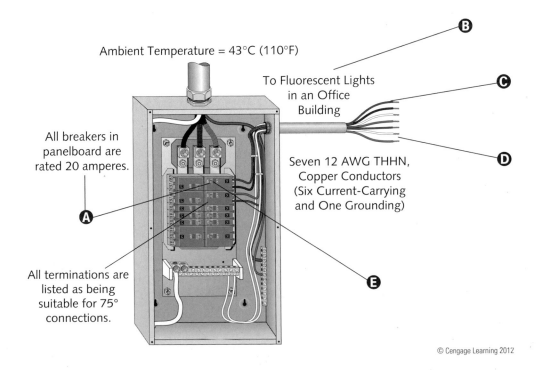

Ambient Temperature = 43°C (110°F)

To Fluorescent Lights in an Office Building

All breakers in panelboard are rated 20 amperes.

Seven 12 AWG THHN, Copper Conductors (Six Current-Carrying and One Grounding)

All terminations are listed as being suitable for 75° connections.

© Cengage Learning 2012

Support Requirements for Conductors in Vertical Raceways

Ⓐ Conductors in vertical raceways must be supported if the vertical rise exceeds Table 300.19(A) values ≫ *300.19(A)* ≪. Although three specific methods for supporting cables are listed in 300.19(C)(1), (2), and (3), other equally effective methods are also permitted.

Ⓑ Clamping devices constructed of or employing insulating wedges inserted in raceway ends are one type of conductor support. If clamping the insulation does not adequately support the cable, the conductor must also be clamped ≫ *300.19(C)(1)* ≪.

Ⓒ One cable support must be provided at the top of the vertical raceway, or as close to the top as practical. Intermediate supports must be provided so that supported conductor lengths do not exceed those specified in Table 300.19(A) ≫ *300.19(A)* ≪.

Ⓓ Covered boxes can be inserted at the required intervals in which insulating supports are installed and secured in a satisfactory manner to withstand the weight of the conductors attached to them ≫ *300.19(C)(2)* ≪.

Ⓔ Cables can be supported in junction boxes, by deflecting the cables at least 90° and carrying them horizontally to a distance not less than twice the diameter of the cable, provided the cables are carried on at least two insulating supports and are secured to them by tie wires (if desired) ≫ *300.19(C)(3)* ≪.

Ⓕ In using an insulator to deflect cable, cables must be supported at intervals not greater than 20% of the values in Table 300.19(A) ≫ *300.19(C)(3)* ≪.

Table 300.19(A) Maximum Distance between Conductor Supports in Vertical Raceways (in ft)

AWG or Circular-Mil Size of Conductor	Aluminum or Copper-Clad Aluminum	Copper
18 AWG through 8 AW	100	100
6 AWG through 1/0 AW	200	100
2/0 AWG through 4/0 AW	180	80
250 through 350 kcmil	135	60
400 through 500 kcmil	120	50
600 through 750 kcmil	95	40
Over 750 kcmil	85	35

© Cengage Learning 2012

Ⓒ

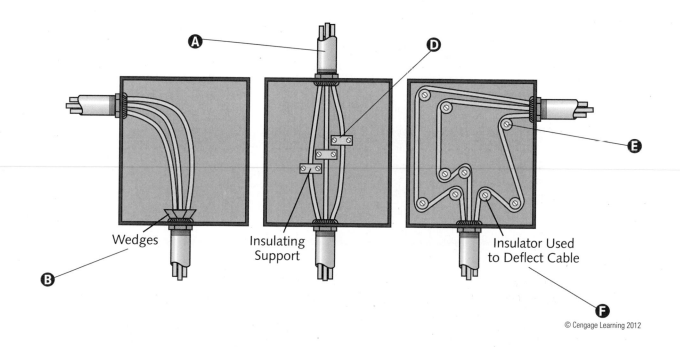

Wedges

Insulating Support

Insulator Used to Deflect Cable

Ⓐ　Ⓓ　Ⓔ　Ⓑ　Ⓕ

© Cengage Learning 2012

Summary

- Raceways shall not be supported by ceiling grid support wires, but can be supported by independent (additional) support wires.
- Raceways shall not support other raceways, cables, conductors, or nonelectric equipment.
- The maximum total of conduit bends, between pull points, is four quarter bends (360°).
- RMC is a heavy-walled metal raceway that can be threaded.
- IMC has thinner walls than does RMC, and it can also be threaded.
- Unless made up with threaded couplings, RMC and IMC support requirements are the same as EMT.
- EMT, the thinnest-walled classification of metal nonflexible raceways, provides protection from all but severe physical damage.
- The distance between PVC supports increases as conduit size increases.
- Certain raceways, permitted as luminaire whips, do not require support if installed in lengths of 6 ft or less.

- Wireways are troughs with hinged or removable covers for housing and protecting conductors and cables.
- Wireways can pass transversely through walls if the length passing through the wall is unbroken and if conductors can be accessed on both sides of the wall.
- Raceway fill can be calculated from Tables 4 and 5, located in Chapter 9.
- Conduit (and tubing) nipples can be filled to 60% of their cross-sectional area.
- Certain conductor properties are listed in Table 310.104(A).
- Conductor temperature limitations must be considered when determining overcurrent protection.
- Ambient temperature, the number of current-carrying conductors, and continuous loads can alter maximum conductor ampacity or overcurrent protection.

Unit 5 Competency Test

NEC Reference	Answer
230	B
230	45°
230	10.54
342	3/4
380	C
360	75%

1. Where exposed to the weather, raceways shall be:
    a) rainproof and arranged to drain.
    b) watertight and arranged to drain.
    c) weatherproof and arranged to drain.
    d) suitable for use in wet locations and arranged to drain.

2. Most conduit and tubing type raceways have a maximum equivalent of four _____ bends between pull points, for example, conduit bodies and boxes.

3. The cross-sectional area of 2-in. EMT is _____ in.²

4. All cut ends of conduits shall be reamed or otherwise finished to remove rough edges. Where conduit is threaded in the field, a standard cutting die with a _____-in. taper per ft (1 in 16) shall be used.

5. Multioutlet assemblies shall not:
    I. be installed in hoistways.
    II. run within dry partitions.
    III. run through dry partitions.
       a) I only   b) III only   c) I and II only   d) I, II, and III

6. Splices and taps shall be permitted within a wireway provided they are accessible. The conductors, including splices and taps, shall not fill the wireway to more than _____ of its area at that point.

NEC Reference	Answer
372	header

7. A _____ shall be defined as a transverse raceway for electric conductors, providing access to predetermined cells of a cellular metal floor, thereby permitting the installation of electric conductors from a distribution center to the cells.

| 300 | 1¼ |

8. In both exposed and concealed locations, where a cable or raceway-type wiring method is installed through bored holes in joist, rafters, or wood members, holes shall be bored so that the edge of the hole is not less than _____ in. from the nearest edge of the wood member.

| 300 | 2 |

9. How many 14 AWG THHN conductors, including an equipment grounding conductor, can be installed in a ⅜-in. flexible metal conduit using inside fittings?

| 300 | 3 |

10. What is the maximum distance (in ft) between supports for a straight run of 2-in. IMC made up with threaded couplings?

| 300 | 40% |

11. The sum of cross-sectional areas of all contained conductors at any cross section of a wireway shall not exceed _____ fill.

| 300 | induction |

12. Where conductors carrying alternating current are installed in ferrous metal raceways, they shall be so arranged as to avoid heating the surrounding metal by _____.

| 590 | trench |

13. A _____ is a type of surface, flush, or freestanding raceway designed to hold conductors and receptacles, assembled in the field or at the factory.

| 344 | galvanic reation |

14. Where practicable, contact between dissimilar metals anywhere in the system shall be avoided to eliminate the possibility of _____.

| 300 | 2 |

15. What is the maximum number of 4 AWG THHN conductors permitted in 1½-in. rigid metal conduit?

| 300 | 4 |

16. Where the same 4 AWG conductor enters and then exits a wireway through conduit or tubing, the distance between those raceway entries shall not be less than _____ times the trade diameter of the larger raceway.

| 300 | 14 |

17. All components of an exposed wiring system, installed where walls are frequently washed, must be mounted with a minimum of _____ in. airspace between components (boxes, fittings, conduits, etc.) and the wall (supporting surface).

| 590 | 6 |

18. Where an equipment bonding jumper is installed on the outside of a raceway, the length shall not exceed _____ ft and shall be routed with the raceway or enclosure.

| 390 | C |

19. Connections from headers to cabinets and other enclosures, in cellular concrete floor raceways, shall be made by means of:

    I. listed metal raceways.

    II. listed nonmetallic raceways.

    III. listed fittings.

        a) I only   b) I and II only   c) I and III only   d) I, II, and III

| 390 | 6 |

20. ENT shall be secured at least every _____ ft.

| 390 | wet |

21. IMC encased in a concrete slab, on the first floor of an office building, is considered a _____ location.

| 390 | C |

22. When an outlet from an underfloor raceway is discontinued, the sections of circuit conductors supplying the outlet:

    a) shall be permitted to be spliced.

    b) shall be removed from the raceway.

    c) shall be permitted to be reinsulated.

    d) shall be handled like abandoned outlets on loop wiring.

NEC Reference	Answer
390	insulated, etc.
390	4
390	3
390	4
300	2
390	3/2
390	1.74 in
390	EBJ
390	3
390	6
390	.5
390	45°

23. RMC, elbows, couplings, and fittings shall be permitted to be installed in concrete, in direct contact with the earth, or in areas subject to severe corrosive influences where protected by _____ and judged suitable for the condition.

24. Where installed in raceways, conductors _____ and larger, not specifically permitted or required elsewhere in the *NEC* to be solid, shall be stranded.

25. In vertical raceways, 3/0 AWG aluminum conductors must be supported if the vertical rise is greater than _____ ft.

26. Where raceways contain _____ or larger insulated circuit conductors, and these conductors enter a cabinet, a box, an enclosure, or a raceway, the conductors shall be protected by an identified fitting providing a smoothly rounded insulating surface, unless the conductors are separated from the fitting or raceway by identified insulating materials that are securely fastened in place.

27. Nonmetallic wireways shall be supported where run horizontally at intervals not to exceed _____ ft, and at each end or joint, unless listed for other support intervals.

28. Horizontal runs of IMC supported by openings through framing members at intervals not exceeding _____ ft and securely fastened within _____ ft of termination points shall be permitted.

29. If a 1¼-in. EMT raceway containing three conductors is already filled to 20%, what is the raceway's cross-sectional area?

30. Expansion fittings and telescoping sections of metal raceways shall be made electrically continuous by _____ or other means.

31. RMC shall be securely fastened within _____ ft of each outlet box, junction box, device box, cabinet, conduit body, or other conduit termination.

32. The maximum distance between supports for 4-in. PVC is _____ ft.

33. When calculating the maximum number of conductors or cables permitted in a conduit or tubing, all of the same size (total cross-sectional area including insulation), the next-higher whole number shall be used to determine the maximum number of conductors permitted when the calculation results in a decimal greater than or equal to _____.

34. Where insulated conductors are deflected within a wireway, either at the ends or where conduits, fittings, or other raceways or cables enter or leave the wireway, or where the direction of the wireway is deflected greater than _____, dimensions corresponding to one wire per terminal in Table 312.6(A) shall apply.

ONE-FAMILY DWELLINGS

UNIT 6 General Provisions

Objectives

After studying this unit, the student should:

▶ be able to calculate the minimum number of 15- and 20-ampere branch circuits in a one-family dwelling.

▶ know the requirements for single receptacles on individual branch circuits.

▶ understand the branch-circuit ratings allowed for general-purpose receptacles.

▶ know how to lay out general-purpose receptacles in a dwelling.

▶ know the receptacle ratings allowed on various size branch circuits.

▶ understand the requirements for receptacle boxes.

▶ have a good understanding of split-wire receptacles.

▶ know the requirements for wet bar receptacles.

▶ be familiar with the provisions concerning the minimum length of free conductors inside boxes.

▶ know the receptacle replacement requirements.

▶ understand the general requirements for lighting and switches.

▶ be familiar with the use of a white conductor as an ungrounded conductor.

▶ understand general requirements for devices and luminaire boxes.

▶ know the provisions for outdoor receptacles.

▶ be familiar with the provisions concerning illuminating outdoor entrances and exits.

▶ understand the requirements for attaching receptacles or lighting to vegetation (such as trees).

Introduction

Unit 6 contains provisions concerning general areas both inside and outside of one-family dwellings. A one-family dwelling is a building consisting solely of one dwelling unit »Article 100«. Learning codes that pertain to one-family dwellings is of great importance because these codes are the foundation for all residential wiring. One-family dwellings are built in all shapes and sizes. Some general codes apply to every dwelling, such as the placement of receptacles in habitable rooms.

Nonmetallic-sheathed cable is the most common wiring type installed in one-family dwellings. Article 334 covers the use, installation, and construction specifications of nonmetallic-sheathed cable (Types NM, NMC, and NMS). Uses permitted for nonmetallic-sheathed cable are covered in 334.10 and uses not permitted are in 334.12. (See Unit 4 of this text for *Code* requirements pertaining to nonmetallic-sheathed cable.)

Prior to the 2008 edition of the *NEC*, only 120-volt, single-phase, 15- and 20-ampere branch circuits supplying outlets installed in dwelling unit bedrooms were required to be arc-fault circuit-interrupter protected. Now all 120-volt, single-phase, 15- and 20-ampere branch circuits supplying outlets or devices installed in dwelling unit kitchens, family rooms, dining rooms, living rooms, parlors, libraries, dens, bedrooms, sunrooms, recreation rooms, closets, hallways, laundry areas, or similar rooms or areas shall be protected by any of the means described in 210.12(A)(1) through (6).

Over the last several editions of the NEC, requirements for GFCI and AFCI protection have expanded. See 210.8(A) for GFCI protection for personnel requirements and 210.12 for AFCI protection requirements.

ELECTRICAL FLOOR PLAN (BLUEPRINT)

Electrical blueprints (floor plans) may be optional on a one-family dwelling. Sometimes it is the responsibility of the electrician to design and install the electrical system. The accompanying electrical floor plan shows receptacles, lighting, switches, and other items that might be found on a residential plan. The *NEC* references (shown here in red), are, of course, not present on general plans.

This chapter covers everything in a one-family dwelling, from a single receptacle in an individual branch circuit to illumination of entrances and exits. Branch circuits are explained briefly in this unit and in greater detail in later units. Receptacles have both general and specific codes. General codes are explained in this unit, while specific codes for rooms and areas with special receptacle requirements are explained in Unit 7. This unit also includes general requirements for lighting and switches, while specific requirements can be found in Unit 7.

> ## CAUTION
> Some state and local jurisdictions may not allow the use of nonmetallic-sheathed cable in any dwelling. Obtain a copy of any additional rules and regulations for your area.

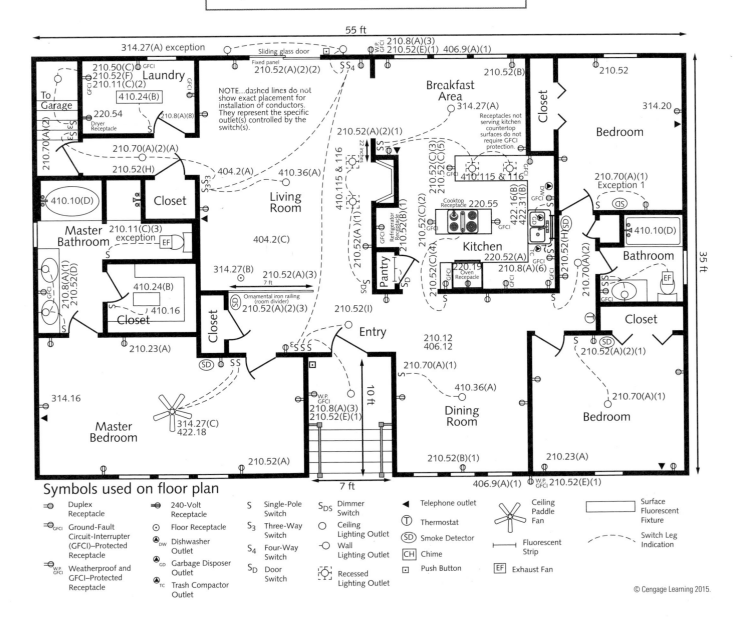

BRANCH CIRCUITS

AFCI Protection

A An arc-fault circuit interrupter (AFCI) is a device intended to provide protection from the effects of arc faults by recognizing characteristics unique to arcing and by functioning to de-energize the circuit when an arc fault is detected ≫ *Article 100* ≪.

B All 120-volt, single-phase, 15- and 20-ampere branch circuits supplying outlets or devices installed in dwelling unit kitchens, family rooms, dining rooms, living rooms, parlors, libraries, dens, bedrooms, sunrooms, recreation rooms, closets, hallways, laundry areas, or similar rooms or areas shall be protected by any of the means described in 210.12(A)(1) through (6):

(1) A listed combination-type arc-fault circuit interrupter, installed to provide protection of the entire branch circuit

(2) A listed branch/feeder-type AFCI installed at the origin of the branch-circuit in combination with a listed outlet branch-circuit type arc-fault circuit interrupter installed at the first outlet box on the branch circuit. The first outlet box in the branch circuit shall be marked to indicate that it is the first outlet of the circuit.

(3) A listed supplemental arc protection circuit breaker installed at the origin of the branch circuit in combination with a listed outlet branch-circuit type arc-fault circuit interrupter installed at the first outlet box on the branch circuit where all of the following conditions are met:

   a. The branch-circuit wiring shall be continuous from the branch-circuit overcurrent device to the outlet branch-circuit arc-fault circuit interrupter.

   b. The maximum length of the branch-circuit wiring from the branch-circuit overcurrent device to the first outlet shall not exceed 50 ft (15.2 m) for a 14 AWG conductor or 70 ft (21.3 m) for a 12 AWG conductor.

   c. The first outlet box in the branch circuit shall be marked to indicate that it is the first outlet of the circuit.

(4) A listed outlet branch-circuit type arc-fault circuit interrupter installed at the first outlet on the branch circuit in combination with a listed branch-circuit overcurrent protective device where all of the following conditions are met:

   a. The branch-circuit wiring shall be continuous from the branch-circuit overcurrent device to the outlet branch-circuit arc-fault circuit interrupter.

   b. The maximum length of the branch-circuit wiring from the branch-circuit overcurrent device to the first outlet shall not exceed 50 ft (15.2 m) for a 14 AWG conductor or 70 ft (21.3 m) for a 12 AWG conductor.

   c. The first outlet box in the branch circuit shall be marked to indicate that it is the first outlet of the circuit.

   d. The combination of the branch-circuit overcurrent device and outlet branch-circuit AFCI shall be identified as meeting the requirements for a system combination-type AFCI and shall be listed as such.

(5) If RMC, IMC, EMT, Type MC, or steel-armored Type AC cables meeting the requirements of 250.118, metal wireways, metal auxiliary gutters, and metal outlet and junction boxes are installed for the portion of the branch circuit between the branch-circuit overcurrent device and the first outlet, it shall be permitted to install a listed outlet branch-circuit type AFCI at the first outlet to provide protection for the remaining portion of the branch circuit.

(6) Where a listed metal or nonmetallic conduit or tubing or Type MC cable is encased in not less than 2 in. (50 mm) of concrete for the portion of the branch circuit between the branch-circuit overcurrent device and the first outlet, it shall be permitted to install a listed outlet branch-circuit type AFCI at the first outlet to provide protection for the remaining portion of the branch circuit ≫ *210.12(A)* ≪.

C In any of the areas specified in 210.12(A), where branch-circuit wiring is modified, replaced, or extended, the branch circuit shall be protected by one of the following: (1) a listed combination-type AFCI located at the origin of the branch circuit, or (2) a listed outlet branch-circuit type AFCI located at the first receptacle outlet of the existing branch circuit ≫ *210.12(B)* ≪.

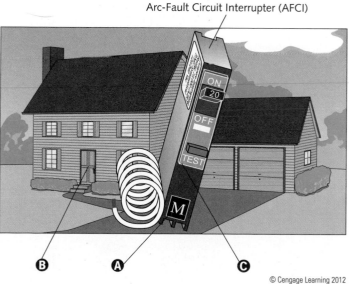

Arc-Fault Circuit Interrupter (AFCI)

© Cengage Learning 2012

> **NOTE**
>
> AFCI protection shall not be required where the extension of the existing conductors is not more than 6 ft (1.8 m) and does not include any additional outlets or devices ≫ *210.12(B) Exception* ≪.

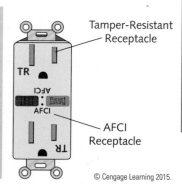

Tamper-Resistant Receptacle

AFCI Receptacle

© Cengage Learning 2015.

Tamper-Resistant Receptacles in Dwelling Units

A In all areas specified in 210.52, all nonlocking-type 125-volt, 15- and 20-ampere receptacles shall be listed tamper-resistant receptacles ≫*406.12(A)*≪.

B Tamper-resistant receptacles are identified by the letters "TR" or the words "Tamper Resistant." After the receptacle has been installed, the identification is only required to be visible with the cover plate removed. See UL (Underwriter Laboratories) White Book, category: Receptacles for Plugs and Attachment Plugs (RTRT).

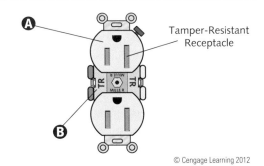
Tamper-Resistant Receptacle

© Cengage Learning 2012

> **NOTE**
>
> It is not required to install tamper-resistant receptacles in the following areas:
> 1. Receptacles located more than 5½ ft (1.7 m) above the floor.
> 2. Receptacles that are part of a luminaire or appliance.
> 3. A single receptacle or a duplex receptacle for two appliances located within dedicated space for each appliance that, in normal use, is not easily moved from one place to another and that is cord-and-plug connected in accordance with 400.7(A)(6), (A)(7), or (A)(8).
> 4. Nongrounding receptacles used for replacements as permitted in 406.4(D)(2)(a) ≫*406.12 Exception*≪.

Required Branch Circuits

A A minimum of two 20-ampere small appliance branch circuits are required ≫*210.11(C)(1) and 210.52(B)(1)*≪.

B At least one 20-ampere laundry branch circuit is required ≫*210.11(C)(2) and 210.52(F)*≪.

C A minimum of one 120-volt, 20-ampere bathroom branch circuit is required ≫*210.11(C)(3) and 210.52(D)*≪.

General-Purpose Branch Circuit

A general-purpose branch circuit may feed lights, receptacles, or both.

A General-purpose branch-circuit overcurrent protection shall not exceed 15 amperes for 14 AWG copper; 20 amperes for 12 AWG copper; and 30 amperes for 10 AWG copper ≫*240.4(D)*≪.

B Branch-circuit conductors shall have an ampacity not less than the maximum load to be served ≫*210.19(A)(1)*≪.

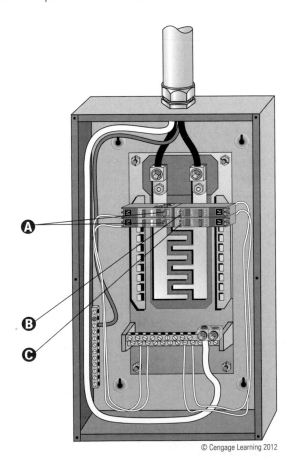

© Cengage Learning 2012

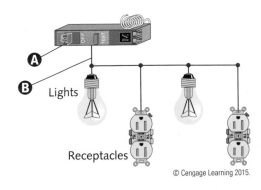
Lights
Receptacles

© Cengage Learning 2015.

Calculating 15-Ampere Branch Circuits

Ⓐ Calculation to find the minimum number of 15-ampere branch circuits:

- Calculate the general lighting and receptacle load by using Table 220.12.
- First, multiply 1400 ft² by 3 volt-amperes:

  1400 × 3 = 4200 volt-amperes

- Next, divide the total volt-amperes by 120 volts to find the total amperes:

  4200 ÷ 120 = 35 amperes

- Finally, divide 35 amperes by 15 (for a 15-ampere circuit) to find the minimum number of circuits: 35 ÷ 15 = 2.33. (Round up to the next whole number if the result of the calculation is not a whole number.)
- If 15-ampere circuits are installed for general lighting and receptacles, then at least three circuits are required.

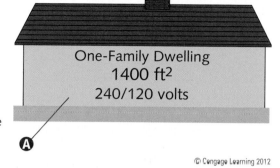

One-Family Dwelling
1400 ft²
240/120 volts

Ⓐ

© Cengage Learning 2012

Calculating 20-Ampere Branch Circuits

Ⓐ Calculation to find the minimum number of 20-ampere branch circuits:

- Calculate the general lighting and receptacle load by using Table 220.12.
- First, multiply 1400 ft² by 3 volt-amperes:

  1400 × 3 = 4200 volt-amperes

- Next, divide the total volt-amperes by 120 volts to find the total amperes: 4200 ÷ 120 = 35 amperes
- Finally, divide 35 amperes by 20 (for a 20-ampere circuit) to find the minimum number of circuits: 35 ÷ 20 = 1.75. (Round up to the next whole number if the result of the calculation is not a whole number.)
- If 20-ampere circuits are installed for general lighting and receptacles, then at least two circuits are required.

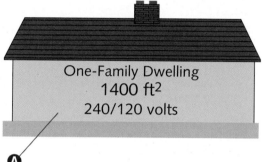

One-Family Dwelling
1400 ft²
240/120 volts

Ⓐ

© Cengage Learning 2012

NOTE

At least four 20-ampere branch circuits are required (in addition to any installed for general lighting and receptacles): two small appliance branch circuits, one laundry branch circuit, and one bathroom branch circuit.

Individual Branch Circuit for a Water Heater

Ⓐ Calculation to find the maximum overcurrent protective device:

- The overcurrent protection shall not exceed the rating marked on the appliance. If the rating is not marked and the appliance is rated over 13.3 amperes, the overcurrent protection shall not exceed 150% of the rated current ⟫ *422.11(E)(3)* ⟪: 18.75 × 150% = 28.13.
- If the calculated rating does not correspond to a standard size fuse or breaker, as found in 240.6(A), the next higher standard rating shall be permitted ⟫ *422.11(E)(3)* ⟪. The next standard size fuse or breaker higher than the calculated rating of 28.13 amperes is 30 amperes.
- The maximum overcurrent protective device (fuse or breaker) is 30 amperes.

Ⓑ The minimum conductor size is 10 AWG copper ⟫ *Table 310.15(B)(16)* ⟪. Because the overcurrent protection shall not exceed 20 amperes for 12 AWG copper, 12 AWG conductors are not permitted ⟫ *240.4(D)* ⟪.

Ⓒ A branch circuit supplying a fixed storage-type water heater (120-gallon capacity or less) shall be considered a continuous load and, therefore, must have a rating not less than 125% of the water heater nameplate rating ⟫ *422.13 and 422.10(A)* ⟪.

Ⓓ Calculation to find the minimum circuit ampacity:

- First, divide the wattage by the voltage to find amperage:

  4500 ÷ 240 = 18.75

- Next, multiply the amperage by 125% as required by 422.13: 18.75 × 125% = 23.44
- The minimum ampacity required for this water heater is 23.44 amperes.

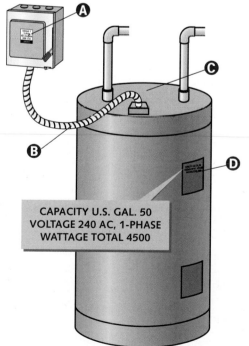

CAPACITY U.S. GAL. 50
VOLTAGE 240 AC, 1-PHASE
WATTAGE TOTAL 4500

© Cengage Learning 2012

Ampere Rating of Receptacles

Receptacle branch circuits must have a rating of 20 amperes for laundry areas, bathrooms, kitchens (except for refrigeration equipment supplied from an individual branch circuit rated 15 amperes or greater), dining rooms, pantries, breakfast rooms, and similar areas. Throughout the remainder of a one-family dwelling, 15-ampere receptacle branch circuits are allowed.

Lighting and receptacles can share the same branch circuit, except for small-appliance receptacles, bathroom receptacles (unless the circuit supplies a single bathroom), and laundry receptacles ≫ *210.23(A)* ≪. See Unit 7 for additional information concerning receptacles and lighting in these specific areas.

(A) Receptacles installed on 15-ampere circuits must have an ampere rating of not more than 15 amperes ≫ *Table 210.21(B)(3)* ≪.

(B) A 20-ampere duplex receptacle shall not be installed on a 15-ampere circuit.

(C) Receptacles installed on 20-ampere circuits can have an ampere rating of either 15 or 20 amperes ≫ *Table 210.21(B)(3)* ≪.

> ### NOTE
>
> Branch circuits for receptacles shall not be smaller than 14 AWG ≫ *210.19(A)(4)* ≪. Generally, 15-ampere circuits require 14 AWG and 20-ampere circuits require 12 AWG copper conductors. Larger conductors may be needed to compensate for voltage drop or ambient temperature, or where more than three current-carrying conductors exist in a raceway or cable.

A Single Receptacle on an Individual Branch Circuit

(D) A single receptacle installed on a branch circuit with no other device or outlet shall have an ampere rating not less than the rating of that circuit ≫ *210.21(B)(1)* ≪.

(E) Installing two or more 20-ampere receptacles or outlets on a 15-ampere circuit is not permitted, but this requirement does not apply to a single receptacle. [See 210.21(B)(1)] Note, while there is no provision that prohibits installing a 20-ampere single receptacle on a 15-ampere circuit, there is a provision that does prohibit the load from exceeding the branch-circuit ampere rating. [See 210.22]

(F) A 15-ampere single receptacle shall not be installed on a 20-ampere individual branch circuit.

(G) A duplex receptacle is not a single receptacle. As defined in Article 100, a single receptacle is a single contact device with no other contact device on the same yoke.

(H) Receptacles on 20-ampere circuits can have a rating of either 15 or 20 amperes ≫ *Table 210.21(B)(3)* ≪.

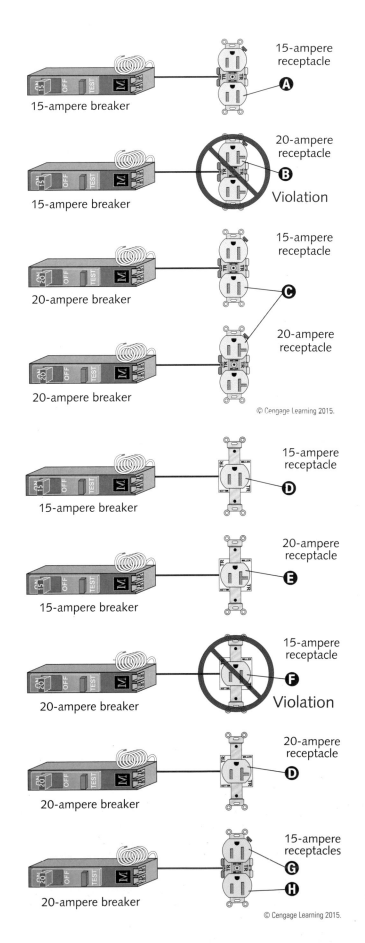

15-ampere breaker — 15-ampere receptacle **(A)**

15-ampere breaker — 20-ampere receptacle **(B)** Violation

20-ampere breaker — 15-ampere receptacle **(C)**

20-ampere breaker — 20-ampere receptacle

© Cengage Learning 2015.

15-ampere breaker — 15-ampere receptacle **(D)**

15-ampere breaker — 20-ampere receptacle **(E)**

20-ampere breaker — 15-ampere receptacle **(F)** Violation

20-ampere breaker — 20-ampere receptacle **(D)**

20-ampere breaker — 15-ampere receptacles **(G)** **(H)**

© Cengage Learning 2015.

RECEPTACLES

General Receptacle Placement

Wall space determines the minimum number of receptacles in a given dwelling. Receptacle outlets shall be installed in kitchens, family rooms, dining rooms, living rooms, parlors, libraries, dens, sunrooms, bedrooms, recreation rooms, or similar rooms or areas in accordance with the provisions specified in 210.52(A).

A Receptacles shall be installed so that no point measured horizontally along the floor line of any wall space is more than 6 ft (1.8 m) from a receptacle outlet ≫ *210.52(A)(1)* ≪.

B The maximum distance between receptacles is 12 ft (3.6 m).

C General receptacle placement requirements for 125-volt, 15- and 20-ampere receptacle outlets are in 210.52.

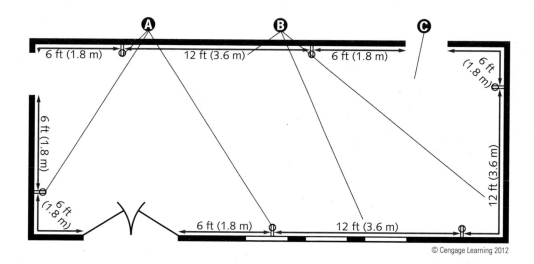

© Cengage Learning 2012

CAUTION

The receptacles required by 210.52(A) through (H) shall be in addition to any receptacle that is (1) part of a luminaire or appliance; (2) controlled by a wall switch in accordance with 210.70(A) Exception No. 1; (3) located within cabinets or cupboards; or (4) located more than 5½ ft (1.7 m) above the floor ≫ *210.52* ≪.

NOTE

Dwelling unit receptacle outlet general provisions specified in 210.52 apply to any part of a basement containing habitable rooms, such as a den, recreational room, etc.

Wall Spaces 2 ft (600 mm) in Width

A A receptacle is required for any wall space 2 ft (600 mm) or more in width (including space measured around corners), unbroken along the floor line by doorways and similar openings, fireplaces, and fixed cabinets ≫ *210.52(A)(2)(1)* ≪.

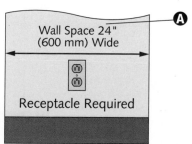

© Cengage Learning 2012

Maximum Distance to a Receptacle

Ⓐ An easy way to understand the placement of dwelling receptacles is to imagine having a floor lamp with a 6-ft (1.8-m) cord. Anywhere this lamp is placed around the wall, there should be a receptacle within reach, without using an extension cord. Even when placed beside a door opening, an outlet should be within reach. If the lamp is placed next to a wall that is at least 24 in. (600 mm) wide, an outlet should be available within that wall space.

Ⓑ The maximum distance to any receptacle, along the floor line of any wall space, measured horizontally, shall be 6 ft (1.8 m) ≫ *210.52(A)(1)* ≪.

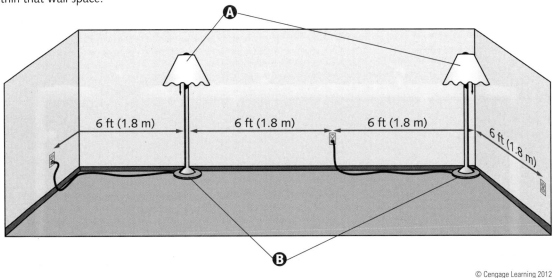

© Cengage Learning 2012

Space Measured around Corners

Ⓐ The maximum distance **between receptacles** along the floor line of any wall space measured horizontally shall be 12 ft (3.6 m) ≫ *210.52(A)(1)* ≪.

Ⓑ Wall space includes space measured around corners ≫ *210.52(A)(2)(1)* ≪.

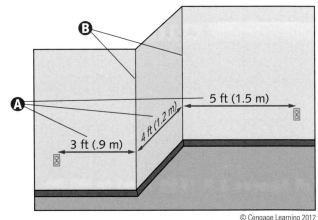

© Cengage Learning 2012

Fixed Panels

Ⓐ Glass door fixed panels in exterior walls are counted as wall space ≫ *210.52(A)(2)(2)* ≪.

Ⓑ Sliding panels in exterior walls are not counted as wall space ≫ *210.52(A)(2)(2)* ≪.

Ⓒ The maximum distance to any receptacle along the floor line measured horizontally shall be 6 ft (1.8 m) ≫ *210.52(A)(1)* ≪.

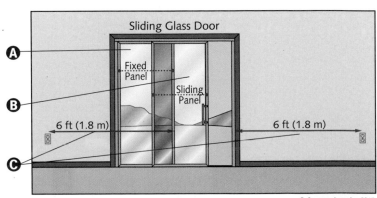

© Cengage Learning 2015.

Fixed Room Dividers

Ⓐ The space afforded by fixed room dividers, such as railings or freestanding bar-type counters, shall be counted as wall space ≫ *210.52(A)(2)(3)* ≪ .

Ⓑ Receptacle outlets in or on floors are permitted. Receptacles located more than 18 in. (450 mm) from the wall (or room divider) may not be counted as required receptacles ≫ *210.52(A)(3)* ≪ .

> ### N O T E
>
> Not all types of receptacle boxes are suitable for floor installation. Different types of boxes may be required for different types of floor construction, that is, wood or concrete. Receptacle floor boxes must be listed specifically for the type of floor in which they are installed ≫ *314.27(B)* ≪.

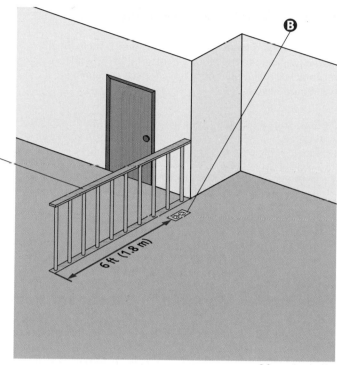

© Cengage Learning 2012

Miscellaneous Receptacle Requirements

Luminaires or appliance(s) with built-in receptacle(s) are permitted, but these receptacles do not count as required receptacles ≫ *210.52* ≪ .

Ⓐ Although receptacles located more than 5½ ft (1.7 m) above the floor are permitted, they are not counted as required receptacles ≫ *210.52* ≪ .

Ⓑ Although receptacle outlets within cabinets or cupboards are permitted, they are not counted as required receptacles ≫ *210.52* ≪ .

> ### N O T E
>
> Height requirements for wall receptacles are not defined.

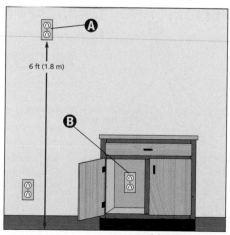

© Cengage Learning 2012

Balcony Handrail

Ⓐ A receptacle floor box must be listed specifically for the type of floor in which it is installed ≫ *314.27(B)* ≪ .

Ⓑ A floor receptacle may be required if a balcony handrail is longer than 6 ft (1.8 m) and the area is one listed in 210.52(A).

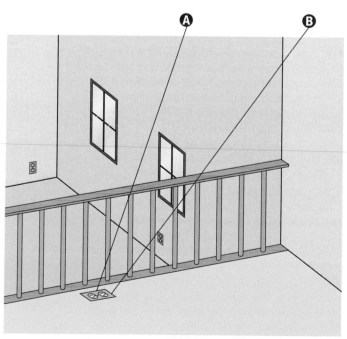

© Cengage Learning 2012

Electric Baseboard Heaters

Ⓐ A baseboard heater more than 12 ft (3.6 m) in length does not eliminate the requirements of 210.52(A)(1). A receptacle is still required for a wall space containing a baseboard heater.

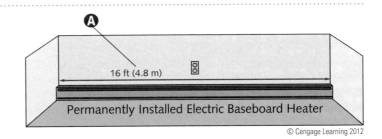

Permanently Installed Electric Baseboard Heater

© Cengage Learning 2012

NOTE

Listed or labeled equipment must be installed (and used) per any instructions included in the listing or labeling **》110.3(B)《**.

CAUTION

Listed baseboard heaters include instructions that may not permit their installation below receptacle outlets **》210.52** *Informational Note* and *424.9 Informational Note*《.

Receptacles Mounted in Baseboard Heaters

Ⓐ Permanently installed electric baseboard heaters equipped with factory-installed outlets or outlets provided as a separate assembly by the manufacturer shall be permitted as the required outlet(s) for the wall space utilized by such permanently installed heaters **》210.52** and *424.9*《.

Ⓑ The receptacle outlet(s) shall not be connected to the heater circuit **》210.52** and *424.9*《.

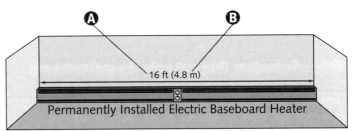

Permanently Installed Electric Baseboard Heater

© Cengage Learning 2012

Receptacle Boxes

Receptacle boxes shall comply with Article 314. An expanded description of boxes, including box fill calculations, can be found in Unit 3 of this guidebook.

Ⓐ Metal raceways, cable trays, cable armor, cable sheath, enclosures, frames, fittings, and other metal non-current-carrying parts that serve as equipment grounding conductors (with or without the use of supplementary equipment grounding conductors) shall be bonded where necessary to ensure electrical continuity and the capacity to conduct safely any fault current likely to be imposed on them **》250.96(A)《**.

Ⓑ A connection shall be made between the equipment grounding conductor(s) and a metal box by means of a grounding screw used for no other purpose. Other grounding devices or equipment listed for grounding are allowed **》250.148(C)《**.

Ⓒ The arrangement of grounding connections shall be such that the disconnection or removal of a receptacle, luminaire, or other device fed from the box will not interfere with (or interrupt) the grounding continuity **》250.148(B)《**.

Grounding Clip

Grounding Screw

Thread-Forming Screw that Engages Less than Two Threads

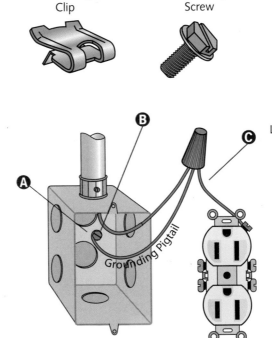

Grounding Pigtail

CAUTION

Thread-forming machine screws that do not engage at least two threads in the enclosure shall not be used as grounding screws **》250.8《**.

© Cengage Learning 2012

Grounding-Type Receptacles

Whereas 14 AWG copper conductors are required for 15-ampere circuits, 20-ampere circuits require 12 AWG ≫*240.4(D)*≪.

A Nongrounding-type receptacles may only be used to replace existing nongrounding receptacles (see "Receptacle Replacements" later in this unit).

B Grounding-type receptacles shall be used on 15- and 20-ampere branch circuits except as provided in 406.4(D) ≫*406.4(A)*≪.

C For 15-, 20-, and 30-ampere circuits, the equipment grounding conductor shall be the same size as the current-carrying conductors ≫*Table 250.122*≪. The size grounding conductor for a 15-ampere circuit is 14 AWG copper conductor; for a 20-ampere circuit is 12 AWG; and for a 30-ampere circuit is 10 AWG.

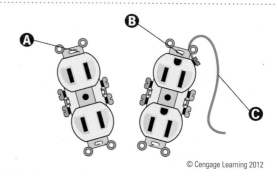

© Cengage Learning 2012

> **NOTE**
>
> Where receptacles have equipment grounding conductor contacts, those contacts shall be grounded by connection to the equipment grounding conductor of the circuit that supplies the receptacle ≫*406.4(B) and (C)*≪.

Split-Wire Receptacles

A A switch has been installed to control one of the multiwire branch circuits.

B Each multiwire branch circuit must be provided with a means that simultaneously disconnects all ungrounded (hot) conductors at the point where the branch circuit originates ≫*210.4(B)*≪. This is accomplished through the use of either one double-pole breaker or two single-pole breakers with identified handle ties.

C A split-wire receptacle receiving power from a single source (one breaker or one fuse) is not a multiwire receptacle.

D Switch-controlled, split-wire duplex receptacle(s) are sometimes installed in lieu of a lighting outlet ≫*210.70(A)(1) Exception No. 1*≪.

E The tab has been removed so one receptacle can be controlled from the switch and the other receptacle can remain live whether the switch is on or off.

F One-half of the duplex receptacle is controlled by a wall switch, while the other half is a typical receptacle.

> **WARNING**
>
> In multiwire branch circuits, the continuity of a grounded conductor shall not be dependent on the device ≫*300.13(B)*≪. If breaking the grounded conductor at the receptacle breaks the circuit down the line, then the grounded conductors must *not* be connected to the receptacle. Simply splice the grounded conductors and install a jumper wire to the receptacle.

> **NOTE**
>
> For additional information concerning multiwire branch circuits, see the definition located in Unit 2.

> **NOTE**
>
> Installing a grounded circuit conductor at the switch location is not required where the switch controls a receptacle load ≫*404.2(C)(7)*≪.

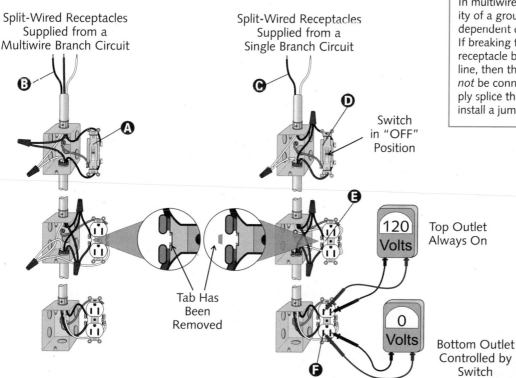

Split-Wired Receptacles Supplied from a Multiwire Branch Circuit

Split-Wired Receptacles Supplied from a Single Branch Circuit

Switch in "OFF" Position

Tab Has Been Removed

120 Volts — Top Outlet Always On

0 Volts — Bottom Outlet Controlled by Switch

© Cengage Learning 2012

Wet Bar Receptacles

If located outside the kitchen, pantry, breakfast room, dining room, or similar area, wet bar countertop receptacle placement is not specified ⟫*210.52(C)*⟪.

A Regardless of whether the receptacle serves countertop space or not, ground-fault circuit-interrupter (GFCI) protection is required if the receptacle is within 6 ft (1.8 m) of the outside edge of the sink ⟫*210.8(A)(7)*⟪.

B GFCI protection is required where receptacles are installed within 6 ft (1.8 m) of the outside edge of the sink ⟫*210.8(A)(7)*⟪.

C For cord-and plug-connected appliances, an accessible, separable connector or an accessible plug and receptacle can serve as the disconnecting means. Where the separable connector (or plug and receptacle) are not readily accessible, cord-and plug-connected appliances shall have a disconnecting means in accordance with 422.31 ⟫*422.33(A)*⟪.

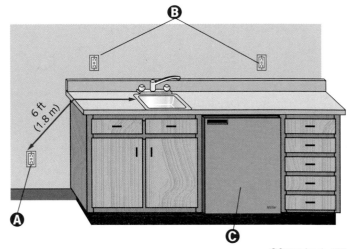

© Cengage Learning 2012

Room Air Conditioner Receptacle

A If supplied by an individual branch circuit, the rating of a cord- and attachment plug–connected room air conditioner can be no more than 80% of the circuit's ampere rating ⟫*210.23(A)(1)* and *110.62(B)*⟪.

B Where other receptacles, lighting units, or appliances are also supplied, the rating of a cord- and attachment plug–connected room air conditioner shall not exceed 50% of the branch-circuit ampere rating ⟫*210.23(A)(2)* and *440.62(C)*⟪.

C The attachment plug can serve as the disconnecting means ⟫*440.13*⟪.

D The length of the supply cord shall not exceed 10 ft (3.0 m) for a 120-volt or 6 ft (1.8 m) for a 208- or 240-volt room air conditioner ⟫*440.64*⟪.

> **NOTE**
>
> For additional provisions pertaining to room air conditioners, see Part VII of Article 440.

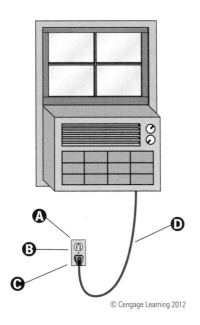

© Cengage Learning 2012

Length of Free Conductor

A At least 6 in. (150 mm) of free conductor, measured from the point in the box where it emerges from its raceway or cable sheath, shall be left at each outlet, junction, and switch point ⟫*300.14*⟪.

B Conductors must extend a minimum of 3 in. (75 mm) outside of the box opening where the opening to the box is less than 8 in. (200 mm) in any dimension ⟫*300.14*⟪.

> **NOTE**
>
> No minimum length is specified for conductors that are not spliced or terminated at the outlet, junction, or switch point
> ⟫*300.14 Exception*⟪.

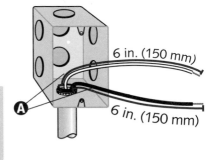
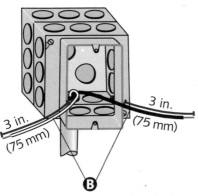

© Cengage Learning 2012

RECEPTACLE REPLACEMENTS

Grounding Means Present or Installed in the Enclosure

A A grounding-type receptacle shall be installed in any receptacle enclosure having a grounding means ≫ *406.4(D)(1)* ≪ .

B If an equipment grounding conductor is installed and connected to the grounding electrode system in accordance with 250.130(C), a grounding-type receptacle shall be installed ≫ *406.4(D)(1)* ≪ .

C A grounding-type receptacle shall be connected to the equipment grounding conductor in accordance with 406.4(C) or 250.130(C).

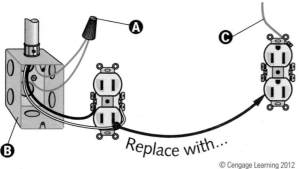

© Cengage Learning 2012

> **NOTE**
>
> An existing two-wire receptacle may be attached to a metal receptacle enclosure. Just because the box is metal does not guarantee that it is grounded to the grounding electrode system.

No Grounding Means Present . . . Nongrounding-Type Replaces Nongrounding-Type

A Where attachment to an equipment grounding conductor does not exist in the receptacle enclosure, a nongrounding-type receptacle can replace an existing nongrounding-type receptacle ≫ *406.4(D)(2)(a)* ≪ .

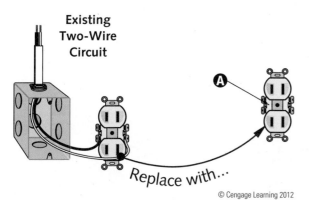

Existing Two-Wire Circuit

Replace with…

© Cengage Learning 2012

Receptacles Requiring GFCI Protection

A Because of changes in GFCI-protection requirements, some old receptacles must now be replaced with GFCI-protected receptacles ≫ *406.4(D)(3)* ≪ . Arc-fault circuit-interrupter type and ground-fault circuit-interrupter type receptacles shall be installed in a readily accessible location ≫ *406.4(D)* ≪ .

B All receptacles that serve kitchen countertop surfaces shall have GFCI protection ≫ *210.8(A)(6)* ≪ .

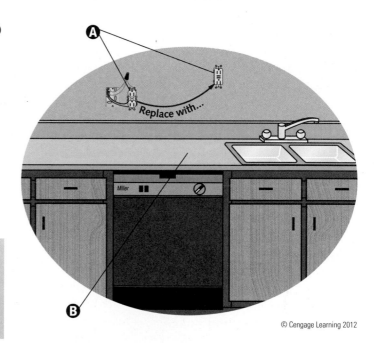

© Cengage Learning 2012

> **NOTE**
>
> Where replacement of the receptacle type is impracticable, such as where the outlet box size will not permit the installation of the GFCI receptacle, the receptacle shall be permitted to be replaced with a new receptacle of the existing type, where GFCI protection is provided and the receptacle is marked "GFCI protected" and "no equipment ground," in accordance with 406.4(D)(2)(a), (b), or (c) ≫ *406.4(D)(3) Exception* ≪ .

No Grounding Means Present . . . GFCI Replaces Nongrounding-Type

A Where attachment to an equipment grounding conductor does not exist in the receptacle enclosure, a GFCI receptacle can replace the old receptacle ≫ *406.4(D)(2)(b)* ≪.

B This GFCI receptacle must be labeled **"No Equipment Ground"** ≫ *406.4(D)(2)(b)* ≪.

No Grounding Means Present . . . GFCI and Grounding-Type Replaces Nongrounding-Type

A This GFCI receptacle must be labeled **"No Equipment Ground"** ≫ *406.4(D)(2)(b)* ≪.

B Where attachment to an equipment grounding conductor does not exist in the receptacle enclosure, a GFCI receptacle can replace an existing nongrounding-type receptacle ≫ *406.4(D)(2)(b)* ≪.

C A grounding-type receptacle may replace a nongrounding-type receptacle where the circuit is ground-fault protected. These receptacles shall be labeled **"GFCI Protected"** and **"No Equipment Ground"** ≫ *406.4(D)(2)(c)* ≪.

D Do not connect an equipment grounding conductor between the nongrounding-type receptacle and the grounding-type receptacle ≫ *406.4(D)(2)(c)* ≪.

> **CAUTION**
>
> An equipment grounding conductor shall not be connected from the GFCI-type receptacle to any outlet supplied from the GFCI receptacle ≫ *406.4(D)(2)(b)* ≪.

> **NOTE**
>
> Arc-fault circuit-interrupter type and ground-fault circuit-interrupter type receptacles shall be installed in a readily accessible location ≫ *406.4(D)* ≪.

No Grounding Means Present . . . Grounding-Type Replaces Nongrounding-Type

A A grounding-type receptacle may replace a nongrounding-type receptacle where the circuit is ground-fault protected.

B These receptacles shall be labeled **"GFCI Protected"** and **"No Equipment Ground"** ≫ *406.4(D)(2)(c)* ≪.

C Do not connect an equipment grounding conductor between the nongrounding-type receptacle and the grounding-type receptacle ≫ *406.4(D)(2)(c)* ≪.

> **NOTE**
>
> Consumers should be alerted to the fact that a receptacle is not grounded. This is especially important as more goods (such as personal computers) require grounded receptacles.

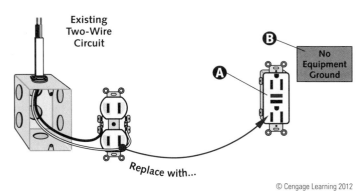

© Cengage Learning 2012

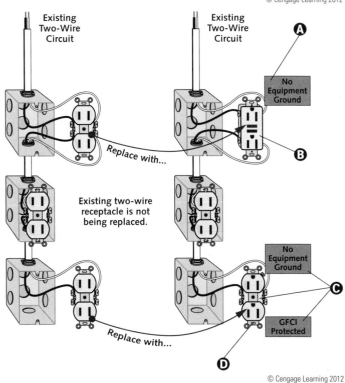

© Cengage Learning 2012

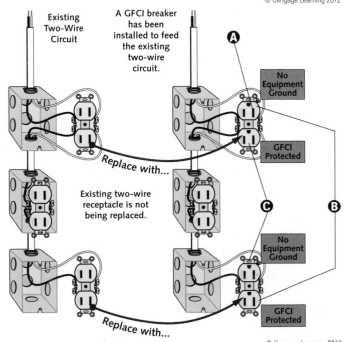

© Cengage Learning 2012

RECEPTACLES REQUIRING AFCI PROTECTION

Ⓐ Where a receptacle outlet is supplied by a branch circuit that requires arc-fault circuit-interrupter (AFCI) protection as specified elsewhere in the *NEC,* a replacement receptacle at this outlet shall be one of the following:

(1) A listed outlet branch-circuit type arc-fault circuit-interrupter receptacle.

(2) A receptacle protected by a listed outlet branch-circuit type arc-fault circuit interrupter–type circuit receptacle.

(3) A receptacle protected by a listed combination-type arc-fault circuit interrupter–type circuit breaker.

This requirement becomes effective January 1, 2014 》 *406.4(D)(4)* 《.

Ⓑ Requirements for arc-fault circuit-interrupter protection are in 210.12.

Ⓒ Listed tamper-resistant receptacles shall be provided where replacements are made at receptacle outlets that are required to be tamper-resistant elsewhere in the NEC 》 *406.4(D)(5)* 《.

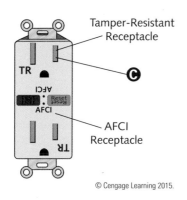

Tamper-Resistant Receptacle

Ⓒ

AFCI Receptacle

© Cengage Learning 2015.

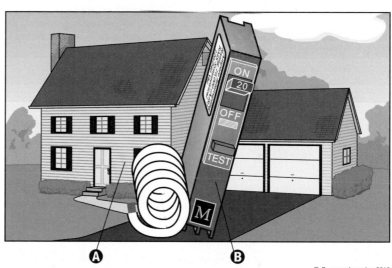

Ⓐ   **Ⓑ**

© Cengage Learning 2012

OTHER CONSIDERATIONS WHEN REPLACING RECEPTACLES

Ⓐ Listed tamper-resistant receptacles shall be provided where replacements are made at receptacle outlets that are required to be tamper-resistant elsewhere in the *NEC* 》 *406.4(D)(5)* 《. In all areas specified in 210.52, all nonlocking-type 125-volt, 15- and 20-ampere receptacles shall be listed tamper-resistant receptacles 》 *406.12(A)* 《.

Ⓑ Tamper-resistant receptacles are identified by the letters "TR" or the words "Tamper Resistant." After the receptacle has been installed, the identification is only required to be visible with the cover-plate removed. See UL (Underwriter Laboratories) White Book, category: Receptacles for Plugs and Attachment Plugs (RTRT).

Ⓒ Weather-resistant receptacles shall be provided where replacements are made at receptacle outlets that are required to be so protected elsewhere in the *NEC* 》 *406.4(D)(6)* 《. All 15- and 20-ampere, 125- and 250-volt nonlocking-type receptacles installed in damp and wet locations shall be listed weather-resistant type 》 *406.9(A) and (B)* 《.

Ⓓ Weather-resistant receptacles are identified by the letters "WR" or the words "Weather Resistant." After the receptacle has been installed, the identification must be visible with the cover plate secured in place. See UL (Underwriter Laboratories) White Book, category: Receptacles for Plugs and Attachment Plugs (RTRT).

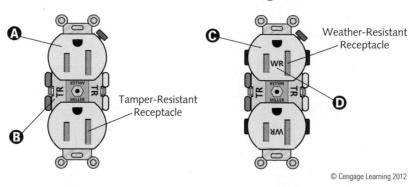

Ⓐ

TR   MILLER   TR

Tamper-Resistant Receptacle

Ⓑ

Ⓒ

WR

TR   MILLER   TR

Weather-Resistant Receptacle

Ⓓ

WR

© Cengage Learning 2012

LIGHTING AND SWITCHING

Outdoor Entrances and Exits

A At least one wall switch–controlled lighting outlet shall be installed in every habitable room of a dwelling ≫210.70(A)(1)≪.

B At least one wall switch–controlled lighting outlet is required to provide exterior side illumination of outdoor entrances or exits having grade-level access ≫210.70(A)(2)(b)≪.

> **NOTE**
>
> Remote, central, or automatic lighting control is permitted for outdoor entrances ≫**210.70(A)(2) Exception**≪.

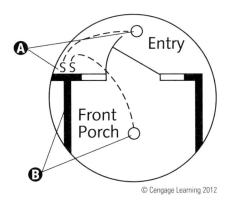

© Cengage Learning 2012

General Lighting

All switches shall be located so that they may be operated from a readily accessible place, not more than 6 ft, 7 in. (2.0 m), from the floor to the center of the switch operating handle ≫404.8(A)≪. No other height requirements are stipulated for wall switches.

A Dwelling hallways require a minimum of one wall switch–controlled lighting outlet ≫210.70(A)(2)(a)≪.

B At least one wall switch–controlled lighting outlet shall be installed in every habitable room and bathroom of a dwelling ≫210.70(A)(1)≪.

C Except for kitchens and bathrooms, every habitable room is permitted one or more wall switch–controlled receptacle(s) in lieu of a lighting outlet ≫210.70(A)(1) Exception No. 1≪. Switch–controlled receptacles allow for cord-connected lighting.

D At least one wall switch–controlled lighting outlet shall be installed in every bathroom of a dwelling ≫210.70(A)(1)≪.

> **NOTE**
>
> Occupancy sensors are permitted to control lighting outlets in lieu of wall switches. One of two requirements must be met in order to use occupancy sensors: either a regular wall switch must be present in addition to the sensor, or the sensor must be equipped with a manual override that allows the sensor to function as a switch. Either type of wall switch shall be located at a customary location ≫**210.70(A)(1) Exception No. 2**≪.

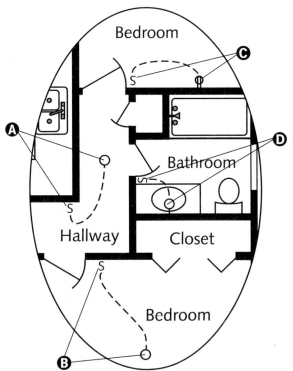

© Cengage Learning 2012

Switches Controlling Lighting Loads

Ⓐ Where switches control lighting loads supplied by a grounded general-purpose branch circuit, the grounded circuit conductor for the controlled lighting circuit shall be provided at the switch location ≫*404.2(C)*≪.

Ⓑ The provision for a (future) grounded conductor is to complete a circuit path for electronic lighting control devices ≫*404.2(C) Informational Note*≪.

> ### NOTE
>
> The grounded circuit conductor shall be permitted to be omitted from the switch enclosure for the following
> (1) Where conductors enter the box enclosing the switch through a raceway, provided that the raceway is large enough for all contained conductors, including a grounded conductor
> (2) Where the box enclosing the switch is accessible for the installation of an additional or replacement cable without removing finish materials
> (3) Where snap switches with integral enclosures comply with 300.15(E)
> (4) Where a switch does not serve a habitable room or bathroom
> (5) Where multiple switch locations control the same lighting load such that the entire floor area of the room or space is visible from the single or combined switch locations
> (6) Where lighting in the area is controlled by automatic means
> (7) Where a switch controls a receptacle load ≫**404.2(C)**≪.

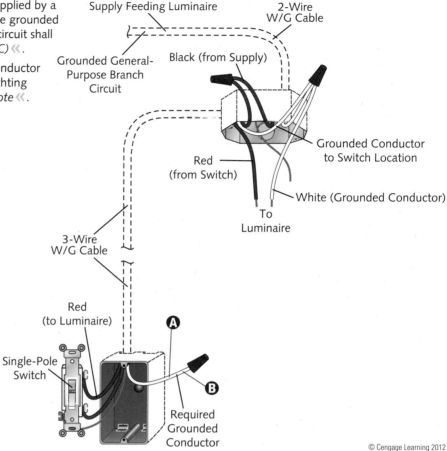

© Cengage Learning 2012

Ceiling-Suspended (Paddle) Fans

Include any light kit or accessory when determining the total weight of the fan.

Ⓐ Outlet boxes or outlet box systems used as the sole support of a ceiling-suspended (paddle) fan shall be listed, shall be marked by their manufacturer as suitable for this purpose, and shall not support ceiling-suspended (paddle) fans that weigh more than 70 pounds (32 kg). For outlet boxes or outlet box systems designed to support ceiling-suspended (paddle) fans that weigh more than 35 pounds (16 kg), the required marking must include the maximum weight that can be supported ≫*314.27(C)*≪.

Ⓑ In a completed installation, each outlet box shall be provided with a cover unless covered by means of a ceiling-suspended (paddle) fan canopy ≫*422.20*≪. Any combustible ceiling finish exposed between the edge of a ceiling-suspended (paddle) fan canopy or pan and an outlet box shall be covered with noncombustible material ≫*422.21*≪.

Ⓒ Canopies of ceiling-suspended (paddle) fans and outlet boxes taken together shall provide sufficient space so that conductors and their connecting devices are capable of being installed in accordance with 314.16 ≫*422.19*≪.

> ### WARNING
>
> Even if a ceiling-suspended (paddle) fan will not be installed initially, it may be necessary to install a box (ceiling fan box) that is listed to support a ceiling-suspended (paddle) fan. In accordance with 314.27(C), where spare, separately switched, ungrounded conductors are provided to a ceiling-mounted outlet box, in a location acceptable for a ceiling-suspended (paddle) fan in single-family, two-family, or multifamily dwellings, the outlet box or outlet box system shall be listed for sole support of a ceiling-suspended (paddle) fan.

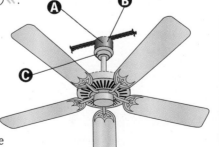

> ### NOTE
>
> Ceiling-suspended (paddle) fans must be supported independently of an outlet box or by listed outlet box or outlet box systems identified for the use and installed in accordance with 314.27(C) ≫**422.18**≪.

© Cengage Learning 2015.

Cables Installed to Feed Switches

No provision permits the installation of a smaller conductor as a switch loop. Switch loops (legs) that are part of a 20-ampere branch circuit cannot be smaller than 12 AWG copper conductors. Table 210.24 requires 12 AWG conductors throughout a 20-ampere branch circuit, except for taps. Taps and switch loops are not the same.

A white conductor in a cable can be used as an ungrounded conductor when supplying power to a switch, but not as a return conductor from the switch to the outlet. The conductor shall be permanently reidentified to indicate its use by marking tape, painting, or other effective means at its terminations and at each location where the conductor is visible and accessible ≫ *200.7(C)(2)* ≪. **The reidentified conductor can be any color except white, gray, or green.**

Ⓐ Metal switch boxes shall be connected to an equipment grounding conductor as specified in Part IV of Article 250 ≫ *404.12* ≪.

Ⓑ Where switches control lighting loads supplied by a grounded general-purpose branch circuit, the grounded circuit conductor for the controlled lighting circuit shall be provided at the switch location ≫ *404.2(C)* ≪.

Ⓒ Snap switches, including dimmer and similar control switches, shall be connected to an equipment grounding conductor and shall provide a means to connect metal faceplates to the equipment grounding conductor, whether or not a metal faceplate is installed. Snap switches shall be considered to be part of an effective ground-fault current path if either of the following conditions is met: (1) the switch is mounted with metal screws to a metal box or metal cover that is connected to an equipment grounding conductor or to a nonmetallic box with integral means for connecting to an equipment grounding conductor; or (2) an equipment grounding conductor or equipment bonding jumper is connected to an equipment grounding termination of the snap switch ≫ *404.9(B)* ≪.

Ⓓ Where circuit conductors are spliced within a box, or terminated on equipment within or supported by a box, any equipment grounding conductor(s) associated with those circuit conductors shall be connected within the box or to the box with devices suitable for the use in accordance with 250.148(A) through (E) ≫ *250.148* ≪.

> **N O T E**
>
> An outlet box used exclusively for lighting can support a luminaire or lampholder weighing no more than 50 pounds (23 kg), unless the box is listed and marked for the maximum weight to be supported. A separate means for supporting a luminaire (independent of the outlet box) is also permitted ≫ *314.27(A)(2)* ≪.

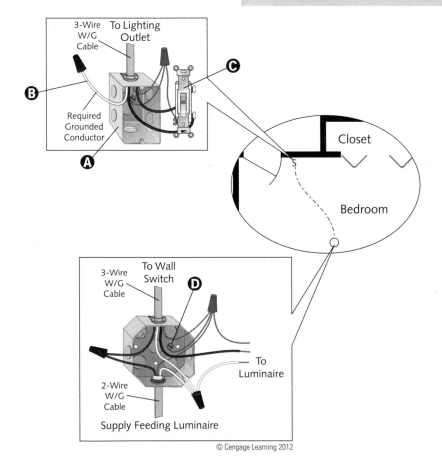

OUTDOOR RECEPTACLES AND LIGHTING

Required Outdoor Receptacles

(A) One-family dwellings require at least one receptacle outlet installed at the front and back of the dwelling. Each receptacle must be readily accessible from grade and located not more than 6 ft, 6 in. (2.0 m) above grade level ≫ *210.52(E)(1)* ≪.

(B) *Each* grade level unit of a two-family dwelling requires at least one receptacle outlet installed at the front and back of the dwelling. Each receptacle must be readily accessible from grade and located not more than 6 ft, 6 in. (2.0 m) above grade level ≫ *210.52(E)(1)* ≪.

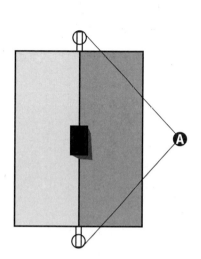
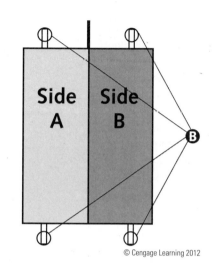

© Cengage Learning 2012

Balconies, Decks, and Porches

(A) Balconies, decks, and porches that are attached to the dwelling unit and are accessible from inside the dwelling unit shall have at least one receptacle outlet accessible from the balcony, deck, or porch. The receptacle outlet shall not be located more than 6-½ ft (2.0 m) above the balcony, deck, or porch walking surface ≫ *210.52(E)(3)* ≪.

(B) Receptacles of 15 and 20 amperes installed in wet location shall have an enclosure that is weatherproof whether or not the attachment plug cap is inserted. An outlet box hood installed for this purpose shall be listed and shall be identified as "extra duty." All 15- and 20-ampere, 125- and 250-volt nonlocking-type receptacles shall be listed weather-resistant type ≫ *406.9(B)(1)* ≪.

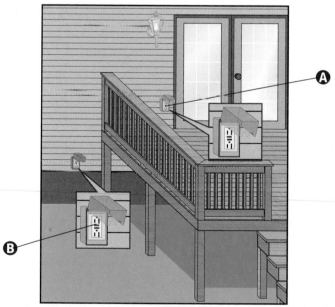

© Cengage Learning 2015.

> ### CAUTION
>
> All balconies, decks, and porches that are accessible from inside the dwelling unit must have the required receptacle outlet regardless of size. There is no exception for small balconies, decks, or porches. For example, a balcony with an area of 10 ft² is on the second floor of a new dwelling. The balcony is outside, but it is accessible from inside the dwelling. Although this balcony is small, a receptacle outlet must be installed.

Receptacles in Damp and Wet Locations

Ⓐ An outdoor receptacle, in a location protected from the weather or in other damp locations, requires an enclosure for the receptacle that is weatherproof when the receptacle is covered (attachment plug cap not inserted and receptacle covers closed). Receptacle locations protected from the weather include areas such as under roofed open porches, canopies, marquees, etc., that are not subjected to a beating rain or water runoff ≫ *406.9(A)* ≪.

Ⓑ The enclosure for a receptacle installed in a flush-mounted outlet box in a finished surface requires the use of a weatherproof faceplate assembly that provides a watertight seal between the plate and the finished surface ≫ *406.9(E)* ≪.

Ⓒ An installation suitable for wet locations shall also be considered suitable for damp locations ≫ *406.9(A)* ≪.

Ⓓ Receptacles of 15 and 20 amperes installed in a wet location shall have an enclosure that is weatherproof whether or not the attachment plug cap is inserted. An outlet box hood

installed for this purpose shall be listed and shall be identified as "extra duty" ≫ *406.9(B)(1)* ≪.

Ⓔ All 15- and 20-ampere, 125- and 250-volt nonlocking receptacles installed in damp and wet locations shall be a listed weather-resistant type ≫ *406.9(A)* and *(B)* ≪.

Ⓕ All nonlocking-type 125-volt, 15- and 20-ampere receptacles, in all areas specified in 210.52, shall be listed tamper-resistant receptacles ≫ *406.12(A)* ≪.

WARNING

When replacing receptacles, compliance with 406.4(D)(1) through (6) is required. In accordance with 406.12, tamper-resistant receptacles are required. In accordance with 406.9(A) and (B)(1), listed weather-resistant type receptacles are required. In older dwellings, it may be necessary to install a ground-fault circuit-interrupter protected receptacle.

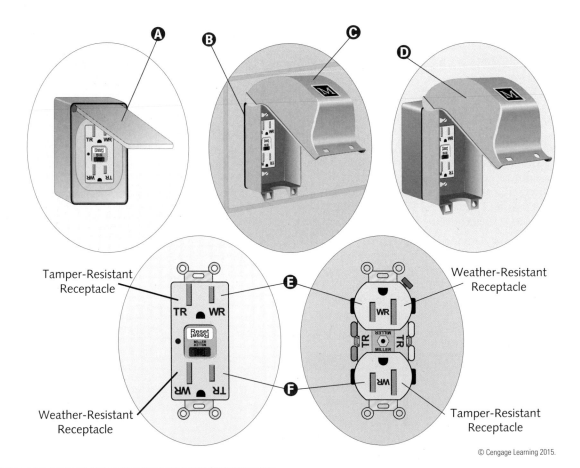

Tamper-Resistant Receptacle

Weather-Resistant Receptacle

Weather-Resistant Receptacle

Tamper-Resistant Receptacle

© Cengage Learning 2015.

NOTE

All 15- and 20-ampere, 125- through 250-volt receptacles installed in a wet location and subject to routine high-pressure spray washing shall be permitted to have an enclosure that is weatherproof when the attachment plug is removed ≫*406.9(B)(1) Exception*≪.

Receptacles More than 6 ft 6 in. (2.0 m) Above Grade

Ⓐ For a one-family dwelling and each unit of a two-family dwelling that is at grade level, at least one receptacle outlet readily accessible from grade and not more than 6½ ft (2.0 m) above grade level shall be installed at the front and back of the dwelling »*210.52(E)(1)*«.

Ⓑ Balconies, decks, and porches that are attached to the dwelling unit and are accessible from inside the dwelling unit shall have at least one receptacle outlet accessible from the balcony, deck, or porch. The receptacle outlet shall not be located more than 6½ ft (2.0 m) above the balcony, deck, or porch walking surface »*210.52(E)(3)*«.

> **N O T E**
>
> GFCI protection is not required for receptacles that are not readily accessible and are supplied by a branch circuit dedicated to electric snow melting–, deicing–, or pipeline- and vessel-heating equipment. The installation must comply with 426.28 or 427.22, as applicable »**210.8(A) (3) Exception**«. (See 426.28 and 427.22 for ground-fault protection of equipment.) Outdoor receptacles that are not readily accessible (such as under an eave) require GFCI protection.

> **CAUTION**
>
> Outdoor receptacles require GFCI protection »*210.8(A)(3)*«.

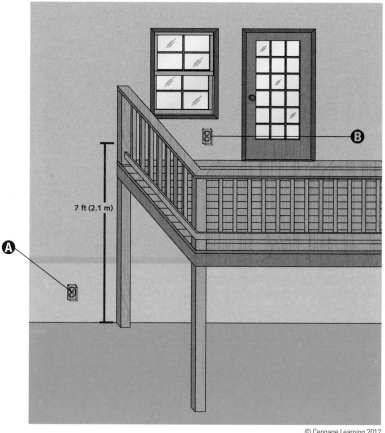

7 ft (2.1 m)

© Cengage Learning 2012

Equipment Outlet

Ⓐ A 125-volt, single-phase, 15- or 20-ampere-rated receptacle outlet shall be installed at an accessible location for the servicing of heating, air-conditioning, and refrigeration equipment. The receptacle shall be located on the same level and within 25 ft (7.5 m) of the heating, air-conditioning, and refrigeration equipment »*210.63*«.

Ⓑ At least one receptacle outlet, readily accessible from grade and not more than 6½ ft (2.0 m) above grade level, shall be installed at the front and back of the dwelling »*210.52(E)(1)*«.

Ⓒ If installed in accordance with both 210.52(E) and 210.63, only one receptacle outlet is required.

> **N O T E**
>
> A receptacle outlet shall not be required at one- and two-family dwellings for the service of evaporative coolers »**210.63 Exception**«.

> **CAUTION**
>
> The receptacle outlet shall not be connected to the load side of the equipment disconnecting means »*210.63*«.

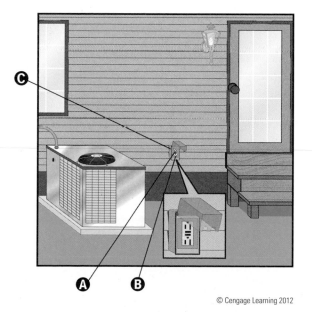

© Cengage Learning 2012

Support for Receptacle Enclosures

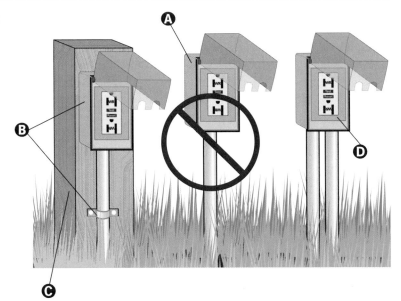

A As a general rule, a single conduit shall not support an enclosure that contains a device 》314.23(F)《.

B Rigidly and securely fasten in place all boxes and raceways 》300.11(A) and 314.23(A)《.

C Supporting an enclosure from grade requires rigid support in the form of conduit or a brace (metal, polymeric, or wood) 》314.23(B)《. Wood brace cross sections must be at least 1 in. × 2 in. (25 mm × 50 mm). Wood braces must be treated to withstand the conditions when used in wet locations 》314.23(B)(2)《.

D An enclosure containing devices can be supported by two conduits under the conditions found in 314.23(F): (1) the box shall not exceed 100 in.³ (1650 cm³); (2) the box has either threaded entries or identified hubs; (3) the box shall be supported by two or more conduits threaded wrenchtight into the enclosure or hubs; and (4) each conduit shall be secured within 18 in. (450 mm) of the enclosure.

© Cengage Learning 2012

N O T E

Dwelling-unit outdoor receptacles require GFCI protection for personnel 》**210.8(A)(3)**《.

Illumination of Entrances and Exits

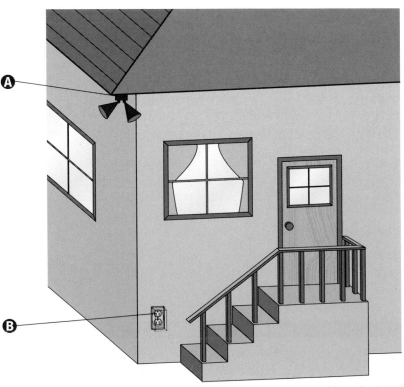

A Install at least one wall switch–controlled lighting outlet to provide illumination on the exterior side of outdoor entrances and exits (doorways) 》210.70(A)(2)(b)《. It is not required that the luminaire be located adjacent to the doorway, as long as illumination around the doorway is provided.

B Install at least one receptacle outlet at the front and back of the dwelling 》210.52(E)(1)《. Outdoor receptacles are required to be GFCI protected 》210.8(A)(3)《.

© Cengage Learning 2012

Wiring in Trees

Ⓐ Outdoor luminaires and associated equipment can be supported by trees ⟫*410.36(G)*⟪.

Ⓑ Vegetation (such as trees) shall not be used to support overhead conductor spans ⟫*225.26*⟪.

> ### NOTE
>
> Direct-buried conductors and cables emerging from grade and specified in columns 1 and 4 of Table 300.5 shall be protected by enclosures or raceways that extend from the minimum cover distance below grade required by 300.5(A) to a point 8 ft (2.5 m) or more above finished grade ⟫*300.5(D)(1)*⟪.

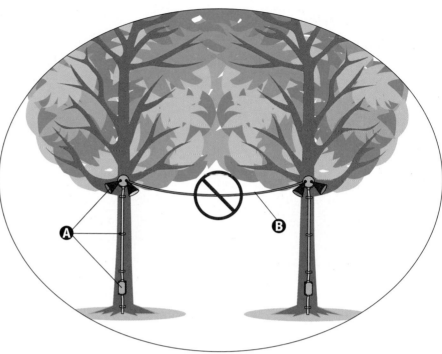

© Cengage Learning 2012

Summary

- All 120-volt, single-phase, 15- and 20-ampere branch circuits supplying outlets or devices installed in dwelling-unit kitchens, family rooms, dining rooms, living rooms, parlors, libraries, dens, bedrooms, sunrooms, recreation rooms, closets, hallways, laundry areas, or similar rooms or areas shall be protected by any of the arc-fault circuit-interrupter types described in 210.12(A)(1) through (6).

- All nonlocking-type 125-volt, 15- and 20-ampere receptacles, in all areas specified in 210.52, shall be listed tamper-resistant receptacles.

- Balconies, decks, and porches that are attached to the dwelling unit and are accessible from inside the dwelling unit shall have at least one receptacle outlet accessible from the balcony, deck, or porch.

- General-purpose branch circuits may feed lights, receptacles, or any combination of those items.

- An individual branch circuit feeds only one piece of equipment.

- The maximum distance to any receptacle measured horizontally along the floor line is 6 ft (1.8 m).

- A receptacle is required for any wall space 2 ft (600 mm) or wider.

- Where connected to a branch circuit supplying two or more receptacles or outlets, receptacles rated 20 amperes are not permitted on a 15-ampere branch circuit.

- Both 15- and 20-ampere receptacles are permitted on 20-ampere branch circuits.

- GFCI-protected receptacles may replace nongrounding-type receptacles where the receptacle enclosure is without a grounding means.

- GFCI-protected receptacles on nongrounded systems must be marked in accordance with 406.4(D)(2) provisions.

- A means to simultaneously disconnect all ungrounded conductors must be provided on multiwire branch circuits.

- A switch-controlled lighting outlet shall be installed in every habitable room of a dwelling, including bathrooms, hallways, and stairways.

- Where switches control lighting loads supplied by a grounded general-purpose branch circuit for other than the items described in 404.2(C)(1) though (7), the grounded circuit conductor for the controlled lighting circuit shall be provided at the switch location.

- Boxes are permitted to support luminaires weighing 50 pounds (23 kg) or less.

- Listed boxes are permitted to support ceiling fans weighing no more than 35 pounds (16 kg).

- One-family dwellings require at least one exterior GFCI-protected receptacle outlet at the front and back of the dwelling.

- Outdoor entrances and exits must have at least one wall switch-controlled lighting outlet.

- Although lighting and receptacles can be supported by trees, trees shall not support overhead conductor spans.

Unit 6 Competency Test

NEC **Reference**   **Answer**

_____ _____ 1. A duplex receptacle rated 20 amperes can be installed on a _____ branch circuit.

       a) 15-ampere   b) 20-ampere

       c) 15- or 20-ampere   d) 20- or 25-ampere

_____ _____ 2. The continuity of a grounded conductor shall not be dependent on the device in _____.

       a) branch circuits not having a grounded conductor

       b) multiwire branch circuits

       c) individual branch circuits

       d) branch circuits

_____ _____ 3. Receptacles located more than _____ ft above the floor are not counted in the required number of receptacles along the wall.

_____ _____ 4. At least _____ in. of free conductor, measured from the point in the box where it emerges from its raceway or cable sheath, shall be left at each outlet, junction, and switch point for splices or the connection of luminaires or devices.

_____ _____ 5. In nonmetallic-sheathed cable, the equipment grounding conductor for 15-, 20-, and 30-ampere branch circuits _____.

       a) is required only with aluminum or copper-clad aluminum cable

       b) must be the same size as the insulated circuit conductors

       c) may be one size smaller than the insulated circuit conductors

       d) may be two sizes smaller than the insulated circuit conductors

_____ _____ 6. Receptacles installed for the attachment of portable cords shall be rated at not less than 15 amperes, 125 volts, or _____ amperes, 250 volts, and shall be of a type not suitable for use as lampholders.

_____ _____ 7. A luminaire that weighs more than 6 pounds (3 kg) or exceeds _____ in. in any dimension shall not be supported by the screw shell of a lampholder.

_____ _____ 8. Which of the following is not a standard classification for a branch circuit supplying several loads?

       a) 20 amperes   b) 25 amperes   c) 30 amperes   d) 50 amperes

_____ _____ 9. A cord connector that is supported by a permanently installed cord pendant shall be considered a(n) _____ outlet.

_____ _____ 10. Grounding-type receptacles shall be installed only on circuits of the _____ for which they are rated, except as provided in Tables 210.21(B)(2) and (B)(3).

       I. voltage class

       II. wattage

       III. current

       a) I only   b) I and III only   c) I and II only   d) I, II, and III

_____ _____ 11. Outlet boxes or outlet box systems used as the sole support of a ceiling-suspended (paddle) fan shall be listed, shall be marked by their manufacturer as suitable for this purpose, and shall not support ceiling-suspended (paddle) fans that weigh more than _____ pounds.

_____ _____ 12. Receptacle outlets in or on floors shall not be counted as part of the required number of receptacle outlets unless located within _____ in. of the wall.

_____   _____   13. Luminaires shall be wired with conductors having insulation suitable for the environmental conditions, _____, _____, and _____ to which the conductors will be subjected.

_____   _____   14. A luminaire that weighs more than _____ pounds shall be supported independently of the outlet box, unless the outlet box is listed and marked on the interior of the box to indicate the maximum weight the box shall be permitted to support.

_____   _____   15. A duplex receptacle rated 15 amperes can be installed on a _____ ampere general-purpose branch circuit.

_____   _____   16. A branch circuit that supplies only one utilization equipment is known as a(n) _____ branch circuit.

_____   _____   17. The maximum distance between receptacles in a one-family dwelling is _____ ft.

_____   _____   18. A nongrounding-type receptacle has been replaced with a new grounding-type duplex receptacle. The 15-ampere circuit breaker feeding the circuit has been replaced with a 15-ampere ground-fault circuit breaker. The new receptacle must be marked _____.

      I. "Two-Wire Circuit"

     II. "No Equipment Ground"

   III. "GFCI Protected"

      a) III only   b) I and III only   c) II and III only   d) I, II, and III

_____   _____   19. An outlet where one or more receptacles are installed is called a(n) _____.

_____   _____   20. What is the maximum distance that a receptacle wall box can be set back from the finished surface of a ¼-in. wood paneling wall?

      a) 0.0625 in.   b) 0.125 in.   c) 0.25 in.   d) 0.0 in.

_____   _____   21. Switch-controlled receptacles (in lieu of a lighting outlet) are permitted in all but which of the following?

      a) Bedroom   b) Library   c) Dining room   d) Hallway

_____   _____   22. A single receptacle rated 50 amperes can be installed on a _____-ampere individual branch circuit.

_____   _____   23. A(n) _____ is a system or circuit conductor that is intentionally grounded.

_____   _____   24. The definition of a branch circuit is _____.

      a) the circuit conductors between the service and the subpanel

      b) the circuit conductors prior to the final overcurrent device protecting the circuit

      c) the circuit conductors between the final overcurrent device protecting the circuit and the outlet(s)

      d) the circuit conductors between the final overload device protecting the circuit and the outlet(s)

_____   _____   25. In dwelling units, a duplex receptacle fed from a multiwire branch circuit must have _____ as a means of disconnect.

      I. two single-pole circuit breakers without identified handle ties

     II. one double-pole circuit breaker

   III. two single-pole circuit breakers with identified handle ties

      a) I only   b) I or II only   c) II only   d) II or III only

UNIT 7

Specific Provisions

Objectives

After studying this unit, the student should:

▶ know the required ampere rating for receptacles and branch circuits in kitchens, pantries, dining rooms, breakfast rooms, and similar areas.

▶ be familiar with the requirements for kitchen countertop receptacle placement, including island countertop spaces and peninsular countertop spaces.

▶ know the minimum number of circuits required for kitchens, pantries, dining rooms, breakfast rooms, and similar areas.

▶ understand requirements pertaining to permanently connected, as well as cord- and plug-connected, appliances.

▶ be able to perform load calculations for appliance branch circuits.

▶ know the specific provisions for placement of GFCI-protected receptacles.

▶ understand the requirements for hallway and stairway receptacles and lighting.

▶ be able to identify which luminaires are, and which are not, permitted in closets, and the placement thereof.

▶ understand the comprehensive definition of a bathroom and know the requirements for receptacles, lighting, and fans.

▶ know the requirements for receptacles and lighting in attached garages, detached garages, and basements.

▶ understand the requirements for laundry (including clothes dryers) receptacles and branch circuits.

▶ be familiar with the requirements for attic and crawl space lighting and receptacles.

▶ understand lighting and receptacle requirements around HVAC equipment.

Introduction

Unit 6 contains provisions concerning general areas both inside and outside of one-family dwellings. The material in Unit 7 addresses more complex issues, requiring additional provisions for specific areas (such as kitchens, hallways, clothes closets, bathrooms, garages, basements, etc.). For instance, a kitchen may require the general provisions found in Unit 6, as well as specific provisions found in this unit. A dining room having no countertop surface requires receptacle placement by the general rule, but it must be fed from a 20-ampere branch circuit. Some areas require only one receptacle outlet, while bathroom receptacle requirements depend on the number and location of sinks. This unit also includes specific requirements for lighting outlets and switches. The pages that follow provide significant information on specific provisions for areas such as kitchens, dining rooms, breakfast rooms, hallways, stairways, clothes closets, bathrooms, attached as well as detached garages, basements, attics, and crawl spaces.

KITCHENS, DINING ROOMS, AND BREAKFAST ROOMS

Receptacle and Branch-Circuit Rating

Receptacle placement in a kitchen is determined by 210.52(A) and (C). All of the general receptacle placement provisions found in 210.52(A) apply, as well as additional requirements found in 210.52(C) for countertops.

A Receptacles installed on the 20-ampere small-appliance circuits may be rated either 15 or 20 amperes ≫ *Table 210.21(B)(3)* ≪.

All receptacles installed to serve kitchen countertop surfaces shall have GFCI protection ≫ *210.8(A)(6)* ≪.

B The maximum distance to any receptacle, measured horizontally along the wall line, is 24 in. (600 mm) ≫ *210.52(C) (1)* ≪. An easy way to understand countertop receptacle placement is to imagine a toaster with a 24-in. (600-mm) cord. Anywhere this toaster is placed around the countertop wall, there should be a receptacle within reach.

> **NOTE**
>
> In kitchens, pantries, breakfast rooms, dining rooms, or similar areas of a dwelling unit, circuits that serve countertop surface receptacles and general-purpose receptacles shall be 20-ampere small-appliance branch circuits ≫*210.52(B)(1)*≪.

> **NOTE**
>
> Receptacle outlets are not required on a wall directly behind a range, counter-mounted cooking unit, or some sinks. Receptacle outlets in this location shall not be counted as required countertop outlets ≫*210.52(C)(1) Exception*≪.

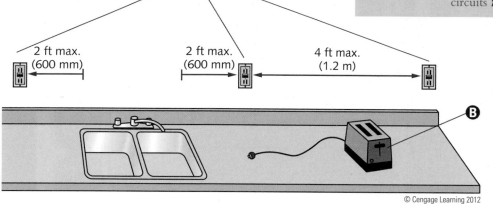

© Cengage Learning 2012

Other Outlets Fed from Small-Appliance Branch Circuits

A A dedicated clock outlet, installed to provide power and support, is permitted on a small-appliance branch circuit ≫ *210.52(B)(2) Exception No. 1* ≪.

B A hood fan is not permitted on a small-appliance branch circuit ≫ *210.52(B)(2)* ≪.

C Outdoor receptacles are not permitted on small-appliance branch circuits ≫ *210.52(B)(2)* ≪.

D A lighting outlet is not permitted on a small-appliance branch circuit ≫ *210.52(B)(2)* ≪.

E Circuits feeding receptacles in kitchens, pantries, breakfast rooms, dining rooms, or similar areas shall not feed receptacles outside of these areas ≫ *210.52(B)(2)* ≪.

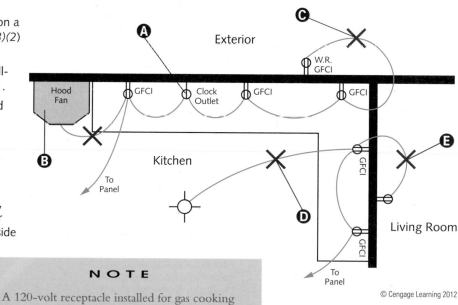

© Cengage Learning 2012

> **NOTE**
>
> A 120-volt receptacle installed for gas cooking equipment can be fed from a small-appliance branch circuit ≫*210.52(B)(2) Exception No. 2*≪.

Kitchen Countertop Receptacle Placement

All receptacles installed to serve kitchen countertop surfaces shall have GFCI protection ⟫*210.8(A)(6)*⟪.

Branch circuits in kitchens shall be 20-ampere circuits ⟫*210.52(B)(1)*⟪.

A minimum of two small-appliance branch circuits are required for receptacles serving kitchen countertops ⟫*210.52(B)(3)*⟪. Either or both of these small-appliance branch circuits may feed other receptacles in the kitchen, pantry, breakfast room, or dining area.

Receptacles located more than 20 in. (500 mm) above countertops are allowed but do not count as required receptacles ⟫*210.52(C)(5)*⟪.

Receptacles inside appliance garages are permitted, but do not count as required countertop receptacles ⟫*210.52(C)(5)*⟪.

A The maximum distance to any receptacle along the wall line measured horizontally is 24 in. (600 mm) ⟫*210.52(C)(1)*⟪.

B A receptacle installed for refrigeration equipment may be fed from one of the 20-ampere small-appliance branch circuits, or from an individual branch circuit rated 15 amperes or greater ⟫*210.52(B)(1) Exception No. 2*⟪.

C Receptacles not serving kitchen countertops do not require GFCI protection.

D A receptacle installed behind a refrigerator does not count as a required countertop receptacle ⟫*210.52(C)(5)*⟪.

E Each wall counter space 12 in. (300 mm) or wider requires a receptacle ⟫*210.52(C)(1)*⟪.

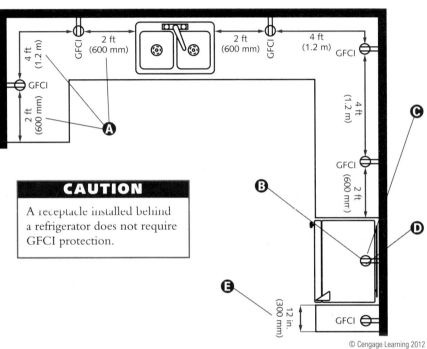

CAUTION

A receptacle installed behind a refrigerator does not require GFCI protection.

© Cengage Learning 2012

Kitchen Countertop Receptacles

A As a general rule, receptacle outlets shall be located on or above the countertop ⟫*210.52(C)(5)*⟪. Installation of receptacles below the countertop is allowed by some exceptions (see 210.52(C)(5) Exception).

B GFCI protection is not needed for receptacles installed along the wall to meet 210.52(A) requirements.

C GFCI protection is required for receptacles installed to serve kitchen countertops ⟫*210.8(A)(6)*⟪.

D Receptacles located inside cabinets or cupboards, which do not serve kitchen countertop surfaces, do not require GFCI protection unless the receptacle is supplying power to a dishwasher ⟫*210.8(A)(6)* and *210.8(D)*⟪.

E Receptacles located inside cabinets or cupboards do not count as required kitchen countertop receptacles ⟫*210.52*⟪.

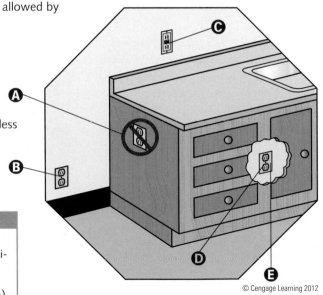

© Cengage Learning 2012

NOTE

Receptacle outlets are permitted to be installed in countertops, but only if the outlet assemblies are listed for installation in countertops ⟫*210.52(C)(5)*⟪.

WARNING

Receptacles installed for countertop surfaces as specified in 210.52(C) shall not be considered as the receptacles required by 210.52(A) ⟫*210.52(A)(4)*⟪.

Island Receptacle Placement

At least one receptacle outlet shall be installed at each island counter space with a long dimension of 24 in. (600 mm) or more and a short dimension of 12 in. (300 mm) or more **》210.52(C)(2)《**.

A receptacle is required for each island counter space with at least a 12- × 24-in. (300- × 600-mm) area that is separated from other counter space because of range tops, refrigerators, or sinks **》210.52(C)(4)《**.

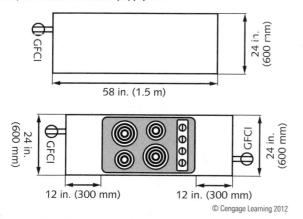

© Cengage Learning 2012

NOTE

Where a range, counter-mounted cooking unit, or sink is installed in an island or peninsular countertop and the countertop's depth behind the range, counter-mounted cooking unit, or sink is less than 12 in. (300 mm), the range, counter-mounted cooking unit, or sink is considered to divide the island into two separate countertop spaces. Each separate countertop space shall comply with the applicable requirements in 210.52(C) **》210.52(C)(4)《**.

Receptacles Mounted below Countertops

Where an island or peninsular countertop is flat across its entire surface, receptacle installation below the countertop is permitted under certain conditions. If a wall, backsplash, overhead cabinet, or similar area is available above the counter, the receptacle must be mounted above the countertop. Receptacles installed more than 20 in. (500 mm) above the countertop, such as in an overhead cabinet, are permitted but are not included as required countertop receptacles **》210.52(C)(5) Exception《**.

Ⓐ Receptacles installed faceup in the work surface or countertop are not allowed unless the receptacle outlet assembly is listed for countertop applications **》210.52(C)(5) and 406.5(E)《**.

Ⓑ GFCI protection is required for all receptacles installed to serve kitchen countertop surfaces **》210.8(A)(6)《**.

Ⓒ Although receptacles located more than 12 in. (300 mm) below countertops are permitted, they are not to be included as required countertop receptacles **》210.52(C)(5) Exception《**.

Ⓓ Receptacles mounted in the cabinet are permitted under an overhanging countertop but are not counted as required countertop receptacles where the countertop extends more than 6 in. (150 mm) beyond its support base **》210.52(C)(5) Exception《**.

Peninsular Receptacle Placement

Counter spaces separated by range tops, refrigerators, or sinks require a receptacle for each 12- × 24-in. (300- × 600-mm) area **》210.52(C)(4)《**.

A receptacle is required for each peninsular counter space with at least a 12- × 24-in. (300- × 600-mm) area that is separated from other counter space because of range tops, refrigerators, or sinks **》210.52(C)(3)《**.

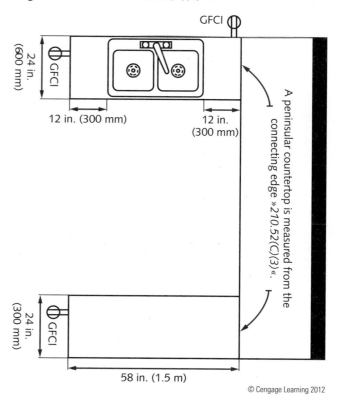

© Cengage Learning 2012

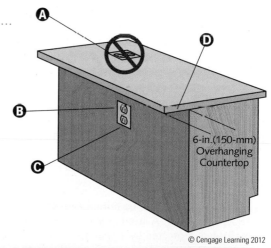

6-in.(150-mm) Overhanging Countertop

© Cengage Learning 2012

CAUTION

A peninsular counter side without doors or drawers might be considered wall space by the authority having jurisdiction (AHJ). If so, and if longer than 6 ft (1.8 m), a receptacle is required.

Permanently Connected Appliances

Ⓐ For permanently connected appliances rated over 300 volt-amperes, a disconnecting means is not required if the branch-circuit switch or circuit breaker is within sight from the appliance or is lockable in accordance with 110.25 》 *422.31(B)* 《.

Ⓑ A disconnecting means is not required if the branch-circuit switch or circuit breaker can be locked in the open (off) position. The provision for locking or adding a lock to the disconnecting means must be installed on or at the switch or circuit breaker used as the disconnecting means and shall remain in place with or without the lock installed 》 *422.31(B)* and *110.25* 《.

Ⓒ Water heaters shall not be connected with flexible cords 》 *422.16(A)* 《.

Ⓓ Grounding must meet 250.134 requirements for equipment fastened in place or connected by permanent wiring methods.

Ⓔ Adequately protect conductors where subject to physical damage 》 *300.4* 《.

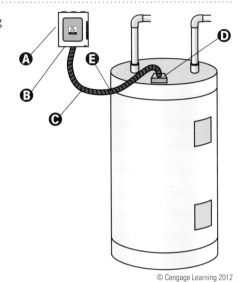

© Cengage Learning 2012

> ### N O T E
>
> A means must be provided to disconnect all ungrounded (hot) conductors from the appliance **》422.30《**.

A fastened-in-place appliance (dishwasher, in-sink waste disposer, etc.) shall not be connected to a small-appliance branch circuit 》 *210.52(B)(2)* 《.

Ⓐ A permanently connected appliance rated over 300 volt-amperes or ⅛ horsepower must have a means to disconnect all ungrounded (hot) conductors from the appliance 》 *422.30* 《.

Ⓑ The disconnecting means shall comply with 430.109 and 430.110. For permanently connected motor-operated appliances with motors rated over ⅛ horsepower, the disconnecting means shall meet 422.31(C)(1) or (2).

(1)   The branch-circuit switch or circuit breaker shall be permitted to serve as the disconnecting means where the switch or circuit breaker is within sight from the appliance.

(2)   The disconnecting means shall be installed within sight of the appliance. 》 *422.31(C)* 《.

Ⓒ Grounding must meet 250.134 requirements for equipment fastened in place or connected by permanent wiring methods.

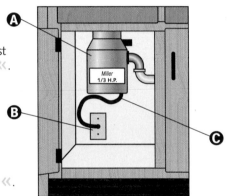

© Cengage Learning 2012

Kitchen In-Sink Waste Disposers

Some appliances (dishwashers, trash compactors, and in-sink waste disposers) can be cord and plug connected. This arrangement provides a means of disconnect where accessible 》 *422.33(A)* 《.

Electrically operated in-sink waste disposers can be cord and plug connected under certain conditions 》 *422.16(B)(1)* 《. The flexible cord must be identified as suitable in the manufacturer's installation instructions and must meet *all* of the following conditions:

• The flexible cord shall be terminated with a grounding-type attachment plug. A listed in-sink waste disposer distinctly marked as protected by a system of double insulation or its equivalent does not require termination with a grounding-type attachment plug.

• The cord must be between 18 and 36 in. (450 and 900 mm) in length.

• The receptacle must be located so the flexible cord does not become damaged.

• The receptacle shall be accessible.

Ⓐ The switch controlling an in-sink waste disposer serves as a disconnecting means. Also see 430.109(C).

Ⓑ Appliance receptacles installed inside cabinets for in-sink waste disposers and built-in trash compactors, as permitted by 422.16, do not require GFCI protection.

Ⓒ Grounding must meet 250.138 requirements for cord- and plug-connected equipment.

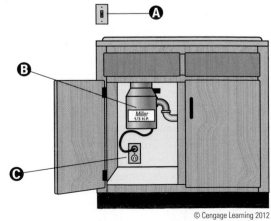

© Cengage Learning 2012

Built-In Dishwashers and Trash Compactors

Built-in dishwashers and trash compactors can be cord and plug connected under certain conditions ≫*422.16(B)(2)*≪. The flexible cord must be identified as suitable for the purpose by the appliance manufacturer's installation instructions and must meet *all* of the following conditions:

- The flexible cord shall be terminated with a grounding-type attachment plug. A listed dishwasher or trash compactor distinctly marked as protected by a system of double insulation, or its equivalent, does not require termination with a grounding-type attachment plug.

- The cord must be between 3 and 4 ft (0.9 and 1.2 m) in length when measured from the face of the attachment plug to the plane of the rear of the appliance.

- The receptacle must be located so the flexible cord does not become damaged.

- The receptacle must be located in the space occupied by the appliance or adjacent thereto.

- The receptacle shall be accessible.

A Where accessible, cord and plug arrangements may serve as a disconnecting means ≫*422.33(A)*≪. A receptacle installed under the sink (inside the cabinet) may serve as a means of disconnect.

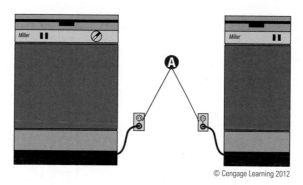

© Cengage Learning 2012

Appliance Branch-Circuit Rating

Branch circuits, rated 15 or 20 amperes, may feed lighting units, utilization equipment, or a combination of both. The rating of any one cord- and plug-connected utilization equipment shall not exceed 80% of the branch-circuit ampere rating ≫*210.23(A)*≪. Cord- and plug-connected equipment may be portable or may be fastened-in-place equipment. Fastened-in-place utilization equipment shall not exceed 50% of the branch-circuit ampere rating where the circuit also feeds lighting units or non-fastened-in-place utilization equipment ≫*210.23(A)(2)*≪.

A The full-load current for a single-phase motor is found in Table 430.248.

B This receptacle could be split-wired to accommodate two individual branch circuits.

C The total load shall not exceed the branch-circuit rating ≫*220.18*≪. For motor-operated, fastened-in-place appliances with a motor larger than 1/8 horsepower, multiply the largest motor by 125% and add 100% of the other loads ≫*220.18(A)*≪.

> **NOTE**
>
> Local codes may require a separate branch circuit for the dishwasher and the in-sink waste disposer, that is, one branch circuit for each.

> **CAUTION**
>
> GFCI protection shall be provided for outlets that supply dishwashers installed in dwelling unit locations ≫*210.8(D)*≪.

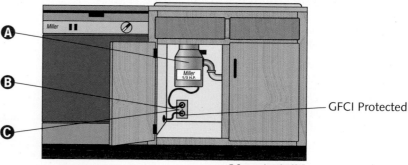

GFCI Protected

© Cengage Learning 2015.

Calculation for One Counter-Mounted Cooking Unit (Cooktop)

Branch-circuit calculation for one cooktop with a nameplate rating of 6.5 kW at 240 volts:

- Use the nameplate rating to determine the branch-circuit conductor size and the overcurrent protection for one counter-mounted cooking unit »*Table 220.55, Note 4*«
- 6.5 kW = 6500 watts
  6500 watts ÷ 240 = 27.1 amperes
- Overcurrent protective device = 30 amperes »*240.4(B)*«
- Conductor size = 10 AWG »*Table 310.15(B)(16)*«

Calculation for One Wall-Mounted Oven

Branch-circuit calculation for one oven with a nameplate rating of 4 kW at 240 volts:

- Use the nameplate rating to determine the branch-circuit conductor size and the overcurrent protection for one wall-mounted oven »*Table 220.55, Note 4*«
- 4 kW = 4000 watts
  4000 watts ÷ 240 = 16.7 amperes
- Overcurrent protective device = 20 amperes »*240.4(B)*«
- Conductor size = 12 AWG »*Table 310.15(B)(16)*«

Counter-Mounted Cooking Units (Cooktops) and Wall-Mounted Ovens

A The nameplate rating of a counter-mounted cooking unit or a wall oven shall be the minimum branch-circuit load »*Table 220.55, Note 4*«.

B Wall-mounted ovens and counter-mounted cooking units may be connected permanently or by cord and plug »*422.16(B)(3)*«.

C A disconnecting means is required »*422.30*«.

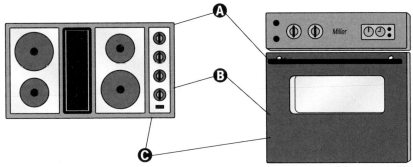

© Cengage Learning 2012

Cooktop and Wall-Mounted Oven(s) on Same Branch Circuit

A Tap conductors from a 50-ampere branch circuit must have an ampacity of at least 20 and not less than the ampacity of the individual unit served. These tap conductors include any conductors that are part of the pigtail supplied with the appliance that are smaller than the branch-circuit conductors. The taps shall not be longer than necessary for servicing the appliance »*210.19(A)(3) Exception No. 1*«.

B Where a single branch circuit feeds one cooktop and one or two ovens (in the same room), add the nameplate ratings of the appliances, treating the total kW as one range to be found in Table 220.55 »*Table 220.55, Note 4*«.

C No disconnecting means is required if the branch-circuit switch or circuit breaker can be locked in the open (off) position »*422.31(B)*«.

D It is permissible to feed one cooktop and one or two wall-mounted ovens with one branch circuit where all of the units are in the same room »*Table 220.55, Note 4*«.

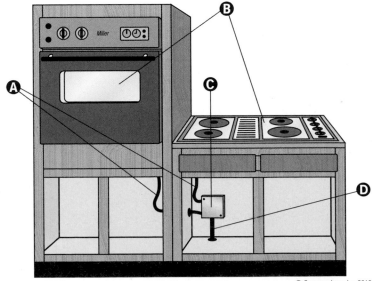

© Cengage Learning 2012

Electric Ranges

Ⓐ Frames of electric ranges, wall-mounted ovens, counter-mounted cooking units, clothes dryers, and outlet or junction boxes that are part of the circuit for these appliances shall be connected to the equipment grounding conductor in the manner specified by 250.134 or 250.138 »*250.140*«.

Ⓑ It is permissible to compute the branch-circuit load for one range in accordance with Table 220.55 »*Table 220.55, Note 4*«.

Ⓒ A disconnecting means is required »*422.30*«. An accessible separable connector or an accessible plug and receptacle can serve as the disconnecting means »*422.33(A)*«.

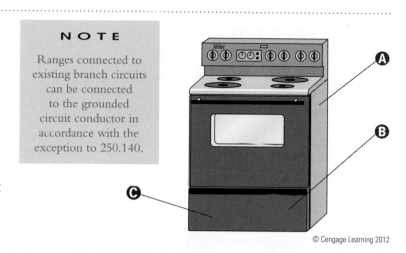

NOTE

Ranges connected to existing branch circuits can be connected to the grounded circuit conductor in accordance with the exception to 250.140.

© Cengage Learning 2012

Ⓐ Most household electric ranges are cord and plug connected. The attachment plug and receptacle can serve as the disconnecting means if they are accessible by removing a drawer on the front of the range »*422.33(B)*«.

Ⓑ New branch-circuit conductors installed for a range must have an insulated grounded (neutral) conductor and a grounding means. Grounding must meet 250.138 requirements for cord- and plug-connected equipment.

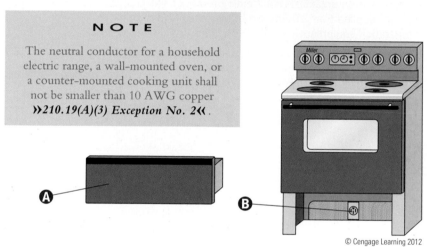

NOTE

The neutral conductor for a household electric range, a wall-mounted oven, or a counter-mounted cooking unit shall not be smaller than 10 AWG copper »*210.19(A)(3) Exception No. 2*«.

© Cengage Learning 2012

Calculation for Ranges Rated 8¾ kW through 12 kW

Branch-circuit calculation for one range with a nameplate rating of 12 kW at 240 volts:

- Compute the branch-circuit load for one range in accordance with Table 220.55.
- 12 kW = 8 kW (Column C)

  8 kW = 8000 watts

  8000 watts ÷ 240 = 33.3 amperes
- Overcurrent protective device = 40 amperes »*240.4(B)*«
- Conductor size = 8 AWG »*Table 310.15(B)(16)*«

NOTE

The minimum branch-circuit rating for a range rated 8¾ kW or more is 40 amperes »*210.19(A)(3)*«.

Calculation for Ranges Rated More than 12 kW

Branch-circuit calculation for one range with a nameplate rating of 15 kW at 240 volts:

- Compute the branch-circuit load for one range in accordance with Table 220.55.
- Column C is for 12-kW and under ranges; Note 1 applies to ranges over 12 kW
- First, subtract 12 kW from the 15-kW range:

  15 kW – 12 kW = 3 kW
- Next, multiply 3 kW by 5% = 15%
- Then, multiply 8 kW (Column A) by 15%:

  8 kW × 15% = 1.2 kW
- Finally, add 1.2 kW to 8 kW:

  1.2 kW + 8 kW = 9.2 kW = 9200 watts
- 9200 watts ÷ 240 volts = 38.3 amperes
- Overcurrent protective device = 40 amperes »*240.4(B)*«
- Conductor size = 8 AWG »*Table 310.15(B)(16)*«

Kitchen, Dining Room, and Breakfast Room

Ⓐ A minimum of two small-appliance branch circuits are required for receptacles serving kitchen countertops ≫ *210.52(B)(3)* ≪. Either or both of these small-appliance branch circuits may feed other receptacles in the kitchen, pantry, breakfast room, or dining room.

Ⓑ Noncountertop receptacle placement is determined by wall space ≫ *210.52(A)* ≪.

Ⓒ The receptacle outlet installed for refrigeration equipment can be supplied from an individual branch circuit rated 15 amperes or greater ≫ *210.52(B)(1) Exception No. 2* ≪.

Ⓓ Range hoods and lights are not permitted on small-appliance branch circuits ≫ *210.52(B)(2)* ≪.

Ⓔ All wall and floor receptacle outlets covered by 210.52(A) and all countertop outlets covered by 210.52(C) in the kitchen, pantry, breakfast room, dining room, or similar area shall be supplied from 20-ampere branch circuits (except for refrigeration equipment) ≫ *210.52(B)(1)* ≪.

Ⓕ A minimum of two circuits is required for receptacles located in the kitchen, pantry, breakfast room, dining room, or similar areas ≫ *210.52(B)(1)* ≪.

Ⓖ At least one wall switch-controlled lighting outlet shall be installed in dining rooms, kitchens, and breakfast rooms ≫ *210.70(A)(1)* ≪. One or more receptacles controlled by a wall switch are permitted in lieu of a lighting outlet in dining rooms and breakfast rooms, but not in kitchens ≫ *210.70(A)(1) Exception No. 1* ≪.

> **N O T E**
>
> All 120-volt, single-phase, 15- and 20-ampere branch circuits supplying outlets or devices installed in dwelling unit kitchens, family rooms, dining rooms, living rooms, parlors, libraries, dens, bedrooms, sunrooms, recreation rooms, closets, hallways, laundry areas, or similar rooms or areas shall be arc-fault circuit-interrupter (AFCI) protected by any of the means described in 210.12(A)(1) through (6) ≫ *210.12(A)* ≪.

© Cengage Learning 2012

HALLWAYS AND STAIRWAYS

Hallway and Stairway Lighting

A minimum of one lighting outlet is required in each hallway and stairway ≫ *210.70(A)(2)(a)* ≪. **The light must be wall switch–controlled, except where controlled by remote, central, or automatic means** ≫ *210.70(A)(2) Exception* ≪. **A receptacle controlled by a wall switch is not permitted.**

Ⓐ Interior stairways of six or more risers (steps) shall have a wall switch located at each floor level and landing level that includes an entryway ≫ *210.70(A)(2)(c)* ≪.

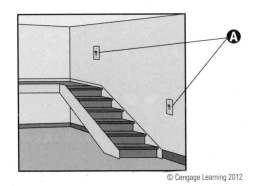

© Cengage Learning 2012

Hallway Receptacle Placement

Receptacles in hallways are not subject to the general provisions for placement determined by wall space.

A If length is less than 10 ft (3.0 m), no receptacle is required.

B Hallways measuring 10 ft (3.0 m) or more in length must have at least one receptacle outlet »*210.52(H)*«. Hallway length is determined by measuring along the hallway centerline, without passing through a doorway.

C If length is 10 ft (3.0 m) or more, a receptacle is required.

> **CAUTION**
>
> Foyers that are not part of a hallway in accordance with 210.52(H) and that have an area that is greater than 60 ft² (5.6 m²) shall have a receptacle(s) located in each wall space 3 ft (900 mm) or more in width. Doorways, door-side windows that extend to the floor, and similar openings shall not be considered wall space »*210.52(I)*«.

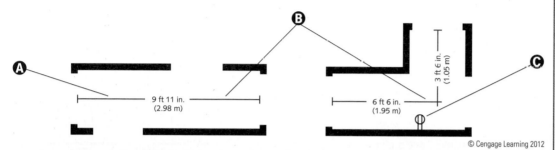

9 ft 11 in. (2.98 m)   6 ft 6 in. (1.95 m)   3 ft 6 in. (1.05 m)

© Cengage Learning 2012

CLOTHES CLOSETS

Luminaire Types *Not* Permitted in Clothes Closets »*410.16(B)*«

A Pendant luminaires or lampholders
B Incandescent luminaires with open or partially enclosed lamps

> **NOTE**
>
> The volume is bound by the sides and back closet walls and planes extending from the closet floor vertically to a height of 6 ft (1.8 m) or to the highest clothes-hanging rod and parallel to the walls at a horizontal distance of 24 in. (600 mm) from the sides and back of the closet walls, respectively, and continuing vertically to the closet ceiling parallel to the walls at a horizontal distance of 12 in. (300 mm) or the width of the shelf, whichever is greater; for a closet that permits access to both sides of a hanging rod, this space includes the volume below the highest rod extending 12 in. (300 mm) on either side of the rod on a plane horizontal to the floor extending the entire length of the rod »*410.2*«. For an illustrated view of this definition, see "Dedicated Space in Clothes Closets."

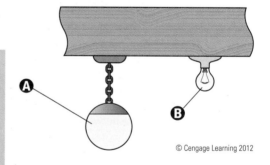

© Cengage Learning 2012

Luminaire Types Permitted in Clothes Closets »*410.16(A)*«

A Recessed incandescent or light-emitting diode (LED) luminaire with a completely enclosed light source
B Surface-mounted incandescent or LED luminaire with a completely enclosed light source
C Surface-mounted fluorescent luminaire
D Recessed fluorescent luminaire

> **NOTE**
>
> Also permitted in clothes closets are surface-mounted fluorescent or light-emitting diode (LED) luminaires that are identified as suitable for installation within the closet storage space »*410.16(A)(3)*«.

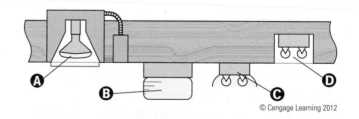

© Cengage Learning 2012

> **NOTE**
>
> All luminaires, lampholders, and retrofit kits shall be listed »*410.6*«.

Dedicated Space in Clothes Closets

Receptacles in clothes closets are permitted, but are not required.

Lighting outlets are permitted in closets, but are not required.

A Surface-mounted incandescent or LED luminaires with a completely enclosed light source shall have a minimum 12-in. (300-mm) clearance between the luminaire and the nearest point of a closet storage space ≫410.16(C)(1)≪.

B The blue shading represents closet storage space.

C The depth of the closet storage space above either 6 ft (1.8 m) or to the highest clothes-hanging rod is 12 in. (300 mm) or the width of the shelf, whichever is greater ≫410.2 Closet Storage Space≪.

D The closet storage space extends up from the floor to a height of 6 ft (1.8 m) or to the highest clothes-hanging rod ≫410.2 Closet Storage Space≪.

E The depth of the closet storage space is 24 in. (600 mm) ≫410.2≪.

F Surface-mounted fluorescent luminaires shall have a minimum 6-in. (150-mm) clearance between the luminaire and the nearest point of a closet storage space ≫410.16(C)(2)≪.

G The blue shading represents closet storage space.

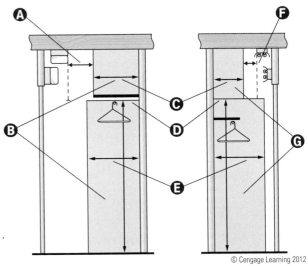

© Cengage Learning 2012

A The closet storage space extends up from the floor to a height of 6 ft (1.8 m) or to the highest clothes-hanging rod ≫410.2≪.

B The depth of the closet storage space is 24 in. (600 mm) ≫410.2≪.

C Recessed incandescent or LED luminaires with a completely enclosed light source shall have a minimum 6-in. (150-mm) clearance between the luminaire and the nearest point of a storage space ≫410.16(C)(3)≪.

D Surface-mounted incandescent or LED luminaires with a completely enclosed light source shall have a minimum 12-in. (300-mm) clearance between the luminaire and the nearest point of a storage space ≫410.16(C)(1)≪.

E Surface-mounted fluorescent luminaires shall have a minimum 6-in. (150-mm) clearance between the luminaire and the nearest point of a storage space ≫410.16(C)(2)≪.

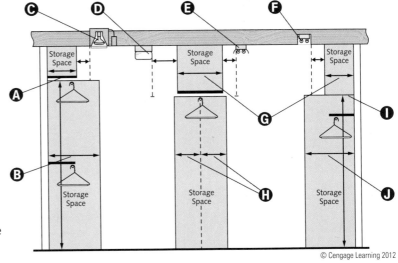

© Cengage Learning 2012

F Recessed fluorescent luminaires shall have a minimum 6-in. (150-mm) clearance between the luminaire and the nearest point of a storage space ≫410.16(C)(4)≪.

G The depth of the closet storage space above either 6 ft (1.8 m) or to the highest clothes-hanging rod is 12 in. (300 mm) or the width of the shelf, whichever is greater ≫410.2≪.

H For a closet with access to both sides of a hanging rod, the storage space shall include the volume below the highest rod extending 12 in. (300 mm) on either side of the rod on a plane horizontal to the floor and extending the entire length of the rod ≫410.2≪.

I The closet storage space extends up from the floor to a height of 6 ft (1.8 m) or to the highest clothes-hanging rod ≫410.2≪.

J The depth of the closet storage space is 24 in. (600 mm) ≫410.2≪.

NOTE

Surface-mounted fluorescent or LED luminaires shall be permitted to be installed within the closet storage space where identified for this use ≫410.16(C)(5)≪.

CAUTION

Overcurrent devices shall not be located in the vicinity of easily ignitible material, such as in clothes closets ≫240.24(D)≪.

BATHROOMS

Bathroom Area

The general provisions for receptacle placement by wall space do not apply to bathrooms.

Ⓐ GFCI protection for personnel is required for every 125-volt, single-phase 15- and 20-ampere receptacle located in the bathroom area »*210.8(A)(1)*«.

> ### NOTE
>
> Service disconnecting means shall not be located in bathrooms »*230.70(A)(2)*« . Overcurrent devices, other than supplementary overcurrent protection, shall not be installed in bathrooms »*240.24(E)*« . Supplementary overcurrent protection is not branch-circuit overcurrent protection. It is an additional overcurrent protection usually installed within luminaires, appliances, and other equipment. It is not required that supplementary overcurrent protection be readily accessible »*240.10*«.

Ⓑ **Bathroom Definition:** An area including a basin (lavatory or sink) with one or more of the following: a toilet, a urinal, a tub, a shower, a bidet, or similar plumbing fixtures »*Article 100—Definitions*« . A bathroom is an **area** and not necessarily a single room.

Ⓒ One or more lighting outlets controlled by a wall switch are required in bathrooms »*210.70(A)(1)*«.

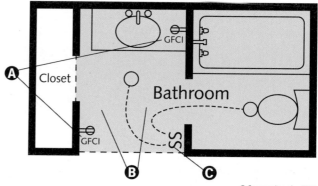

© Cengage Learning 2012

Receptacle within 36 in. (900 mm) of Sink

Ⓐ At least one receptacle shall be located within 3 ft (900 mm) of the outside edge of each basin (lavatory or sink) »*210.52(D)*« .

> ### NOTE
>
> In no case shall the receptacle be located more than 12 in. (300 mm) below the top of the basin »*210.52(D)*«.

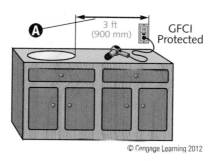

© Cengage Learning 2012

Bathroom Branch-Circuit Rating

Ⓐ Bathroom receptacle outlets shall be supplied by at least one 120-volt, 20-ampere branch circuit »*210.11(C)(3)*«.

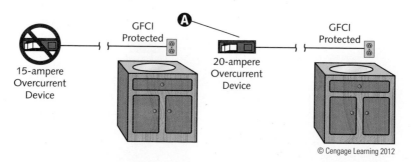

© Cengage Learning 2012

Bathroom Receptacles

Receptacles, unless listed as receptacle assemblies for countertop applications, shall not be installed in a face-up position in countertops or similar work surfaces »*406.5(E)*«.

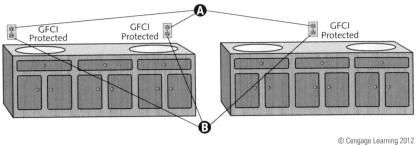

© Cengage Learning 2012

A At least one receptacle shall be located within 36 in. (900 mm) of the outside edge of each basin (lavatory or sink) »*210.52(D)*«.

B Receptacles may have an individual rating of either 15 or 20 amperes, but must be supplied from a 20-ampere branch circuit »*210.11(C)(3)*«. GFCI protection for personnel is required for all bathroom receptacles. The ground-fault circuit interrupter must be installed in a readily accessible location »*210.8(A)(1)*«.

NOTE

Receptacle outlet assemblies listed for the application shall be permitted to be installed in the countertop »*210.52(D)*«.

CAUTION

The receptacle outlet shall be located on a wall or partition that is adjacent to the basin or basin countertop, located on the countertop, or installed on the side or face of the basin cabinet. In no case shall the receptacle be located more than 12 in. (300 mm) below the top of the basin »*210.52(D)*«.

Bathroom

A A branch circuit providing power to a bathroom receptacle may also provide power to other bathroom receptacles, whether in the same bathroom or in different bathrooms »*210.11(C)(3)*«.

B A branch circuit providing power to a bathroom receptacle may also provide power to other equipment, such as lighting and exhaust fans, but only within the same bathroom »*210.11(C)(3) Exception*«. If the branch circuit provides power to a bathroom receptacle and other equipment, the circuit shall not provide power to any other bathroom.

C A branch circuit providing power to bathroom receptacles shall not provide power to any receptacle or lighting outside of bathrooms »*210.11(C)(3)*«.

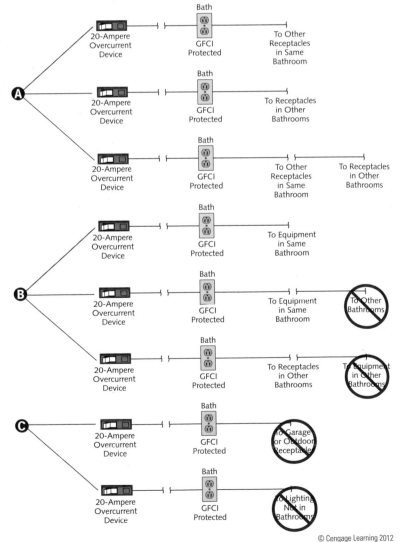

© Cengage Learning 2012

Bathtub and Shower Zone

A Luminaire permitted (not in zone)

B Luminaire not permitted

C No parts of cord-connected luminaires, chain-, cable-, or cord-suspended luminaires, lighting track, pendants, and ceiling-suspended (paddle) fans are permitted within a certain zone of a bathtub ≫*410.10(D)*≪. The bathtub zone measures 3 ft (900 mm) horizontally and 8 ft (2.5 m) vertically from the top of the bathtub rim or shower stall threshold and includes the space directly over the tub.

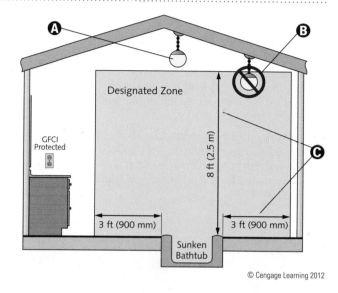

© Cengage Learning 2012

A Recessed lights, surface-mounted lights, and exhaust fans are permitted within the bathtub zone. Exposed, normally non-current-carrying metal parts must be connected to an equipment grounding conductor in accordance with 250.110.

B Chain-, cable-, or cord-suspended luminaires, cord-connected luminaires, lighting track, pendants, and ceiling-suspended (paddle) fans are permitted only *outside* the zone.

C The bathtub zone measures 3 ft (900 mm) horizontally and 8 ft (2.5 m) vertically from the top of the bathtub rim or shower stall threshold. The zone is all-encompassing and includes the space directly over the tub or shower stall ≫*410.10(D)*≪.

NOTE

All 125-volt, single-phase, 15- and 20-ampere receptacles installed within 6 ft (1.8 m) of the outside edge of the bathtub or shower stall shall have ground-fault circuit-interrupter protection for personnel ≫*210.8(D)*≪. Since receptacles installed in bathrooms are already required to have ground-fault circuit-interrupter protection for personnel, this provision applies to bathtubs and shower stalls that are not located in bathrooms.

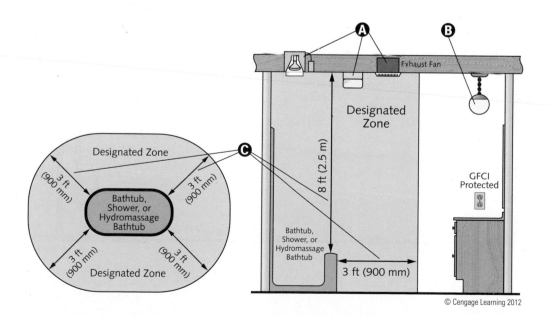

© Cengage Learning 2012

BASEMENTS AND GARAGES

Garage Receptacles

Ⓐ In each attached garage and in each detached garage with electric power, at least one receptacle outlet shall be installed. At least one receptacle outlet shall be installed for each car space ≫ *210.8(A)(2)* ≪ .

Ⓑ Laundry equipment receptacles shall not be counted as required garage receptacles ≫ *210.52(G)* ≪ . In dwelling units, at least one receptacle outlet shall be installed in areas designated for the installation of laundry equipment ≫ *210.52(F)* ≪ .

Ⓒ Receptacles not readily accessible must have GFCI protection for personnel ≫ *210.8(A)(2)* ≪ .

Ⓓ If a duplex receptacle is installed for two appliances, GFCI protection is required ≫ *210.8(A)(2)* ≪ .

Ⓔ A single receptacle installed on an individual 20-ampere branch circuit must have a rating of 20 amperes ≫ *210.21(B)(1)* ≪ .

NOTE

The branch circuit supplying garage receptacle(s) shall not supply outlets outside of the garage ≫ ***210.52(G)(1)*** ≪ .

WARNING

All 125-volt, single-phase, 15- and 20-ampere receptacles installed in garages shall have ground-fault circuit-interrupter protection for personnel ≫ *210.8(A)(2)* ≪ . Note, there are no exceptions to this rule.

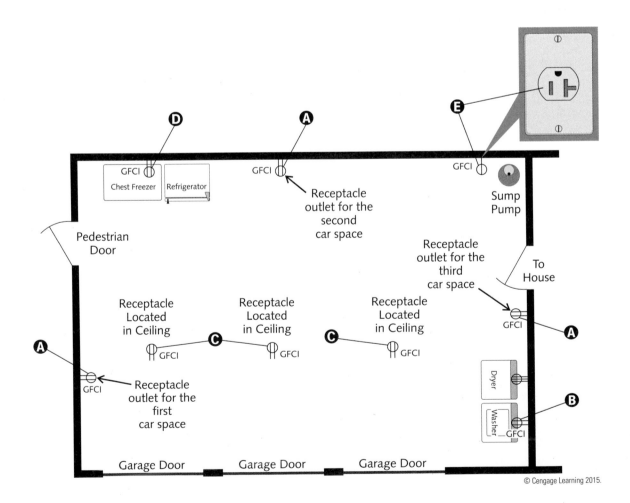

© Cengage Learning 2015.

Lighting

(A) Basements and attached garages require at least one receptacle each. The branch circuit supplying the garage receptacle(s) shall not supply outlets outside of the garage. At least one receptacle outlet shall be installed for each car space in the garage »*210.52(G)(1) and (3)*«. (Receptacle placement is not determined by wall space.)

(B) At least one wall switch–controlled lighting outlet is required to provide illumination on the exterior side of an outside entrance or exit with grade-level access »*210.70(A)(2)(b)*«.

(C) A vehicle garage door is not considered an outside entrance or exit and therefore does not require a wall switch–controlled lighting outlet »*210.70(A)(2)(b)*«.

(D) A lighting outlet shall be installed at, or near, equipment that may require servicing »*210.70(A)(3)*«.

(E) At least one wall switch–controlled lighting outlet is required for basements and attached garages »*210.70(A) (2) and (3)*«.

(F) Interior stairways with six or more steps shall have a wall switch located at each floor level and landing that includes an entryway »*210.70(A)(2)(c)*«.

(G) A lighting outlet installed to provide illumination for the general area (basement, garage, etc.) may also serve as the equipment lighting, provided it is located at or near the equipment.

(H) All 125-volt, single-phase, 15- and 20-ampere receptacles installed in unfinished basements shall have GFCI protection for personnel »*210.8(A)(5)*«.

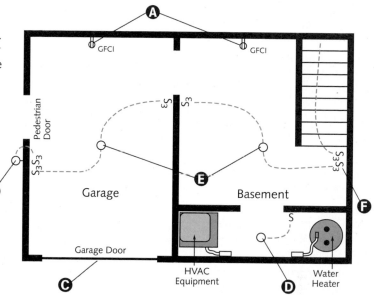

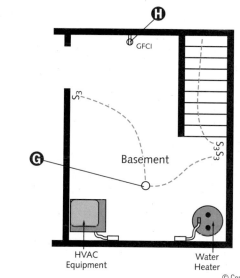

© Cengage Learning 2012

NOTE

A receptacle in an unfinished basement supplying only a permanently installed fire alarm or burglar alarm system shall not be required to have GFCI protection »*210.8(A)(5) Exception*«. Receptacles installed under this exception shall not be considered as meeting the requirements of 210.52(G).

NOTE

Unfinished basements are defined as portions or areas of the basement not intended as habitable rooms and limited to storage areas, work areas, and the like »*210.8(A)(5)*«.

CAUTION

At least one receptacle outlet is required in each separate unfurnished portion of a basement »*210.52(G)(3)*«.

Disconnecting Means

(A) A disconnecting means is not required if the branch-circuit switch or circuit breaker is within 50 ft (15 m) and can be seen while working on the appliance ≫ 422.31(B) ≪.

(B) A disconnecting means is required unless the branch-circuit switch or circuit breaker can be locked in the open (off) position. The provision for locking or adding a lock to the disconnecting means shall be installed on or at the switch or circuit breaker used as the disconnecting means and must remain in place whether the lock is installed or not ≫ 422.31(B) and 110.25 ≪.

(C) In this illustration, the branch-circuit switch or circuit breaker cannot be seen while working on the appliance.

(D) A permanently connected appliance rated over 300 volt-amperes requires a means to simultaneously disconnect all ungrounded (hot) conductors from the appliance ≫ 422.30 ≪.

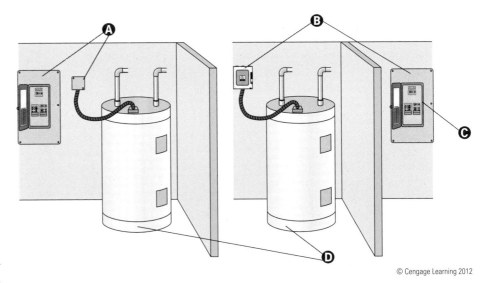

© Cengage Learning 2012

Detached Garages

(A) At least one receptacle is required in each detached garage with electric power ≫ 210.52(G)(1) ≪.

(B) All receptacles installed in detached garages must have GFCI protection, unless installed under one of the exceptions found in 210.8(A)(2).

(C) At least one wall switch–controlled lighting outlet is required for detached garages with electric power ≫ 210.70(A)(2)(a) ≪.

(D) A detached garage with electric power requires at least one wall switch–controlled lighting outlet to provide illumination on the exterior side of an outside entrance or exit with grade-level access ≫ 210.70(A)(2)(b) ≪.

> **NOTE**
>
> Lighting and receptacles are not required in detached garages. There are no provisions to provide electric power to detached garages. However, when electric power is provided, the provisions for attached garages must be met ≫ **210.52(G)(2)** ≪.

© Cengage Learning 2012

> **CAUTION**
>
> In all areas specified in 210.52, all nonlocking-type 125-volt, 15- and 20-ampere receptacles shall be listed tamper-resistant receptacles ≫ 406.12(A) ≪.

Accessory Buildings

A All receptacles installed in accessory buildings having a floor located at or below grade level not intended as habitable rooms and limited to storage areas, work areas, and areas of similar use must have GFCI protection ≫*210.8(A)(2)* ≪.

B Lighting and receptacles are not required in accessory buildings. There are no provisions to provide electric power to detached garages or accessory buildings.

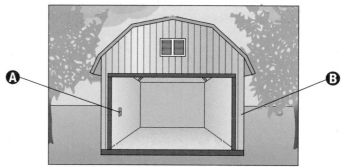

© Cengage Learning 2012

LAUNDRY AREAS

Receptacles and Lighting

A The clothes dryer outlet must have an insulated grounded (neutral) conductor and an equipment grounding conductor ≫*250.138* ≪.

B A lighting outlet is required for spaces containing equipment that requires servicing ≫*210.70(A)(3)* ≪. A lighted area immediately adjacent to a laundry area (such as a hallway) may provide sufficient light for both areas.

C Laundry receptacle outlet(s) must be fed from a 20-ampere branch circuit ≫*210.11(C)(2)* ≪.

D At least one receptacle outlet shall be installed in areas designated for the installation of laundry equipment ≫*210.52(F)* ≪.

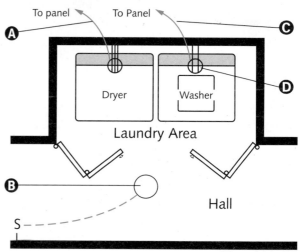

© Cengage Learning 2012

> **NOTE**
>
> Electric clothes dryers connected to existing branch circuits are permitted to follow 250.140.

Receptacles and Lighting

A Laundry receptacle outlet(s) must be fed from a 20-ampere branch circuit ≫*210.11(C)(2)* ≪.

B In dwelling units, at least one receptacle outlet shall be installed in areas designated for the installation of laundry equipment ≫*210.52(F)* ≪.

C Multiple receptacles are permitted on the laundry circuit as long as all of the outlet(s) are designated for laundry equipment ≫*210.11(C)(2)* ≪.

D All 125-volt, single-phase, 15- and 20-ampere receptacles installed in laundry areas shall have ground-fault circuit-interrupter protection for personnel ≫*210.8(A)(10)* ≪.

E Outlet(s) not designated for laundry equipment are not permitted on a laundry branch circuit ≫*210.11(C)(2)* ≪.

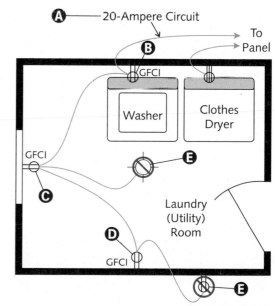

© Cengage Learning 2015.

> **NOTE**
>
> A laundry room/area is not subject to the general provisions for receptacle placement found under 210.52(A). Although additional laundry area receptacles are permitted, only one receptacle outlet is required for the installation of laundry equipment ≫**210.52(F)**≪.

Receptacles and Lighting *(continued)*

A receptacle outlet installed for a specific appliance, such as laundry equipment, shall be installed within 6 ft (1.8 m) of the intended location of the appliance ≫*210.50(C)*≪.

A The clothes dryer outlet must have an insulated grounded (neutral) conductor *and* an equipment grounding conductor ≫*250.138*≪.

B Laundry receptacle outlet(s) must be fed from a 20-ampere branch circuit ≫*210.11(C)(2)*≪. A duplex receptacle installed on an individual 20-ampere branch circuit may have either a 15- or 20-ampere rating ≫*210.21(B)(2)*≪.

C Height requirements for laundry receptacle outlet(s) are not specified, except that they shall not be located more than 5½ ft (1.7 m) above the floor ≫*210.52*≪.

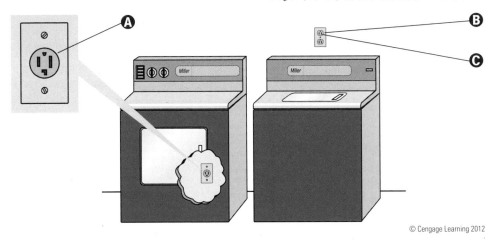

© Cengage Learning 2012

ATTIC AND CRAWL SPACES

Attic Spaces

A receptacle is not required for attic spaces that do not contain equipment that may require servicing.

Attic spaces not used for storage or attics that do not contain equipment that may require servicing do not require lighting outlets.

A Attic spaces used for storage or containing equipment that may require servicing must have at least one lighting outlet ≫*210.70(A)(3)*≪.

B At least one switch shall be located at the usual entry and exit to the attic space ≫*210.70(A)(3)*≪. If the luminaire contains a switch and is located at the usual point of entry, a separate switch may not be required.

C Cables not installed within framing members may need protection up to a height of at least 7 ft (2.1 m) ≫*320.23(A), 330.23,* and *334.23*≪.

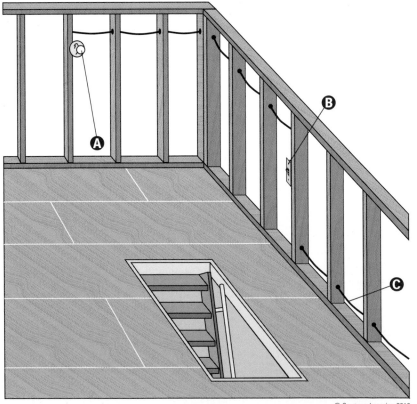

© Cengage Learning 2012

Crawl and Underfloor Spaces

A Crawl spaces with lighting outlet(s) require at least one switch located at the usual entry and exit to the crawl space *»210.70(A)(3)«*. If the luminaire contains a switch and is located at the usual point of entry, a separate switch may not be required.

B A 125-volt, single-phase, 15- or 20-ampere rated receptacle outlet shall be installed at an accessible location for the servicing of equipment. The receptacle shall be located on the same level and within 25 ft (7.5 m) of the equipment. The receptacle shall not be connected to the load side of the equipment disconnecting means *»210.63«*.

C Receptacle outlet(s) require GFCI protection where installed in a crawl space that is not above grade level *»210.8(A)(4)«*.

D Underfloor spaces used for storage containing equipment that may require servicing must have at least one lighting outlet. A lighting outlet shall be located at or near equipment that may require servicing *»210.70(A)(3)«*.

E A disconnecting means must be provided for HVAC equipment *»440.3(B), 422.30, and 424.19«*.

No lighting outlets are required for underfloor spaces that are not used for storage or contain no equipment that may require servicing.

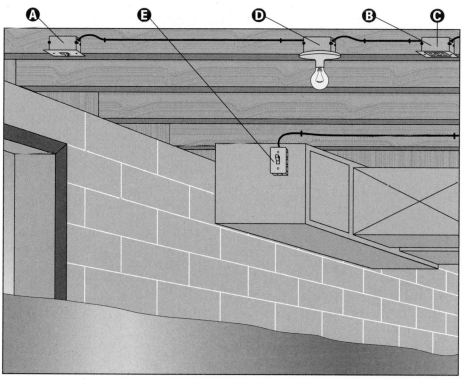

© Cengage Learning 2012

CAUTION

Where cable is run at angles with joists in crawl spaces, it shall be permissible to secure cables not smaller than two 6 AWG or three 8 AWG conductors directly to the lower edges of the joists. Smaller cables must be run either through bored holes in joists or on running boards *»334.15(C)«*.

Summary

- Receptacle branch circuits rated 20 amperes are required for small-appliance branch circuits in kitchens, pantries, dining rooms, breakfast rooms, and similar areas.
- The maximum distance to any countertop receptacle, along the wall line measured horizontally, is 24 in. (600 mm).
- General provisions for countertop receptacle placement do not apply to islands and peninsulars.
- A minimum of two circuits is required for receptacles serving kitchen countertops.
- Receptacles located in kitchens, pantries, dining rooms, breakfast rooms, and similar areas also require a minimum of two circuits.
- All receptacles installed to serve kitchen countertops shall have GFCI protection.

- Some appliances can be cord and plug connected.
- Permanently connected as well as cord- and plug-connected appliances require a disconnecting means.
- Branch-circuit conductors installed for ranges and clothes dryers must have an insulated grounded (neutral) conductor and a grounding means.
- Hallways measuring 10 ft (3.0 m) or more in length require only one receptacle.
- Clothes closets have dedicated space where luminaires are not permitted.
- Only certain types of luminaires are permitted in the dedicated space within bathrooms.
- Receptacles in bathrooms must be GFCI protected and be supplied by 20-ampere branch circuit(s).
- All garage receptacles require GFCI protection.

- Lighting is required at or near equipment that may require servicing.
- All 125-volt, single-phase, 15- and 20-ampere receptacles installed in laundry areas shall have GFCI protection.
- Under certain conditions, attics and crawl spaces may require a lighting outlet and receptacle.

- A receptacle is required on the same level and within 25 ft (7.5 m) of HVAC equipment that may require servicing.
- GFCI protection shall be provided for outlets that supply dishwashers installed in dwelling unit locations.

Unit 7 Competency Test

NEC Reference	Answer	
_____	_____	1. The maximum distance between kitchen countertop receptacles is _____ ft.
_____	_____	2. Hanging luminaires located directly above any part of the bathtub shall be installed so that the luminaire is not less than _____ ft above the top of the bathtub rim.
_____	_____	3. A 15-ampere rated duplex receptacle may be installed on a _____ branch circuit.

3.
a) 15-ampere
b) 20-ampere
c) 15- or 20-ampere
d) 15-, 20-, or 25-ampere

| _____ | _____ | 4. The minimum size THHN copper neutral conductor permitted for an 8¾ kW, 240-volt, 3-wire household electric range is _____. |

4. a) 12 AWG b) 10 AWG c) 8 AWG d) 6 AWG

_____	_____	5. Luminaires shall be constructed or installed so that adjacent combustible material will not be subjected to temperatures in excess of _____ °Fahrenheit.
_____	_____	6. A one-family dwelling has a front entrance on the east side of the house, an attached garage with two 10-ft wide doors on the south side, and a back entrance door on the west side of the house. This one-family dwelling requires _____ lighting outlet(s) for these outdoor entrances.
_____	_____	7. Overcurrent devices shall not be located in the vicinity of easily ignitible material such as in _____.
_____	_____	8. A single receptacle installed on an individual branch circuit shall have an ampere rating not less than _____% of the rating of the branch circuit.

8. a) 125 b) 100 c) 80 d) 50

_____	_____	9. Appliance receptacle outlets installed for specific appliances, such as laundry equipment, shall be placed within _____ ft of the intended location of the appliance.
_____	_____	10. Tap conductors for household cooking equipment supplied from a 50-ampere branch circuit shall have an ampacity of not less than _____ amperes.
_____	_____	11. One cooktop and two ovens are connected to one branch circuit in a kitchen of a one-family dwelling. The cooktop has a nameplate rating of 4 kW at 240 volts, while each oven has a nameplate rating of 5.5 kW at 240 volts. The minimum load calculation for this branch circuit is _____ watts.
_____	_____	12. The rating of a single cord- and plug-connected utilization equipment shall not exceed _____% of the branch-circuit ampere rating.
_____	_____	13. A recessed incandescent luminaire with a completely enclosed light source installed in the ceiling of a clothes closet must have a minimum clearance of _____ in. between the luminaire and the nearest point of the closet storage space.
_____	_____	14. Where the NEC specifies that one equipment shall be "within sight of" another equipment, the specified equipment shall be visible and not more than _____ ft distant from the other.

NEC Reference	Answer

15. In a dwelling unit, a hallway measuring 21½ ft in length requires a minimum of _____ receptacle outlet(s).

16. A permanently connected dishwasher has a nameplate rating of 780 volt-amperes. The circuit breaker is permitted to serve as the disconnecting means where it is:

    I. within sight from the dishwasher.

    II. lockable in accordance with 110.25.

    III. capable of being locked in the closed position.

        a) I only   b) I or III only   c) I or II only   d) I, II, or III

17. Interior stairway lighting outlets shall be controlled by a wall switch at each floor level, where the difference between levels is _____ steps or more.

18. Supplementary overcurrent devices shall not be required to be _____.

19. The maximum length cord permitted on an electrically operated trash compactor is _____ in.

20. For a one-family dwelling, at least one receptacle outlet shall be installed in each:

    I. basement.

    II. attached garage.

    III. detached garage.

        a) I only   b) I and II only   c) II and III only   d) I, II, and III

21. A refrigerator receptacle can be fed from a:

    I. 15-ampere individual branch circuit.

    II. 20-ampere individual branch circuit.

    III. 20-ampere small appliance branch circuit.

        a) II only   b) III only   c) II or III   d) I, II, or III

22. A receptacle installed in an area designated for the installation of laundry equipment and where there are no other receptacles can be a:

    I. 15-ampere single receptacle.

    II. 15-ampere duplex receptacle.

    III. 20-ampere duplex receptacle.

        a) III only   b) I or III   c) II or III   d) I, II, or III

23. At least one GFCI-protected receptacle shall be located within _____ ft from the outside edge of each bathroom sink.

24. Receptacles are permitted under an overhanging countertop, but are not counted as required countertop receptacles, where the countertop extends more than _____ in. beyond its support base.

25. A branch circuit providing power to a bathroom receptacle may also provide power to:

    I. two other receptacles, two luminaires, and an exhaust fan in the same bathroom.

    II. two other receptacles in the same bathroom and receptacles in two additional bathrooms.

    III. one luminaire in the same bathroom and one luminaire in another bathroom.

        a) I only   b) I or II only   c) II or III only   d) I, II, or III

Load Calculations

Objectives

After studying this unit, the student should:

▶ be able to calculate the general lighting load in a one-family dwelling.

▶ know the minimum volt-ampere (VA) requirements for small-appliance and laundry branch circuits.

▶ know how to apply demand factors to the general lighting load.

▶ be able to apply demand factors to fastened-in-place appliances.

▶ be able to calculate feeder-demand loads for household clothes dryers.

▶ know how to calculate feeder-demand loads for household cooking equipment.

▶ be able to calculate heating and air-conditioning feeder-demand loads.

▶ be able to calculate a one-family dwelling service or feeder using the standard method.

▶ be able to calculate a one-family dwelling service or feeder using the optional method.

▶ know how to size service and feeder conductors.

▶ be able to calculate and choose the appropriate size neutral conductor.

▶ understand how the grounding electrode conductor is selected.

Introduction

Load calculations must be performed in order to determine the size of services and feeders. Service conductors are the conductors between the power provider and the disconnecting means, whether a main breaker or main lug panel. A feeder is defined as all circuit conductors between the service disconnecting means and the panelboard containing fuses or breakers that feed branch circuits. Additional load calculations are needed for dwellings with panelboards that are not part of the service. For example, a separate panelboard might be installed near the middle of the house to help eliminate voltage drop in branch-circuit conductors. Since the remote panelboard (subpanel) does not carry the total load, an additional calculation is needed to size the feeder and panelboard.

This unit simplifies the standard and optional load calculation methods for one-family dwellings. A blank form is provided at the beginning of each method. The floor plan of the one-family dwelling being calculated is found on page 108 while the specifications are on page 157. Both calculation methods use the same specifications. Each line on each form is explained in detail using that information. The completed form follows the explanations.

COMPILING INFORMATION ESSENTIAL TO LOAD CALCULATIONS

Total Floor Area Square Footage

A Open porches are not included in the calculated floor area of dwelling units ≫*220.12*≪. Although this porch has a roof, it is considered open and, therefore, is not included in the calculation.

B Scale rulers are also useful in determining the location of outlets (receptacles, lights, switches, etc.).

C Before a scale ruler can be used to determine dimensions, the scale used to produce the blueprint (floor plan) must be known. (Then the area of an exterior porch, for example, is simple to calculate.)

D Scale rulers are useful in determining the total square footage of a structure (house, building, etc.) that has been drawn on a set of blueprints (floor plans).

E A regular tape measure (ruler, yard stick, etc.) can be used, but the use of a scale ruler makes the job much easier.

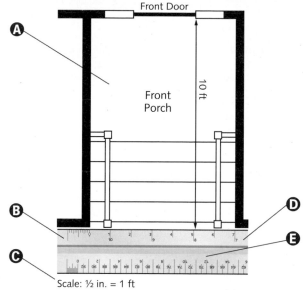

Scale: ½ in. = 1 ft

© Cengage Learning 2012

Gathering Information

Gathering certain information is necessary to accurately perform load calculations. All the essential data can be identified by simply answering the questions found on the load calculation form, either standard or optional. Enter the information on the appropriate line of the load calculation forms provided with this textbook.

A What is the dwelling's total square-foot (ft²) area, using the outside dimensions?

B What fastened-in-place appliances will be installed and what is the volt-ampere load of each?

C How many small appliance branch circuits will be installed?

D How many laundry circuits will be installed?

E If an electric clothes dryer will be installed, what is the volt-ampere load?

F What household cooking equipment will be installed and what is the volt-ampere load of each?

G Will there be an air-conditioning system? If so, what is the total volt-ampere load, including the compressor and fan motors? If an electric heating system will be installed, what is the total volt-ampere load, including the strip heat and blower motor?

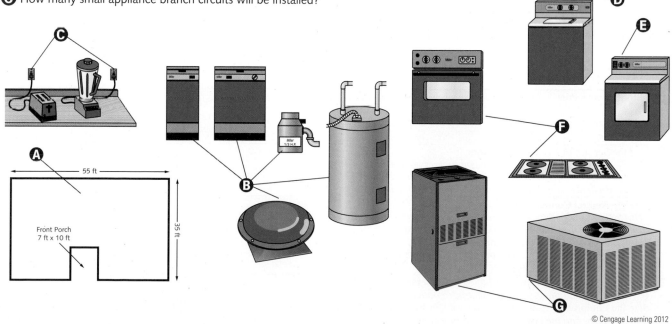

© Cengage Learning 2012

STANDARD METHOD: ONE-FAMILY DWELLINGS

Standard Method Load Calculation for One-Family Dwellings

1	General Lighting and Receptacle Loads *220.12* *Do not include open porches, garages, or unused or* *unfinished spaces not adaptable for future use.*	$3 \times$ _____ $=$ (sq ft using outside dimensions)	**1**
2	Small-Appliance Branch Circuits *220.52(A)* At least **two** small-appliance branch circuits must be included. *210.11(C)(1)*	$1500 \times$ _____ $=$ (minimum of two)	**2**
3	Laundry Branch Circuit(s) *220.52(B)* At least **one** laundry branch circuit must be included. *210.11(C)(2)*	$1500 \times$ _____ $=$ (minimum of one)	**3**

4 Add lines 1, 2, and 3	**4**		Lines 5 through 8 utilize the demand factors found in *Table 220.42.*
5 _____ $- 3000 =$ (line 4)	**5**	(if 117,000 or less, skip to line 8)	**6** _____ $- 117,000 =$ **6** (line 5, if more than 117,000)
7 _____ $\times 25\% =$ **7** (line 6)			**8** _____ $\times 35\% =$ **8** (smaller of line 5 or 117,000)

9 Total General Lighting and Receptacle Load	$3000 +$ _____ $+$ _____ $=$ (line 7) (line 8)	**9**

10 Fastened-In-Place Appliances *220.53*
Use the nameplate rating.
Do not include electric ranges,
clothes dryers, space-heating
equipment, or air-conditioning
equipment.

water heater /	dishwasher /	disposer /
compactor /	/	/
/	/	/

If fewer than four units, put total volt-amperes on line 10.
If four or more units, multiply total volt-amperes by 75%. _____ $\times 75\% =$ **10**
(volt-amps of four or more)

11 Clothes Dryers *220.54* *(If present; otherwise skip to line 12.) Use 5000 watts or the nameplate rating, whichever is larger.* The neutral demand load is 70% for feeders. *220.61(B)*		**11**
12 Ranges, Ovens, Cooktops, and Other Household Cooking Appliances over 1750 Watts *220.55* *(If present; otherwise skip to line 13.) Use Table 220.55 and all of the applicable Notes.* The neutral demand load is 70% for feeders. *220.61(B)*		**12**
13 Heating or Air-Conditioning System (Compare the heat and A/C, and omit the smaller.) *220.60* *Include the air handler when using either one. For heat pumps, include the compressor and the* *maximum amount of electric heat that can be energized while the compressor is running.*		**13**
14 Largest Motor (one motor only) *220.50* and *430.24* *Multiply the volt-amperes of the largest motor by 25%.*	_____ $\times 25\% =$ (volt-amps of largest motor)	**14**
15 Total Volt-Ampere Demand Load: *Add lines 9 through 14 to find the minimum required volt-amperes.*		**15**

16 Minimum Amperes *Divide the total* *volt-amperes by* *the voltage.*	**16** _____ $\div$ _____ $=$ _____ (line 15) (voltage) (minimum amperes)	**Minimum Size** **17 Service or** **Feeder** *240.6(A)*	**17**
18 Size the Service or Feeder Conductors. *Use 310.15(B)(7) to find the service conductors up to 400 amperes.* *Ratings in excess of 400 amperes shall comply with Table 310.15(B)(16)* *310.15(B)(7) also applies to feeder conductors supplying the entire load.*		**Minimum Size** **Conductors**	**18**
19 Size the Neutral Conductor. *220.61* *310.15(B)(7) states that the neutral service or feeder conductor* *can be smaller than the ungrounded ("hot") conductors, provided the* *requirements of 215.2, 220.61, and 230.42 are met.* *250.24(C)(1) states that the grounded conductor shall not be smaller* *than specified in Table 250.102(C)(1).*		**Minimum Size** **Neutral** **Conductor**	**19**
20 Size the Grounding Electrode Conductor (for Service). *250.66* *Use line 18 to find the grounding electrode conductor in Table 250.66.* Size the Equipment Grounding Conductor (for Feeder). *250.122* *Use line 17 to find the equipment grounding conductor in Table 250.122.* *Equipment grounding conductor types are listed in 250.118.*		**Minimum Size** **Grounding** **Electrode** **Conductor**	**20**

Sample One-Family Dwelling Load Calculation

Assume water heater, clothes dryer, counter-mounted cooking unit, wall-mounted oven, and electric heat kilowatt (kW) ratings equivalent to kilovolt-ampere (kVA).

> **NOTE**
>
> The information to the right will be used throughout the remainder of this unit to demonstrate the use of the load calculation forms, both standard and optional.

Dwelling outside dimension . . . . 35 ft × 55 ft
Front porch (included within outside dimensions) . . . . . 7 ft × 10 ft
Small-appliance branch circuits . . . . four
Laundry branch circuits . . . . two
Water heater . . . . 4.5 kW, 240 volt
Dishwasher . . . . 10 amperes, 120 volt
In-sink waste disposer . . . . ½HP, 115 volt
Trash compactor . . . . 7.5 amperes, 120 volt
Three attic fans (4.2 amperes each) . . . . 12.6 amperes, 120 volt
Clothes dryer . . . . 5.5 kW, 240 volt
Counter-mounted cooking unit (cooktop) . . . . 7 kW, 240 volt
Wall-mounted oven . . . . 6 kW, 240 volt
Electric heat (three banks at 5 kW each) . . . . 15 kW, 240 volt
Air handler (blower motor) . . . . 3.2 amperes, 115 volt
Air-conditioner compressor . . . . 16.6 amperes, 230 volt
Condenser fan motor . . . . 2 amperes, 115 volt

Line 1—General Lighting and Receptacle Loads

(A) In Table 220.12 the general lighting loads are listed according to the type of occupancy and are determined by requiring a minimum lighting load for each square foot of floor area.

(B) The lighting load for residential dwelling units is 3 volt-amperes (VA) per square foot.

(C) The dwelling floor plan on page 108 has an outside dimension of 35 ft × 55 ft (1925 ft²), but there is an open porch with a dimension of 7 ft × 10 ft (70 ft²). Since open porches are not included in the total area, 70 is deducted from 1925, leaving 1855 ft².

(D) The floor area for each floor shall be computed using the outside dimensions. The total floor area does not include open porches, garages, or unused (including unfinished) spaces that are not adaptable for future use ≫*220.12*≪.

> **NOTE**
>
> All 15- or 20-ampere general-use receptacle outlets (except for small-appliance and laundry outlets) are considered part of the general lighting load. Footnote "a" at the bottom of Table 220.12 refers to 220.14(J), which states that no additional load calculations are required for these outlets.

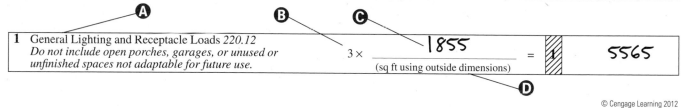

| 1 | General Lighting and Receptacle Loads *220.12* *Do not include open porches, garages, or unused or unfinished spaces not adaptable for future use.* | 3 × ___1855___ (sq ft using outside dimensions) | = | 5565 |

© Cengage Learning 2012

Line 2—Small-Appliance Branch Circuits

(A) Small-appliance branch circuits must be included when calculating a dwelling unit service ≫*220.52(A)*≪.

(B) Each small-appliance branch circuit is calculated at no less than 1500 volt-amperes ≫*220.52(A)*≪.

(C) A dwelling unit must have at least two small-appliance branch circuits ≫*210.11(C)(1)*≪.

(D) These loads can be included with the general lighting load and subjected to the demand factors provided in Table 220.42 ≫*220.52(A)*≪.

> **NOTE**
>
> An individual branch circuit for refrigeration equipment (permitted by 210.52(B)(1) Exception No. 2) can be excluded from this calculation ≫*220.52(A) Exception*≪. If a dwelling has more than one feeder, a separate load calculation is needed for each. It is not necessary to include this step if a feeder does not supply small-appliance branch circuits. A load calculation for a feeder supplying small-appliance branch circuits must include at least 1500 volt-amperes for each 2-wire circuit ≫*220.52(A)*≪.

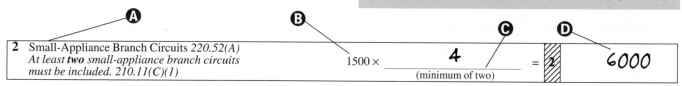

| 2 | Small-Appliance Branch Circuits *220.52(A)* *At least **two** small-appliance branch circuits must be included. 210.11(C)(1)* | 1500 × ___4___ (minimum of two) | = | 6000 |

© Cengage Learning 2012

Line 3—Laundry Branch Circuits

Ⓐ Laundry branch circuits must be included when calculating a dwelling unit service ≫ *220.52(B)* ≪.

Ⓑ Each laundry branch circuit is calculated at no less than 1500 volt-amperes ≫ *220.52(B)* ≪.

Ⓒ At least one laundry branch circuit is required per dwelling unit ≫ *210.11(C)(2)* ≪.

Ⓓ These loads can be included with the general lighting load and subjected to the demand factors provided in Table 220.42 ≫ *220.52(B)* ≪.

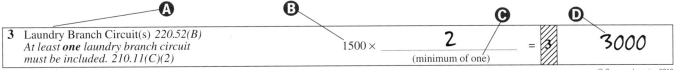

© Cengage Learning 2012

Lines 4 through 8—Applying Demand Factors Found in Table 220.42

Ⓐ Line 4 is simply the total of lines 1 through 3.

Ⓑ Lines 5 through 8 outline the procedure that derates by using demand factors provided in Table 220.42.

Ⓒ Line 5 is the result of subtracting 3000 from the number in line 4.

Ⓓ The first 3000 (or less) volt-amperes must remain at 100% ≫ *Table 220.42* ≪.

Ⓔ If line 5 is 117,000 or less, then lines 6 and 7 can be skipped.

Ⓕ If Line 4 is 120,000 or less, then lines 6 and 7 will remain empty.

Ⓖ Insert the smaller of line 5 or 117,000.

Ⓗ The next 117,000 (from 3001 to 120,000) is multiplied by 35% ≫ *Table 220.42* ≪.

Ⓘ The result found on line 8 has been rounded up from 4047.75 to the next whole number, thus eliminating the decimal.

> **NOTE**
>
> Generally, one-family dwelling load calculations will not use lines 6 and 7. (For example, a one-family dwelling with a total floor area of 25,000 ft², twenty small-appliance circuits and ten laundry branch circuits will not need lines 6 and 7.)

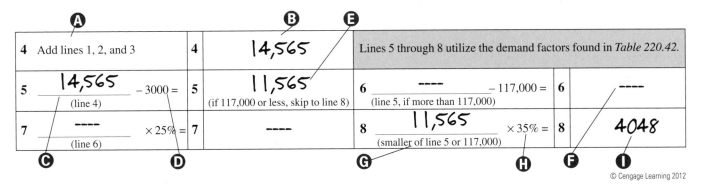

© Cengage Learning 2012

Line 9—Total General Lighting (and Receptacle) Load

Ⓐ Line 9 is the total of general lighting and general-use receptacles, small-appliance branch circuits, and laundry branch circuits after derating.

Ⓑ If line 7 was skipped (left blank), leave this space blank also.

Ⓒ This shaded block draws attention to the first of six lines (boxes) that will be added together to give the total volt-ampere demand load for the one-family dwelling.

Ⓓ If performing a load calculation for a feeder (and panel) that supplies less than 1000 ft² of floor area and does not supply small-appliance or laundry branch circuits, the result could be less than 3000. For example, if a feeder is installed to supply only 800 ft² of floor area and will not supply any small appliance or laundry branch circuits, then line 9 will be 2400 (3 VA per ft² × 800 ft² = 2400).

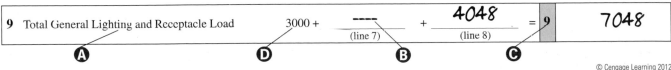

© Cengage Learning 2012

Line 10—Fastened-In-Place Appliances

Ⓐ The term **appliance** designates utilization equipment commonly built in standardized types and sizes, and installed as a unit to perform specific function(s) ≫ *Article 100* ≪.

Ⓑ For the purpose of this load calculation, a kW rating will be the same as a kVA rating.

Ⓒ Multiplying voltage by amperage produces volt-amperes (VA).

Ⓓ Horsepower ratings must be converted to volt-amperes (VA) before they can be added to this form. The full load current ratings for single-phase motors are found in Table 430.248. Multiply the amperes (9.8) by the rated voltage (115) to find the volt-amperes (1127).

Ⓔ No derating is allowed when there are only one, two, or three fastened-in-place appliances.

Ⓕ If the feeder being calculated contains at least four fastened-in-place appliances, the combined volt-ampere rating of those appliances is multiplied by 75% and placed on line 10 ≫ *220.53* ≪.

Ⓖ Space is provided for additional fastened-in-place appliances.

> **N O T E**
>
> Electric ranges, dryers, space-heating equipment, and air-conditioning equipment are not included as fastened-in-place appliances **≫220.53≪**.
> Household cooking appliances individually rated in excess of 1750 watts are derated under 220.55 provisions; see Line 12.

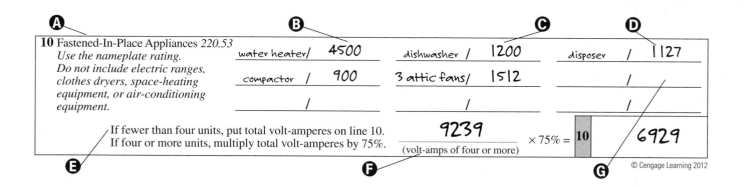

© Cengage Learning 2012

Line 11—Clothes Dryers

Ⓐ A clothes dryer is not a requirement for a load calculation. Skip this line if there is no clothes dryer.

Ⓑ If one of the circuits on the feeder being calculated is a clothes dryer, the rating for the dryer must be at least 5000 watts (volt-amperes) ≫ *220.54* ≪.

Ⓒ A clothes dryer neutral load on a feeder can be calculated at 70% ≫ *220.61(B)* ≪.

Ⓓ If the nameplate rating is more than 5000 watts (volt-amperes), the larger number is used ≫ *220.54* ≪.

Ⓔ 5.5 kW is equal to 5500 watts (volt-amperes).

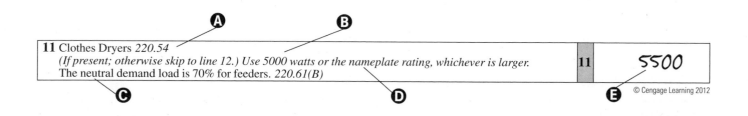

© Cengage Learning 2012

Line 12—Household Cooking Appliances

Ⓐ Household cooking appliances (ranges, wall-mounted ovens, counter-mounted cooking units, etc.) are not required in a load calculation. Skip this line if there are no cooking appliances rated over 1¾ kW.

Ⓑ Individual household cooking appliances rated more than 1750 watts (volt-amperes) can be derated by using Table 220.55 demand factors (and all of the applicable notes).

Ⓒ The neutral load on a feeder can be calculated at 70% for household electric ranges, wall-mounted ovens, and counter-mounted cooking units ≫ *220.61(B)* ≪ .

Ⓓ A 7-kW counter-mounted cooking unit (cooktop) and a 6-kW wall-mounted oven are included in this calculation. Both units fall within the parameters of Column B. The demand factor percentage found in that column for 2 units is 65%. Multiply the total kW (13) by the demand factor percentage (65%) to find the derated load (8.45 kW); 8.45 kW is equal to 8450 watts (volt-amperes).

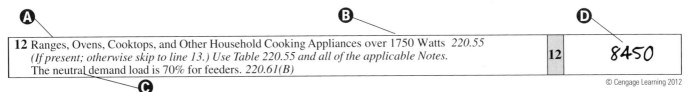

Ⓐ   **Ⓑ**   **Ⓓ**

12 Ranges, Ovens, Cooktops, and Other Household Cooking Appliances over 1750 Watts *220.55*
(If present; otherwise skip to line 13.) Use Table 220.55 and all of the applicable Notes.
The neutral demand load is 70% for feeders. *220.61(B)* | **12** | *8450*

Ⓒ

© Cengage Learning 2012

Line 13—Heating or Air-Conditioning System

Ⓐ The smaller of two (or more) **noncoincident loads** can be omitted, as long as they are never energized simultaneously ≫ *220.60* ≪ . By definition, noncoincident means not occurring at the same time. A dwelling's electric heating system and air-conditioning system may be noncoincident loads.

Ⓑ The air handler (blower motor) has a rating of 368 volt-amperes (3.2 × 115 = 368). Since both loads are energized simultaneously, add the air handler to the 15 kW of electric heat.

Ⓒ Fixed electric space heating loads are calculated at 100% of the total connected load ≫ *220.51* ≪ .

Ⓓ The air handler (blower) works with either the heating or the air conditioning. Add it to both calculations.

Ⓔ The air-conditioner compressor (16.6 × 230 = 3818 VA), condenser fan motor (2 × 115 = 230 VA), and blower motor (3.2 × 115 = 368 VA) have a combined load of 4416 volt-amperes. Because this total is less than the heating load, omit the air-conditioner compressor and the condenser fan motor.

> **N O T E**
>
> A heat pump with supplementary heat is not considered a noncoincident load. Add the compressor full-load current to the maximum amount of heat that can be energized while the compressor is running. Remember to include all associated motors—blower, condenser, etc.

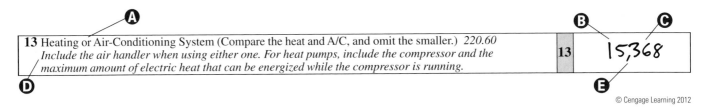

Ⓐ   **Ⓑ**   **Ⓒ**

13 Heating or Air-Conditioning System (Compare the heat and A/C, and omit the smaller.) *220.60*
Include the air handler when using either one. For heat pumps, include the compressor and the
maximum amount of electric heat that can be energized while the compressor is running. | **13** | *15,368*

Ⓓ   **Ⓔ**

© Cengage Learning 2012

Line 14—Largest Motor

Ⓐ Multiply the largest motor in the dwelling calculation by 25% ≫ *220.50* and *430.24* ≪ .

Ⓑ The largest motor is usually the air-conditioner compressor. If the air-conditioner load has been omitted, use the second largest motor. In this case, use the ½-horsepower in-sink waste disposer motor.

Ⓒ The result of 281.75 has been rounded up, thus eliminating the decimal.

Ⓐ   **Ⓑ**   **Ⓒ**

14 Largest Motor (one motor only) *220.50* and *430.24*
Multiply the volt-amperes of the largest motor by 25%. | *1127* (volt-amps of largest motor) × 25% = | **14** | *282*

© Cengage Learning 2012

Line 15—Total Demand Load

A Add lines 9 through 14 to find the service's minimum volt-ampere load.

A

15 Total Volt-Ampere Demand Load: *Add lines 9 through 14 to find the minimum required volt-amperes.*	**15**	**43,577**

© Cengage Learning 2012

Lines 16 and 17—Minimum Service or Feeder

A Place the total volt-ampere amount found in line 15 here.

B Write down the source voltage that will supply the feeder.

C It is permissible to round calculations to the nearest whole ampere. Fractions of an ampere 0.5 and higher are rounded up, while fractions less than 0.5 are dropped ≫ *220.5(B)* ≪ .

D The service (or feeder) overcurrent protection must be higher than the number found in line 16. A list of standard ampere ratings for fuses and circuit breakers is provided in 240.6(A).

E Find the amperage by dividing the volt-amperes by the voltage.

F This is the minimum amperage rating required for the service or feeder being calculated.

G The next standard size fuse or breaker above 182 amperes is 200 ≫ *240.6(A)* ≪ .

A   **B**   **C**   **D**

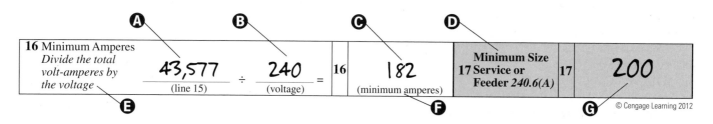

© Cengage Learning 2012

Line 18—Minimum Size Conductors

A For a service rated 100 through 400 A, the service conductors supplying the entire load associated with a one-family dwelling, or the service conductors supplying the entire load associated with an individual dwelling unit in a two-family or multifamily dwelling, shall be permitted to have an ampacity not less than 83 percent of the service rating ≫ *310.15(B)(7)(1)* ≪ .

B Section 310.15(B)(7) also applies to feeder conductors that supply the entire load associated with a one-family dwelling ≫ *310.15(B)(7)(2)* ≪ .

C The maximum service or feeder rating permitted when using 310.15(B)(7) is 400 amperes.

D Multiply the minimum size service or feeder rating (Line 17) by 83 percent and then select a conductor from Table 310.15(B)(16).

200 × 83% = 166 amperes

In accordance with the 75°C column of Table 310.15(B)(16), the ampacity of a 2/0 AWG copper conductor is 175 amperes.

E 200 × 83% = 166 amperes

In accordance with the 75°C column of Table 310.15(B)(16), the ampacity of a 4/0 AWG aluminum conductor is 180 amperes.

A   **B**   **C**   **D**

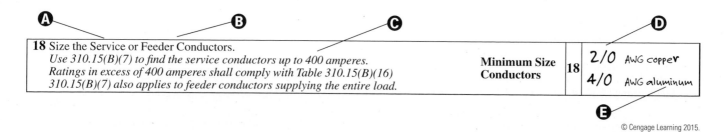

© Cengage Learning 2015.

Line 19—Neutral Conductor

A No neutral conductor is connected to the water heater in this calculation.

B All fastened-in-place appliances utilizing a grounded conductor must be included.

C The electric heat and air-conditioner compressor are 240-volt loads (no neutral).

D If the largest motor is 120 (115) volts, use the motor load in line 14.

E For the purpose of this load calculation, the terms **neutral** and **grounded** are synonymous.

F The feeder (or service) neutral load is the maximum unbalance of the load determined by Article 220 ≫ 220.61(A) ≪.

G The neutral (grounded) conductor can be smaller than the ungrounded (hot) conductors provided certain requirements in 215.2, 220.61 and 230.42 are met ≫ 310.15(B)(7)(4) ≪.

H All numbers are shown as volt-amperes.

I Since all of the general lighting and receptacle loads have a rating of 120 volts, the total volt-ampere rating (from line 9) must be included in the neutral calculation.

J A demand factor of 75% can be applied to the nameplate rating load of four or more fastened-in-place appliances ≫ 220.53 ≪.

K Multiply the clothes dryer load (line 11) by 70% to determine the neutral load ≫ 220.61(B) ≪.

L Multiplying line 12 by 70% provides the neutral cooking appliances load ≫ 220.61(B) ≪.

M Because both of these motors are energized simultaneously, thereby contributing to the neutral load, they must be included in the neutral calculation.

N The total neutral volt-ampere load (21,247) divided by the voltage (240) renders the minimum neutral amperes (89).

O Size the neutral by finding the maximum unbalanced load.

P A 3 AWG (75°C [167°F]) copper conductor is required for 89 amperes ≫ Table 310.15(B)(16) ≪.

Q A 2 AWG (75°C [167°F]) aluminum conductor is required for 89 amperes ≫ Table 310.15(B)(16) ≪.

```
General lighting and receptacles (line 9) . . . . . . . . . .7048
Fastened-in-place appliances
       Water heater. . . . . . . . . . . . . . . . . . . .0
       Dishwasher. . . . . . . . . . . . . . . . . .1200
       In-sink waste disposer . . . . . . . . .1127
       Trash compactor . . . . . . . . . . . . . . .900
       Three attic fans. . . . . . . . . . . . . . .1512
       Total . . . . . . . . . . . . . . . . . . . . . . .4739
Total fastened-in-place appliances (4739 × 75%). . .3554
Clothes dryer (line 11 × 70%). . . . . . . . . . . . . . . .3850
Cooking appliances (line 12 × 70%). . . . . . . . . . . .5915
Electric heat . . . . . . . . . . . . . . . . . . . . . . . . . . . . . . .0
Air-conditioner compressor . . . . . . . . . . . . . . . . . . . .0
Air handler (blower motor). . . . . . . . . . . . . . . . . . .368
Condenser fan motor. . . . . . . . . . . . . . . . . . . . . . . .230
Largest motor. . . . . . . . . . . . . . . . . . . . . . . . . . . . .282
TOTAL. . . . . . . . . . . . . . . . . . . . . . . . . . . . . . . 21,247
```

19 Size the Neutral Conductor *220.61*
310.15(B)(7) states that the neutral service or feeder conductor can be smaller than the ungrounded ("hot") conductors, provided the requirements of 215.2, 220.61, and 230.42 are met. 250.24(C)(1) states that the grounded conductor shall not be smaller than specified in Table 250.102(C)(1).

| Minimum Size Neutral Conductor | 19 | 3 copper |
| | | 2 aluminum |

© Cengage Learning 2015.

Line 20—Grounding Electrode Conductor

A Use 310.15(B)(7) to find service entrance and service lateral conductors.

B The minimum size grounding electrode conductor is 4 AWG copper (or 2 AWG aluminum) ≫ Table 250.66 ≪.

C In outdoor applications, aluminum (or copper-clad aluminum) grounding electrode conductors must not be installed within 18 in. (450 mm) of the earth ≫ 250.64(A) ≪.

20 Size the Grounding Electrode Conductor (for Service). *250.66*
Use line 18 to find the grounding electrode conductor in Table 250.66.
Size the Equipment Grounding Conductor (for Feeder). *250.122*
Use line 17 to find the equipment grounding conductor in Table 250.122.
Equipment grounding conductor types are listed in 250.118.

| Minimum Size Grounding Electrode Conductor | 20 | 4 copper |
| | | 2 aluminum |

© Cengage Learning 2012

The following page shows the completed one-family dwelling load calculation according to the standard method.

Standard Method Load Calculation for One-Family Dwellings

1	General Lighting and Receptacle Loads *220.12* *Do not include open porches, garages, or unused or unfinished spaces not adaptable for future use.*	$3 \times \underline{\quad 1855 \quad}$ (sq ft using outside dimensions) $= $	**1**		**5565**

2 Small-Appliance Branch Circuits *220.52(A)*
*At least **two** small-appliance branch circuits must be included. 210.11(C)(1)* $1500 \times \underline{\quad 4 \quad} = $ (minimum of two) **2** **6000**

3 Laundry Branch Circuit(s) *220.52(B)*
*At least **one** laundry branch circuit must be included. 210.11(C)(2)* $1500 \times \underline{\quad 2 \quad} = $ (minimum of one) **3** **3000**

| **4** | Add lines 1, 2, and 3 | **4** | **14,565** | Lines 5 through 8 utilize the demand factors found in *Table 220.42*. | | |

5 $\underline{\quad 14,565 \quad}$ (line 4) $- 3000 =$ **5** **11,565** (if 117,000 or less, skip to line 8) **6** $\underline{\quad ---- \quad}$ (line 5, if more than 117,000) $- 117,000 =$ **6** ----

7 $\underline{\quad ---- \quad}$ (line 6) $\times 25\% =$ **7** ---- **8** $\underline{\quad 11,565 \quad}$ (smaller of line 5 or 117,000) $\times 35\% =$ **8** **4048**

9 Total General Lighting and Receptacle Load $3000 + \underline{\quad ---- \quad}$ (line 7) $+ \underline{\quad 4048 \quad}$ (line 8) $=$ **9** **7048**

10 Fastened-In-Place Appliances *220.53*
Use the nameplate rating.
Do not include electric ranges, clothes dryers, space-heating equipment, or air-conditioning equipment.

water heater / **4500**   dishwasher / **1200**   disposer / **1127**

compactor / **900**   3 attic fans / **1512**   /

/   /   /

If fewer than four units, put total volt-amperes on line 10.
If four or more units, multiply total volt-amperes by 75%. $\underline{\quad 9239 \quad}$ (volt-amps of four or more) $\times 75\% =$ **10** **6929**

11 Clothes Dryers *220.54*
(If present; otherwise skip to line 12.) Use 5000 watts or the nameplate rating, whichever is larger.
The neutral demand load is 70% for feeders. 220.61(B) **11** **5500**

12 Ranges, Ovens, Cooktops, and Other Household Cooking Appliances over 1750 Watts *220.55*
(If present; otherwise skip to line 13.) Use Table 220.55 and all of the applicable Notes.
The neutral demand load is 70% for feeders. 220.61(B) **12** **8450**

13 Heating or Air-Conditioning System (Compare the heat and A/C, and omit the smaller.) *220.60*
Include the air handler when using either one. For heat pumps, include the compressor and the maximum amount of electric heat that can be energized while the compressor is running. **13** **15,368**

14 Largest Motor (one motor only) *220.50* and *430.24*
Multiply the volt-amperes of the largest motor by 25%. $\underline{\quad 1127 \quad}$ (volt-amps of largest motor) $\times 25\% =$ **14** **282**

15 Total Volt-Ampere Demand Load: *Add lines 9 through 14 to find the minimum required volt-amperes.* **15** **43,577**

16 Minimum Amperes
Divide the total volt-amperes by the voltage $\underline{\quad 43,577 \quad}$ (line 15) $\div \underline{\quad 240 \quad}$ (voltage) $=$ **16** **182** (minimum amperes) **17** **Minimum Size Service or Feeder** *240.6(A)* **17** **200**

18 Size the Service or Feeder Conductors.
Use 310.15(B)(7) to find the service conductors up to 400 amperes.
Ratings in excess of 400 amperes shall comply with Table 310.15(B)(16)
310.15(B)(7) also applies to feeder conductors supplying the entire load. **Minimum Size Conductors** **18** **2/0** AWG copper
4/0 AWG aluminum

19 Size the Neutral Conductor *220.61*
310.15(B)(7) states that the neutral service or feeder conductor can be smaller than the ungrounded ("hot") conductors, provided the requirements of 215.2, 220.61, and 230.42 are met.
250.24(C)(1) states that the grounded conductor shall not be smaller than specified in Table 250.102(C)(1). **Minimum Size Neutral Conductor** **19** **3** copper
2 aluminum

20 Size the Grounding Electrode Conductor (for Service). *250.66*
Use line 18 to find the grounding electrode conductor in Table 250.66.
Size the Equipment Grounding Conductor (for Feeder). *250.122*
Use line 17 to find the equipment grounding conductor in Table 250.122.
Equipment grounding conductor types are listed in 250.118. **Minimum Size Grounding Electrode Conductor** **20** **4** copper
2 aluminum

OPTIONAL METHOD: ONE-FAMILY DWELLINGS

Optional Method Load Calculation for One-Family Dwellings

1 General Lighting and Receptacle Loads *220.82(B)(1)*
Do not include open porches, garages, or unused or
unfinished spaces not adaptable for future use.

$3 \times$ _____ = **1**
(sq ft using outside dimensions)

2 Small-Appliance Branch Circuits *220.82(B)(2)*
*At least **two** small-appliance branch circuits*
must be included. 210.11(C)(1)

$1500 \times$ _____ = **2**
(minimum of two)

3 Laundry Branch Circuit(s) *220.82(B)(2)*
*At least **one** laundry branch circuit*
must be included. 210.11(C)(2)

$1500 \times$ _____ = **3**
(minimum of one)

4 Appliances *220.82(B)(3)* and *(4)*
*Use the nameplate rating of **all**
appliances (fastened in place,
permanently connected, or
connected to a specific circuit),
ranges, ovens, cooktops, motors,
and clothes dryers.
Convert any nameplate rating
given in amperes to volt-amperes
by multiplying the amperes
by the rated voltage.*

*Do not include any heating
or air-conditioning
equipment in this section.*

Total volt-amperes
of all appliances
LISTED BELOW **4**

water heater/ _____	clothes dryer/ _____	range / _____
dishwasher / _____	disposer / _____	/ _____
/ _____	/ _____	/ _____
/ _____	/ _____	/ _____
/ _____	/ _____	/ _____

5 Apply *220.82(B)* demand factor to the total of lines 1 through 4.

5

_____ $- 10,000 =$ _____ $\times 40\% =$ _____ $+ 10,000 =$
(total of lines 1 through 4)

6 Heating or Air-Conditioning System *220.82(C)*
*Use the nameplate ratings in volt-amperes for
all applicable systems in lines **a** through **e**.*

a) Air-conditioning and cooling systems, including heat pumps without
any supplemental electric heating:

_____ $\times 100\% =$ **a)**

b) Electric thermal storage and other heating systems where
the usual load is expected to be continuous at full nameplate
value. *Systems qualifying under this selection shall not be
figured under any other selection in 220.82(C).*

_____ $\times 100\% =$ **b)**

c) Supplemental electric heating equipment for heat-pump systems. Include the
heat-pump compressor(s) at 100%. *If the heat-pump compressor is prevented
from operating with the supplemental heat,
omit the compressor.*

_____ $\times 65\% =$ **c)**

d) Electric space-heating equipment, if fewer than four
separately controlled units:

_____ $\times 65\% =$ **d)**

e) Electric space-heating equipment, if four or more
separately controlled units:

_____ $\times 40\% =$ **e)**

7 Total Volt-Ampere
Demand Load:

_____ $+$ _____ = **7**
(largest VA rating from lines 6a through 6e)   (line 5)

8 Minimum Amperes
*Divide the total
volt-amperes
by the voltage.*

_____ $\div$ _____ = **8**
(line 7)   (voltage)   (minimum amperes)

9 Minimum Size
Service or
Feeder *240.6(A)* **9**
(minimum is 100 amperes)

10 Size the Service or Feeder Conductors.
*Use 310.15(B)(7) to find the service conductors up to 400 amperes.
Ratings in excess of 400 amperes shall comply with Table 310.15(B)(16)
310.15(B)(7) also applies to feeder conductors supplying the entire load.*

**Minimum Size
Conductors** **10**

11 Size the Neutral Conductor. *220.61*
Note: There is no optional method for calculating the neutral conductor.
*310.15(B)(7) states that the neutral service or feeder conductor
can be smaller than the ungrounded ("hot") conductors, provided the
requirements of 215.2, 220.61, and 230.42 are met.
250.24(C)(1) states that the grounded conductor shall not be smaller
than specified in Table 250.102(C)(1).*

**Minimum Size
Neutral
Conductor** **11**

12 Size the Grounding Electrode Conductor.
Use line 10 to find the grounding electrode conductor in Table 250.66.
Size the Equipment Grounding Conductor (for Feeder). *250.122*
*Use line 9 to find the equipment grounding conductor in Table 250.122.
Equipment grounding conductor types are listed in 250.118.*

**Minimum Size
Grounding
Electrode
Conductor** **12**

Lines 1 through 3—General Lighting and Receptacle Loads

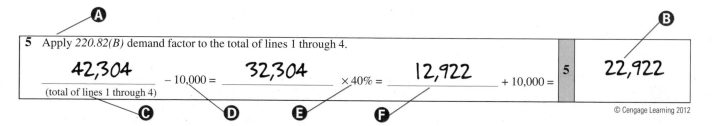

A In the **optional method,** lines 1, 2, and 3 are calculated exactly as described in the **standard method.** (See explanations on pages 157 and 158.)

1 General Lighting and Receptacle Loads *220.82(B)(1)* *Do not include open porches, garages, or unused or unfinished spaces not adaptable for future use.*	3 × ___**1855**___ (sq ft using outside dimensions)	=	**1**	**5565**
2 Small-Appliance Branch Circuits *220.82(B)(2)* *At least **two** small-appliance branch circuits must be included. 210.11(C)(1)*	1500 × ___**4**___ (minimum of two)	=	**2**	**6000**
3 Laundry Branch Circuit(s) *220.82(B)(2)* *At least **one** laundry branch circuit must be included. 210.11(C)(2)*	1500 × ___**2**___ (minimum of one)	=	**3**	**3000**

© Cengage Learning 2012

Line 4—Appliances

A This section includes appliances (fastened in place, permanently connected, or connected to a specific circuit), ranges, wall-mounted ovens, counter-mounted cooking units, clothes dryers, and water heaters »*220.82(B)(3)*«. Include all permanently connected motors not already included »*220.82(B)(4)*«.

B Heating and air-conditioning systems are *not* included in this calculation.

C Use the exact nameplate rating of the clothes dryer, even if less than 5000 volt-amperes.

D Do not apply the demand factors in Table 220.55 to household cooking equipment (ranges, cooktops, and ovens). List the nameplate ratings as they appear.

4 Appliances *220.82(B)(3) and (4)* *Use the nameplate rating of **all** appliances (fastened in place, permanently connected, or connected to a specific circuit), ranges, ovens, cooktops, motors, and clothes dryers. Convert any nameplate rating given in amperes to volt-amperes by multiplying the amperes by the rated voltage.*	*Do not include any heating or air-conditioning equipment in this section.* water heater/ **4500** dishwasher / **1200** compactor / **900** ___ / ___ /	Total volt-amperes of all appliances LISTED BELOW clothes dryer/ **5500** disposer / **1127** 3 attic fans/ **1512** ___ / ___ /	**4** range / cooktop / **7000** oven / **6000** ___ / ___ /	**27,739**

© Cengage Learning 2012

Line 5—Applying Demand Factors to the Total of Lines 1 through 4

A This demand factor is found in 220.82(B).

B This is the newly calculated load including everything except the heating and air-conditioning systems.

C (5565 + 6000 + 3000 + 27,739)

D Because the first 10,000 volt-amperes are calculated at 100%, subtract 10,000 here.

E The remainder of the load is calculated at 40%.

F The result on line 5 has been rounded up from 12,921.6.

5 Apply *220.82(B)* demand factor to the total of lines 1 through 4.					
___**42,304**___ (total of lines 1 through 4)	− 10,000 =	___**32,304**___	× 40% =	___**12,922**___	+ 10,000 =	**5**	**22,922**

© Cengage Learning 2012

Line 6—Heating or Air-Conditioning Systems

Ⓐ The air-conditioner compressor (16.6 × 230 = 3818 VA), condenser fan motor (2 × 115 = 230 VA), and blower motor (3.2 × 115 = 368 VA) have a combined load of 4416 volt-amperes.

Ⓑ The electric heat (3 × 5000 = 15,000) added to the blower motor (3.2 × 115 = 368) is 15,368 volt-amperes.

> **NOTE**
>
> The total volt-ampere rating of one, two, or three separately controlled heating units is multiplied by 65%. The total volt-ampere rating of four (or more) separately controlled heating units is multiplied by 40%.

Ⓐ

Ⓑ

6	Heating or Air-Conditioning System *220.82(C)* *Use the nameplate ratings in volt-amperes for all applicable systems in lines **a** through **e.***		**c)**	Supplemental electric heating equipment for heat-pump systems. Include the heat-pump compressor(s) at 100%. *If the heat-pump compressor is prevented from operating with the supplemental heat, omit the compressor.* _____ × 65% =	**c)**
	a) Air-conditioning and cooling systems, including heat pumps without any supplemental electric heating: _____ **4416** _____ × 100% =	**a)** **4416**	**d)**	Electric space-heating equipment, if fewer than four separately controlled units: _____ **15,368** _____ × 65% =	**d)** **9989**
	b) Electric thermal storage and other heating systems where the usual load is expected to be continuous at full nameplate value. *Systems qualifying under this selection shall not be figured under any other selection in 220.82(C).* _____ × 100% =	**b)**	**e)**	Electric space-heating equipment, if four or more separately controlled units: _____ × 40% =	**e)**

Line 7—Total Volt-Ampere Demand Load

Ⓐ The largest volt-ampere rating of the heating or air-conditioning system(s), after application of demand factors.

Ⓑ The volt-ampere demand load from line 5.

Ⓒ The total volt-ampere load calculated by the optional method. (For comparison, this one-family dwelling, calculated by the standard method, is 43,577 volt-amperes.)

Ⓐ **Ⓑ** **Ⓒ**

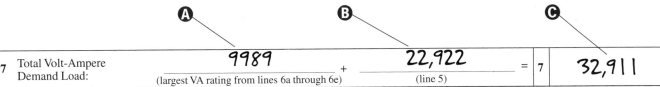

| 7 | Total Volt-Ampere Demand Load: | **9989** (largest VA rating from lines 6a through 6e) | + | **22,922** (line 5) | = | 7 | **32,911** |

Lines 8 and 9—Minimum Service or Feeder

Ⓐ The volt-ampere load from line 7

Ⓑ The source voltage

Ⓒ The fraction of an ampere (.13) has been dropped.

Ⓓ The overcurrent protection chosen for the service (or feeder) must be higher than the number found on line 8.

Standard ampere ratings for fuses and circuit breakers are listed in 240.6(A).

Ⓔ The next standard size fuse or breaker above 137 amperes is 150. (By the standard method, this same dwelling requires a 200-ampere service.)

Ⓐ **Ⓑ** **Ⓒ** **Ⓓ** **Ⓔ**

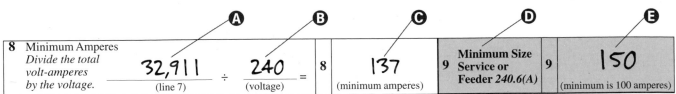

| 8 | Minimum Amperes *Divide the total volt-amperes by the voltage.* | **32,911** (line 7) | ÷ | **240** (voltage) | = | 8 | **137** (minimum amperes) | 9 | **Minimum Size Service or Feeder 240.6(A)** | 9 | **150** (minimum is 100 amperes) |

Line 10—Minimum Size Conductors

Ⓐ For a service rated 100 through 400 A, the service conductors supplying the entire load associated with a one-family dwelling, or the service conductors supplying the entire load associated with an individual dwelling unit in a two-family or multifamily dwelling, shall be permitted to have an ampacity not less than 83 percent of the service rating ≫*310.15(B)(7)(1)*≪.

Ⓑ Section 310.15(B)(7) also applies to feeder conductors that supply the entire load associated with a one-family dwelling ≫*310.15(B)(7)(2)*≪.

Ⓒ The maximum service or feeder rating permitted when using 310.15(B)(7) is 400 amperes.

Ⓓ Multiply the minimum size service or feeder rating (Line 8) by 83 percent and then select a conductor from Table 310.15(B)(16).

150 × 83% = 124.5 = 125 amperes

In accordance with the 75°C column of Table 310.15(B)(16), the ampacity of a 1 AWG copper conductor is 130 amperes.

Ⓔ 150 × 83% = 124.5 = 125 amperes

In accordance with the 75°C column of Table 310.15(B)(16), the ampacity of a 2/0 AWG aluminum conductor is 135 amperes.

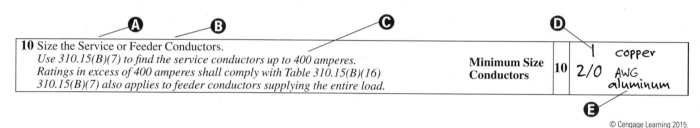

© Cengage Learning 2015.

Line 11—Neutral Conductor

Ⓐ The neutral conductor must be calculated by the standard method.

Ⓑ Review the complete neutral conductor explanation on page 162, since the calculation method is the same.

> **NOTE**
>
> There is no optional method for calculating the neutral conductor.

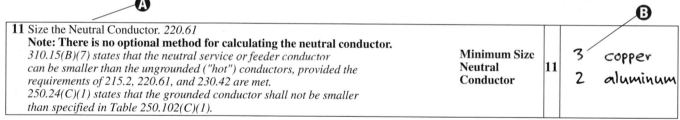

© Cengage Learning 2015.

Line 12—Grounding Electrode Conductor

Ⓐ The minimum size grounding electrode conductor is 6 AWG copper (or 4 AWG aluminum) ≫*Table 250.66*≪.

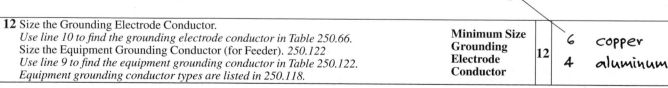

© Cengage Learning 2012

The following page shows the completed one-family dwelling load calculation according to the optional method.

Optional Method Load Calculation for One-Family Dwellings

1	General Lighting and Receptacle Loads *220.82(B)(1)* *Do not include open porches, garages, or unused or* *unfinished spaces not adaptable for future use.*	$3 \times$ __1855__ (sq ft using outside dimensions)	= **1**	**5565**
2	Small-Appliance Branch Circuits *220.82(B)(2)* *At least **two** small-appliance branch circuits* *must be included. 210.11(C)(1)*	$1500 \times$ __4__ (minimum of two)	= **2**	**6000**
3	Laundry Branch Circuit(s) *220.82(B)(2)* *At least **one** laundry branch circuit* *must be included. 210.11(C)(2)*	$1500 \times$ __2__ (minimum of one)	= **3**	**3000**

| **4** | Appliances *220.82(B)(3)* and *(4)*
*Use the nameplate rating of **all**
appliances (fastened in place,
permanently connected, or
connected to a specific circuit),
ranges, ovens, cooktops, motors,
and clothes dryers.
Convert any nameplate rating
given in amperes to volt-amperes
by multiplying the amperes
by the rated voltage.* | *Do not include any heating
or air-conditioning
equipment in this section.* | Total volt-amperes
of all appliances
LISTED BELOW | **4** | **27,739** |

water heater / 4500   clothes dryer / 5500   range /

dishwasher / 1200   disposer / 1127   cooktop / 7000

compactor / 900   3 attic fans / 1512   oven / 6000

____ / ____   ____ / ____   ____ / ____

____ / ____   ____ / ____   ____ / ____

5 Apply *220.82(B)* demand factor to the total of lines 1 through 4.

__42,304__ $- 10,000 =$ __32,304__ $\times 40\% =$ __12,922__ $+ 10,000 =$ **5**   **22,922**
(total of lines 1 through 4)

6 Heating or Air-Conditioning System *220.82(C)* *Use the nameplate ratings in volt-amperes for all applicable systems in lines **a** through **e**.*	**c)** Supplemental electric heating equipment for heat-pump systems. Include the heat-pump compressor(s) at 100%. *If the heat-pump compressor is prevented from operating with the supplemental heat, omit the compressor.* _____ $\times 65\% =$ **c)**

a) Air-conditioning and cooling systems, including heat pumps without any supplemental electric heating:

__4416__ $\times 100\% =$ **a)** **4416**

d) Electric space-heating equipment, if fewer than four separately controlled units:

__15,368__ $\times 65\% =$ **d)** **9989**

b) Electric thermal storage and other heating systems where the usual load is expected to be continuous at full nameplate value. *Systems qualifying under this selection shall not be figured under any other selection in 220.82(C).*

_____ $\times 100\% =$ **b)**

e) Electric space-heating equipment, if four or more separately controlled units:

_____ $\times 40\% =$ **e)**

7 Total Volt-Ampere Demand Load: __9989__ + __22,922__ = **7** **32,911**
(largest VA rating from lines 6a through 6e)   (line 5)

8 Minimum Amperes
Divide the total volt-amperes by the voltage. __32,911__ $\div$ __240__ = **8** **137**
(line 7)   (voltage)   (minimum amperes)

9 Minimum Size Service or Feeder *240.6(A)* **9** **150**
(minimum is 100 amperes)

10 Size the Service or Feeder Conductors.
Use 310.15(B)(7) to find the service conductors up to 400 amperes.
Ratings in excess of 400 amperes shall comply with Table 310.15(B)(16)
310.15(B)(7) also applies to feeder conductors supplying the entire load.
Minimum Size Conductors **10** **1 copper**
2/0 AWG aluminum

11 Size the Neutral Conductor. *220.61*
Note: There is no optional method for calculating the neutral conductor.
310.15(B)(7) states that the neutral service or feeder conductor
can be smaller than the ungrounded ("hot") conductors, provided the
requirements of 215.2, 220.61, and 230.42 are met.
250.24(C)(1) states that the grounded conductor shall not be smaller
than specified in Table 250.102(C)(1).
Minimum Size Neutral Conductor **11** **3 copper**
2 aluminum

12 Size the Grounding Electrode Conductor.
Use line 10 to find the grounding electrode conductor in Table 250.66.
Size the Equipment Grounding Conductor (for Feeder). 250.122
Use line 9 to find the equipment grounding conductor in Table 250.122.
Equipment grounding conductor types are listed in 250.118.
Minimum Size Grounding Electrode Conductor **12** **6 copper**
4 aluminum

Summary

- A dwelling unit load calculation, for general lighting and receptacles, is determined by total square footage.
- Table 220.12 requires a unit load of 3 volt-amperes per square foot.
- At least one laundry and two small-appliance branch circuits are required when performing a service (or main power feeder) calculation.
- Laundry and small-appliance branch circuits are calculated using a rating of at least 1500 volt-amperes each.
- Table 220.42 demand factors are applied to the general lighting, small appliance, and laundry loads.
- The total volt-ampere rating of four or more fastened-in-place appliances is multiplied by 75%.

- Demand loads for household cooking appliances (over 1¾ kW rating) are found in Table 220.55.
- The heating and air-conditioning loads are compared, and the larger of the two is used.
- For heat pumps, include the compressor and the maximum amount of electric heat that can be energized while the compressor is running.
- Two load calculation methods are provided in Article 220, standard and optional.
- The neutral conductor can be smaller than the ungrounded (hot) conductors, provided the requirements of 215.2, 220.61, and 230.42 are met.
- Table 250.66 is used to size grounding electrode conductors.

Unit 8 Competency Test

NEC Reference	Answer	
_____	_____	1. It shall be permissible to apply a demand factor of _____% to the nameplate rating load of _____ or more appliances fastened in place, other than electric ranges, clothes dryers, space-heating equipment, or air-conditioning equipment, that are served by the same feeder in a one-family dwelling.
_____	_____	2. What is the lighting demand load (before derating) for a house with outside dimensions of 30 ft × 48 ft on the first floor and 22 ft × 42 ft on the second floor?
_____	_____	3. A homeowner wants to add twelve general-purpose receptacle outlets to the electrical drawings before the house is built. What additional load, in volt-amperes, will they add to the service-load calculation?
_____	_____	4. In each dwelling unit, the feeder load shall be computed at _____ volt-amperes for each 2-wire small-appliance branch circuit required by 210.11(C)(1).
_____	_____	5. What is the demand load for the service for two 3½-kW wall-mounted ovens and one 5-kW counter-mounted cooking unit? All of the appliances are supplied from a single branch circuit and are located in the same room of a one-family dwelling. a) 12 kW b) 9.6 kW c) 8 kW d) 6.6 kW
_____	_____	6. A 4.5-kVA, 240-volt clothes dryer contributes _____ amperes to the neutral load, when calculating the service by the standard method.
_____	_____	7. A one-family dwelling contains the following appliances: a 1-kVA, 115-volt dishwasher; a ¼-HP, 115-volt in-sink waste disposer; and a ⅓-HP, 115-volt trash compactor. Using the standard method, what is the load contribution, in volt-amperes, for these appliances?
_____	_____	8. For one-family dwellings, the service disconnecting means must have a rating of at least _____ amperes, 3-wire.
_____	_____	9. A one-family dwelling has a 175-ampere, 240-volt service that is fed with 75°C (167°F) conductors. What is the minimum size copper grounding electrode conductor required?

NEC Reference   Answer

_____   _____   10. A one-family dwelling has six 3-kW electric wall heaters (with individual thermostats) and five room air conditioners. Two air conditioners are rated 10.5 amperes at 230 volts, and the others are rated 7.2 amperes at 115 volts. A service calculation is being performed, using the optional method. How many volt-amperes should be included (after demand factors) for the heating and air conditioning? (Assume heater kW ratings equivalent to kVA.)

_____   _____   11. In each dwelling unit, a feeder load of at least _____ volt-amperes must be included for each 2-wire laundry branch circuit installed as required by 210.11(C)(2).

_____   _____   12. A 12-kW, 240-volt range contributes _____ watts to the load when calculating the service by the standard method.

Questions 13 through 21 are based on a 120/240-volt, 3-wire, single-phase one-family dwelling containing the following:

Floor area . . . . 2500 ft²		Clothes dryer . . . . 4 kW, 240-volt	
Range. . . . . 12 kW, 240-volt		Electric heat (two banks at 3 kW each). . . . 6 kW, 240-volt	
Water heater . . . . 4 kW, 240-volt		Air handler (blower motor) . . . . ¼ HP, 115-volt	
Dishwasher . . . . 1.5 kW, 120-volt		Air-conditioner compressor . . . . 5 HP, 230-volt	
In-sink waste disposer . . . . ⅓ HP, 115-volt		Condenser fan motor. . . . . ⅙ HP, 115-volt	

Assume water heater, clothes dryer, range, and electric heat kW ratings equivalent to kVA.

Using Part B (the standard method) of Article 220:

_____   _____   13. What is the minimum ampere rating for the service?

_____   _____   14. What is the minimum size service?

_____   _____   15. What is the minimum size THWN copper conductor that can be installed as the service (ungrounded) conductor?

_____   _____   16. What is the minimum size grounding electrode conductor?

Using the Optional Method:

_____   _____   17. What is the minimum ampere rating for the service?

_____   _____   18. What is the minimum size service?

_____   _____   19. What is the minimum size THWN copper conductor that can be installed as the service (ungrounded) conductor?

_____   _____   20. What is the minimum size grounding electrode conductor?

Finding the Neutral Load:

_____   _____   21. What is the minimum size THWN copper conductor that can be installed as the neutral conductor?

Services and Electrical Equipment

Objectives

After studying this unit, the student should:

▶ be able to determine adequate strength for a mast supporting service-drop conductors.

▶ understand service-entrance cable provisions.

▶ know the definition of a service lateral and understand the applicable provisions.

▶ be familiar with minimum wiring clearance requirements (vertical, horizontal, etc.) for service and outside wiring.

▶ have a good understanding of the amount of three-dimensional working space required around service equipment, panelboards, electric equipment, etc.

▶ thoroughly understand panelboards, cabinets, and cutout boxes.

▶ be familiar with the multitude of provisions relating to circuit breakers and fuses.

▶ know and understand different aspects of service equipment and panelboard installations.

▶ know how to calculate conductor sizes in accordance with 310.15(B)(7).

▶ understand that electric equipment must be installed in a neat and professional manner.

▶ have a detailed understanding of grounding and bonding procedures.

▶ be able to properly position grounded and grounding conductors in remote panelboards (subpanels).

▶ have an extensive understanding of the grounding system as a whole.

▶ understand how to ground and bond panelboards on the supply side as well as the load side of the service disconnecting means.

▶ be able to prevent objectionable current flow in grounding conductors.

▶ understand the grounding requirements for panelboards installed in separate buildings or structures.

Introduction

Because the service is the heart of the electrical system, the importance of understanding this unit is enormous. Many of the regulations learned in this unit are applicable in all types of occupancies, not merely single-family dwellings. Unit 9 is divided into five sections. The first, "Service-Entrance Wiring Methods," addresses both overhead and underground service entrance provisions. The section entitled "Service and Outside Wiring Clearances" covers the clearances required above roofs and final grade. Other clearances (horizontal, vertical, and diagonal) are shown for windows, doors, balconies, platforms, etc. Next, the "Working Space around Equipment" section illustrates common working space requirements. "Service Equipment and Panelboards" explains panelboards, cutout boxes, and cabinets. Specific provisions pertaining to circuit breakers, fuses, and circuit breaker panels are also covered. "Grounding," the last section, explains in detail Article 250 provisions relating to service equipment and panelboards. Refer to Unit 9 often when studying (or installing) services, whether one-family or multifamily dwellings, or commercial and even industrial buildings.

Since many local jurisdictions and electric utility companies have supplemental requirements, exercise caution in compiling a complete set of service installation requirements. Examples of these requirements include, but are not limited to, restrictions on the type of conduit used for service mast; minimum conduit trade size; specific strapping methods; restrictions on the use of service-entrance cables; special underground service provisions; limitations on the maximum distance service-entrance conductors can run within a building or structure; and various grounding provisions. Obtain a copy of these additional rules (if any) for your area.

SERVICE-ENTRANCE WIRING METHODS

Service Mast as Support

A Service-entrance and overhead service conductors shall be arranged so that water shall not enter the service raceway or equipment ≫ *230.54(G)* ≪.

B It is not required (where approved) that rigid (and intermediate) metal conduit be securely fastened within 3 ft (900 mm) of the service head for above-the-roof termination of a mast ≫ *342.30(A)(3) and 344.30(A)(3)* ≪.

C Open conductors must be attached to fittings identified for use with service conductors or to noncombustible, nonabsorbent insulators securely attached to the building or other structure ≫ *230.27* ≪.

D The service mast shall be of adequate strength or be supported by braces or guys to withstand safely the strain imposed by the service-drop or overhead service conductors ≫ *230.28(A)* ≪.

E Hubs intended for use with a conduit that serves as a service mast shall be identified for use with service-entrance equipment ≫ *230.28(A)* ≪.

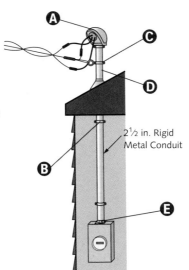

2½ in. Rigid Metal Conduit

© Cengage Learning 2015.

> ### CAUTION
>
> Service-drop or overhead service conductors shall not be attached to a service mast between a weatherhead or the end of the conduit and a coupling, where the coupling is located above the last point of securement to the building or other structure or is located above the building or other structure ≫*230.28(B)*◄.

Service Heads (Weather Heads)

A Service raceways must be equipped with a service head at the point of connection to service-drop or overhead service conductors. The service head shall be listed for use in wet locations ≫ *230.54(A)* ≪.

B Service-entrance and overhead service conductors shall be arranged so that water shall not enter the service raceway or equipment ≫ *230.54(G)* ≪.

C Conductors of different potential must be brought out of service heads through separately bushed openings ≫ *230.54(E)* ≪.

D An insulated grounded conductor 4 AWG or larger must be identified either by a continuous white or gray outer finish or by three continuous white or gray stripes on other than green insulation along its entire length or by a distinctive white or gray marking at its terminations during installation. This marking must completely encircle the conductor or insulation ≫ *200.6(B)* ≪.

E Conductors must have adequate mechanical strength and sufficient ampacity to carry the current for the load (as computed in accordance with Article 220) ≫ *230.23(A)* ≪.

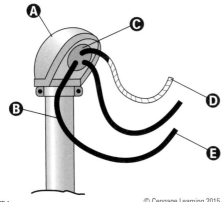

© Cengage Learning 2015.

Service Mast Attachments

A Service drop is defined as the overhead conductors between the utility electric supply system and the service point ≫ *Article 100* ≪.

B Nothing except power service-drop or overhead service conductors can be attached to the service mast ≫ *230.28* ≪.

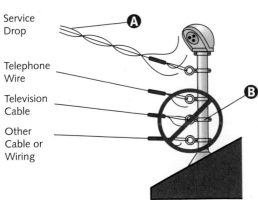

Service Drop

Telephone Wire

Television Cable

Other Cable or Wiring

© Cengage Learning 2012

Service Drops Attached to Buildings

Ⓐ Open conductors must be attached to fittings identified for use with service conductors, or to noncombustible, nonabsorbent insulators that are securely attached to the building (or other structure) ≫230.27≪.

Ⓑ Electrical metallic tubing (EMT) couplings and connectors shall be made up tight. If installed in wet locations, they shall be listed for use in wet locations ≫358.42 and 314.15≪.

Ⓒ Service heads must be located above the point where service-drop or overhead service conductors are attached to the building or other structure ≫230.54(C)≪.

Ⓓ Where it is impracticable to locate the service head above the point of attachment, the service head must be located within 24 in. (600 mm) of that point ≫230.54(C) Exception≪.

Ⓔ Because this service mast does not support service-drop conductors, another wiring method (EMT) is used. Other allowable service-entrance methods for 1000 volts, nominal, or less are listed in 230.43.

Ⓕ Electric equipment must be firmly secured to the surface on which it is mounted. Wooden plugs driven into holes in masonry, concrete, plaster, etc., are not acceptable ≫110.13(A)≪.

> **NOTE**
>
> Overhead service conductors are defined as the overhead conductors between the service point and the first point of connection to the service-entrance conductors at the building or other structure ≫**Article 100**≪.

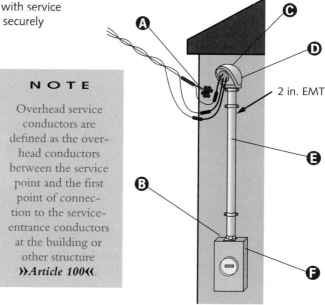

2 in. EMT

© Cengage Learning 2012

Support of Overhead Conductors

Ⓐ Vegetation (such as trees) shall not be used to support overhead conductor spans, except for temporary wiring in accordance with Article 527 ≫225.26≪.

Ⓑ Drip loops must be formed on each individual conductor. To prevent moisture penetration, service-entrance conductors are connected to the service-drop or overhead service conductors either (1) below the level of the service head or (2) below the level of the termination of the service-entrance cable sheath ≫230.54(F)≪.

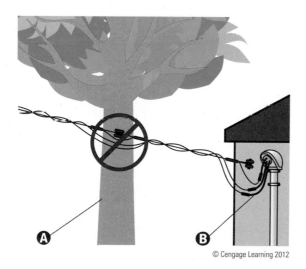

© Cengage Learning 2012

Goosenecks in Service-Entrance Cables

Ⓐ Bends in service-entrance cable must be made so that the cable is not damaged. In addition, the radius of the curve of the inner edge of any bend must be at least five times the diameter of the cable ≫338.24≪.

Ⓑ Goosenecks in service-entrance cables must be located above the point of attachment of the service-drop or overhead service conductors to the building (or other structure) ≫230.54(C)≪.

Ⓒ Type SE cable can be formed in a gooseneck and taped with a self-sealing weather-resistant thermoplastic ≫230.54(B) Exception≪.

Ⓓ Drip loops must be formed on all individual conductors ≫230.54(F)≪.

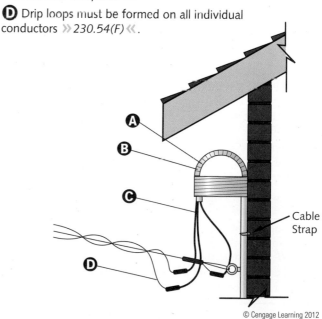

Cable Strap

© Cengage Learning 2012

Service-Entrance Cable

A Service-entrance cables must be fitted with a raintight service head, except for Type SE cable that has been formed into a gooseneck and taped with a self-sealing, weather-resistant thermoplastic ⟫230.54(B)⟪.

B If Type SE or USE cable consists of at least two conductors, one can be uninsulated ⟫338.100⟪.

C Service-entrance cable is a single conductor or multiconductor assembly (with or without an overall covering), primarily used for services. Type SE has a flame-retardant, moisture-resistant covering ⟫338.2⟪.

D Service-entrance cables must be held securely in place ⟫230.54(D)⟪.

E Service head conductors of different potential must be brought out through separately bushed openings ⟫230.54(E)⟪.

F Service-entrance cables must be supported by straps (or other approved means) within 12 in. (300 mm) of every service head or gooseneck ⟫230.51(A)⟪.

G Entrance cables must be supported at intervals not to exceed 30 in. (750 mm) ⟫230.51(A)⟪.

H Entrance cables must be supported within 12 in. (300 mm) of every raceway or enclosure connection ⟫230.51(A)⟪.

I Service-entrance and overhead service conductors must be arranged so that water will not enter service raceway or equipment ⟫230.54(G)⟪.

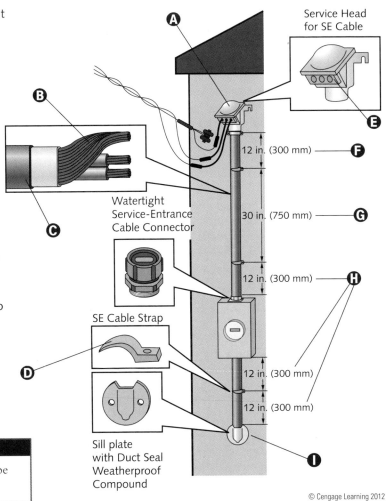

Service Head for SE Cable

12 in. (300 mm) — **F**

30 in. (750 mm) — **G**

12 in. (300 mm) — **H**

12 in. (300 mm)

12 in. (300 mm)

Watertight Service-Entrance Cable Connector

SE Cable Strap

Sill plate with Duct Seal Weatherproof Compound

© Cengage Learning 2012

CAUTION
Where subject to physical damage, service cables must be appropriately protected ⟫230.50⟪.

Conductors Considered Outside of a Building

A Building or structure interior wall

B An interior service disconnecting means must be installed as close as possible to the point of entrance of the service conductors and must be readily accessible. A raceway beneath a 2-in. (50-mm) concrete slab lies outside the building. Under this condition, the service disconnecting means could be installed on (or in) an interior wall of the building. (See Caution.)

C Conductors installed under 2 in. (50 mm) or more of concrete beneath a building (or other structure) are considered outside ⟫230.6(1)⟪.

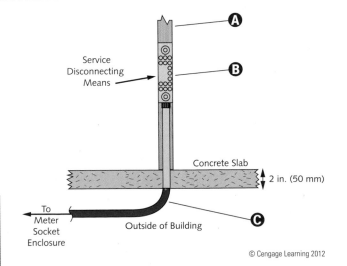

Service Disconnecting Means

Concrete Slab

2 in. (50 mm)

To Meter Socket Enclosure

Outside of Building

© Cengage Learning 2012

CAUTION
Although not stipulated in the *NEC*, some local jurisdictions have adopted specific distance requirements for service conductors. (For example, the length for each service conductor between the meter socket enclosure and the service disconnecting means could be limited to only 6 ft [1.8 m].)

Underground Services

Ⓐ Underground service conductors shall be insulated for the applied voltage ≫*230.30(A)* ≪. Grounded conductors qualifying under 230.30(A) Exception can be uninsulated (bare).

Ⓑ Metal raceways containing service conductors must be connected to the grounded system conductor if the electrical system is grounded ≫*250.80* ≪.

Ⓒ Wiring methods permitted for services of 600 volts, nominal, or less are listed in 230.43.

Ⓓ Underground service conductors are defined as the underground conductors between the service point and the first point of connection to the service-entrance conductors in a terminal box, meter, or other enclosure, inside or outside the building wall ≫*Article 100* ≪. Where there is no terminal box, meter, or other enclosure, the point of connection is considered to be the point of entrance of the service conductors into the building ≫*Informational Note to Service Conductors, Underground* ≪.

Ⓔ Direct-buried conduit (or other raceways) must be installed in adherence with Table 300.5 minimum cover requirements ≫*300.5(A)* ≪.

> **N O T E**
>
> Underground service conductors must have adequate mechanical strength and sufficient ampacity to carry the current for the load (as calculated in accordance with Article 220) ≫***230.31(A)***≪.

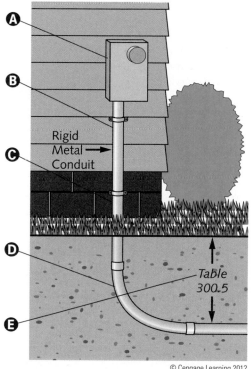

Rigid Metal Conduit

Table 300.5

© Cengage Learning 2012

Underground Service-Entrance Cable

Although Type USE cable identified for underground use has a moisture-resistant covering, a flame-retardant covering is not required ≫*338.2* ≪.

If Type USE cable contains at least two conductors, one can be uninsulated ≫*338.100* ≪.

Ⓐ Underground service conductors must be protected against damage in accordance with 300.5. Service conductors entering a building or other structure must be installed in accordance with 230.6 or must be protected by an approved raceway wiring method identified in 230.43 ≫*230.32* ≪.

Ⓑ Where the conductors or cables emerge as a direct-burial wiring method, a bushing (or fitting) must be installed to protect conductors from abrasion on the end of the conduit (or tubing) that terminates underground. A seal providing the same level of protection can be used instead of a bushing ≫*300.5(H)* and *300.15(C)* ≪.

Ⓒ Type USE cable used for service laterals can emerge aboveground at outside terminations (in meter bases or other enclosures) provided 300.5(D) protection requirements are met ≫*338.12(B)(2)* ≪.

Ⓓ When installed, direct-buried cable must meet Table 300.5 minimum cover requirements ≫*300.5(A)* ≪.

Ⓔ Cabled, single-conductor, Type USE constructions, recognized for underground use, may have a bare copper conductor cabled with the assembly. Underground-approved Type USE single, parallel, or cabled conductor assemblies may have a bare copper concentric conductor applied. An outer overall covering is not required for these constructions ≫*338.100* ≪.

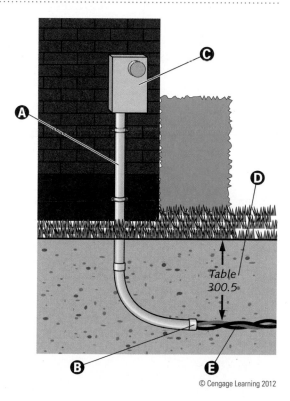

Table 300.5

© Cengage Learning 2012

Wiring Methods

A Service-entrance conductors must be installed according to applicable *Code* requirements for the type of wiring method used and limited to the following methods: (1) open wiring on insulators; (2) Type IGS cable; (3) rigid metal conduit (RMC); (4) intermediate metal conduit (IMC); (5) electrical metallic tubing (EMT); (6) electrical nonmetallic tubing (ENT); (7) service-entrance cables; (8) wireways; (9) busways; (10) auxiliary gutters; (11) rigid polyvinyl chloride conduit (PVC); (12) cablebus; (13) Type MC cable; (14) mineral-insulated, metal-sheathed cable, Type MI; (15) flexible metal conduit (FMC) or liquidtight flexible metal conduit (LFMC) not longer than 6 ft (1.8 m) between raceways, or between raceway and service equipment, with a supply-side bonding jumper routed with the FMC or LFMC according to 250.102(A), (B), (C), and (E) provisions; (16) liquidtight flexible nonmetallic conduit (LFNC); (17) high-density polyethylene conduit (HDPE); (18) nonmetallic underground conduit with conductors (NUCC); or (19) reinforced thermosetting resin conduit (RTRC) »*230.43*«.

> ### NOTE
>
> The wiring method of choice must be installed according to provisions for that particular method. For example, Type rigid metal conduit (RMC) must be installed according to Article 344 provisions.

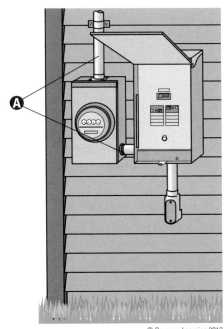

© Cengage Learning 2012

SERVICE AND OUTSIDE WIRING CLEARANCES

Accessibility to Pedestrians

A The point of attachment of the overhead service conductors to a building (or other structure) must meet 230.9 and 230.24 minimum clearances. In all cases, this point of attachment must be at least 10 ft (3.0 m) above finished grade »*230.26*«.

B Overhead service conductors, accessible only to pedestrian traffic, must have a minimum 10-ft (3.0-m) vertical clearance above finished grade, sidewalks, platforms, or projections from which they might be reached. The voltage must not exceed 150 to ground »*230.24(B)(1)* and *225.18(1)*«.

C The 3-ft (900-mm) minimum requirement does not apply to conductors run above the top level of a window »*230.9(A) Exception* and *225.19(D)(1) Exception*«.

D The 3-ft (900-mm) clearance from windows found in 230.9 does not apply to raceways and service equipment.

E A drip loop's lowest point must have a minimum vertical clearance of 10 ft (3.0 m) »*230.24(B)(1)*«.

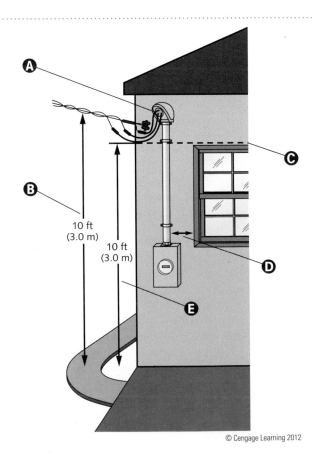

10 ft (3.0 m)

10 ft (3.0 m)

© Cengage Learning 2012

Vertical Clearances

Ⓐ A 10-ft (3.0-m) minimum vertical clearance is required from any platform or projection from which overhead conductors (of 150 volts, or less, to ground) might be reached ≫ *230.24(B)(1) and 225.18(1)* ≪ .

Ⓑ A minimum 12-ft (3.7-m) vertical clearance is required over residential property and driveways, and over commercial areas (not subject to truck traffic) provided conductor voltage does not exceed 300 volts to ground ≫ *230.24(B)(2) and 225.18(2)* ≪ .

Ⓒ Conductor voltage exceeding 300 volts to ground has a minimum vertical clearance of 15 ft (4.5 m) ≫ *230.24(B)(3) and 225.18(3)* ≪ .

Ⓓ Public streets, alleys, roads, parking areas subject to truck traffic, driveways on nonresidential property, and other land traversed by vehicles (such as cultivated, grazing, forest, and orchard) require a minimum 18-ft (5.5-m) vertical clearance ≫ *230.24(B)(4) and 225.18(4)* ≪ .

> **NOTE**
>
> Swimming pool overhead conductor clearances are listed in 680.8.

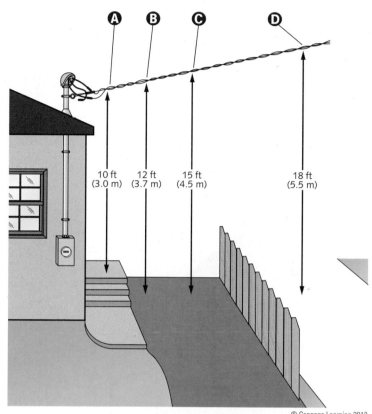

© Cengage Learning 2012

Clearance from Windows and Accessible Areas

Ⓐ Service conductors installed as open conductors or multi-conductor cable (having no overall outer jacket) must have a clearance of at least 3 ft (900 mm) from windows (designed to open), doors, porches, balconies, ladders, stairs, fire escapes, and similar locations ≫ *230.9(A) and 225.19(D)(1)* ≪ .

Ⓑ The 3-ft (900-mm) clearance requirement does not apply to conductors run above the top level of a window ≫ *230.9(A) Exception and 225.19(D)(1) Exception* ≪ .

Ⓒ The 3-ft (900-mm) clearance is not required for windows that are not designed to open.

Ⓓ Except for the area above the window, the 3-ft (900-mm) clearance applies to the entire perimeter of windows, doors, etc.

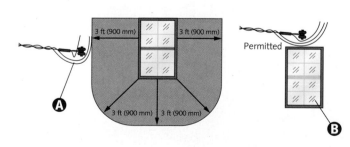
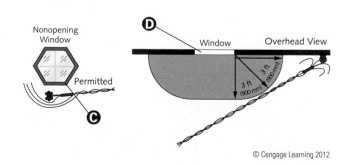

© Cengage Learning 2012

Conductors Obstructing Openings

A Overhead conductors must not obstruct building entrances and must not be installed beneath openings designed for passage of materials, such as openings in farm and commercial buildings ≫230.9(C) and 225.19(D)(3)≪.

> **NOTE**
>
> A minimum 18-ft (5.5-m) vertical clearance is required over land routinely traversed by vehicles (such as cultivated, grazing, forest, and orchard) ≫**230.24(B)(4)** and **225.18(4)**≪.

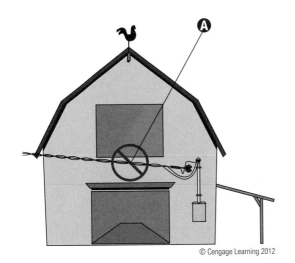

© Cengage Learning 2012

Conductors above Overhanging Portions of Roofs

A A maximum of 6 ft (1.8 m) of overhead service conductors pass over the roof.

B Conductors terminate in a through-the-roof raceway (or other approved support).

C The vertical clearance for overhead service conductors can be reduced to 18 in. (450 mm) provided all 230.24(A) Exception No. 3 requirements are met.

D The voltage between conductors does not exceed 300 volts.

E The horizontal overhead service conductor measurement shall not exceed 4 ft (1.2 m).

F This reduction in clearance applies to conductors passing above roof overhang only.

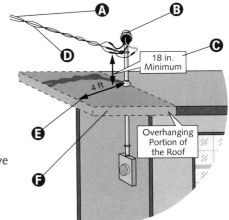

© Cengage Learning 2012

> **NOTE**
>
> If all the requirements of 225.19(A) Exception No. 3 are met, the vertical clearance for the overhead spans of open conductors, including multiconductor cables, can be reduced to 18 in. (450 mm).

Conductor Clearance above Flat Roofs

A Conductors must have a vertical clearance of at least 8 ft (2.4 m) above the roof surface ≫230.24(A) and 225.19(A)≪.

B Although overhead service conductors must not be readily accessible, they must comply with 230.24(A) through (E) for services not over 1000 volts, nominal ≫230.24≪.

C The service mast shall be of adequate strength or be supported by braces or guys to withstand safely the strain imposed by the service-drop or overhead service conductors ≫230.28≪.

D The vertical clearance above roof level must be maintained for a distance of at least 3 ft (900 mm) from all edges of the roof ≫230.24(A) and 225.19(A)≪.

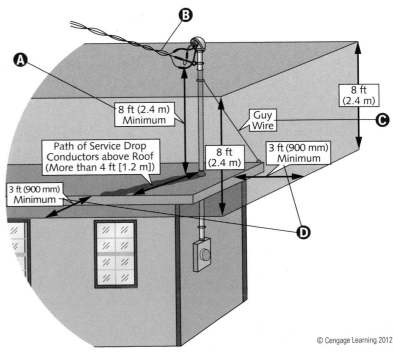

© Cengage Learning 2012

Conductors above Sloped Roofs

Ⓐ The voltage between the ungrounded (hot) conductors is 240 volts.

Ⓑ Where the voltage between conductors does not exceed 300, and the roof has a slope of not less than 4 in. in 12 in. (100 mm in 300 mm), a 3-ft (900-mm) clearance is permitted ≫ *230.24(A) Exception No. 2* and *225.19(A) Exception No. 2* ≪.

Ⓒ This roof, as denoted by this symbol, has a slope (or slant) of 6 in. in 12 in., which is greater than 4 in. in 12 in.

Ⓓ Unlike 230.24(B), which stipulates that "vertical clearance" also includes the drip loop, 230.24 provisions mention only the service-drop conductor. Because the end of the service-drop conductors could be part (or all) of the drip loop, caution is advised.

Ⓔ The horizontal service-drop measurement is less than 4 ft (1.2 m). However, the conductors pass over more than just the overhanging portion of this roof. Therefore, the clearance shall not be reduced to 18 in. (450 mm) ≫ *230.24(A) Exception No. 3* and *225.19(A) Exception No. 3* ≪.

Ⓕ Service equipment rated at 1000 volts or less must be marked as suitable for use as service equipment. All service equipment shall be listed. Individual meter socket enclosures are not considered service equipment ≫ *230.66* ≪.

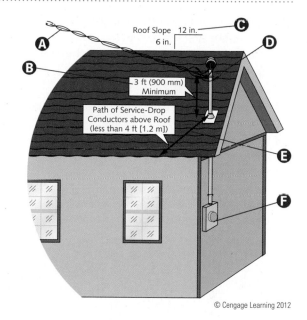

© Cengage Learning 2012

Various Roof Designs

Ⓐ Overhead service conductors above a roof with a slope of at least 4 in. in 12 in. (100 mm in 300 mm) have a minimum clearance of 3 ft (900 mm) ≫ *230.24(A) Exception No. 2* ≪.

Ⓑ This roof has a slope of 4 in. in 12 in.

Ⓒ The point of attachment is usually the lowest point of the service drop. If, however, the conductors run above the roof's highest point (ridge or peak), a taller service mast may be required.

Ⓓ This roof has a slope of 6 in. in 12 in. (more than 4 in. in 12 in.).

Ⓔ The supply voltage for all three overhead service conductor illustrations is 240 volts.

Ⓕ A service mast supporting service-drop conductors must be of adequate strength or be supported by braces (or guys) to safely withstand the strain imposed by the service drop ≫ *230.28* ≪.

Ⓖ Overhead service conductors above a roof with a slope less than 4 in. in 12 in. (100 mm in 300 mm) shall have a minimum clearance of 8 ft (2.5 m) ≫ *230.24(A)* ≪.

Ⓗ This roof has a slope of 3 in. in 12 in. (less than 4 in. in 12 in.).

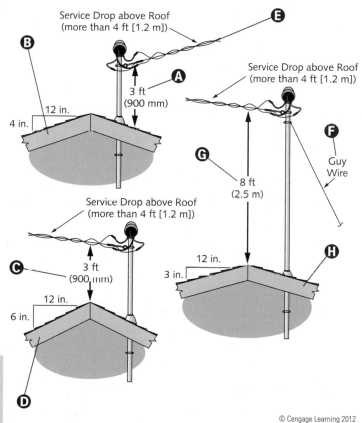

© Cengage Learning 2012

NOTE

Section 225.19 lists the vertical clearance for the overhead span of open conductors and open multiconductor cables. Because each of the illustrated service drops has a horizontal measurement of more than 4 ft (1.2 m), the amount of slope determines the minimum clearance requirement (either 3 ft [900 mm] or 8 ft [2.5 m]).

Clearance Reduction

Ⓐ No more than 6 linear ft (1.8 m) of overhead service conductors pass over the roof.

Ⓑ Voltage between conductors does not exceed 300 volts.

Ⓒ The horizontal overhead service conductor measurement does not exceed 4 ft (1.2 m).

Ⓓ Conductors terminate in a through-the-roof raceway (or other approved support).

Ⓔ The conductor's minimum vertical clearance is 18 in. (450 mm) where all 230.24(A) Exception No. 3 requirements have been met.

Ⓕ The reduction in clearance (to 18 in. [450 mm]) applies to sloped roofs as well as flat roofs where the service drop passes above the overhang portion of the roof only.

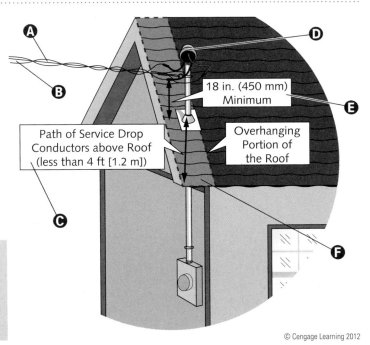

© Cengage Learning 2012

> ### NOTE
>
> If all the requirements of 225.19(A) Exception No. 3 are met, the vertical clearance for the overhead spans of open conductors, including multiconductor cables, can be reduced to 18 in. (450 mm).

WORKING SPACE AROUND EQUIPMENT

Headroom and Height

Ⓐ Overcurrent devices shall not be located in the vicinity of easily ignitible material (such as in clothes closets) »240.24(D)«. Although mentioned specifically, clothes closets are only one type of area that may contain easily ignitible materials.

Ⓑ The minimum working space height for equipment operating at 600 volts, nominal, or less to ground and likely to require examination, adjustment, servicing, or maintenance while energized is 6½ ft (2.0 m) »110.26(A)(3)«. Service equipment and panelboards are examples of equipment likely to require examination, adjustment, servicing, or maintenance while energized.

Ⓒ A clear workspace must extend from the grade, floor, or platform to the height of 6½ ft (2.0 m) or the height of the equipment, whichever is greater »110.26(A)(3)«.

Ⓓ Workspace illumination must be provided for all indoor installations of service equipment, panelboards, etc. Supplemental luminaires are not required where the workspace is illuminated by an adjacent light source »110.26(D)«.

Ⓔ Access and working space must be provided and maintained around all electric equipment to permit ready and safe equipment operation and maintenance »110.26«.

Ⓕ Working space required by 110.26 must not be used for storage »110.26(B)«.

> ### NOTE
>
> Compliance with the minimum headroom of working space is optional for 200 amperes or less service equipment or panelboards in existing dwellings »**110.26(A)(3)** *Exception No. 1*«. Electric equipment workspace must have at least one entrance of sufficient area to provide access »**110.26(C)**«.

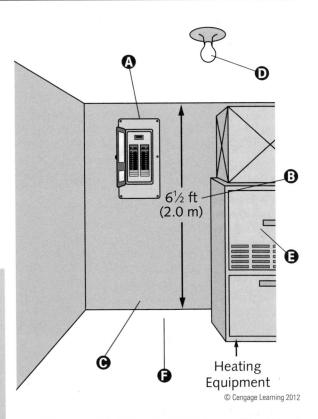

© Cengage Learning 2012

> ### CAUTION
>
> Working space illumination shall not be controlled by automatic means only »110.26(D)«.

Maximum Depth of Associated Equipment

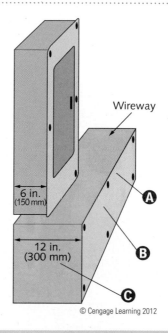

Wireway

6 in.
(150 mm)

12 in.
(300 mm)

Ⓐ

Ⓑ

Ⓒ

© Cengage Learning 2012

Ⓐ Wireways are troughs (sheet metal or nonmetallic) with hinged or removable covers used to house and protect electric wires and cables. Conductors are laid in place after the raceway has been installed as a complete system ≫*376.2* and *378.2*≪.

Ⓑ Article 376 outlines metal wireway provisions.

Ⓒ In an electrical installation, other associated equipment located above or below the electric equipment shall not extend more than 6 in. (150 mm) beyond the front of that electrical equipment ≫*110.26(A)(3)*≪.

CAUTION

All switchboards, switchgear, panelboards, and motor control centers shall be located in dedicated spaces and protected from damage ≫*110.26(E)*≪. Surrounding space (equal to the width and depth of the equipment) must be clear of foreign systems, unless protection is provided against damage from condensation, leaks, or breaks in such foreign systems. This dedicated equipment space extends from the floor to a height of 6 ft (1.8 m) above the equipment or to the structural ceiling, whichever is lower. A dropped, suspended, or similar ceiling that does not strengthen the building is not considered a structural ceiling. Sprinkler protection shall be permitted for the dedicated space, but the piping shall not be located in the dedicated equipment space ≫*110.26(E)(1)(b)–(d)*≪.

NOTE

Meters that are installed in meter sockets shall be permitted to extend beyond the other equipment. The meter socket must be installed in accordance with the provisions in 110.26 ≫***110.26(A)(3) Exception No. 2***≪.

Working Space Width

Ⓐ The frontal workspace width of electric equipment must be the greater of the width of the equipment or 30 in. (762 mm) ≫*110.26(A)(2)*≪.

Ⓑ Section 110.26 does not stipulate that the panel (or equipment) be located in the center of the workspace. It can be located anywhere within the required space.

Ⓒ A panel can be located in the middle, with equal amounts of space on each side, or . . .

Ⓓ A panel can be located on the far left side, with the additional space on the right, or . . .

Ⓔ A panel can be located on the far right, with the additional space on the left side.

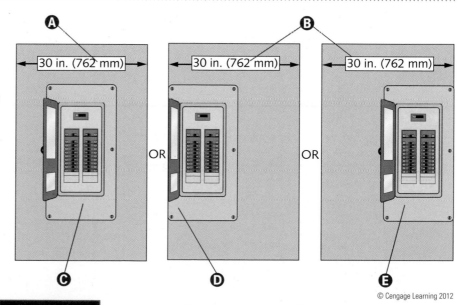

Ⓐ 30 in. (762 mm) **Ⓑ** 30 in. (762 mm) 30 in. (762 mm)

OR OR

Ⓒ **Ⓓ** **Ⓔ**

© Cengage Learning 2012

CAUTION

The workspace extends to a height of 6½ ft (2.0 m) or the height of the equipment, whichever is greater. Above 6½ ft (2.0 m) or the height of the equipment, the dedicated space must equal the width and depth of the equipment ≫*110.26(A)(3)* and *110.26(E)*≪.

NOTE

The working space must always permit at least a 90° opening of equipment doors or hinged panels ≫***110.26(A)(2)***≪.

Overlapping Workspace

Ⓐ The workspace width required in front of electric equipment is the greater of the width of the equipment or 30 in. (762 mm) ›› *110.26(A)(2)* ‹‹.

Ⓑ In this configuration, it is not necessary that each panel have its own individual working space.

Ⓒ A panel (or equipment) can share the working space of other equipment associated with the electrical installation.

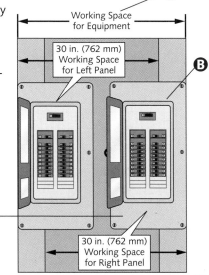

Working Space for Equipment — **Ⓐ**

30 in. (762 mm) Working Space for Left Panel

Ⓑ

Ⓒ

30 in. (762 mm) Working Space for Right Panel

© Cengage Learning 2012

Equipment behind Doors

Ⓐ Equipment can be located behind a door provided a 3-ft (914-mm) working space is achieved by closing the door.

Ⓑ Overcurrent devices must be readily accessible unless one of the conditions in 240.24(A)(1) through (4) apply ›› *240.24(A)* ‹‹. (The term "readily accessible" is defined in Unit 2 of this book.)

© Cengage Learning 2012

Working Space Depth

Ⓐ Concrete, brick, or tile walls are recognized as grounded ›› *Table 110.26(A) Note 2* ‹‹.

Ⓑ Table 110.26(A) specifies working space depth requirements in the direction of access to live parts. Distances are measured from the live parts (if exposed) or from the front or opening (if enclosed) ›› *110.26(A)(1)* ‹‹.

Ⓒ The minimum workspace depth for equipment operating at 150 volts, nominal (or less) to ground, is 3 ft (914 mm), regardless of the conditions found in Table 110.26(A). For equipment operating between 151 and 600 volts, nominal to ground, refer to Table 110.26(A).

Ⓓ Electric equipment must be firmly secured to the surface on which it is mounted. Wooden plugs driven into holes in masonry, concrete, plaster, or similar materials are not acceptable ›› *110.13(A)* ‹‹.

Ⓔ Except as elsewhere required or permitted, the workspace area for equipment operating at 600 volts, nominal (or less) to ground, which is likely to require examination, adjustment, servicing, or maintenance while energized, must comply with (1), (2), and (3) of 110.26(A) ›› *110.26(A)* ‹‹.

Ⓕ The minimum working space height is 6½ ft (2.0 m), except for existing dwellings whose service equipment or panelboard does not exceed 200 amperes ›› *110.26(A)(3)* ‹‹.

Ⓖ The minimum acceptable width of working space is 30 in. (762 mm) ›› *110.26(A)(2)* ‹‹.

NOTE

Installation and working space requirements for equipment operating over 600 volts, nominal, are found in 110.30 through 110.40.

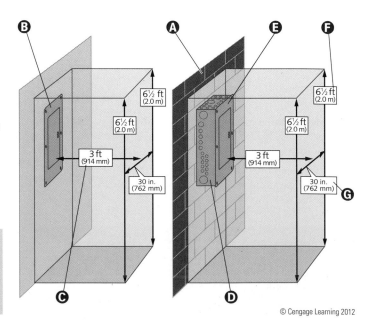

© Cengage Learning 2012

Dedicated Space

A The space equal to the width and depth of the equipment and extending from the floor to a height of 6 ft (1.8 m) above the equipment or to the structural ceiling, whichever is lower, must be dedicated to the electrical system. A dropped, suspended, or similar ceiling that does not strengthen the building is not considered a structural ceiling ≫ *110.26(E)(1)(A) and (D)* ≪.

B All switchboards, switchgear, panelboards, distribution boards, and motor control centers must be located in dedicated spaces and must be protected from damage ≫ *110.26(E)* ≪.

> **NOTE**
>
> Outdoor clear workspace includes the zone described in 110.26(A). No architectural appurtenances or other equipment can occupy space in this zone ≫*110.26(E)(2)(a)*≪.

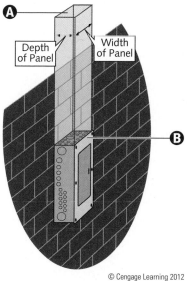

Depth of Panel / Width of Panel

© Cengage Learning 2012

Outdoor Service Equipment

A The space equal to the width and depth of the equipment, and extending from grade to a height of 1.8 m (6 ft) above the equipment, shall be dedicated to the electrical installation. No piping or other equipment foreign to the electrical installation shall be located in this zone ≫ *110.26(E)(2)(a)* ≪.

B Panelboards shall be mounted in cabinets, cutout boxes, or identified enclosures and shall be dead-front ≫ *408.38* ≪. "Dead front" means having no exposed live parts on the operating side of the equipment ≫ *Article 100* ≪.

C Outdoor electrical equipment shall be installed in suitable enclosures and shall be protected from accidental piping system spills and leaks and from accidental contact by unauthorized personnel or vehicular traffic. The zone described in 110.26(A) must be included in the minimum clear working space. No architectural appurtenance or other equipment can occupy space in this zone ≫ *110.26(E)(2)(a)* ≪.

> **NOTE**
>
> Panelboards in damp or wet locations must comply with 312.2 ≫*408.37*≪. In damp or wet locations, surface-type enclosures must be installed so moisture or water will not enter or accumulate within the enclosure. Unless the enclosure is nonmetallic, it must be mounted so there is at least ¼-in. (6-mm) airspace between the enclosure and the wall or other supporting surface. Enclosures installed in wet locations shall be weatherproof. For enclosures in wet locations, raceways or cables entering above the level of uninsulated live parts shall use fittings listed for wet locations ≫*312.2*≪.

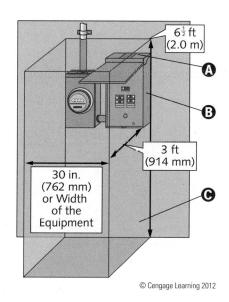

6½ ft (2.0 m)

3 ft (914 mm)

30 in. (762 mm) or Width of the Equipment

© Cengage Learning 2012

Electrical Equipment

Ⓐ All electric equipment must be surrounded by sufficient unobstructed space to allow ready and safe equipment operation and maintenance ≫ *110.26* ≪. Working space is mandatory for equipment likely to require examination, adjustment, servicing, or maintenance while energized ≫ *110.26(A)* ≪.

Ⓑ Article 440 details air-conditioning and refrigerating equipment provisions. General requirements, disconnecting means, overcurrent protection, conductor sizing, and room air conditioners are a few of the covered topics.

Ⓒ The disconnecting means for air-conditioning or refrigerating equipment must be located within sight from the equipment and must also be readily accessible ≫ *440.14* ≪.

Ⓓ Installation of the disconnecting means can be on, or within, the air-conditioning or refrigerating equipment ≫ *440.14* ≪. If installed within the equipment, it must be readily accessible. ("Accessible, Readily" is defined in Article 100 of the *NEC* and Unit 2 of this book.)

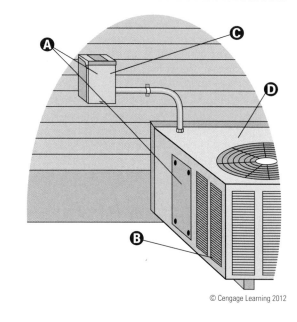

© Cengage Learning 2012

NOTE

Equipment grounding and equipment grounding conductor provisions are found in Article 250, Part VI (250.110 through 250.126).

SERVICE EQUIPMENT AND PANELBOARDS

Panelboards

Ⓐ A panelboard (1) consists of one or more panel units designed for assembly, thereby forming a single panel; (2) includes buses and automatic overcurrent devices; (3) may contain switches for controlling circuits (light, heat, or power); (4) is designed for mounting in a cabinet or cutout box; (5) is designed for placement in or against a wall, partition, or other support; and (6) is accessible only from the front ≫ *Article 100* ≪. (Panelboards include, but are not limited to, circuit breaker panels and fused or nonfused switches [disconnects].)

Ⓑ Panelboards include overcurrent devices.

Ⓒ Panelboards include buses.

Ⓓ Panelboards are mounted in cabinets or cutout boxes that are accessible only from the front. Panelboard requirements are in Article 408. Article 312 covers the installation and construction specifications of cabinets.

Ⓔ Panelboards can be single units or multipanel units.

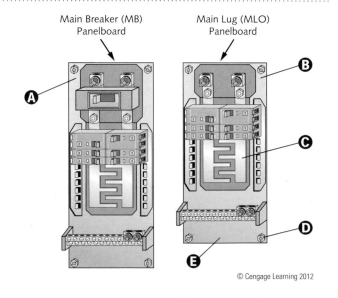

Main Breaker (MB) Panelboard Main Lug (MLO) Panelboard

© Cengage Learning 2012

Enclosures

(A) A cabinet is an enclosure that by design may be surface or flush mounted, having a frame, mat, or trim attached to which are (or can be) swinging door(s) ≫ *Article 100* ≪.

(B) Cabinets can be either flush or surface mounted.

(C) Cutout boxes have swinging doors or covers attached directly to the enclosure.

(D) Cabinets include trims with swinging doors.

(E) **Type 1** enclosures, intended for indoor use, primarily provide protection against contact with the enclosed equipment in locations not subject to unusual service conditions.

(F) Another type of enclosure, a cutout box, is designed for surface mounting and has swinging doors (or covers) secured directly to and extending from one of the enclosure walls ≫ *Article 100* ≪.

(G) Cutout boxes are surface mounted.

(H) "Dead front" is defined as having no exposed live parts on the operating side of the equipment ≫ *Article 100* ≪.

(I) **Type 3R** enclosures, intended for outdoor use, primarily provide protection against falling precipitation and must remain undamaged by external formation of ice. They are not intended to provide protection against conditions such as dust, internal condensation, or internal icing.

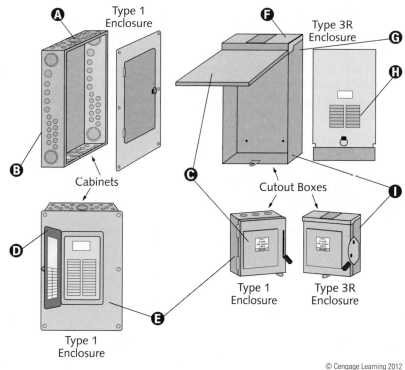

© Cengage Learning 2012

> **N O T E**
>
> Enclosure types are listed in Table 110.28.

Unused Openings and Circuit Directory

(A) The clear, evident, and specific purpose of all panelboard circuits (and circuit modifications) must be legibly identified on a circuit directory affixed to the face or inside of the panel door(s). The identification shall include an approved degree of detail that allows each circuit to be distinguished from all others. Spare positions that contain unused overcurrent devices or switches shall be described accordingly. No circuit shall be described in a manner that depends on transient conditions of occupancy ≫ *408.4(A)* ≪.

(B) Busbars, both insulated and bare, must be rigidly mounted ≫ *408.51* ≪.

(C) Unused openings, other than those intended for the operation of equipment, those intended for mounting purposes, or those permitted as part of the design for listed equipment, shall be closed to afford protection substantially equivalent to the wall of the equipment ≫ *110.12(A)* ≪.

(D) Each disconnecting means shall be legibly marked to indicate its purpose unless located and arranged so the purpose is evident. The marking must be durable enough to withstand the surrounding environment ≫ *110.22(A)* ≪.

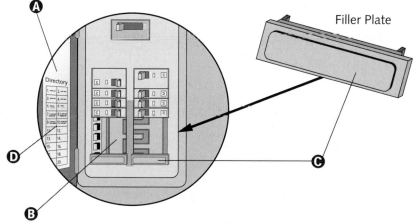

© Cengage Learning 2012

Circuit Breakers

The standard ampere ratings for fuses and inverse-time circuit breakers are 15, 20, 25, 30, 35, 40, 45, 50, 60, 70, 80, 90, 100, 110, 125, 150, 175, 200, 225, 250, 300, 350, 400, 450, 500, 600, 700, 800, 1000, 1200, 1600, 2000, 2500, 3000, 4000, 5000, and 6000 amperes. Fuses and inverse-time circuit breakers with nonstandard ampere ratings can be used. Additional standard ampere ratings for fuses are 1, 3, 6, 10, and 601 ≫ *240.6(A)* ≪.

Ⓐ A **circuit breaker** is a device designed to both open and close a circuit by nonautomatic means and to open the circuit automatically when subjected to a predetermined overcurrent without being damaged (when properly applied within its rating) ≫ *Article 100* ≪.

Ⓑ All circuit breakers having an interrupting rating other than 5000 amperes must be appropriately marked. The interrupting rating is not required on circuit breakers used for supplementary protection ≫ *240.83(C)* ≪. (Note: The ampere interrupting rating [AIR] is not the current rating of the circuit breaker.)

Ⓒ Circuit breakers must have a voltage rating (clearly marked) no less than the nominal system voltage that indicates their ability to interrupt fault current (between phases or phase to ground) ≫ *240.83(E)* ≪.

Ⓓ General-use and motor-circuit switches, circuit breakers, and molded case switches, where mounted in an enclosure as described in 404.3, shall clearly indicate whether they are in the open (off) or closed (on) position ≫ *404.7* ≪.

Ⓔ For vertically operated circuit breakers, the up position of the handle shall be the closed (on) position ≫ *240.81* and *404.7* ≪.

Ⓕ A circuit breaker's ampere rating must be durably marked and remain visible after installation. When necessary to ensure visibility, removal of a trim or cover is permitted ≫ *240.83(A)* ≪.

Ⓖ A circuit breaker's **inverse time** is a qualifying term indicating an intentional delay in the tripping action of the breaker. The length of the delay decreases as the magnitude of the current increases ≫ *Article 100* ≪.

Ⓗ A **nonadjustable** circuit breaker does not allow the value of current that triggers the tripping action, or the time required for its operation, to be altered ≫ *Article 100* ≪.

Ⓘ A circuit breaker's open (off) or closed (on) positions must be clearly identified ≫ *240.81* and *404.7* ≪.

Ⓙ The ampere rating of any circuit breaker rated at 100 amperes or less, and 1000 volts or less, must be molded, stamped, etched, or similarly marked into the handle or escutcheon area ≫ *240.83(B)* ≪.

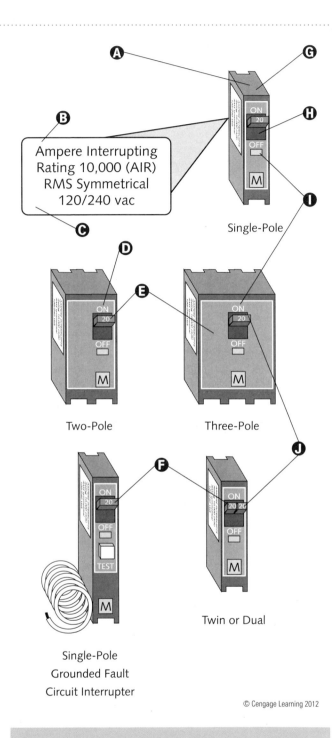

Ampere Interrupting Rating 10,000 (AIR) RMS Symmetrical 120/240 vac

Single-Pole

Two-Pole

Three-Pole

Single-Pole
Grounded Fault
Circuit Interrupter

Twin or Dual

© Cengage Learning 2012

CAUTION

Circuit breakers and fuses can be connected in parallel provided they are factory assembled and listed as a unit. Individual fuses, circuit breakers, or combinations shall not otherwise be connected in parallel ≫ *240.8* ≪.

NOTE

Circuit breakers must be trip free as well as manually operable in both the closed and opened positions. Although a circuit breaker may normally operate by electrical or pneumatic means, a means of manual operation must also be provided ≫ *240.80* ≪.

Circuit breakers used as switches in 120-volt and 277-volt fluorescent lighting circuits must be listed and be marked "SWD" or "HID" ≫ *240.83(D)* ≪.

Arc-Fault Circuit Interrupter (AFCI)

Ⓐ An AFCI is a device intended to provide protection from the effects of arc faults by recognizing characteristics unique to arcing and by functioning to de-energize the circuit when an arc fault is detected ≫ *Article 100* ≪.

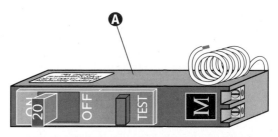

Arc-Fault Circuit Interrupter (AFCI)

© Cengage Learning 2012

> **WARNING**
>
> All 120-volt, single-phase, 15- and 20-ampere branch circuits supplying outlets or devices installed in dwelling unit kitchens, family rooms, dining rooms, living rooms, parlors, libraries, dens, bedrooms, sunrooms, recreation rooms, closets, hallways, laundry areas, or similar rooms or areas shall be protected by any of the means described in 210.12(A)(1) through (6) ≫ *210.12(A)* ≪. See 210.12(A)(1) through (6) for the complete list of permitted types of AFCI protection.

> **NOTE**
>
> In any of the areas specified in 210.12(A), where branch-circuit wiring is modified, replaced, or extended, the branch circuit shall be protected by one of the following: (1) a listed combination-type AFCI located at the origin of the branch circuit; or (2) a listed outlet branch circuit–type AFCI located at the first receptacle outlet of the existing branch circuit ≫ *210.12(B)* ≪.

Back-Fed Devices

Ⓐ Since a bolt-on breaker is not a plug-in-type overcurrent protection device, no additional fastener (screw, clip, etc.) is needed.

Ⓑ These remote panelboard (subpanel) feeder conductors originated from a main panelboard. The conductors are connected to the load side terminals of the circuit breaker, instead of connecting to the main lugs. In this configuration (the reverse of normal operation), the conductor feeds the breaker, thus the term "back-fed device."

Ⓒ Back-fed plug-in-type overcurrent protection devices or plug-in-type main lug assemblies used to terminate field-installed ungrounded supply conductors must be secured to the panelboard by an additional fastener mounting means that requires other than a pull to release the device from the mounting means on the panel ≫ *408.36(D)* ≪.

> **NOTE**
>
> A back-fed device functioning as a service disconnect must be permanently marked to identify it as a service disconnect ≫ *230.70(B)* ≪.

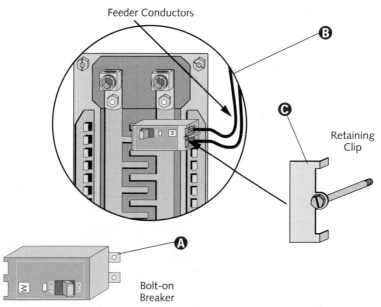

Feeder Conductors

Retaining Clip

Bolt-on Breaker

© Cengage Learning 2012

Installation

A Enclosures installed in wet locations must be weatherproof ≫ *312.2* ≪ .

B Cabinets and cutout boxes must have approved space to prevent crowding of installed conductors ≫ *312.7* ≪ .

C Panelboards in damp or wet locations must be installed in compliance with 312.2 ≫ *408.37* ≪ .

D Concrete, brick, and tile walls are considered to be grounded ≫ *Table 110.26(A) Condition 2* ≪ .

E The service conductors must be attached to the service disconnecting means by pressure connectors, clamps, or other approved means ≫ *230.81* ≪ .

F In damp or wet locations, surface-type enclosures (within the scope of Article 312) must be located or equipped so that moisture or water does not enter or accumulate within the cabinet or cutout box. A minimum of $\frac{1}{4}$-in. (6-mm) airspace between the enclosure and the wall or other supporting surface is required ≫ *312.2* ≪ .

G Electric equipment must be firmly secured to the surface on which it is mounted. Wooden plugs driven into holes in masonry, concrete, plaster, or similar materials are not acceptable ≫ *110.13(A)* ≪ .

> **NOTE**
>
> For enclosures in wet locations, raceways or cables entering above live parts shall use fittings listed for wet locations ≫**312.2**≪.

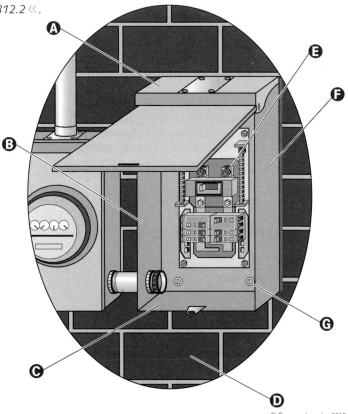

© Cengage Learning 2012

Grouping of Service Disconnects

A Energized parts of service equipment must be enclosed as specified in 230.62(A) or guarded as specified in 230.62(B) ≫ *230.62* ≪ .

B Two to six circuit breakers or sets of fuses are acceptable as the overcurrent device to provide overload protection. The sum of the ratings of the circuit breakers or fuses shall be permitted to exceed the ampacity of the service conductors, provided the calculated load does not exceed the ampacity of the service conductors ≫ *230.90(A) Exception No. 3* ≪ .

C A service disconnecting means consisting of more than one switch or circuit breaker, as permitted by 230.71, must have a combined rating of all switches or circuit breakers used, not less than the rating required by 230.79 ≫ *230.80* ≪ .

D Multiple service disconnects (from two to six) as permitted in 230.71 must be grouped. Each disconnect shall be marked to indicate the load served ≫ *230.72(A)* ≪ .

> **NOTE**
>
> Equipment whose purpose is to interrupt current at fault levels shall have an interrupting rating at nominal circuit voltage sufficient for the current that is available at the line terminals of the equipment. Equipment intended to interrupt current at other than fault levels shall have an interrupting rating at nominal circuit voltage sufficient for the current that must be interrupted ≫**110.9**≪.

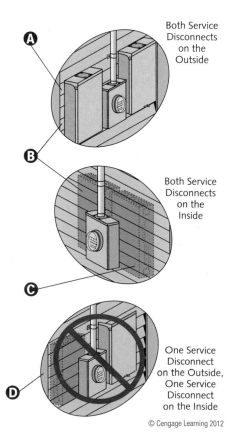

Both Service Disconnects on the Outside

Both Service Disconnects on the Inside

One Service Disconnect on the Outside, One Service Disconnect on the Inside

© Cengage Learning 2012

Location of Service Disconnecting Means

A The service disconnecting means shall be installed at a readily accessible location either outside of a building or structure or inside as near as possible to the point of entrance of the service conductors 》*230.70(A)(1)* 《.

> ### NOTE
>
> Maximum service conductor lengths are not specified within a building or structure. The provision merely states that the disconnecting means must be located nearest the point of service conductor entrance. Since the length of service conductors can alter available short-circuit current, the length should be kept to a minimum. Some local jurisdictions define a maximum acceptable length for service conductors.

> ### CAUTION
>
> Neither service disconnecting means nor overcurrent devices (other than supplementary overcurrent protection) can be located in bathrooms 》*230.70(A)(2)* and *240.24(E)*《.

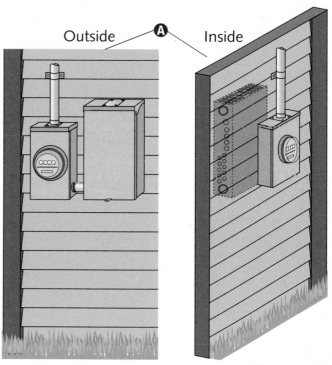

Outside — **A** — Inside

© Cengage Learning 2012

Service Equipment

A Because individual meter socket enclosures are not considered service equipment, a "suitable for use as service equipment" marking is not required 》*230.66*《.

B Service equipment rated 1000 volts or less must be clearly marked as suitable for use as service equipment. All service equipment shall be listed 》*230.66*《.

C Each service disconnect must be permanently marked, identifying it as a service disconnect 》*230.70(B)*《.

D A one-family dwelling service disconnecting means must have a minimum rating of 100 amperes, 3-wire 》*230.79(C)*《.

E A means to disconnect all conductors in a building (or other structure) from the service-entrance conductors must be provided 》*230.70*《.

> ### WARNING
>
> Local utility companies often set parameters (height, grounding, etc.) for metering equipment installation. Local jurisdictions may also have special regulations concerning service equipment. Obtain copies of all applicable provisions for the area in which the equipment will be installed.

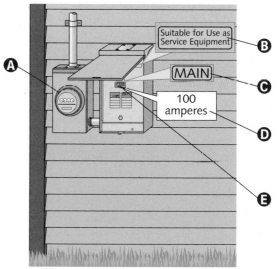

Suitable for Use as Service Equipment — **B**

MAIN — **C**

100 amperes — **D**

A

E

© Cengage Learning 2012

Multiple Service Disconnects

A Section 230.90 requires that each ungrounded (hot) service conductor have overload protection. Such protection shall be provided by an overcurrent device in series with each ungrounded service conductor having a rating (or setting) not higher than the allowable conductor ampacity ≫ 230.90(A) ≪. Under 230.90(A), five exceptions allow more than one overcurrent device for overload protection.

B Two to six circuit breakers or sets of fuses can serve as the overcurrent device providing overload protection ≫ 230.90(A) Exception No. 3 ≪.

C The total circuit breaker (or fuse) ratings can exceed the service conductors ampacity, provided the calculated load (in accordance with Article 220) does not exceed the ampacity of the service conductors ≫ 230.90(A) Exception No. 3 ≪.

D The service disconnecting means shall have a rating not less than the calculated load to be carried, determined in accordance with Parts III, IV, or V of Article 220, as applicable. In no case shall the rating be lower than specified in 230.79(A), (B), (C), or (D) ≫ 230.79 ≪.

E Each service disconnect must be permanently marked as a service disconnect ≫ 230.70(B) ≪. Each individual service disconnect must be marked. A single marking for multiple service disconnects is not permitted.

F The service disconnecting means for each service (permitted by 230.2) or for each set of service-entrance conductors (permitted by 230.40, Exception Nos. 1, 3, 4, or 5) shall consist of not more than six switches or six circuit breakers, whether mounted in a single enclosure, in a group of separate enclosures, or in or on a switchboard or in switchgear. No more than six disconnects per service can be grouped in any one location ≫ 230.71(A) ≪.

G Two or three single-pole switches or breakers, individually operable, are permitted on multiwire circuits (one pole for each ungrounded conductor). The multipole disconnect must be equipped with identified handle ties or a master handle to disconnect all service conductors. Six operations of the hand are the maximum allowed ≫ 230.71(B) ≪.

H Single-pole circuit breakers grouped in accordance with 230.71(B) constitute one protective device ≫ 230.90(A) ≪.

> **NOTE**
>
> A set of fuses shall be considered all the fuses required to protect all of a circuit's ungrounded conductors
> ≫ **230.90(A)** ≪.

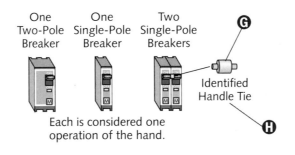

One Two-Pole Breaker   One Single-Pole Breaker   Two Single-Pole Breakers   **G**
Identified Handle Tie
Each is considered one operation of the hand.   **H**

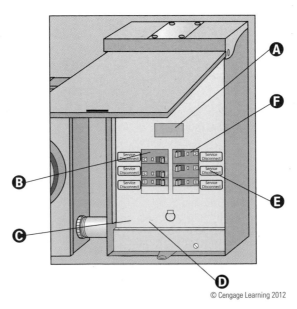

© Cengage Learning 2012

Panelboards

A All panelboards shall have a rating not less than the minimum feeder capacity required for the load calculated in accordance with Part III, IV, or V of Article 220, as applicable ≫ *408.30* ≪.

B In addition to the requirement of 408.30, a panelboard shall be protected by an overcurrent protective device having a rating not greater than that of the panelboard unless meeting one of the exceptions in 408.36.

C An individual overcurrent protective device is not required for a panelboard used as service equipment with multiple disconnecting means in accordance with 230.71 ≫ *408.36 Exception No. 1* ≪.

> **NOTE**
>
> In panelboards protected by three or more main circuit breakers or sets of fuses, the circuit breakers or sets of fuses shall not supply a second bus structure within the same panelboard assembly ≫**408.36 Exception No. 1**≪.

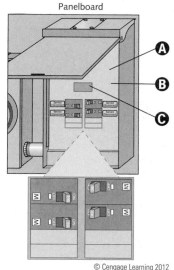

Panelboard

© Cengage Learning 2012

A In addition to the requirement of 408.30, a panelboard shall be protected by an overcurrent protective device having a rating not greater than that of the panelboard. This overcurrent protective device shall be located within or at any point on the supply side of the panelboard ≫ *408.36* ≪.

B The front edge of cabinets (panelboards) situated in walls constructed of noncombustible material (concrete, tile, etc.) must be within ¼ in. (6 mm) of the finished surface ≫ *312.3* ≪.

C Panelboards equipped with snap switches rated at 30 amperes or less shall have overcurrent protection of 200 amperes or less ≫ *408.36(A)* ≪.

D A grounded conductor shall not be connected to normally non-current-carrying metal parts of equipment, to equipment grounding conductor(s), or be reconnected to ground on the load side of the service disconnecting means except as otherwise permitted in Article 250 ≫ *250.24(A)(5)* ≪.

> **NOTE**
>
> Individual protection shall not be required for a panelboard protected on its supply side by two main circuit breakers or two sets of fuses having a combined rating not greater than that of the panelboard. A panelboard constructed or wired under this exception shall not contain more than 42 overcurrent devices. For the purposes of determining the maximum of 42 overcurrent devices, a 2-pole or a 3-pole circuit breaker shall be considered as two or three overcurrent devices, respectively ≫**408.36 Exception No. 2**≪.

Raceway containing feeder conductors

Ungrounded (hot) Conductors

Equipment Grounding Conductor

Neutral Conductor

Panelboard

© Cengage Learning 2012

> **CAUTION**
>
> Cabinets in walls constructed of combustible material (wood, etc.) must be flush with or extend beyond the finished surface ≫*312.3*≪.

> **NOTE**
>
> For existing panelboards, individual protection shall not be required for a panelboard used as service equipment for an individual residential occupancy ≫**408.36 Exception No. 3**≪.

Common-Trip and Single-Pole Circuit Breakers

Ⓐ Circuit breakers must open all ungrounded conductors of the circuit both manually and automatic unless permitted in 240.15(B)(1) through (4) ≫ *240.15(B)* ≪.

Ⓑ Individual single-pole circuit breakers, with identified handle ties, are permitted as protection for each ungrounded conductor of multiwire branch circuits serving only single-phase, line-to-neutral loads ≫ *240.15(B)(1)* ≪. In accordance with 210.4(B), each multiwire branch circuit must be provided with a means that will simultaneously disconnect all ungrounded (hot) conductors at the point where the branch circuit originates. There are no exceptions to this rule.

Ⓒ In grounded systems, individual single-pole circuit breakers, rated 240 volts alternating current (ac), with identified handle ties are permitted as protection for each ungrounded conductor for line-to-line connected loads for single-phase circuits ≫ *240.15(B)(2)* ≪.

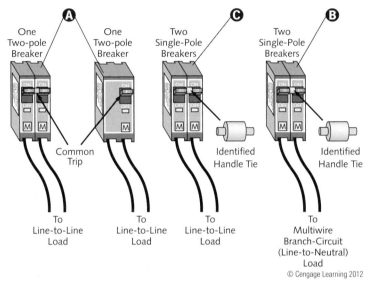

One Two-pole Breaker — Ⓐ — Common Trip — To Line-to-Line Load

One Two-pole Breaker — To Line-to-Line Load

Two Single-Pole Breakers — Ⓒ — Identified Handle Tie — To Line-to-Line Load

Two Single-Pole Breakers — Ⓑ — Identified Handle Tie — To Multiwire Branch-Circuit (Line-to-Neutral) Load

© Cengage Learning 2012

> **NOTE**
>
> For line-to-line loads in 4-wire, 3-phase systems or 5-wire, 2-phase systems, individual single-pole circuit breakers rated 120/240 volts ac with identified handle ties shall be permitted as the protection for each ungrounded conductor, if the systems have a grounded neutral point and the voltage to ground, does not exceed 120 volts ≫ *240.15(B)(3)* ≪.

Cartridge Fuses and Fuseholders

Cartridge fuses and fuseholders are classified according to voltage and amperage ranges. Fuses rated 1000 volts, nominal, or less can be used for voltages at or below their ratings ≫ *240.61* ≪.

Ⓐ Fuseholders are designed so that it is difficult to put a fuse of any given class into a fuseholder intended for a current lower, or voltage higher, than that of the class to which the fuse belongs ≫ *240.60(B)* ≪.

Ⓑ Fuses must be plainly marked (by printing on or a label attached to the fuse barrel), showing the following: (1) ampere rating, (2) voltage rating, (3) interrupting rating if other than 10,000 amperes, (4) "current-limiting" where applicable, and (5) the name or trademark of the manufacturer. The interrupting rating is not required on fuses used for supplementary protection ≫ *240.60(C)* ≪.

Ⓒ Fuseholders for current-limiting fuses must not permit insertion of non-current-limiting fuses ≫ *240.60(B)* ≪.

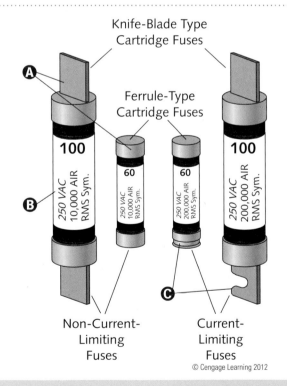

Knife-Blade Type Cartridge Fuses

Ⓐ

Ferrule-Type Cartridge Fuses

Ⓑ

100 250 VAC 10,000 AIR RMS Sym.

60 250 VAC 10,000 AIR RMS Sym.

60 250 VAC 200,000 AIR RMS Sym.

100 250 VAC 200,000 AIR RMS Sym.

Ⓒ

Non-Current-Limiting Fuses

Current-Limiting Fuses

© Cengage Learning 2012

> **NOTE**
>
> Cartridge fuses and fuseholders with a 300-volt rating can be used in the following: (1) circuits not exceeding 300 volts between conductors and (2) single-phase, line-to-neutral circuits supplied from a 3-phase, 4-wire solidly grounded neutral source with line-to-neutral voltage not exceeding 300 volts ≫ *240.60(A)* ≪.

> **NOTE**
>
> Class H cartridge fuses of the renewable type are only permitted as replacements in existing installations where there is no evidence of overfusing or tampering ≫ *240.60(D)* ≪.

Conductor Sizing

A For one-family dwellings and the individual dwelling units of two-family and multifamily dwellings, service and feeder conductors supplied by a single-phase, 120/240-volt system shall be permitted be sized in accordance with 310.15(B)(7)(1) through (4).

(1) For a service rated 100 through 400 A, the service conductors supplying the entire load associated with a one-family dwelling, or the service conductors supplying the entire load associated with an individual dwelling unit in a two-family or multifamily dwelling, shall be permitted to have an ampacity not less than 83 percent of the service rating »*310.15(B)(7)*«.

B Multiply the service rating by 83 percent and then select a conductor from Table 310.15(B)(16).

200 × 83% = 166 amperes

The minimum size conductor is 2/0 AWG copper or 4/0 AWG aluminum.

> ### N O T E
>
> The grounded (neutral) conductor can be smaller than the ungrounded (hot) conductors, provided 215.2, 220.61, and 230.42 requirements are met »***310.15(B)(7)(4)***«. The grounded conductor shall not be smaller than specified in Table 250.102(C)(1) »***250.24(B)(1)***«. The feeder conductors to a dwelling unit shall not be required to have an allowable ampacity rating greater than specified in 310.15(B)(7)(1) or (2) »***310.15(B)(7)(3)***«.

C It shall also be permissible to use 310.15(B)(7) to size feeder conductors if they supply the entire load associated with the one-family dwelling. In this illustration, it is not permissible to use 310.15(B)(7) to size these feeder conductors because they do not supply the entire load associated with this dwelling.

D Article 440 outlines provisions relating to disconnecting means, branch-circuit overload protection, and branch-circuit conductors for air-conditioning and refrigeration equipment.

E Conductor ampacities are found in Table 310.15(B)(16).

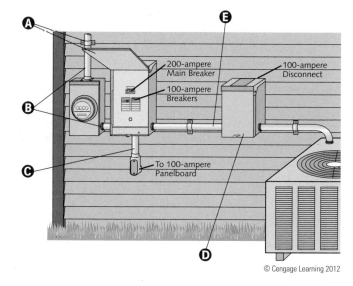

200-ampere Main Breaker
100-ampere Disconnect
100-ampere Breakers
To 100-ampere Panelboard

© Cengage Learning 2012

Installation

A Electric equipment must be installed in a neat and professional manner »*110.12*«.

B All electric equipment internal parts (busbars, wiring terminals, insulators, etc.) must remain undamaged and uncontaminated by foreign materials (paint, plaster, cleaners, abrasives, corrosive residues, etc.). Equipment parts that are broken, bent, cut, corroded, impaired due to chemical action or heat, or otherwise damaged in such a way as to adversely affect safe operation or mechanical strength must not be used »*110.12(B)*«.

C Sections 312.10 and 312.11 itemize construction specifications for cabinets, cutout boxes, and meter socket enclosures.

D Cabinets and cutout boxes must have approved space to accommodate (without crowding) all installed conductors »*312.7*«.

E Conductor sizes are expressed in American Wire Gage (AWG) or in circular mils »*110.6*«.

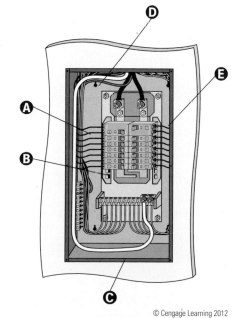

© Cengage Learning 2012

> ### N O T E
>
> The wiring space of enclosures for switches or overcurrent devices shall be permitted for conductors feeding through, spliced, or tapping off to other enclosures, switches, or overcurrent devices where all of the following conditions are met: (1) the total of all conductors installed at any cross section of the wiring space does not exceed 40% of the cross-sectional area of that space; (2) the total area of all conductors, splices, and taps installed at any cross section of the wiring space does not exceed 75% of the cross-sectional area of that space; and (3) a warning label complying with 110.21(B) is applied to the enclosure that identifies the closest disconnecting means for any feed-through conductors »***312.8***«.

Plug Fuses, Fuseholders, and Adapters

A Plug fuses of 15-ampere and lower rating are identified by a hexagonal configuration in a prominent location (such as window cap or cap) to distinguish them from fuses of higher ampere ratings »240.50(C)«.

B Type S adapters must fit Edison-base fuseholders »240.54(A)«.

C Once inserted into a fuseholder, Type S adapters cannot be removed »240.54(C)«.

D Type S fuses are classified at not over 125 volts, and 0 to 15 amperes, 16 to 20 amperes, or 21 to 30 amperes »240.53(A)«.

E Fuseholders of the Edison-base type shall be installed only where they are made to accept Type S fuses by the use of adapters »240.52«.

F Plug fuses rated 16 to 30 amperes are identified by means other than a hexagonal configuration.

G Type S fuseholders and adapters are designed so that either the fuseholder itself or the fuseholder with a Type S adapter inserted can be used only for a Type S fuse »240.54(B)«.

H Type S fuses of an ampere classification specified in 240.53(A) are not interchangeable with a lower-ampere classification. They can be used only in a Type S fuseholder or a fuseholder with a Type S adapter inserted »240.53(B)«.

I Type S fuses, fuseholders, and adapters are designed to make tampering or shunting (bridging) difficult »240.54(D)«.

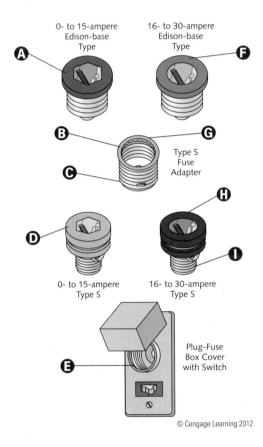

0- to 15-ampere
Edison-base
Type

16- to 30-ampere
Edison-base
Type

Type S
Fuse
Adapter

0- to 15-ampere
Type S

16- to 30-ampere
Type S

Plug-Fuse
Box Cover
with Switch

© Cengage Learning 2012

NOTE

The screw shell of a plug-type fuseholder must be connected to the circuit's load side »**240.50(E)**«.

WARNING

Plug fuses of the Edison-base type can be used only as replacements in existing installations exhibiting no evidence of overfusing or tampering »240.51(B)«.

GROUNDING

Grounded (Neutral) Conductor in the Service Disconnecting Means

A Equipment grounding conductors, grounding electrode conductors, and bonding jumpers shall be connected by one or more of the following means: (1) listed pressure connectors; (2) terminal bars; (3) pressure connectors listed as grounding and bonding equipment; (4) exothermic welding process; (5) machine screw-type fasteners that engage not less than two threads or are secured with a nut; (6) thread-forming machine screws that engage not less than two threads in the enclosure; (7) connections that are part of a listed assembly; or (8) other listed means ›› *250.8* ‹‹.

B An ac system grounded at any point and operating at 1000 volts or less must have a grounded conductor that is routed with the ungrounded conductors to each service disconnecting means and connected to each service disconnecting means grounded conductor(s) terminal or bus. A main bonding jumper shall connect the grounded conductor(s) to each service disconnecting means enclosure ›› *250.24(C)* ‹‹.

C A grounded system must have an unspliced main bonding jumper connecting the equipment grounding conductor(s) and the service-disconnect enclosure to the grounded conductor within each service disconnect enclosure in accordance with 250.28 ›› *250.24(B)* ‹‹.

D A single bar in the service-disconnect enclosure may serve as the neutral bar (or bus) and the equipment grounding terminal bar (or bus).

E The grounded conductor shall not be smaller than specified in Table 250.102(C)(1). ›› *250.24(C)(1)* ‹‹.

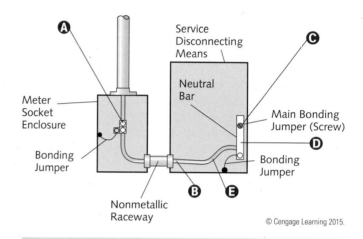

© Cengage Learning 2015.

N O T E

The main bonding jumper (screw, strap, etc.) is installed only in service equipment and not in subpanels.

Main Bonding Jumper

A A main bonding jumper is the connection between the grounded circuit conductor (neutral) and the equipment grounding conductor at the service ›› *Article 100* ‹‹.

B Main bonding jumpers and system bonding jumpers must be connected in the manner specified by the applicable provisions of 250.8 ›› *250.28(C)* ‹‹.

C A grounded system must have an unspliced main bonding jumper connecting the equipment grounding conductor(s) and the service-disconnect enclosure to the grounded conductor within each service disconnect enclosure ›› *250.24(B)* ‹‹.

D In cases where the main bonding jumper or a system bonding jumper is a screw only, it must be identified with a green finish that remains visible after installation ›› *250.28(B)* ‹‹.

E Main bonding jumpers and system bonding jumpers must be made of copper or other corrosion-resistant material; a wire, bus, screw, or similar suitable conductor is acceptable ›› *250.28(A)* ‹‹.

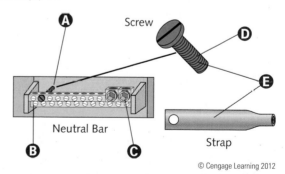

© Cengage Learning 2012

N O T E

Main bonding jumpers and system bonding jumpers shall not be smaller than specified in Table 250.102(C)(1). ›› *250.28(D)(1)* ‹‹. (Listed main bonding jumpers furnished with panelboards meet the provisions of 250.28.)

CAUTION

Panelboards listed as suitable for use as service equipment are furnished with a main bonding jumper that has not yet been installed. When using the panelboard as service equipment, install the main bonding jumper (screw, strap, etc.). When the panelboard is used for other purposes, such as a subpanel, do not install the main bonding jumper.

Remote Panelboards (Subpanels)

A Unless grounded by connection to the grounded circuit conductor (permitted by 250.32, 250.140, and 250.142), non-current-carrying metal parts of equipment, raceways, and other enclosures if grounded, must be connected to an equipment grounding conductor by one of the following methods: (A) by connecting to any of the equipment grounding conductors permitted by 250.118, or (B) by connecting it to an equipment grounding conductor contained within the same raceway or cable or otherwise run with the circuit conductors ≫ *250.134(A)* and *(B)* ≪.

B Equipment grounding conductors can be bare, covered, or insulated. Individually covered or insulated equipment grounding conductors must have a continuous outer finish, either green or green with yellow stripe(s), except as otherwise permitted by Article 250, Part VI ≫ *250.119* ≪.

C Only in service disconnecting means can the neutral terminal bar be connected to the enclosure or equipment grounding terminal bar. In this illustration, the neutral bar is isolated from the equipment grounding system.

D A grounded circuit conductor shall not be used for grounding non-current-carrying metal parts of equipment on the load side of the service disconnecting means, unless an exception has been met ≫ *250.142(B)* ≪.

> **NOTE**
>
> The bonding jumper (screw, strap, etc.) is installed only in the service disconnecting means and not in subpanels.

E Panelboard cabinets and panelboard frames, if metal, must be in physical contact with one another and must be connected to an equipment grounding conductor. Equipment grounding conductors must not be connected to a terminal bar provided for grounded conductors or neutral conductors unless the bar is identified for that purpose and is located where interconnection between equipment grounding conductors and grounded circuit conductors comply with Article 250 ≫ *408.40* ≪.

F A grounding connection must *not* be connected to normally non-current-carrying metal parts of equipment or to equipment grounding conductor(s) or be reconnected to ground on the load side of the service disconnecting means except as otherwise allowed in Article 250 ≫ *250.24(A)(5)* ≪.

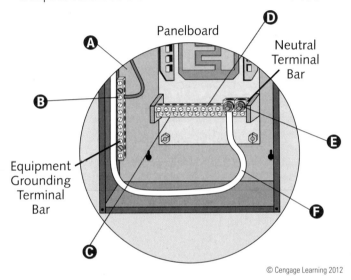

© Cengage Learning 2012

Grounding Electrode Conductor Connections

A The size of the grounding electrode conductor at the service, at each building or structure where supplied by a feeder(s) or branch circuit(s), or at a separately derived system of a grounded or ungrounded ac system shall not be less than given in Table 250.66, except as permitted in 250.66(A) through (C) ≫ *250.66* ≪.

B A premises wiring system supplied by a grounded ac service must have a grounding electrode conductor connected to the grounded service conductor, at each service, in accordance with 250.24(A)(1) through (5) ≫ *250.24(A)* ≪.

C A grounding electrode conductor must connect the equipment grounding conductors, the service-equipment enclosures, and, where the system is grounded, the grounded service conductor to the grounding electrode(s) as required by Part III of Article 250 ≫ *250.24(D)* ≪.

> **NOTE**
>
> The grounding electrode conductor connection shall be made at an accessible point from the load end of the overhead service conductors, service drop, underground service conductors, or service lateral to, including the terminal (or bus) to which the grounded service conductor is connected at the service disconnecting means ≫**250.24(A)(1)**≪.

D Splices in wire-type grounding electrode conductors are acceptable only if by means of irreversible compression-type connectors (listed as grounding and bonding equipment) or by the exothermic welding process ≫ *250.64(C)* ≪.

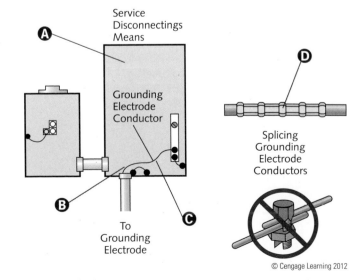

© Cengage Learning 2012

Methods of Bonding at the Service

(A) One method is to bond equipment to the grounded service conductor in a 250.8-approved manner 》*250.92(B)(1)*《.

(B) Connections using threaded couplings or threaded hubs on enclosures are permitted if made up wrenchtight 》*250.92(B)(2)*《. .

(C) Listed devices, such as bonding-type locknuts, bushings, or bushings with bonding jumpers are also permitted to ensure electrical continuity at service equipment 》*250.92(B)(4)*《.

(D) Standard locknuts or bushings shall not be the sole means for bonding as required by 250.92(B), but shall be permitted to be installed to make a mechanical connection of the raceway(s) 》*250.92(B)*《.

(E) Threadless couplings and connectors are approved if made up tight for metal raceways and metal-clad cables. Standard locknuts and bushings shall not be used to bond these items 》*250.92(B)(3)*《.

(F) Bonding jumpers, which meet other Article 250 requirements, shall be used around impaired connections, such as reducing washers or oversized, concentric, or eccentric knockouts 》*250.92(B)*《.

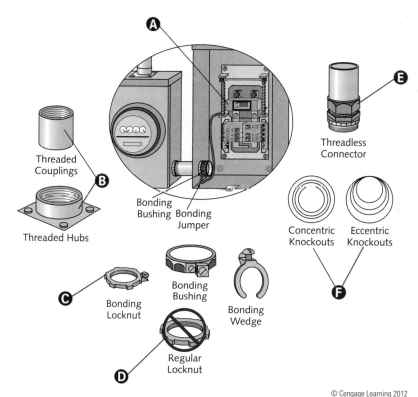

Threaded Couplings

Threaded Hubs

Bonding Bushing Bonding Jumper

Bonding Locknut

Bonding Bushing

Regular Locknut

Bonding Wedge

Threadless Connector

Concentric Knockouts

Eccentric Knockouts

© Cengage Learning 2012

Bonding at the Service

(A) Bonding is connected to establish electrical continuity and conductivity 》*Article 100*《.

(B) Threaded hubs are acceptable when bonding service equipment enclosures to threaded conduits or threadless connectors 》*250.92(B)(2)* and *(3)*《.

(C) All normally non-current-carrying metal parts of service equipment (such as raceways, enclosures containing service conductors, meter fittings, metal raceway, or armor enclosing a grounding electrode conductor) shall be bonded together 》*250.92(A)*《.

(D) Bonding, where necessary, must be provided to ensure not only electrical continuity but also ample capacity to safely conduct any likely fault current 》*250.90*《.

(E) Electrical continuity at service equipment, service raceways, and service conductor enclosures must be accomplished by one of the approved methods in 250.92(B).

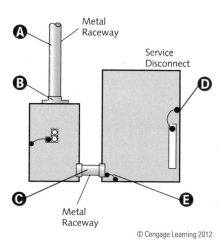

Metal Raceway

Service Disconnect

Metal Raceway

© Cengage Learning 2012

Grounding Electrode System

Grounding electrodes covered in 250.52 include: (1) metal underground water pipe, (2) metal frame of the building or structure, (3) concrete-encased electrode, (4) ground ring, (5) rod and pipe electrodes, (6) other listed electrodes, (7) plate electrodes, and (8) other local metal underground systems or structures. Although some of these grounding electrodes are described on this page, other electrodes from 250.52 are described on the following pages.

Exposed structural metal that is interconnected to form a metal building frame and is not intentionally grounded or bonded and is likely to become energized shall be bonded to the service equipment enclosure; the grounded conductor at the service; the disconnecting means for buildings or structures supplied by a feeder or branch circuit; the grounding electrode conductor, if of sufficient size; or to one or more grounding electrodes used. The bonding jumper(s) shall be sized in accordance with Table 250.66 and installed in accordance with 250.64(A), (B), and (E). The points of attachment of the bonding jumper(s) shall be accessible unless installed in compliance with 250.68(A), Exception No. 2 ≫*250.104(C)* ≪.

Ⓐ The metal frame of the building or structure must be bonded to the grounding electrode system if it is connected to the earth by one or more of the following methods:

(1) At least one structural metal member that is in direct contact with the earth for 10 ft (3.0 m) or more, with or without concrete encasement.

(2) Hold-down bolts that secure the structural steel column and that are connected to a concrete-encased electrode that complies with 250.52(3) and is located in the support footing or foundation. The hold-down bolts shall be connected to the concrete-encased electrode by welding, exothermic welding, the usual steel tie wires, or other approved means ≫*250.52(A)(2)* ≪.

Ⓑ A ground ring, which encircles the building or structure in direct contact with the earth at a depth of at least 2½ ft (750 mm) below the surface, consisting of at least 20 ft (6.0 m) of bare copper conductor not smaller than 2 AWG, is allowed ≫*250.52(A)(4)* and *250.53(F)* ≪.

Ⓒ A concrete-encased electrode shall be encased by at least 2 in. (50 mm) of concrete, located horizontally within that portion of a concrete foundation or footing that is in direct contact with the earth or within vertical foundations or structural components or members that are in direct contact with the earth. The electrode must consist of one or more bare or zinc galvanized or other electrically conductive coated steel reinforcing bars or rods, each at least ½ in. (13 mm) in diameter and 20 ft (6.0 m) continuous length or consisting of at least 20 ft (6.0 m) of bare copper conductor not smaller than 4 AWG. Multiple bars or rods can be connected together by the usual steel tie wires, exothermic welding, welding, or other effective means to create a 20 ft (6.0 m) or greater length ≫*250.52(A)(3)* ≪.

Ⓓ Grounding electrode conductors and bonding jumpers shall be permitted to be connected to the metal structural frame of a building shall be permitted to be used as a conductor to interconnect electrodes that are part of the grounding electrode system, or as a grounding electrode conductor. Grounding electrode conductors and bonding jumpers shall also be permitted to be connected to a concrete-encased electrode of either the conductor type, reinforcing rod or bar installed in accordance with 250.52(A)(3) extended from its location within the concrete to an accessible location above the concrete ≫*250.68(C)(2)* and *(3)* ≪.

NOTE

All grounding electrodes as described in 250.52(A)(1) through (A)(7) that are present at each building or structure served must be bonded together to form the grounding electrode system. Where none of these grounding electrodes exist, one or more of the grounding electrodes specified in 250.52(A)(4) through (A)(8) must be installed and used ≫**250.50**≪.

WARNING

A concrete foundation or footing is not considered to be in "direct contact" with the earth if the concrete is installed with insulation, vapor barriers, films, or similar items that separate the concrete from the earth ≫*250.52(A)(3) Informational Note* ≪.

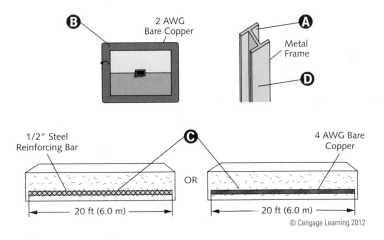

Ⓑ 2 AWG Bare Copper   Ⓐ Metal Frame   Ⓓ

1/2" Steel Reinforcing Bar   Ⓒ   OR   4 AWG Bare Copper

20 ft (6.0 m)   20 ft (6.0 m)

© Cengage Learning 2012

Metal Water Pipe Electrodes

Ⓐ A metal underground water pipe must be in direct contact with the earth for a minimum of 10 ft (3.0 m) (including any metal well casing effectively bonded to the pipe) and electrically continuous (or made so by bonding around insulating joints or insulating pipe) to the points of connection of the grounding electrode conductor and the bonding conductor(s) or jumper(s), if installed ≫ *250.52(A)(1)* ≪.

Ⓑ Grounding path continuity, or the bonding connection to interior piping, must not rely on water meters, filtering devices, or similar equipment ≫ *250.53(D)(1)* ≪.

Ⓒ Bonding jumper(s) shall be installed in accordance with 250.64(A), (B), and (E), shall be sized in accordance with 250.66, and shall be connected as specified in 250.70 ≫ *250.53(C)* ≪.

Ⓓ Although the interior metal water piping system is not a permitted grounding electrode, it can be used as a conductor to interconnect electrodes that are part of the grounding electrode system, but only if located not more than 5 ft (1.52 m) from the point of entrance to the building ≫ *250.68(C)(1)* ≪. There is an exception for industrial, commercial, and institutional buildings or structures.

Ⓔ A metal underground water pipe shall be supplemented by an additional electrode of a type specified in 250.52(A)(2) through (A)(8). If the supplemental electrode is of the rod, pipe, or plate type, it shall comply with 250.52(A). The supplemental electrode shall be bonded to one of the following: (1) grounding electrode conductor; (2) grounded service-entrance conductor; (3) nonflexible grounded service raceway; (4) any grounded service enclosure; and (5) as provided by 250.32(B) ≫ *250.53(D)(2)* ≪.

> **NOTE**
>
> A metal underground water pipe shall be supplemented by an additional electrode of a type specified in 250.52(A)(2) through (A)(8). If the supplemental electrode is of the rod, pipe, or plate type, it shall comply with 250.52(A). The supplemental electrode shall be bonded to one of the following: (1) rod, pipe, or plate electrode; (2) grounding electrode conductor; (3) grounded service-entrance conductor; (4) nonflexible grounded service raceway; or (5) any grounded service enclosure ≫ *250.53(A)(2)* ≪.

> **NOTE**
>
> The grounding electrode conductor can be run to any convenient grounding electrode available in the grounding electrode system where the other electrode(s), if any, is connected by bonding jumpers that are installed in accordance with 250.53(C). Grounding electrode conductor(s) can also be run to one or more grounding electrode(s) individually. It must be sized for the largest grounding electrode conductor required among all the electrodes being connected ≫ *250.64(F)* ≪. Bonding jumper(s) from grounding electrode(s) can also be connected to an aluminum or copper busbar where installed in accordance with 250.64(F)(3).

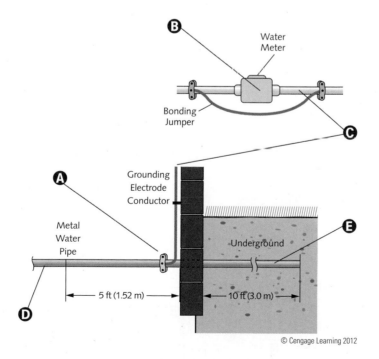

© Cengage Learning 2012

Rod, Pipe, and Plate Electrodes

If none of the grounding electrodes specified in 250.52(A)(1) through (A)(7) exist, one or more of the grounding electrodes specified in 250.52(A)(4) through (A)(8) must be installed and used ≫250.50≪.

Pipe or conduit electrodes shall not be smaller than trade size ¾-in., and, if steel, shall have a galvanized outer surface (or shall be otherwise metal-coated or corrosion-protected) ≫250.52(A)(5)(a)≪.

If practicable, rod, pipe, and plate electrodes shall be embedded below permanent moisture level. Rod, pipe, and plate electrodes shall be free from nonconductive coatings such as paint or enamel ≫250.53(A)(1)≪.

Aluminum grounding electrodes are not permitted ≫250.52(B)(2)≪.

A The electrode's upper end must be flush with or below ground level unless the aboveground end and the grounding electrode conductor attachment are adequately protected against physical damage as specified in 250.10 ≫250.53(G)≪.

B Rod-type grounding electrodes of stainless steel and copper or zinc-coated steel must be at least ⅝ in. (15.87 mm) in diameter unless listed ≫250.52(A)(5)(b)≪.

C In multiple electrode configurations of the type specified in 250.52(A)(5) or (A)(7), each electrode of one grounding system (including that used for strike termination devices) must be at least 6 ft (1.83 m) from any other electrode of another grounding system. Two or more grounding electrodes bonded together are considered a single grounding electrode system ≫250.53(B)≪.

D In accordance with 250.53(A)(2), if a single rod, pipe, or plate electrode is installed, it must be supplemented by an additional electrode. But, if the single rod, pipe, or plate electrode has a resistance to earth of 25 ohms or less, the supplemental electrode is not required ≫250.53(A)(2) Exception≪.

E A metal underground gas piping system *shall not* be used as a grounding electrode ≫250.52(B)≪.

F At least 8 ft (2.44 m) of electrode length must be in contact with the soil ≫250.53(G)≪.

G Where rock bottom is encountered, the electrode can be driven at an oblique angle not to exceed 45° from vertical ≫250.53(G)≪.

H Where rock bottom is encountered, the electrode can be buried in a trench that is at least 2½ ft (750 mm) deep ≫250.53(G)≪.

I Each plate electrode must expose at least 2 ft² (0.186 m²) of surface to exterior soil ≫250.52(A)(7)≪. (One square foot of surface area on each side of a 1 ft × 1 ft plate equals 2 square feet.)

J If multiple rod, pipe, or plate electrodes are installed to meet the requirements of 250.53(A)(2), they shall not be less than 6 ft (1.8 m) apart ≫250.53(A)(3)≪.

The paralleling efficiency of rods is increased by spacing them twice the length of the longest rod ≫250.53(A)(3) Informational Note≪.

K Electrodes of bare or conductively coated iron or steel plates must be at least ¼ in. (6.4 mm) thick, while solid, uncoated nonferrous metal plates must be at least 0.06 in. (1.5 mm) thick ≫250.52(A)(7)≪.

L Plate electrodes must be installed at least 2½ ft (750 mm) below the earth's surface ≫250.53(H)≪.

> ### NOTE
>
> A single rod, pipe, or plate electrode shall be supplemented by an additional electrode of a type specified in 250.52(A)(2) through (A)(8). If the supplemental electrode is of the rod, pipe, or plate type, it shall comply with 250.52(A). The supplemental electrode shall be permitted to be bonded to one of the following: (1) rod, pipe, or plate electrode; (2) grounding electrode conductor; (3) grounded service-entrance conductor; (4) nonflexible grounded service raceway; or (5) any grounded service enclosure ≫**250.53(A)(2)**≪.

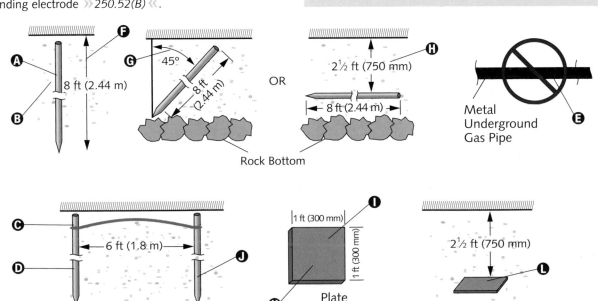

Rock Bottom

Metal Underground Gas Pipe

Plate Electrode

© Cengage Learning 2012

Grounding Electrode Conductor

The grounding electrode conductor must be made of copper, aluminum, copper-clad aluminum, or the items as permitted in 250.68(C). The material selected shall be inherently resistant to corrosive conditions existing at the site of installation or shall be protected against corrosion. Conductors of the wire type shall be solid (or stranded), insulated, covered, or bare ≫ *250.62* ≪.

Bare aluminum or copper-clad aluminum grounding electrode conductors must not be used in direct contact with masonry or earth or where subject to corrosive conditions. If used outside, aluminum or copper-clad aluminum grounding electrode conductors must not be installed within 18 in. (450 mm) of the earth ≫ *250.64(A)* ≪.

Ⓐ A grounding electrode conductor must be used to connect the equipment grounding conductors, the service-equipment enclosures, and, in a grounded system, the grounded service conductor to the grounding electrode(s) as required by Part III of Article 250. The conductor shall be sized in accordance with 250.66 ≫ *250.24(D)* ≪.

Ⓑ The grounding electrode conductor must be installed in one continuous length (without a splice or joint), unless as permitted in 250.30(A)(5) and (A)(6), 250.30(B)(1), and 250.68(C) ≫ *250.64(C)* ≪.

Ⓒ Where exposed, a grounding electrode conductor (or its enclosure) must be securely fastened to the surface on which it is carried ≫ *250.64(B)* ≪.

Ⓓ If a raceway is used as protection for a grounding electrode conductor, the installation must comply with appropriate raceway article requirements ≫ *250.64(E)(4)* ≪.

Ⓔ Ferrous metal raceways and enclosures shall be bonded at each end of the raceway or enclosure to the grounding electrode or grounding electrode conductor ≫ *250.64(E)(1)* ≪.

Ⓕ Ferrous metal raceways and enclosures for grounding electrode conductors shall be electrically continuous from the point of attachment to cabinets or equipment to the grounding electrode and shall be fastened securely to the ground clamp (or fitting) ≫ *250.64(E)(1)* ≪.

Ⓖ The grounding conductor must be connected to the grounding electrode by exothermic welding or by listed means, such as lugs, pressure connectors, clamps, etc. Connections depending on solder shall not be used ≫ *250.70* ≪.

Ⓗ Only one conductor can be connected to the grounding electrode by a single clamp or fitting, unless it is listed for multiple conductors ≫ *250.70* ≪.

Ⓘ This is one variety of ground-rod clamp listed for direct burial.

Ⓙ Ground clamps must be listed for the grounding electrode and grounding electrode conductor materials and, if used on pipe (rod, or other buried electrodes), must also be listed for direct soil burial or concrete encasement ≫ *250.70* ≪.

Ⓚ The bonding jumper for a grounding electrode conductor raceway or cable armor must be the same size (or larger) than the required enclosed grounding electrode conductor ≫ *250.64(E)(3)* ≪.

Ⓛ Grounding electrode conductors and grounding electrode bonding jumpers shall not be required to comply with 300.5 ≫ *250.64(B)* ≪.

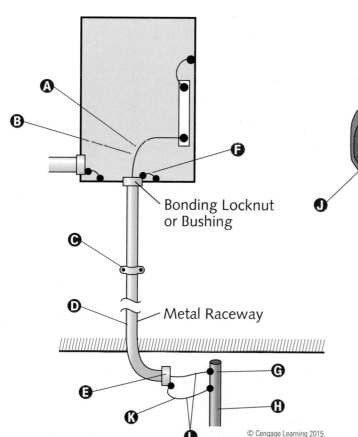

Bonding Locknut or Bushing

Metal Raceway

© Cengage Learning 2015.

Grounding Electrode Conductor Sizing

Where exposed, a grounding electrode conductor or its enclosure shall be securely fastened to the surface on which it is carried. A 4 AWG or larger grounding electrode conductor (copper or aluminum) must be protected if exposed to physical damage. A 6 AWG grounding electrode conductor not exposed to physical damage can run along the surface of the building construction without metal covering or protection, provided it is securely fastened to the construction. Otherwise, it must be protected in RMC, IMC, PVC, RTRC, EMT, or cable armor 》》*250.64(B)*《《.

Grounding electrode conductors smaller than 6 AWG shall be protected in RMC, IMC, PVC, RTRC, EMT, or cable armor 》》*250.64(B)*《《.

A The connection of a grounding electrode conductor at the service, at each building or structure where supplied by a feeder(s) or branch circuit(s), or at a separately derived system and associated bonding jumper(s) shall be made as specified 250.68(A) through (C) 》》*250.68*《《.

B A 4 AWG copper (or 2 AWG aluminum or copper-clad aluminum) grounding electrode conductor is required if the service-entrance conductors are 2/0 AWG or 3/0 AWG copper 》》*Table 250.66*《《.

C The interior metal water piping system can be used as a conductor to interconnect electrodes that are part of the grounding electrode system, but only if located not more than 5 ft (1.52 m) from the point of entrance to the building 》》*250.68(C)(1)*《《.

D To ensure an effective grounding path for a metal piping system used as a grounding electrode, bonding must be provided as necessary around insulated joints as well as around any equipment likely to be disconnected for repairs or replacement. Bonding jumpers shall be of sufficient length to permit removal of such equipment while retaining the integrity of the grounding path 》》*250.68(B)*《《.

E If the grounding electrode conductor is connected to a single or multiple rod, pipe, or plate electrode(s), or any combination thereof, as permitted in 250.52(A)(5) or (A)(7), that portion of the conductor serving as the sole connection to the grounding electrode(s) is not required to be larger than 6 AWG copper wire or 4 AWG aluminum wire 》》*250.66(A)*《《. Grounding electrode conductors and grounding electrode bonding jumpers shall not be required to comply with 300.5 》》*250.64(B)*《《.

F Where the grounding electrode conductor is connected to a single or multiple concrete-encased electrode(s) as permitted in 250.52(A)(3), that portion of the conductor serving as the sole connection to the grounding electrode(s) is not required to be larger than 4 AWG copper wire 》》*250.66(B)*《《.

G If, as 250.52(A)(4) permits, the grounding electrode conductor is connected to a ground ring, that portion of the conductor serving as the sole connection to the grounding electrode does not have to be larger than the ground ring conductor 》》*250.66(C)*《《.

> **NOTE**
>
> Grounding electrode conductors can be installed on or through framing members 》》**250.64(B)**《《.

> **CAUTION**
>
> All mechanical elements used to terminate a grounding electrode conductor or bonding jumper to a grounding electrode shall be accessible unless meeting one of the two exceptions 》》*250.68(A)*《《.

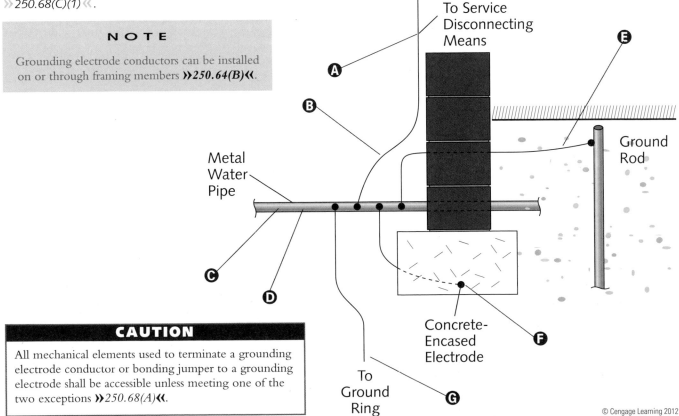

To Service Disconnecting Means

Ground Rod

Metal Water Pipe

Concrete-Encased Electrode

To Ground Ring

© Cengage Learning 2012

Supply-Side Bonding Jumper

A Grounding electrode conductor and equipment bonding jumper sizes on the supply side of the service are determined by service-entrance conductor sizes.

B Unit 8 explains grounded (neutral) conductor sizing.

C The size of the grounding electrode conductor at the service, at each building or structure where supplied by a feeder(s) or branch circuit(s), or at a separately derived system of a grounded or ungrounded ac system shall not be less than given in Table 250.66, except as permitted in 250.66(A) through (C) ≫ *250.66* ≪.

D The supply-side bonding jumper shall not be smaller than specified in Table 250.102(C)(1) ≫ *250.102(C)* ≪.

Service Disconnecting Means

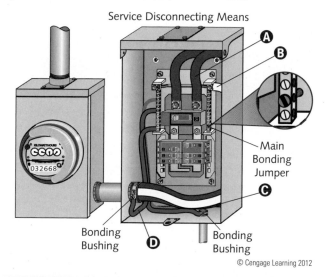

Main Bonding Jumper

Bonding Bushing

Bonding Bushing

© Cengage Learning 2012

CAUTION

Metal piping system(s), including gas piping, with the potential to become energized must be bonded to any of the following: (1) equipment grounding conductor for the circuit that is likely to energize the piping system; (2) service equipment enclosure; (3) grounded conductor at the service; (4) grounding electrode conductor, if of sufficient size; and (5) one or more grounding electrodes used. The bonding conductor(s) or jumper(s) must be sized in accordance with Table 250.122 by using the rating of the circuit that is likely to energize the piping system(s). The points of attachment of the bonding jumper(s) shall be accessible ≫*250.104(B)*≪.

NOTE

The interior metal water piping system must be bonded either to the service equipment enclosure, the grounded conductor at the service, the grounding electrode conductor (if of sufficient size), or to the one or more grounding electrodes used. The bonding jumper must be sized in accordance with Table 250.66 and installed in accordance with 250.64(A), (B), and (E). The bonding jumper's point of attachment must be accessible ≫*250.104(A) and (A)(1)*≪.

Equipment Bonding Jumper on the Load Side of an Overcurrent Device

The equipment bonding jumper on the load side of an overcurrent device(s) shall be sized, in accordance with 250.122 ≫ *250.102(D)* ≪.

A A panelboard (on the load side of an overcurrent device) protected by a 100-ampere overcurrent protection device (circuit breaker or fuse) requires an 8 AWG copper equipment grounding conductor ≫ *Table 250.122* ≪.

B Wire-type copper, aluminum, or copper-clad aluminum equipment grounding conductors must not be less than shown in Table 250.122, but in no case are they required to be larger than the circuit conductors supplying the equipment. Where a raceway, cable armor, or sheath is used as the equipment grounding conductor (as provided in 250.118 and 250.134[A]), it must comply with 250.4(A)(5) or (B)(4) ≫ *250.122(A)* ≪.

C Metal panelboard cabinets and frames must be connected to an equipment grounding conductor and in physical contact with each other. Where the panelboard is used with nonmetallic raceway (or cable), or where separate grounding conductors are provided, a terminal bar for the grounding conductors must be secured within the cabinet. The terminal bar must be bonded to the metal cabinet and panelboard frame. Otherwise it shall be connected to the equipment grounding conductor that is run with the conductors feeding the panelboard ≫ *408.40* ≪.

D Equipment grounding conductor and equipment bonding jumper sizes (on the load side of an overcurrent device) are determined by the size of the overcurrent device that protects the equipment.

E Unit 8 explains grounded (neutral) conductor sizing.

To 100-ampere Feeder

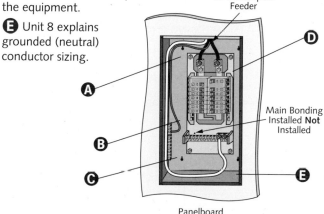

Main Bonding Installed **Not** Installed

Panelboard

© Cengage Learning 2012

NOTE

Where ungrounded conductor sizes are increased in size from the minimum size that has sufficient ampacity for the intended installation, wire-type equipment grounding conductors, where installed, shall be increased in size proportionately according to circular mil area of the ungrounded conductors ≫*250.122(B)*≪.
An equipment grounding conductor run with (or enclosing) the circuit conductors must be one of those listed in 250.118.

Intersystem Bonding Termination

An intersystem bonding termination for connecting intersystem bonding conductors required for other systems shall be provided external to enclosures at the service equipment or metering equipment enclosure and at the disconnecting means for any additional buildings or structures. The intersystem bonding termination shall comply with the following:

(1) Be accessible for connection and inspection.
(2) Consist of a set of terminals with the capacity for connection of not less than three intersystem bonding conductors.
(3) Not interfere with opening the enclosure for a service, building or structure disconnecting means, or metering equipment.
(4) At the service equipment, be securely mounted and electrically connected to an enclosure for the service equipment, to the meter enclosure, or to an exposed nonflexible metallic service raceway, or be mounted at one of these enclosures and be connected to the enclosure or to the grounding electrode conductor with a minimum 6 AWG copper conductor
(5) At the disconnecting means for a building or structure, be securely mounted and electrically connected to the metallic enclosure for the building or structure disconnecting means, or be mounted at the disconnecting means and be connected to the metallic enclosure or to the grounding electrode conductor with a minimum 6 AWG copper conductor.

(6) The terminals shall be listed as grounding and bonding equipment ≫*250.94*≪.

Ⓐ Shall be located at the service equipment and shall not interfere with opening doors or covers.

Ⓑ Shall have at least three terminals for intersystem bonding conductors.

Ⓒ If connected to the grounding conductor, the minimum size is 6 AWG copper.

Ⓓ Shall be accessible for connection and inspection.

Ⓔ The terminals shall be listed as grounding and bonding equipment.

> **NOTE**
>
> Exception: In existing buildings or structures where any of the intersystem bonding and grounding electrode conductors required by 770.100(B)(2), 800.100(B)(2), 810.21(F)(2), 820.100(B)(2), and 830.100(B)(2) exist, installation of the intersystem bonding termination is not required. An accessible means external to enclosures for connecting intersystem bonding and grounding electrode conductors shall be permitted at the service equipment and at the disconnecting means for any additional buildings or structures by at least one of the following means:
> (1) Exposed nonflexible metallic raceways
> (2) An exposed grounding electrode conductor
> (3) Approved means for the external connection of a copper or other corrosion-resistant bonding or grounding electrode conductor to the grounded raceway or equipment.

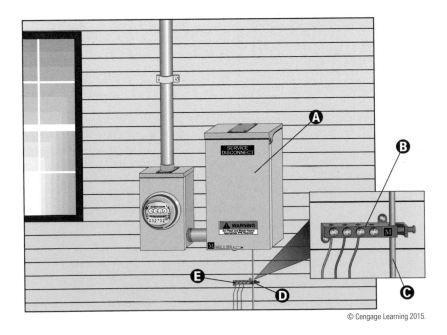

© Cengage Learning 2015.

Separate Building or Structure – Grounding Electrode Not Required

Besides other requirements in the National Electrical Code, installations in separate buildings and structures must meet requirements in Article 225. Requirements in Article 225 cover outside branch circuits and feeders.

Means shall be provided for disconnecting all ungrounded conductors that supply or pass through the building or structure ≫225.31≪. The disconnecting means specified in 225.31 shall be comprised of a circuit breaker, molded case switch, general-use switch, snap switch, or other approved means. Where applied in accordance with 250.32(B), Exception, the disconnecting means shall be suitable for use as service equipment ≫225.36≪.

A building is defined as a structure that stands alone or that is cut off from adjoining structures by fire walls with all openings therein protected by approved fire doors ≫*Article 100*≪.

A A structure is simply defined as that which is built or constructed ≫*Article 100*≪.

B A grounding electrode shall not be required where only a single branch circuit, including a multiwire branch circuit, supplies the building or structure, and the branch circuit includes an equipment grounding conductor for grounding the normally non-current-carrying metal parts of equipment ≫*250.32 Exception*≪.

C If a one-family dwelling has a detached garage or accessory building with electric power, at least one wall switch–controlled lighting outlet shall be installed to provide illumination on the exterior side of outdoor entrances or exits with grade level access. A vehicle door in a garage shall not be considered as an outdoor entrance or exit ≫*210.70(A)(2)(b)*≪.

D In each detached garage with electric power, at least one receptacle outlet shall be installed. The branch circuit supplying this receptacle(s) shall not supply outlets outside of the garage ≫*210.52(G)(1)*≪. In each accessory building with electric power, at least one receptacle outlet shall be installed ≫*210.52(G)(2)*≪.

E At least one receptacle outlet shall be installed for each car space ≫*210.52(G)(1)*≪.

> **NOTE**
>
> At least one wall switch–controlled lighting outlet shall be installed in detached garages with electric power ≫*210.70(A)(2)(a)*≪.

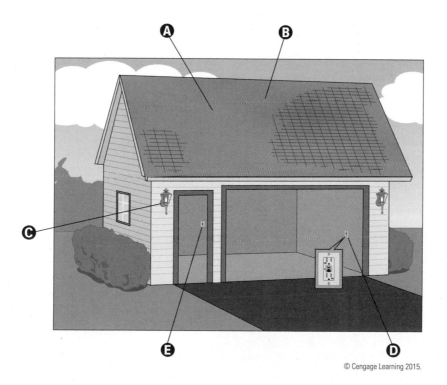

Separate Buildings or Structures

Besides other requirements in the NEC, installations in separate buildings and structures must meet requirements in Article 225. Requirements in Article 225 cover outside branch circuits and feeders.

A Buildings or structures supplied by feeders or branch circuits must have a grounding electrode or grounding electrode system installed in accordance with Part III of Article 250. The grounding electrode conductors must be installed in accordance with 250.32(B) for grounded systems, or (C) for ungrounded systems. If there is no existing grounding electrode, the grounding electrode required in 250.50 must be installed ≫ *250.32(A)* ≪.

B Means shall be provided for disconnecting all ungrounded conductors that supply or pass through the building or structure ≫ *225.31* ≪.

C The disconnecting means shall be installed either inside or outside of the building or structure served or where the conductors pass through the building or structure. The disconnecting means shall be at a readily accessible location nearest the point of entrance of the conductors ≫ *225.32* ≪.

D The disconnecting means specified in 225.31 shall be comprised of a circuit breaker, molded case switch, general-use switch, snap switch, or other approved means ≫ *225.36* ≪.

E The building or structure disconnecting means shall plainly indicate whether it is in the open or closed position ≫ *225.38(D)* ≪.

F An equipment grounding conductor, as described in 250.118, shall be run with the supply conductors and connected to the building or structure disconnecting means and to the grounding electrode(s). The equipment grounding conductor shall be used for grounding or bonding or equipment, structures, or frames required to be grounded or bonded. The equipment grounding conductor must be sized in accordance with 250.122 ≫ *250.32(B)(1)* ≪.

G Any installed grounded conductor shall not be connected to the equipment grounding conductor or to the grounding electrode(s) ≫ *250.32(B)(1)* ≪.

H The size of the grounding electrode conductor to the grounding electrode(s) shall not be smaller than given in 250.66, based on the largest ungrounded supply conductor. The installation shall comply with Part III of Article 250 ≫ *250.32(E)* ≪.

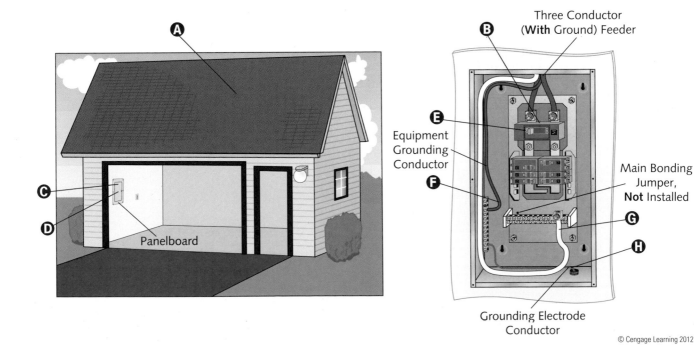

Three Conductor (**With** Ground) Feeder

Equipment Grounding Conductor

Main Bonding Jumper, **Not** Installed

Panelboard

Grounding Electrode Conductor

© Cengage Learning 2012

Preventing Objectionable Current

Normally non-current-carrying electrically conductive materials that are likely to become energized must be connected together and to the electrical supply source in a manner that establishes an effective ground-fault current path ≫*250.4(A)(4)*≪.

Temporary currents resulting from abnormal conditions, such as ground faults, are not classified as objectionable current for the purpose specified in 250.6(A) and (B) ≫*250.6(C)*≪.

Ⓐ Normally non-current-carrying conductive materials enclosing electrical conductors (or equipment), or forming part of such equipment, must be connected to earth so as to limit the voltage to ground on these materials ≫*250.4(A)(2)*≪.

Ⓑ If the use of multiple grounding connections results in objectionable current, one or more of the following alterations in 250.6(B)(1) through (4) are permitted, if the requirements of 250.4(A)(5) or (B)(4) are met: (1) discontinue one or more, but not all, of such grounding connections; (2) change the location of the grounding connections; (3) interrupt the continuity of the conductor (or conductive path) causing the objectionable current; and (4) take other suitable remedial and approved action ≫*250.6(B)*≪.

Ⓒ The grounding of electric systems, circuit conductors, surge arresters, surge-protective devices and conductive normally non-current-carrying materials (and equipment) must be installed and arranged in a manner that prevents objectionable current ≫*250.6(A)*≪.

CAUTION

Connecting a grounded (neutral) and grounding conductor in parallel creates objectionable current on the grounding conductor, because the unbalanced current flows through both the grounded and grounding conductor. Should the grounded conductor lose its continuity (become disconnected), the grounding conductor would then carry all of the unbalanced (neutral) current. A hazard would exist if the grounding conductor's ampacity rating is less than the current flowing through it.

NOTE

Normally non-current-carrying conductive materials enclosing electrical conductors or equipment, or forming part of such equipment, shall be connected together and to the electrical supply source in a manner that establishes an effective ground-fault current path ≫*250.4(A)(3)*≪.

NOTE

Electrical equipment and wiring and other electrically conductive material likely to become energized shall be installed in a manner that creates a low-impedance circuit facilitating the operation of the overcurrent device or ground detector for high-impedance grounded systems. It shall be capable of safely carrying the maximum ground-fault current likely to be imposed on it from any point on the wiring system where a ground fault may occur to the electrical supply source. The earth shall not be considered as an effective ground-fault current path ≫*250.4(A)(5)*≪.

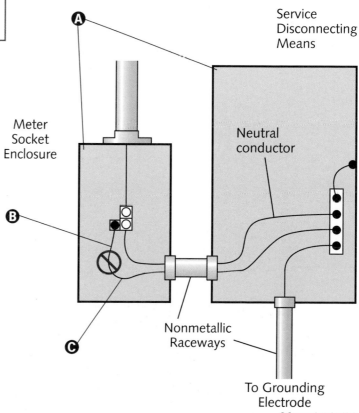

Service Disconnecting Means

Meter Socket Enclosure

Neutral conductor

Nonmetallic Raceways

To Grounding Electrode

Summary

- A service mast must safely withstand the strain imposed by service-drop conductors.
- Nothing except electrical service-drop conductors can be attached to the service mast.
- Wiring methods permitted for services are listed in 230.43.
- Service-drop conductor vertical ground clearances are found in 230.24(B)
- Service-drop conductor roof clearances are in 230.24(A).
- A minimum of 6½ ft (2.0 m) of workspace headroom is required around service equipment and panelboards.
- A minimum frontal workspace width is 30 in. (762 mm) for service equipment and panelboards; the depth of this workspace must be at least 3 ft (914 mm).
- Panelboards are mounted in cabinets or cutout boxes.
- Each service disconnect must be permanently marked identifying it as a service disconnect.
- The service disconnecting means must be installed at a readily accessible location whether inside or outside.
- Two to six circuit breakers or sets of fuses can serve as the overcurrent device, providing overload protection.
- It shall be permissible to use 310.15(B)(7) to size 120/240-volt, single-phase dwelling services and certain feeders rated 100 through 400 amps.
- If the installation does not meet the specifications in 310.15(B)(7), the conductors shall be sized in accordance with Table 310.15(B)(16).
- Electric equipment must be installed in a neat and professional manner.
- Grounding and bonding conductors must be connected by listed means.
- The grounded (neutral) conductor is isolated from the equipment grounding conductor in subpanels.
- Grounding electrode conductors are sized from 250.66 and Table 250.66.
- Use Table 250.102(C)(1) to size main bonding jumpers, system bonding jumpers, and supply side bonding jumpers.
- Equipment grounding conductors and bonding jumpers, on the load side of the service, are sized according to 250.122 and Table 250.122.
- All non-current-carrying metal parts of service equipment must be bonded together.
- Grounding electrode systems (and conductors) are located in Part III of Article 250.
- Panelboards in separate buildings or structures must be installed according to 250.32 provisions.

Unit 9 Competency Test

NEC Reference	Answer	
_____	_____	1. The point of attachment of the service-drop conductors to a building must not be less than _____ ft above finish grade.
_____	_____	2. The supply-side bonding jumper shall not be smaller than the sizes shown in Table _____ for grounding electrode conductors.
_____	_____	3. Service conductors run above the top level of a window shall be:

    a) 8 ft above a window.

    b) accessible.

    c) 3 ft above a window.

    d) considered out of reach.

_____	_____	4. Every circuit breaker having an interrupting rating other than _____ amperes shall have its interrupting rating shown on the circuit breaker.
_____	_____	5. Energized parts of service equipment that are not enclosed shall be installed on a _____ and guarded in accordance with 110.18 and 110.27.

    I. switchboard

    II. panelboard

    III. control board

        a) II only   b) I or II only   c) II or III only   d) I, II, or III

NEC Reference	Answer

6. In addition to the requirement of 408.30, a panelboard shall be protected by an overcurrent protective device having a rating not greater than _____.

7. Service-entrance cables shall be supported by straps or other approved means within _____ of every service head, gooseneck, or connection to a raceway or enclosure.

   a) 1 ft b) 6 ft c) 12 ft d) 30 ft

8. Bonding jumpers meeting the other requirements of this article shall be used around impaired connections, such as reducing washers or oversized _____ knockouts.

9. Residential 240-volt service-drop conductors must have a vertical clearance above a driveway of at least _____ ft.

10. A grounding ring encircling the building, in direct contact with the earth, must consist of at least _____ ft of bare copper conductor not smaller than 2 AWG.

11. Plug fuses of 15-ampere and lower rating shall be identified by a(n) _____ configuration of the window, cap, or other prominent part to distinguish them from fuses of higher ampere ratings.

12. Where more than one grounding electrode is used, each electrode of one grounding system (including that used for strike termination devices) shall not be less than _____ ft from any other electrode of another grounding system.

13. Every circuit and circuit modification in a panelboard shall be legibly identified as to its _____ purpose or use.

14. Service heads and _____ in service-entrance cables shall be located above the point of attachment of the service-drop or overhead service conductors to the building or other structure.

15. The width of the working space in front of the electric equipment shall be the width of the equipment or _____ in., whichever is greater.

16. Which of the following is not a true statement concerning an equipment grounding conductor?

   a) Under certain conditions, equipment grounding conductors shall be required to be larger than circuit conductors supplying the equipment.

   b) One equipment grounding conductor may serve multiple circuits.

   c) Under certain conditions, equipment grounding conductors may be run in parallel.

   d) The size of equipment grounding conductors shall be increased where ungrounded conductors are increased in size.

17. The minimum height of working space for equipment operating at 600 volts, nominal, or less to ground and likely to require examination, adjustment, servicing, or maintenance while energized shall be _____ ft or the height of the equipment, whichever is greater.

18. Service-drop conductors shall have a vertical clearance of not less than _____ ft above the roof surface.

19. In damp or wet locations, surface-type cabinets, cutout boxes, and meter socket enclosures shall be so placed or equipped as to prevent moisture or water from entering and accumulating within them, and mounted so there is at least _____ in. airspace between the enclosure and the wall or other supporting surface.

20. Other equipment associated with the electrical installation located above or below the electrical installation shall be permitted to extend not more than _____ beyond the front of the electrical equipment.

   a) 0 in. b) 6 in. c) 12 in. d) 24 in.

NEC **Reference**   **Answer**

_____   _____   21. What size copper equipment bonding jumper is required for a 200-ampere panelboard located on the load side of the overcurrent device? (The conductors feeding the panelboard are 3/0 AWG copper conductors.)

_____   _____   22. Plug fuses of the Edison-base type shall be classified at not over 125 volts and _____ amperes and below.

_____   _____   23. A _____ is a conductor used to connect the system grounded conductor or the equipment to a grounding electrode or to a point on the grounding electrode system.

_____   _____   24. If a single rod, pipe, or plate grounding electrode has a resistance to earth of _____ or less, a supplemental grounding electrode shall not be required.

_____   _____   25. A disconnecting means serving a hermetic refrigerant motor compressor shall be selected on the basis of the nameplate rated-load current or branch-circuit selection current, whichever is greater, and locked-rotor current, respectively, of the motor compressor. The ampere rating shall be at least _____ of the nameplate rated-load current or branch-circuit selection current, whichever is greater.

_____   _____   26. Circuit breakers rated _____ or less and 1000 volts or less shall have the ampere rating molded, stamped, etched, or similarly marked into their handles or escutcheon areas.

_____   _____   27. A dwelling has a 175-ampere service that is fed with THW copper conductors. What is the minimum size copper grounding electrode conductor?

_____   _____   28. The _____ is the connection between the grounded circuit conductor and the equipment grounding conductor at the service.

_____   _____   29. Which of the following is not a standard ampere rating for fuses?

      a) 25 amperes   b) 50 amperes   c) 75 amperes   d) 601 amperes

_____   _____   30. If multiple rod, pipe, or plate electrodes are installed to meet the installation requirements of the grounding electrode system, they shall not be less than _____ ft apart.

_____   _____   31. Where the supply conductors are larger than 1100 kcmi copper, the main bonding jumper shall have an area not less than _____% of the area of the largest phase conductor.

_____   _____   32. Service conductors are the conductors from the service point to the _____.

_____   _____   33. The minimum depth of working space for a 120/240-volt panelboard, with exposed live parts on one side and grounded parts on the other, shall be at least _____ ft.

_____   _____   34. Each plate electrode shall expose not less than _____ ft^2 of surface to exterior soil and shall be installed not less than _____ ft below the surface of the earth.

_____   _____   35. Where installed on the outside of a raceway, the length of the equipment bonding jumper shall not exceed _____ ft and shall be routed with the raceway.

_____   _____   36. An intersystem bonding termination shall consist of a set of terminals with the capacity for connection of not less than _____ intersystem bonding conductors.

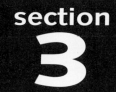
UNIT 10 Comprehensive Provisions

Objectives

After studying this unit, the student should:

▶ be aware that general receptacle and lighting installation provisions for multifamily dwellings are almost identical to those for one-family structures.

▶ know that certain exceptions permit the installation of more than one service on a multifamily building.

▶ understand the provisions limiting the number of service disconnecting means to only six.

▶ be familiar with the provision requiring that each multifamily dwelling occupant have access to the unit's service disconnecting means.

▶ understand that each service supplying a building must be connected to the same grounding electrode.

▶ be familiar with service conductor vertical clearance requirements.

▶ be able to determine when the 83 percent adjustment factor in 310.15(B)(7) can be used to size service and feeder conductors.

▶ know that balconies, decks, and porches that are attached to the dwelling unit and are accessible from inside the dwelling unit shall have at least one receptacle outlet accessible from the balcony, deck, or porch.

▶ be able to choose the appropriate voltage-drop formula, insert the proper data, and calculate the correct answer.

Introduction

A multifamily dwelling is a building containing three or more dwelling units. Apartments or condominiums are examples of multifamily dwellings. They can be any size, from one-story structures containing three units to high-rise buildings containing hundreds of units. One-family dwelling units (covered in Section 2 of this book) contain fundamental provisions essential to all dwelling units, whether one-, two-, or multifamily. Most of the general installation requirements (such as receptacle and switch placement), as well as most of the specific provisions (such as kitchen and bathroom regulations) pertain to all dwelling occupancies. A few requirements in this unit are specific to multifamily dwellings. Service provisions pertaining to multifamily dwellings, such as the maximum number of service disconnecting means, are explained in this unit, which also includes a brief description and explanation of voltage-drop formulas. A thorough understanding of one-family dwelling requirements is essential to understanding those for multifamily dwellings.

PLANS (BLUEPRINTS)

Partial Site Plan

A For individual dwelling units of multifamily dwellings, service and feeder conductors supplied by a single-phase, 120/240-volt system shall be permitted to be sized in accordance with 310.15(B)(7)(1) through (4). For a feeder rated 100 through 400 A, the feeder conductors supplying the entire load associated with an individual dwelling unit in a multifamily dwelling, shall be permitted to have an ampacity not less than 83 percent of the feeder rating 》310.15(B)(7)《.

In a multiple-occupancy building, each occupant must have access to the unit's service disconnecting means, unless electric service and electrical maintenance are provided by building management 》230.72(C)《.

B A laundry receptacle is not required if one of the applicable exceptions is met 》210.52(F)《.

C With few exceptions, general lighting provisions are the same for both one- and multifamily dwellings 》210.70《.

D Use 220.82 to perform optional method load calculations for individual feeders in each unit of multifamily dwellings.

E Clubhouse and swimming pool load calculations are determined in accordance with Article 220, Part III provisions.

The clubhouse is exempt from dwelling unit(s) provisions.

If any portion of the clubhouse is designed or intended for the assembly of more than 99 persons, that portion must comply with Article 518 provisions.

F General receptacle installation provisions for multifamily dwellings are almost identical to those for one-family dwellings 》210.52《.

G Use Part III of Article 220 to perform standard method load calculations for both feeders and services.

Use 220.84 to perform multifamily dwelling optional load calculations for feeders or services consisting of three or more units.

H Swimming pool provisions are contained in Article 680. Swimming pool motor loads are computed in accordance with 430.22 and 430.24 requirements.

I Air-conditioning and refrigeration equipment provisions are found in Article 440.

NOTE

If calculating by the optional method, house loads must be calculated in accordance with Article 220, Part III, and must be in addition to the dwelling unit loads calculated by Table 220.84 demand factors 》220.84(B)《.

CAUTION

At least one 125-volt, single-phase, 15- or 20-ampere-rated receptacle outlet shall be installed within 50 ft (15 m) of the electrical service equipment 》210.64《.

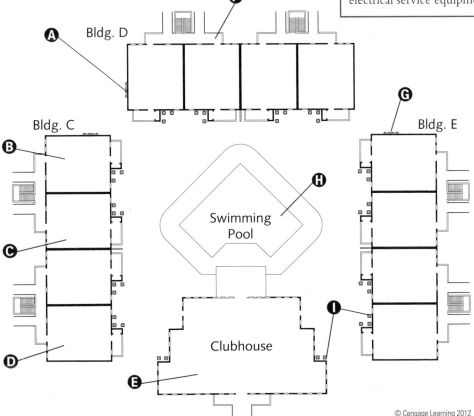

Electrical Plan

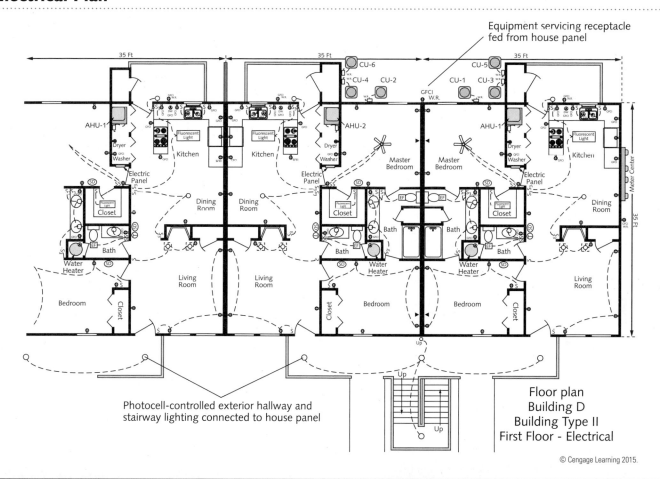

Equipment servicing receptacle fed from house panel

Photocell-controlled exterior hallway and stairway lighting connected to house panel

Floor plan
Building D
Building Type II
First Floor - Electrical

© Cengage Learning 2015.

SERVICES

Maximum Number of Services

Ⓐ Vertical service-drop conductor clearances are covered in Unit 9 of this book.

Ⓑ A building (or other structure) shall be supplied by only one service unless permitted in 230.2(A) through (D) 》230.2 《.

Ⓒ Each service disconnecting means permitted by 230.2, whether for a service or for a set of service-entrance conductors permitted by 230.40 (Exception Nos. 1, 3, 4, or 5) shall consist of not more than six switches (or sets of circuit breakers) mounted in a single enclosure, in a group of separate enclosures, or in a switchboard or in switchgear 》230.71(A) 《.

Ⓓ Auxiliary gutter provisions are located in Article 366.

Ⓔ The conductors and equipment that deliver energy from the electricity supply system to the wiring system of the premises are called the service 》Article 100 《.

Ⓕ Wiring spaces can be supplemented by auxiliary gutters at meter centers, distribution centers, switchboards, and similar points and may enclose conductors or busbars. Auxiliary gutters, however, shall not be used to enclose switches, overcurrent devices, appliances, or similar equipment 》366.10 and 12 《.

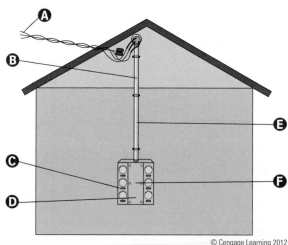

© Cengage Learning 2012

> ### NOTE
>
> Unit 9 explains equipment working space, 110.26.

Maximum Number of Disconnects per Service

A Section 230.40 basic requirements state that each service drop (or lateral) must supply only one set of service-entrance conductors. However, several exceptions exist.

B Two to six service disconnecting means in separate enclosures, grouped at one location and supplying separate loads from one service drop, set of overhead service conductors, sets of underground service conductors, or service lateral, can be supplied by one set of service-entrance conductors to each or several such service equipment enclosures ≫ *230.40 Exception No. 2* ≪.

C There shall be no more than six sets of disconnects per service grouped in a single location ≫ *230.71(A)* ≪.

D The two to six disconnects, as permitted in 230.71, must be grouped ≫ *230.72(A)* ≪.

E Each disconnect must be marked to indicate the load served ≫ *230.72(A)* ≪.

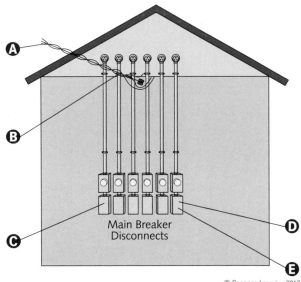

© Cengage Learning 2012

> ### N O T E
>
> For the purpose of 230.71, a disconnecting means used solely for the control circuit of the ground-fault protection system, installed as part of the listed equipment, is not considered a service disconnecting means ≫*230.71(A)*≪.

> ### CAUTION
>
> At least one 125-volt, single-phase, 15- or 20-ampere-rated receptacle outlet shall be installed within 50 ft (15 m) of the electrical service equipment ≫*210.64*≪.

Two Disconnects Supplying Twelve Units

A Conductors must have adequate mechanical strength and sufficient ampacity to carry the current for the load as calculated in accordance with Article 220 ≫ *230.23(A)* ≪.

B Many possible combinations exist for these twelve units: one main disconnect feeding all twelve units; three main disconnects feeding twelve units (four units per disconnect); and six main disconnects feeding twelve units (two units per disconnect), to name a few. The arrangement of service disconnects is not critical, as long as no more than six operations of the hand disconnect all of the service conductors.

C There must be no more than six disconnects per service grouped in a single location ≫ *230.71(A)* ≪.

D If the service disconnecting means consists of more than one switch or circuit breaker, as permitted by 230.71, the combined ratings of all the switches or circuit breakers used must meet the rating requirements of 230.79 ≫ *230.80* ≪.

E In a multiple-occupancy building, each occupant must have access to his or her unit's service disconnecting means ≫ *230.72(C)* ≪.

F Each service disconnect must be permanently marked to identify it as a service disconnect ≫ *230.70(B)* ≪.

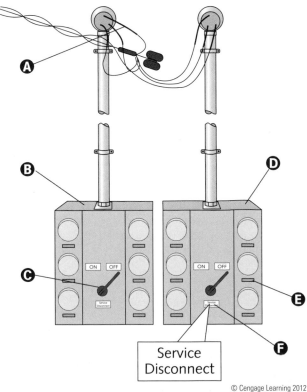

Service Disconnect

© Cengage Learning 2012

> ### N O T E
>
> In a multiple-occupancy building where electric service and electrical maintenance are provided by building management, under continuous supervision, the service disconnecting means supplying more than one occupancy can be accessible only to authorized management personnel ≫*230.72(C) Exception*≪.

More than One Service

A building or other structure can have another service for fire pumps where a separate service is required »*230.2(A)(1)*«.

A building or other structure can have another service for emergency, standby, or parallel power production systems where a separate service is required »*230.2(A)(2) through (5)*«.

By special permission, in multiple-occupancy buildings where there is no space available, service equipment does not have to be accessible to all occupants »*230.2(B)(1)*«. Special permission is the written consent of the AHJ »*Article 100*«.

Multiple services are permitted where the capacity requirements are in excess of 2000 amperes at a supply voltage of 1000 volts or less »*230.2(C)(1)*«.

A If the load requirements of a single-phase installation are greater than the serving agency normally supplies through one service, multiple services are permitted »*230.2(C)(2)*«.

B Where a building or structure is supplied by multiple services or by any combination of branch circuits, feeders, and services, a permanent plaque (or directory) must be installed at each service disconnect location denoting all other services, feeders, and branch circuits supplying that building as well as the area served by each. See 225.37 »*230.2(F)*«.

C Six is the maximum number of service disconnects for any one location »*230.71(A)*«.

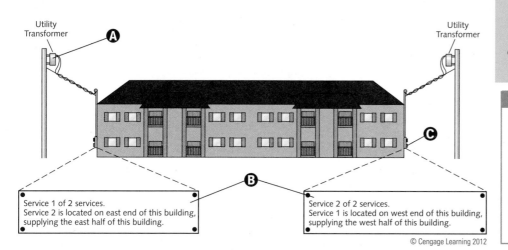

Service 1 of 2 services.
Service 2 is located on east end of this building, supplying the east half of this building.

Service 2 of 2 services.
Service 1 is located on west end of this building, supplying the west half of this building.

© Cengage Learning 2012

> **NOTE**
>
> One set of service-entrance conductors can supply the circuits of a two-family or multifamily dwelling as covered in 230.82(5) or (6) »*230.40 Exception No. 5*«.

> **WARNING**
>
> If a building or structure supplied by separate services requires connection to a grounding electrode, the same grounding electrode must be used for all services. Two or more grounding electrodes that are bonded together constitute a single grounding electrode system in this sense »*250.58*«.

One Service Lateral—One Location

Service lateral is defined as the underground conductors between the utility electric supply system and the service point »*Article 100*«.

A One set of service-entrance conductors can supply two to six service disconnecting means where located in separate enclosures, grouped at one location, and supply separate loads from one service drop or lateral. The one set of service-entrance conductors can supply each or several such service equipment enclosures »*230.40 Exception No. 2*«.

B Only as applied to 230.40, Exception No. 2, underground sets of conductors, 1/0 AWG and larger, running to the same location and connected together at their supply end (but not at their load end) constitute one service »*230.2*«.

> **NOTE**
>
> Underground service conductors are defined as the underground conductors between the service point and the first point of connection to the service-entrance conductors in a terminal box, meter, or other enclosure, inside or outside the building wall »*Article 100*«. Where there is no terminal box, meter, or other enclosure, the point of connection is considered to be the point of entrance of the service conductors into the building »*Informational Note to the definition of Service Conductors, Underground*«.

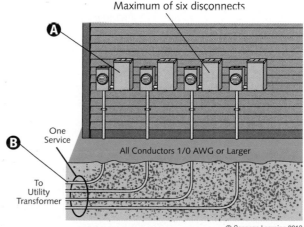

Maximum of six disconnects

One Service

To Utility Transformer

All Conductors 1/0 AWG or Larger

© Cengage Learning 2012

One Service Lateral—Six Disconnects Maximum

Ⓐ Six is the maximum number of disconnects per service grouped in any one location ›› *230.71(A)* ‹‹.

Ⓑ Even though the service conductors are not connected to the same meter socket enclosure, this installation is defined as one service ›› *230.2* ‹‹.

Ⓒ These are size 1/0 AWG and larger conductors running to the same location. Underground conductors are connected together at the supply (in the utility transformer).

Ⓓ Underground service conductor provisions are found in 230.30 through 33.

CAUTION
Since the maximum number of disconnects is six, it is a violation to install two six-gang meter socket enclosures (each without a main service disconnect) in one location. (Two six-gang meter socket enclosures times six service disconnects each equals twelve disconnects.)

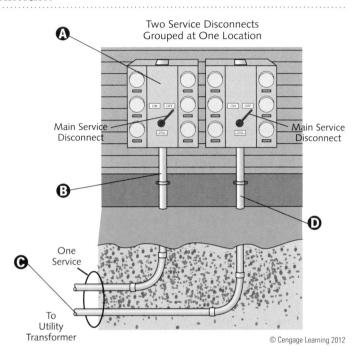

Two Service Disconnects Grouped at One Location

Main Service Disconnect

Main Service Disconnect

One Service

To Utility Transformer

© Cengage Learning 2012

More than Six Disconnects per Building

Ⓐ Here, two services supply one building as permitted by 230.2(C)(2).

Ⓑ The installation of two six-gang meter socket enclosures (each without a main service disconnect) in two locations is not a violation.

Ⓒ Where a building or structure is supplied by multiple services, or by any combination of branch circuits, feeders, and services, a permanent plaque (or directory) must be installed at each service disconnect location denoting all other services, feeders, and branch circuits supplying that building, as well as the area served by each ›› *230.2(E)* ‹‹.

Ⓓ Section 230.71(A) permits a maximum of six disconnects per service, not per building. If more than one service feeds a building, each service can have six disconnects.

NOTE
Each service must be connected to the same grounding electrode ››*250.58*‹‹.

CAUTION
Service conductors supplying a building (or other structure) shall not pass through the interior of another building ››*230.3*‹‹.

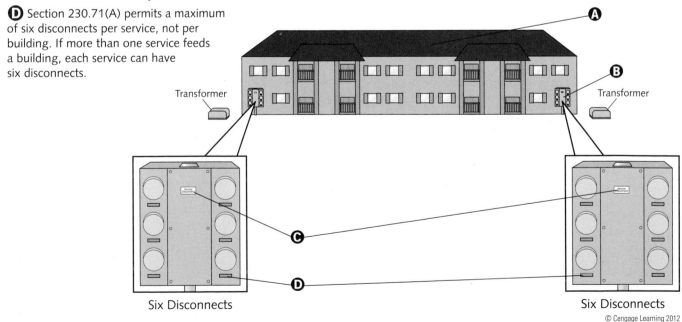

Transformer

Transformer

Six Disconnects

Six Disconnects

© Cengage Learning 2012

Service Conductors

A Section 310.15(B)(7) can be used to size 120/240, single-phase service-entrance conductors and service-lateral conductors, rated 100 through 400 amperes »*310.15(B)(7)*«.

B This book explains standard and optional multifamily load calculations in Unit 11.

C If the service (to each service disconnect) is 400 amperes and the installation meets all the provisions in 310.15(B)(7), it shall be permissible to multiply the service rating by 83 percent and then select a conductor from Table 310.15(B)(16).

400 × 83% = 332 amperes

The minimum size copper conductor is 400 kcmil.

D Table 310.15(B)(16) requires a 500-kcmil conductor to attain the same rating.

E Two main service disconnecting means supplying power to twelve units must be computed on the basis of six units per service. Performing a service calculation on twelve units and then dividing by two is incorrect.

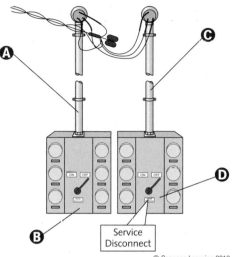

© Cengage Learning 2012

SERVICE WIRING CLEARANCES

Conductor Clearance above Flat Roofs

A The service mast shall be strong enough to safely withstand the strain imposed by the service drop or overhead service conductors. Supplemental braces or guys may be necessary »*230.28*«.

B Overhead service conductors shall not be readily accessible »*230.24*«.

C Conductors must have a minimum vertical clearance of 8 ft (2.5 m) above the roof surface »*230.24(A)*«.

D The vertical clearance above roof level must be maintained for a distance of at least 3 ft (900 mm) from all edges of the roof »*230.24(A)*«.

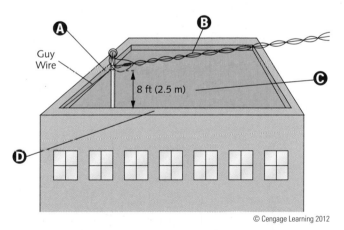

© Cengage Learning 2012

NOTE

Vertical clearances of overhead service conductors, covered in 230.24, pertain to every type of occupancy. Refer to Unit 9 (of this book) for detailed illustrations explaining vertical clearances.

Clearance from Openings

A Open service conductors or multiconductor cable (having no overall outer jacket), must have a clearance of at least 3 ft (900 mm) from windows (designed to open), doors, porches, balconies, ladders, stairs, fire escapes, and similar locations »*230.9(A)*«.

B Public streets, alleys, roads, parking areas subject to truck traffic, and nonresidential driveways require a minimum vertical clearance of 18 ft (5.5 m) »*230.24(B)(4)*«.

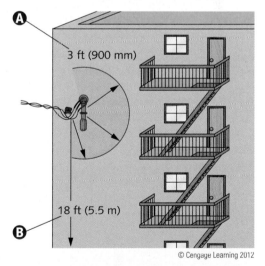

© Cengage Learning 2012

NOTE

Where buildings exceed three stories or 50 ft (15 m) in height, branch-circuit and feeder overhead lines must be arranged, where practicable, so that a clear area at least 6 ft (1.8 m) wide will be left adjacent to the buildings or beginning within 8 ft (2.5 m) to facilitate the raising of fire-fighting ladders »*225.19(E)*«.

Vertical Clearance above Platforms

A A drip loop's lowest point has a minimum vertical clearance of 10 ft (3.0 m) *»230.24(B)(1)«*.

B Overhead service conductors accessible only to pedestrian traffic must have a minimum 10-ft (3.0-m) vertical clearance above final grade or other accessible surface from which they might be reached. The voltage shall not exceed 150 to ground *»230.24(B)(1)«*.

C Vertical clearance of final spans above platforms, projections, or surfaces from which conductors might be reached must be maintained in accordance with 230.24(B) *»230.9(B)«*.

D Here, the 3-ft (900-mm) clearance requirement does not apply to conductors run above the top level of a window *»230.9(A) Exception«*. Because of the fire escape platform, the 10-ft (3.0-m) vertical clearance takes precedence over the window clearance.

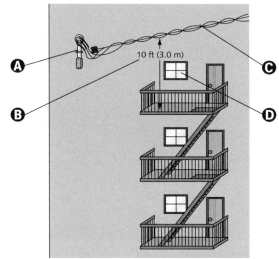

© Cengage Learning 2012

PANELBOARDS AND EQUIPMENT

Access to Overcurrent Devices

Where electric service and electrical maintenance are provided by building management and where these are under continuous building management supervision, service and feeder overcurrent devices supplying more than one occupancy can be accessible only to authorized management personnel in the following: (1) in multiple occupancy buildings and (2) for guest rooms or guest suites *»240.24(B)(1)«*.

A In damp or wet locations, surface-type meter socket enclosures must be situated or equipped so as to prevent moisture from accumulating within the enclosure. There must be at least ¼-in. (6-mm) airspace between the enclosure and the wall *»312.2«*.

B Each occupant must have ready access to all overcurrent devices protecting the conductors supplying that occupancy unless otherwise permitted in 240.24(B)(1) and (B)(2) *»240.24(B)«*.

C It shall be permissible to use 310.15(B)(7) to size 120/240-volt, single-phase dwelling services and certain feeders rated 100 through 400 amps. *»310.15(B)(7)«*.

> **N O T E**
>
> Where the service overcurrent devices are locked, sealed, or otherwise not readily accessible to the occupant, branch-circuit or feeder overcurrent devices must be (1) installed on the load side of the service overcurrent device; (2) mounted in a readily accessible location; and (3) of lower ampere rating than the service overcurrent device *»230.92«*.

> **CAUTION**
>
> Overcurrent devices shall not be located in close proximity to easily ignitible material (such as in clothes closets) *»240.24(D)«*.

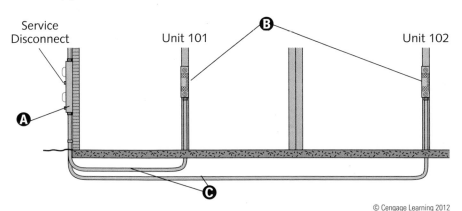

© Cengage Learning 2012

Panelboards Supplying Individual Units

Ⓐ Non-current-carrying metal parts of equipment, raceways, and other enclosures, if grounded, must be connected to an equipment grounding conductor by one of the following methods: (A) by any of the equipment grounding conductors permitted by 250.118, or (B) by an equipment grounding conductor contained within the same raceway or cable, or otherwise run with the circuit conductors 》250.134《.

Ⓑ Install the main bonding jumper (screw, strap, etc.) only in the service disconnecting means, not in subpanels.

Ⓒ At this point, the neutral bar is isolated from the equipment grounding system.

Ⓓ The front edge of cabinets (panelboards) situated in walls constructed of noncombustible material (concrete, tile, etc.) must be within ¼ in. (6 mm) of, or extend beyond, the finished surface 》312.3《.

Ⓔ If all of the qualifications are met, 310.15(B)(7) can be used to size feeder conductors.

Ⓕ A grounded circuit conductor shall not be used for grounding non-current-carrying metal parts of equipment on the load side of the service disconnecting means, unless an exception has been met 》250.142(B)《.

Ⓖ A grounded conductor shall not be connected to normally non-current-carrying metal parts of equipment or to equipment grounding conductor(s) or be reconnected to ground on the load side of the service disconnecting means except as otherwise permitted in Article 250 》250.24(A)(5)《.

WARNING
All switchboards and panelboards supplied by a feeder in other than one- or two-family dwellings shall be marked to indicate the device or equipment where the power supply originates 》408.4(B)《.

CAUTION
Cabinets in walls constructed of combustible material must be flush with or extend beyond the finished surface 》312.3《.

NOTE
A circuit directory must be affixed to the face (or inside) of the panel door(s) indicating the purpose of all circuits and any circuit modifications 》*408.4(A)*《.

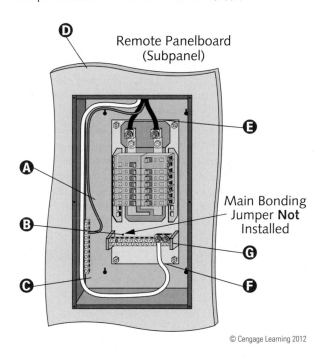

Remote Panelboard (Subpanel)

Main Bonding Jumper **Not** Installed

© Cengage Learning 2012

Rooftop Heating and Air-Conditioning Equipment

Ⓐ All 125-volt, single-phase, 15- and 20-ampere dwelling unit outdoor receptacles must have GFCI protection 》210.8(A)(3)《. Receptacles on rooftops shall not be required to be readily accessible other than from the rooftop.

Ⓑ A 125-volt, single-phase, 15- or 20-ampere-rated receptacle outlet must be installed at an accessible location for the servicing of heating, air-conditioning, and refrigeration equipment 》210.63《.

Ⓒ The receptacle must be located on the same level as, and within 25 ft (7.5 m) of the heating, air-conditioning, and refrigeration equipment 》210.63《. The provision does not stipulate that a receptacle must be located at every unit. One centrally located receptacle could meet the 210.63 requirement for all pieces of equipment located within the 25-ft (7.5-m) radius.

Ⓓ Article 440 contains provisions pertaining to air-conditioning and refrigeration equipment.

CAUTION
The receptacle outlet shall not be connected to the load side of the equipment disconnecting means 》210.63《.

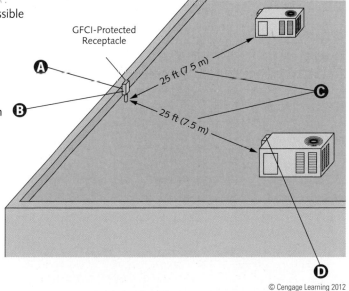

GFCI-Protected Receptacle

25 ft (7.5 m)

25 ft (7.5 m)

© Cengage Learning 2012

Feeder Conductors

A For individual dwelling units of multifamily dwellings, feeder conductors supplied by a single-phase, 120/240-volt system shall be permitted to be sized in accordance with 310.15(B)(7)(2) through (4). For a feeder rated 100 through 400 A, the feeder conductors supplying the entire load associated with an individual dwelling unit in a multifamily dwelling, shall be permitted to have an ampacity not less than 83 percent of the feeder rating ≫ *310.15(B)(7)* ≪ .

B If each feeder is 100 amperes and the installation meets all the provisions in 310.15(B)(7), it shall be permissible to multiply the feeder rating by 83 percent and then select a conductor from Table 310.15(B)(16).

100 × 83% = 83 amperes

The minimum conductor size is 4 AWG copper or 2 AWG aluminum ≫ *310.15(B)(7)* ≪ .

C Load calculations are computed for each individual unit having its own panelboard. Although located in a multifamily dwelling, each unit is calculated by one of the one-family methods. It is incorrect to perform a load calculation on multiple units and then divide by the number of units.

D Standard and optional one-family load calculations are explained in Unit 8 of this book.

E The AHJ may require a diagram showing feeder details prior to installation. Such a diagram must show the area in square feet of the building (or other structure) supplied by each feeder, the total calculated load before applying demand factors, the demand factors used, the calculated load after applying demand factors, and the size and type of conductors that will be used ≫ *215.5* ≪ .

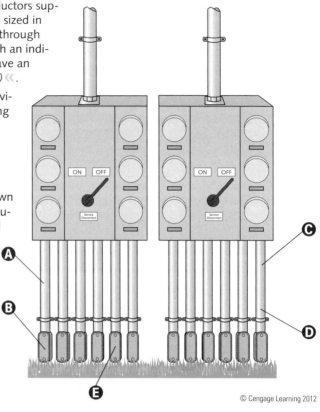

© Cengage Learning 2012

> ### CAUTION
>
> On long runs, it may be necessary to increase the feeder conductors' size to compensate for voltage drop ≫ *215.2(A)(1) Informational Note No. 2* ≪ .

BRANCH CIRCUITS

Receptacles

A For multifamily dwellings, each dwelling unit that is located at grade level and provided with individual exterior entrance/egress shall have at least one receptacle outlet readily accessible from grade and not more than 6½ ft (2.0 m) above grade level ≫ *210.52(E)(2)* ≪ .

B Dwelling-unit branch circuits shall supply only loads within, or associated only with, that dwelling unit ≫ *210.25* ≪ .

C A receptacle for laundry equipment shall not be required in a dwelling unit of a multifamily building where laundry facilities are provided on the premises for use by all building occupants ≫ *210.52(F) Exception No. 1* ≪ .

> ### NOTE
>
> Balconies, decks, and porches that are attached to the dwelling unit and are accessible from inside the dwelling unit shall have at least one receptacle outlet accessible from the balcony, deck, or porch. The receptacle outlet shall not be located more than 6½ ft (2.0 m) above the balcony, deck, or porch walking surface ≫ *210.52(E)(3)* ≪ . There is no exception for balconies, decks, and porches that have small usable areas.

D A receptacle for laundry equipment shall not be required in other than one-family dwellings where laundry facilities are not to be installed or permitted ≫ *210.52(F) Exception No. 2* ≪ .

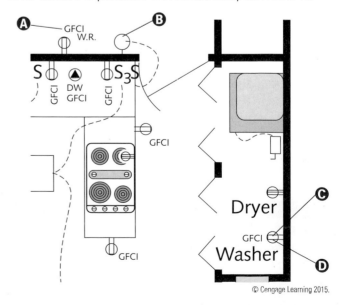

© Cengage Learning 2015.

Receptacles *(continued)*

A The maximum distance to any receptacle, measured horizontally along the wall line, is 24 in. (600 mm) ≫*210.52(C)(1)*≪.

B Dwelling-unit branch circuits shall supply only loads within, or associated only with, that dwelling unit ≫*210.25*≪.

C Receptacles must be installed so that no point measured horizontally along the floor line of any wall space is more than 6 ft (1.8 m) from a receptacle outlet ≫*210.52(A)(1)*≪.

D All 125-volt, 15- and 20-ampere receptacles serving kitchen countertops must have GFCI protection ≫*210.8(A)(6)*≪.

E Kitchen receptacles not serving countertops (such as receptacles behind refrigerators) do not require GFCI protection unless they are installed within 6 ft (1.8 m) of the outside edge of the sink.

F All receptacles in the kitchen, pantry, breakfast room, dining room, or similar areas must be supplied from 20-ampere small-appliance branch circuits (except for refrigeration equipment) ≫*210.52(B)(1)*≪.

G Circuits feeding receptacles in kitchens, pantries, breakfast rooms, dining rooms, or similar areas shall not feed receptacles outside these areas ≫*210.52(B)(2)*≪.

H GFCI protection is required for every 125-volt, 15- and 20-ampere receptacle located in a bathroom area ≫*210.8(A)(1)*≪.

I A receptacle is required for any wall space 2 ft (600 mm) or more in width (including space measured around corners), unbroken along the floor line by doorways and similar openings, fireplaces, and fixed cabinets ≫*210.52(A)(2)(1)*≪.

J Locate at least one receptacle within 36 in. (900 mm) of the outside edge of each basin (lavatory or sink). The receptacle outlet must be located on a wall or partition adjacent to the basin or basin countertop, located on the countertop, or installed on the side or face of the basin cabinet. In no case shall the receptacle be located more than 12 in. (300 mm) below the top of the basin. Receptacle outlet assemblies listed for the application shall be permitted to be installed in the countertop ≫*210.52(D)*≪.

K A receptacle is required for each peninsular counter space (at least 12 by 24 in. [300 × 600 mm] in size) that is separated from other counter space by a range top, refrigerator, or sink ≫*210.52(C)(3) and (4)*≪.

L A circuit providing power to bathroom receptacles shall not provide power to any receptacles (or lighting) outside the bathroom, unless meeting an exception ≫*210.11(C)(3)*≪.

> **NOTE**
>
> Only a few differences exist between one-family and multifamily dwelling receptacle requirements. Refer to Units 6 and 7 for more information pertaining to general as well as specific receptacle provisions.

> **CAUTION**
>
> In all areas specified in 210.52, all nonlocking-type 125-volt, 15- and 20-ampere receptacles shall be listed tamper-resistant receptacles ≫*406.12(A)*≪.

> **WARNING**
>
> Arc-fault circuit interrupter protection shall be provided for all 120-volt, single-phase, 15- and 20-ampere branch circuits supplying outlets or devices installed in dwelling unit kitchens, family rooms, dining rooms, living rooms, parlors, libraries, dens, bedrooms, sunrooms, recreation rooms, closets, hallways, laundry areas, or similar rooms or areas by any of the means described in 210.12(A)(1) through (6). The arc-fault circuit interrupter shall be installed in a readily accessible location ≫*210.12(A)*≪.

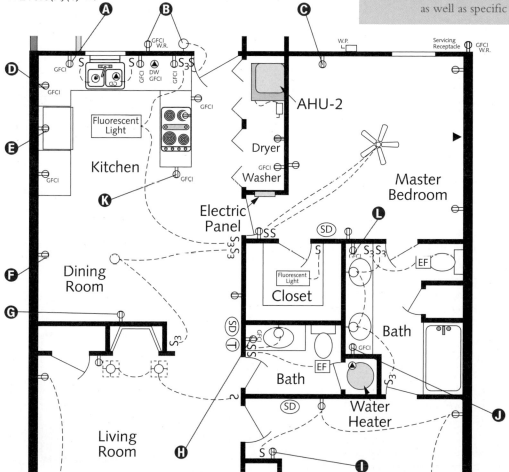

General Lighting

A minimum of one wall switch–controlled lighting outlet is required for dwelling hallways 》*210.70(A)(2)(a)* 《.

Ⓐ At least one wall switch–controlled lighting outlet must be installed in every habitable room of a dwelling 》*210.70(A)(1)* 《.

Ⓑ Clothes closet lighting outlet provisions are found in 410.16.

Ⓒ At least one wall switch–controlled lighting outlet must be installed in every dwelling bathroom 》*210.70(A)(1)* 《.

Ⓓ Except for kitchens and bathrooms, every habitable room is allowed one or more wall switch–controlled receptacle(s) in lieu of a lighting outlet 》*210.70(A)(1) Exception No. 1* 《.

Ⓔ In hallways, stairways, and outdoor entrances, control of lighting can be remote, central, or automatic 》*210.70(A)(2) Exception* 《.

Ⓕ Branch circuits installed for the purpose of lighting, central alarm, signal, communication, or other purposes for public or common areas of a two-family or multifamily dwelling shall not be provided from equipment that supplies an individual dwelling unit or tenant space 》*210.25(B)* 《.

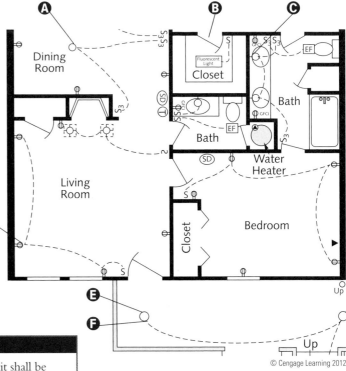

© Cengage Learning 2012

CAUTION

The grounded circuit conductor for the controlled lighting circuit shall be provided at the location where switches control lighting loads that are supplied by a grounded general-purpose branch circuit for other than the following:
(1) Where conductors enter the box enclosing the switch through a raceway, provided that the raceway is large enough for all contained conductors, including a grounded conductor
(2) Where the box enclosing the switch is accessible for the installation of an additional or replacement cable without removing finish materials
(3) Where snap switches with integral enclosures comply with 300.15(E)
(4) Where a switch does not serve a habitable room or bathroom
(5) Where multiple switch locations control the same lighting load such that the entire floor area of the room or space is visible from the single or combined switch locations
(6) Where lighting in the area is controlled by automatic means
(7) Where a switch controls a receptacle load 》*404.2(C)*《.

NOTE

A listed box is required for a ceiling-suspended (paddle) fan where the box provides the only support 》*314.27(C)*《.
Only a few differences exist between one-family and multifamily dwelling lighting requirements. Refer to Units 6 and 7 for more information pertaining to general as well as specific lighting provisions.

Appliance Disconnecting Means

Ⓐ A permanently connected appliance, rated over 300 volt-amperes or ⅛ horsepower, must have a means to simultaneously disconnect all ungrounded (hot) conductors from the appliance 》*422.30 and 422.31* 《.

Ⓑ An appliance switch marked with an "off" position that effectively disconnects all ungrounded conductors can serve as the disconnecting means required by Article 422 where other means for disconnection are provided in 422.34(A) through (D) 》*422.34* 《.

Ⓒ For permanently connected appliances rated over 300 volt-amperes, a disconnecting means is not required if the branch-circuit switch or circuit breaker is lockable in accordance with 110.25 》*422.31(B)* 《.

Ⓓ In multifamily dwellings, the other appliance disconnecting means (as described in 422.34) must be within the dwelling unit or on the same floor as the dwelling unit in which the appliance is installed and can also control lamps and other appliances 》*422.34(A)* 《.

Ⓔ A branch circuit supplying a fixed storage-type water heater with a capacity of 120 gallons (450 L) or less is considered a continuous load and, therefore, must have a rating of at least 125% of the water heater's nameplate rating 》*422.13 and 422.10(A)* 《.

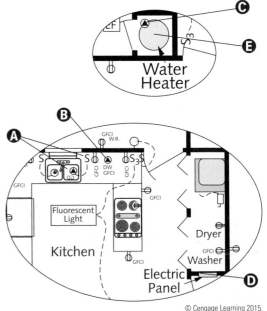

© Cengage Learning 2015.

Heating and Air-Conditioning Equipment

(A) For a hermetic refrigerant motor compressor, the rated-load current marked on the equipment nameplate in which the motor compressor is employed must be used in determining the rating or ampacity of the disconnecting means, the branch-circuit conductors, the controller, the branch-circuit short-circuit and ground-fault protection, and the separate motor overload protection ≫ *440.6(A)* ≪.

(B) The disconnecting means for air-conditioning or refrigerating equipment must be readily accessible and within sight of the equipment ≫ *440.14* ≪.

(C) All electric equipment must be surrounded by sufficient unobstructed space to allow ready and safe equipment operation and maintenance ≫ *110.26* ≪. Working space is mandatory for equipment likely to require examination, adjustment, servicing, or maintenance while energized ≫ *110.26(A)* ≪.

(D) Fixed electric space-heating equipment and motors shall be considered continuous load when sizing the branch circuit ≫ *424.3(B)* ≪.

(E) A hermetic refrigerant motor compressor is defined as a combination consisting of a compressor and motor, both of which are enclosed in the same housing, with no external shaft or shaft seals, with the motor operating in the refrigerant ≫ *Article 100* ≪.

(F) On long runs, it may be necessary to increase the branch-circuit conductor's size to compensate for voltage drop ≫ *210.19(A) Informational Note No. 4* ≪.

(G) Means shall be provided to simultaneously disconnect all ungrounded conductors from the heater, motor controller(s), and supplementary overcurrent protective device(s) of all fixed electric space-heating equipment ≫ *424.19* ≪.

(H) In multifamily dwellings, the other disconnecting means, as described in 424.19(C), must be within the dwelling unit or on the same floor as the dwelling unit in which the fixed heater is installed and can also control lamps and other appliances ≫ *424.19(C)(1)* ≪.

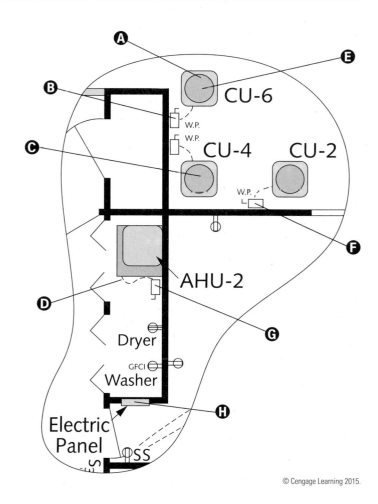

© Cengage Learning 2015.

NOTE

Fixed heater switch(es) marked with an "off" position that effectively disconnects all ungrounded conductors can serve as the disconnecting means required by Article 424, where other means for disconnection are provided in 424.19(C)(1) through (4) ≫***424.19(C)***≪.

CAUTION

The disconnecting means shall be permitted to be installed on or within the air-conditioning or refrigerating equipment. The disconnecting means shall not be located on panels that are designed to allow access to the air-conditioning or refrigeration equipment or to obscure the equipment nameplate(s) ≫*440.14*≪.

VOLTAGE DROP

Conductor Properties (Chapter 9, Tables 8 and 9)

While skin effect and induction change the resistance in alternating-current (ac) circuits, they do not affect direct-current (dc) circuits. The resistance of a conductor is slightly higher in an ac circuit. Table 9 contains the resistance of conductors in PVC, aluminum, and steel raceways. The factors that change resistance in an ac circuit, such as skin effect, are negligible in smaller conductors.

A The larger the circular mil area, the smaller the resistance.

B As listed in Table 8, 6 AWG has an area of 26,240 circular mils.

C Size 3/0 AWG has an area of 167,800 circular mils and contains 19 strands.

D The abbreviated term "kcmil" stands for 1000 (k) circular (c) mills (mil).

E Each conductor size larger than 4/0 AWG is actually the circular mil size. A 250-kcmil conductor has a 250,000 circular mil area. A 500-kcmil conductor has a circular mil area of 500,000.

F Conductor overall diameter is listed in Table 8. This diameter is the actual wire and excludes conductor insulation.

G Conductor strand quantities are listed in Table 8; 6 AWG contains seven strands.

H Table 8 lists dc resistance values (in ohms per 1000 ft) at an ambient temperature of 75°C (167°F).

I Size 500-kcmil conductors have a quantity of 37 strands. The overall diameter is 0.813 in., with each strand having a diameter of 0.116 in.

J The dc resistance for 1000 ft of 6 AWG copper is 0.491 ohms.

K The longer the conductor, the higher the resistance.

L The dc resistance for 100 ft of 6 AWG copper is 0.0491 ohms (0.491 ÷ 1000 = 0.000491 × 100 = 0.0491).

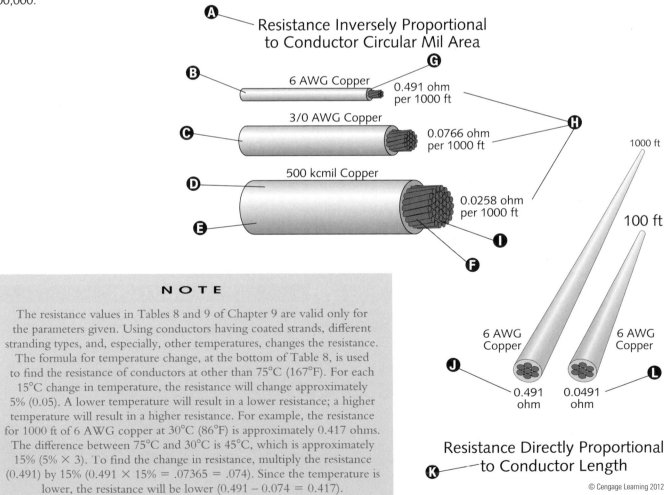

Resistance Inversely Proportional to Conductor Circular Mil Area

6 AWG Copper — 0.491 ohm per 1000 ft

3/0 AWG Copper — 0.0766 ohm per 1000 ft

500 kcmil Copper — 0.0258 ohm per 1000 ft

1000 ft

100 ft

6 AWG Copper — 0.491 ohm

6 AWG Copper — 0.0491 ohm

Resistance Directly Proportional to Conductor Length

© Cengage Learning 2012

NOTE

The resistance values in Tables 8 and 9 of Chapter 9 are valid only for the parameters given. Using conductors having coated strands, different stranding types, and, especially, other temperatures, changes the resistance. The formula for temperature change, at the bottom of Table 8, is used to find the resistance of conductors at other than 75°C (167°F). For each 15°C change in temperature, the resistance will change approximately 5% (0.05). A lower temperature will result in a lower resistance; a higher temperature will result in a higher resistance. For example, the resistance for 1000 ft of 6 AWG copper at 30°C (86°F) is approximately 0.417 ohms. The difference between 75°C and 30°C is 45°C, which is approximately 15% (5% × 3). To find the change in resistance, multiply the resistance (0.491) by 15% (0.491 × 15% = .07365 = .074). Since the temperature is lower, the resistance will be lower (0.491 − 0.074 = 0.417).

Voltage-Drop Recommendation

A Except for sensitive electronic equipment in 647.4(D) and fire pumps in 695.7, voltage-drop considerations are only recommendations; compliance is optional.

B The recommended maximum voltage drop for the combined feeder and branch circuit is 5%. However, the voltage drop for feeders is not necessarily 2%. If the branch circuit is 3%, the feeder is 2%, but if the branch circuit is 1%, the feeder can be 4%. Any combination is possible, as long as the branch circuit does not exceed 3%, and the combined branch circuit and feeder does not exceed 5%.

C There is no recommended maximum voltage drop for service conductors.

D The conductor voltage that is dropped is a percentage of the source voltage.

E The recommended voltage drop for a branch circuit is 3% or less. For example, the maximum allowable voltage drop for a 120-volt branch circuit is 3.6 volts (120 × 3%), and the maximum allowable for a 240-volt branch circuit is 7.2 volts (240 × 3%).

F The higher the resistance through a conductor, the higher the voltage drop. Such a conductor voltage drop could result in a reduction of voltage supplying the load.

G A single-phase feeder (or branch circuit) conductor will have a total resistance twice that of one conductor. For example, the length of one conductor must be doubled to find the total resistance of the circuit. The total length for a three-phase circuit is found by multiplying the length of one conductor by the square root of 3 (1.732).

H Branch-circuit conductors, as defined in Article 100, sized to prevent a voltage drop exceeding 3% at the farthest outlet of power, heating, and lighting loads (or combinations of such loads) and where the maximum total voltage drop on both feeders and branch circuits to the farthest outlet does not exceed 5% will provide reasonable efficient operation ≫ *210.19(A) Informational Note No. 4 and 215.2(A)(1) Informational Note No. 2* ≪.

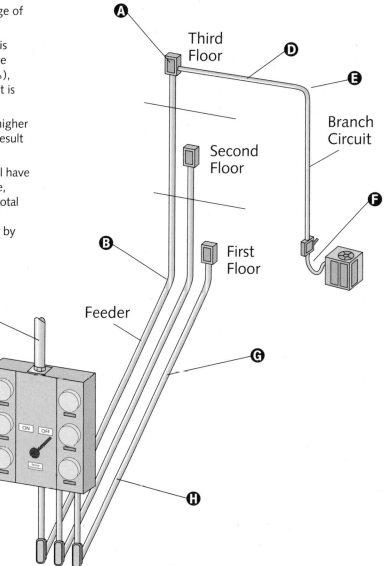

© Cengage Learning 2012

N O T E

Not all electrical installations have feeder conductors.

Single-Phase Voltage-Drop Formulas

Formula Definitions:

V_D = Volts dropped from a circuit.

2 = Multiplying factor for single-phase (the 2 represents the conductor length in a single-phase circuit).

K = Resistivity of the conductor material. The resistance value (measured in ohms) is based on a mil foot at a given temperature. A mil foot is a piece of wire 1 ft long and one mil in diameter. Each conductor resistivity (K) factor can be found by multiplying the circular mil area (cmil) by the resistance (R) for 1 ft. For example, the K factor for a 6 AWG copper conductor at 75°C (167°F) is 12.88384, because 26,240 (area circular mills) × 0.000491 (ohms per ft) = 12.88384. At 75°C (167°F), the approximate K for copper is 12.9 and for aluminum is 21.2.

L = Length, *one way* only: The distance from the voltage source to the load.

I = Actual current used by the load. Do not use 125% for motors and continuous loads.

cmil = Circular mil area of the conductor (see Chapter 9, Table 8).

R = Conductor resistance for 1 ft. *Note:* The resistance values, listed in Chapter 9, Tables 8 and 9, are for 1000 ft. To find the resistance for 1 ft, simply divide the resistance by 1000. Example: The resistance for 1 foot of 6 AWG copper (at 75°C [167°F] is 0.000491 (0.491 ÷ 1000 = 0.000491).

Ⓐ This is one formula used to find voltage drop in a single-phase circuit.

Ⓑ The conductor voltage that is dropped is a percentage of the source voltage. To find the percentage, divide V_D by the source voltage. Example: 2.4 volts dropped from a 120-volt circuit is 2% of the source voltage (2.4 ÷ 120 = 0.02 = 2%).

Ⓒ Here is another formula for finding voltage drop in a single-phase circuit. This method uses the Ohm's law formula. Resistance (for 1 ft of the conductor) is used instead of the K factor and circular mil area.

Ⓓ Use this formula to find the maximum length (distance) from the source to the load. Because of the "2" in the formula, the length is **one way.**

Ⓔ This Ohm's law formula finds the maximum length from the source to the load.

Ⓕ Conductor size can be found using this formula. The calculation provides the minimum circular mil wire size that must be installed to remain below the V_D that was input into the formula.

Ⓖ This formula applies Ohm's law to find the conductor size. The result is the minimum conductor resistance for 1 ft; this keeps the voltage drop below the V_D used in the formula.

Ⓗ This formula calculates the maximum current in amperes.

Ⓘ Use this Ohm's law formula to find the current in amperes.

Ⓙ Since the conductor size is unknown, the exact K cannot be calculated. For copper, at 75°C (167°F), use **12.9** (approximate K). For aluminum, at 75°C (167°F), use **21.2** (approximate K).

Ⓚ V_D represents the actual volts that can drop from the circuit. The maximum (recommended) percentage for a branch circuit is 3%. Be careful not to place "3%" in the V_D position. If the voltage is 120, then 3.6 is the actual volts that can drop in the branch circuit. For 240 voltage, the actual volts that can drop in the branch circuit is 7.2.

$$\text{Ⓐ Ⓑ} \quad V_D = \frac{2 \times K \times L \times I}{cmil}$$

$$\text{Ⓕ Ⓙ Ⓚ} \quad cmil = \frac{2 \times K \times L \times I}{V_D}$$

$$\text{Ⓒ} \quad V_D = 2 \times R \times L \times I$$

$$\text{Ⓖ} \quad R = \frac{V_D}{2 \times L \times I}$$

$$\text{Ⓓ} \quad L = \frac{cmil \times V_D}{2 \times K \times I}$$

$$\text{Ⓗ} \quad I = \frac{cmil \times V_D}{2 \times K \times L}$$

$$\text{Ⓔ} \quad L = \frac{V_D}{2 \times R \times I}$$

$$\text{Ⓘ} \quad I = \frac{V_D}{2 \times R \times L}$$

Three-Phase Voltage-Drop Formulas

Ⓐ The multiplying factor for three-phase is "1.732." The square root of 3 (1.732) represents the conductor length in a three-phase circuit. The only difference between the single-phase and three-phase formulas is that "1.732" has replaced "2."

Ⓑ These formulas are used to find voltage drop in a three-phase circuit.

Ⓒ Use these formulas to find the maximum length (distance) from the source to the load in a three-phase circuit. (Because of the "1.732" in the formula, the length is **one way.**)

Ⓓ These are variations of three-phase formulas.

Ⓔ This formula calculates the conductor size in a three-phase circuit.

Ⓕ This formula is used to determine the resistance of the conductor in a three-phase circuit.

Ⓖ The maximum current in amperes in a three-phase circuit is calculated using this formula.

Ⓗ The single-phase formula can be changed to a three-phase formula by placing "0.866" in the line containing the "2." The purpose of putting "0.866" in the same line as "2" is simple: 2 × 0.866 = 1.732. (Refer to the definitions in the previous section—Single-Phase Voltage-Drop Formulas.)

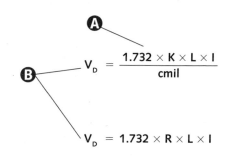

$$V_D = \frac{1.732 \times K \times L \times I}{cmil}$$

$$V_D = 1.732 \times R \times L \times I$$

$$L = \frac{cmil \times V_D}{1.732 \times K \times I}$$

$$L = \frac{V_D}{1.732 \times R \times I}$$

$$cmil = \frac{1.732 \times K \times L \times I}{V_D}$$

$$R = \frac{V_D}{1.732 \times L \times I}$$

$$I = \frac{cmil \times V_D}{1.732 \times K \times L}$$

$$I = \frac{V_D}{1.732 \times R \times L}$$

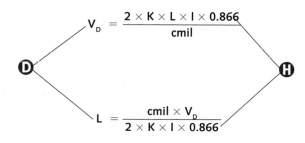

$$V_D = \frac{2 \times K \times L \times I \times 0.866}{cmil}$$

$$L = \frac{cmil \times V_D}{2 \times K \times I \times 0.866}$$

Single-Phase Voltage-Drop Example

At 75°C (167°F), what size conductors are required to feed a 2-hp, 230-volt, single-phase motor that is 150 ft from the source? (Do not exceed *NEC* recommendations.)

A Both the supply and load side of this equipment disconnecting means are branch-circuit conductors; therefore, the maximum recommended voltage drop is 3% of the source voltage.

B Table 430.248 lists full-load current (in amperes) for single-phase ac motors.

C This is the voltage drop formula used to find the minimum size conductor.

D This is the approximate K for copper at 75°C (167°F).

E This is the current in amperes for a 2-hp, 230-volt, single-phase motor ›› *Table 430.248* ‹‹.

F This is the minimum circular mil area required. Notice that a 12 AWG conductor (6530 circular mils) is too small for this installation.

G This is the minimum size conductor recommended for this installation.

H This is the maximum amount of voltage drop on a 230-volt branch circuit that remains within the 3% recommendation (230 × 3% = 6.9).

I This is an alternative formula equally effective in determining the minimum size conductor.

J In multiplying the 1 ft value by 1000, the resistance values will correspond with Tables 8 and 9, which contain ohms per 1000 ft values.

K This is the maximum conductor resistance for 1 ft.

L The conductor chosen shall not exceed this resistance number. Both solid and stranded 12 AWG conductors have a higher resistance; therefore, 10 AWG conductors (solid or stranded) must be used.

NOTE

If the total distance of the branch circuit is less than 145 ft (44.2 m), a 12 AWG copper conductor could be installed.

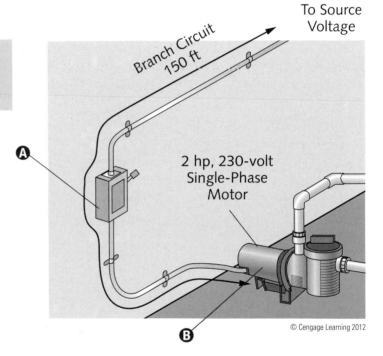

© Cengage Learning 2012

$$\mathbf{C}\ \ \text{cmil} = \frac{2 \times K \times L \times I}{V_D} = \frac{2 \times 12.9 \times 150 \times 12}{6.9} = 6730 = 10\ \text{AWG copper}$$

$$\mathbf{I}\ \ R = \frac{V_D}{2 \times L \times I} = \frac{6.9}{2 \times 150 \times 12} = 0.001917 \times 1000 = 1.917 = 10\ \text{AWG copper}$$

© Cengage Learning 2012

Summary

- Overall receptacle installation requirements, with few exceptions, are the same for both one-family and multi-family dwellings.

- Outdoor receptacle requirements for multifamily dwellings are not the same as requirements for one-family dwellings.

- Under certain conditions, a laundry receptacle is not required.

- A multifamily clubhouse (or portion thereof) designed for the assembly of more than 99 persons must comply with Article 518 provisions.

- Generally, a building is supplied by only one service.

- There can be no more than six disconnects per service grouped in any single location.

- Where separate services supply a building and a grounding electrode is required, the same electrode must be used.

- Separate service laterals, grouped in one location, constitute one service.

- Use 310.15(B)(7) to size 120/240, single-phase service-entrance conductors, service-lateral conductors, and certain feeder conductors rated 100 through 400 amperes.

- Service and open conductor clearances pertain to all dwellings.

- Conductor properties are found in Chapter 9, Tables 8 and 9.

- Voltage-drop compliance is discretionary.

- The recommended voltage drop for a branch circuit is 3% or less.

- The recommended voltage drop for combined feeder and branch circuits is 5% or less.

- Voltage-drop formulas, complete with explanations, are found near the end of this unit.

- At least one 125-volt, single-phase, 15- or 20-ampere-rated receptacle outlet shall be installed within 50 ft (15 m) of the electrical service equipment.

Unit 10 Competency Test

NEC Reference	Answer	
_____	_____	1. _____ is a general term including material, fittings, devices, appliances, luminaires, apparatus, machinery, and the like used as a part of, or in connection with, an electrical installation.
_____	_____	2. Where a building is supplied by more than one service, or combination of branch circuits, feeders and services, a permanent _____ or _____ shall be installed at each service disconnect location denoting all other services, feeders, and branch circuits supplying that building and the area served by each.
_____	_____	3. Where the overcurrent device is rated over _____ amperes, the ampacity of the conductors it protects shall be equal to or greater than the rating of the overcurrent device.
_____	_____	4. A(n) _____ is a device, or group of devices, or other means by which the conductors of a circuit can be disconnected from their source of supply.
_____	_____	5. A(n) _____ is a device intended to provide protection from the effects of arc faults by recognizing characteristics unique to arcing and by functioning to de-energize the circuit when an arc fault is detected.
_____	_____	6. Service-drop conductors shall have a vertical clearance of not less than _____ ft above the roof surface.
_____	_____	7. What size copper equipment grounding conductor is required for a circuit having a 40-ampere overcurrent device?
_____	_____	8. Where a change occurs in the size of the _____, a similar change shall be permitted to be made in the size of the grounded conductor.
_____	_____	9. _____ in each dwelling unit shall supply only loads within that dwelling unit or loads associated only with that dwelling unit.

10. A metal elbow that is installed in an underground installation of nonmetallic raceway and is isolated from possible contact by a minimum cover of _____ in. to any part of the elbow shall not be required to be connected to the grounded system conductor or grounding electrode conductor.

11. Conductors shall be considered outside of a building or other structure where installed under not less than _____ in. of concrete beneath a building or other structure.

12. What is the service demand load (in kW) for an apartment complex containing ten each of the following: 15-kW ranges, 13-kW ranges, 11-kW ranges, and 9-kW ranges? (The service is 120/240 single-phase.)

13. In _____, remote, central, or automatic control of lighting shall be permitted.

14. Service conductors installed as open conductors shall have a clearance of not less than _____ ft from fire escapes.

15. A grounded circuit conductor shall be permitted to ground non-current-carrying metal parts of equipment on the _____ side of the ac service disconnecting means.

16. Each unit of a thirty-unit apartment building has a 4.5-kW clothes dryer. What is the service demand load contribution (in kW) for these clothes dryers? (The service is 120/240 single-phase.)

17. The maximum distance between receptacle outlets in a multifamily dwelling is _____ ft.

18. Overcurrent devices shall not be located in the vicinity of _____, such as in clothes closets.

19. Cabinets and cutout boxes shall have _____ to accommodate all conductors installed in them without crowding.

20. Where conductors carrying alternating current are installed in ferrous metal enclosures or metal raceways, they shall be arranged so as to avoid heating the surrounding metal by _____.

21. _____ is the highest current at rated voltage that a device is identified to interrupt under standard test conditions.

22. Where ungrounded conductors are increased in size to compensate for voltage drop, equipment grounding conductors, where installed, shall be increased in size proportionately according to:

_____

23. Enclosures for overcurrent devices shall be mounted in a _____ position.

24. A metal pole supporting a luminaire, 18 ft above grade, must have a handhole of at least _____ in.

25. Fittings and connectors shall be used only with the specific wiring methods for which they are _____.

26. _____ is a combination consisting of a compressor and motor, both of which are enclosed in the same housing, with no external shaft or shaft seals, with the motor operating in the refrigerant.

27. For permanently connected appliances rated over _____ volt-amperes, the branch-circuit switch or circuit breaker shall be permitted to serve as the disconnecting means where the switch or circuit breaker is within sight from the appliance or is lockable in accordance with 110.25.

NEC Reference	Answer	

28. Fixed electric space heating equipment and motors shall be considered _____ when sizing the branch circuit.

29. _____ is the necessary equipment, usually consisting of a circuit breaker(s) or switch(es) and fuse(s) and their accessories, connected to the load end of service conductors to a building or other structure, or an otherwise designated area, which is intended to constitute the main control and cutoff of the supply.

30. No hanging luminaire parts can be located within a zone measured _____ ft vertically from the top of the bathtub rim and _____ ft vertically from the top of the shower stall threshold.

31. What is the ac resistance of 500 feet of 1000 kcmil aluminum wire in a steel conduit?

32. At 75°C (167°F), what size copper conductors are required (without exceeding *NEC* recommendations) to feed a ¼-horsepower, 115-volt, single-phase motor located 230 ft from the source?

33. Interior metal water piping located not more than _____ ft from the point of entrance to the building shall be permitted to be used as a conductor to interconnect electrodes that are part of the grounding electrode system.

34. What is the minimum neutral demand load (in kW) for twelve apartments, each containing an 8-kW range?

35. What is the voltage-drop percentage on two 10 AWG THW copper, stranded, branch-circuit conductors, 120 ft long, supplying a 21-ampere, 240-volt load?

36. At least one 125-volt, single-phase, 15- or 20-ampere-rated receptacle outlet shall be installed within _____ ft of the electrical service equipment in a multifamily dwelling.

Load Calculations

Objectives

After studying this unit, the student should:

▶ know that the number of dwelling units is essential in performing a multifamily calculation.

▶ be able to calculate the general lighting load in a multifamily dwelling.

▶ correctly identify minimum volt-ampere requirements for small-appliance branch circuits.

▶ thoroughly understand the exceptions pertaining to optional laundry branch circuits.

▶ know how to apply demand factors to the general lighting load, as well as to fastened-in-place appliances.

▶ be able to calculate demand loads for household clothes dryers, cooking equipment, and heating and air-conditioning systems.

▶ understand the standard method for calculating a multifamily dwelling service.

▶ successfully calculate a multifamily dwelling service using the optional method.

▶ be able to calculate a feeder for one unit in a multifamily dwelling using either method.

▶ know how to size service and feeder conductors.

▶ be familiar with paralleled conductor provisions.

▶ correctly calculate and select the appropriate size neutral conductor.

▶ understand how grounding electrode conductors, as well as equipment grounding conductors, are selected.

Introduction

Strong similarities exist between multifamily dwelling load calculations and one-family dwelling load calculations. Both can be performed by a standard method as well as an optional method. In fact, the similarities are so strong that this is reflected in the very design of the load calculation forms. It is, therefore, obviously beneficial to have a clear understanding of one-family dwelling load calculations before studying this unit (see Unit 8). It is absolutely essential to these calculations to know the total number of dwelling units being supplied by the service. To illustrate this point: Two main service disconnecting means supplying power to twelve units must be calculated on the basis of six units per service. Performing a service calculation on twelve units and then dividing the final figure by two would render inaccurate results.

The multifamily dwelling load calculations performed in this unit are based on the floor plan found in Unit 10. The statistical information for each individual dwelling unit remains unchanged for the purpose of load calculation. This data can be found in the floor plan diagram immediately following this introduction. Examples include a multifamily dwelling load calculation for both a twelve-unit and a six-unit service. Each will be calculated by the standard method as well as the optional. Feeder and panelboard sizing for individual units is demonstrated by a standard one-family dwelling calculation. For comparison, the optional method load calculation for one unit of a multifamily dwelling is shown after the optional multifamily dwelling computation.

COMPILING LOAD CALCULATION INFORMATION

Load Calculation Information for Each Unit

A The smaller of two (or more) noncoincident loads can be omitted, as long as they are never energized simultaneously »220.60«.

B Although a clothes dryer is not a requirement for a load calculation, be sure to include a dryer load in units containing clothes dryer outlets.

C A laundry branch circuit is optional in multifamily dwellings »210.52(F) Exception No. 1 and 2«.

D Gather all load calculation information needed to calculate each service. For a service consisting of twelve units (apartment or condominium), compile the essential information for all twelve units. If the service consists of only six units, gather the data for those six units. Since most multifamily dwellings have a limited number of floor plans, compile the information pertaining to each "unit type" (style, floor plan, etc.). Be sure that each unit grouped in a unit type contains the same information, that is, total square feet, small-appliance branch circuits, laundry branch circuits, appliances, etc. For ease of demonstration in this section, each unit is identical in size and content.

E Calculate the floor area using the outside dimensions »220.12«.

F Include household cooking appliances rated over 1¾ kW.

G Include all fastened-in-place appliances.

H Include at least two small-appliance branch circuits »210.11(C)(1)«.

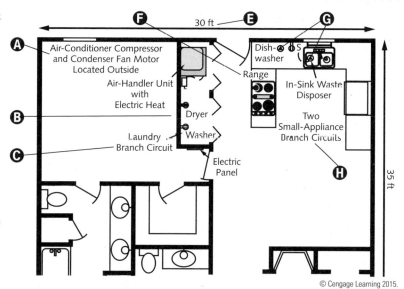

© Cengage Learning 2015.

Dwelling outside dimensions	30′ × 35′
Small-appliance branch circuits	two
Laundry branch circuit	one
Water heater	4.5 kW, 240 volt
Dishwasher	10 amperes, 120 volt
Microwave oven	10 amperes, 120 volt
In-sink waste disposer	7.5 amperes, 120 volt
Clothes dryer	5 kW, 240 volt
Range	12 kW, 240 volt
Electric heat	5 kW, 240 volt
Air handler (blower motor)	3 amperes, 115 volt
Air-conditioner compressor	14 amperes, 230 volt
Condenser fan motor	2 amperes, 115 volt

Assume water heater, clothes dryer, range, and electric heat kW ratings equivalent to kVA.

STANDARD METHOD: MULTIFAMILY DWELLINGS

Load Calculation for Twelve Units

A The first multifamily load calculation form is based on twelve units. Each unit is identical, as previously described.

B In a multiple-occupancy building, each occupant must have access to the unit's service disconnecting means »230.72(C)«.

C There must not be more than six disconnects (switches or circuit breakers) per service grouped in a single location »230.71(A)«.

D One service disconnecting switch controls all twelve units.

E Each unit's feeder (and overcurrent protection) must be calculated individually, using the one-family dwelling form (standard or optional method).

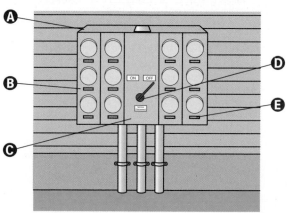

© Cengage Learning 2012

Standard Method Load Calculation for Multifamily Dwellings

1 General Lighting and Receptacle Loads *220.12*
Do not include open porches, garages, or unused or unfinished spaces not adaptable for future use.

$3 \times$ _____ $\times$ _____ = **1**
(sq ft outside dimensions)   (number of units)

2 Small-Appliance Branch Circuits *220.52(A)*
*At least **two** small-appliance branch circuits must be included. 210.11(C)(1)*

$1500 \times$ _____ $\times$ _____ = **2**
(minimum of two)   (number of units)

3 Laundry Branch Circuit(s) *220.52(B)*
*Include at least **one** laundry branch circuit *unless meeting 210.52(F), Exception No. 1 or 2. 210.11(C)(2)*

$1500 \times$ _____ $\times$ _____ = **3**
*(minimum of one)   (number of units)

4 Add lines 1, 2, and 3   **4**

Lines 5 through 8 utilize the demand factors found in *Table 220.42*.

5 _____ $- 3000 =$ **5**
(line 4)

(if 117,000 or less, skip to line 8)

6 _____ $- 117,000 =$ **6**
(line 5, if more than 117,000)

7 _____ $\times 25\% =$ **7**
(line 6)

8 _____ $\times 35\% =$ **8**
(smaller of line 5 or 117,000)

9 Total General Lighting and Receptacle Load

$3000 +$ _____ $+$ _____ = **9**
(line 7)   (line 8)

10 Fastened-In-Place Appliances *220.53*
Use the nameplate rating.
Do not include electric ranges, clothes dryers, space-heating equipment, or air-conditioning equipment.

water heaters/ _____   dishwashers / _____   disposers / _____

_____ /   _____ /   _____ /

_____ /   _____ /   _____ /

If fewer than four units, put total volt-amperes on line 10.
If four or more units, multiply total volt-amperes by 75%.

_____ $\times 75\% =$ **10**
(volt-amps of four or more)

11 Clothes Dryers *220.54*
(If present; otherwise skip to line 12.) Use 5000 watts or the nameplate rating, whichever is larger. The neutral demand load is 70% for feeders. 220.61(B)

11

12 Ranges, Ovens, Cooktops, and Other Household Cooking Appliances over 1750 Watts *220.55*
(If present; otherwise skip to line 13.) Use Table 220.55 and all of the applicable Notes. The neutral demand load is 70% for feeders. 220.61(B)

12

13 Heating or Air-Conditioning System (Compare the heat and A/C, and omit the smaller.) *220.60*
Include the air handler when using either one. For heat pumps, include the compressor and the maximum amount of electric heat that can be energized while the compressor is running.

13

14 Largest Motor (one motor only) *220.50* and *430.24*
Multiply the volt-amperes of the largest motor by 25%.

_____ $\times 25\% =$ **14**
(volt-amps of largest motor)

15 Total Volt-Ampere Demand Load: *Add lines 9 through 14 to find the minimum required volt-amperes.*   **15**

16 Minimum Amperes
Divide the total volt-amperes by the voltage.

_____ $\div$ _____ = **16**
(line 15)   (voltage)   (minimum amperes)

Minimum Size
17 **Service or Feeder** *240.6(A)*   **17**

18 Size the Service or Feeder Conductors.
Use 310.15(B)(7) to find the service conductors up to 400 amperes. Ratings in excess of 400 amperes shall comply with Table 310.15(B)(16). 310.15(B)(7) also applies to feeder conductors supplying the entire load.

Minimum Size Conductors   **18**

19 Size the Neutral Conductor. *220.61*
310.15(B)(7) states that the neutral service or feeder conductor can be smaller than the ungrounded (hot) conductors, provided the requirements of 215.2, 220.61, and 230.42 are met. 250.24(C)(1) states that the grounded conductor shall not be smaller than specified in Table 250.102(C)(1).

Minimum Size Neutral Conductor   **19**

20 Size the Grounding Electrode Conductor (for Service). *250.66*
Use line 18 to find the grounding electrode conductor in Table 250.66.
Size the Equipment Grounding Conductor (for Feeder). *250.122*
Use line 17 to find the equipment grounding conductor in Table 250.122. Equipment grounding conductor types are listed in 250.118.

Minimum Size Grounding Electrode Conductor ... or ... Equipment Grounding Conductor   **20**

Line 1—General Lighting and Receptacle Loads

A The residential dwelling unit's lighting load is 3 volt-amperes (VA) per square foot 》*Table 220.12* 《.

B Each multifamily dwelling unit has an outside dimension of 30 ft × 35 ft (1050 ft²).

C Insert the number of units (apartments or condominiums) that constitute one service. Additional space will be needed for computations if the units have different square footage dimensions.

D Each unit's floor area shall be calculated using the outside dimensions. The total floor area does not include open

porches, garages, or unused (including unfinished) spaces that are not adaptable for future use 》*220.12* 《.

> ### N O T E
> All 15- or 20-ampere general-use receptacle outlets (except for small-appliance and laundry) are considered part of the general lighting load. The "a" footnote at the bottom of Table 220.12 refers to 220.14(J), which states that no additional load calculations are required for these outlets.

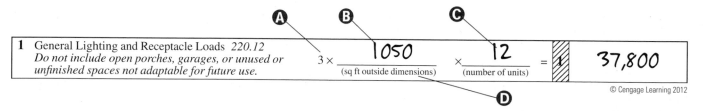

	A	**B**	**C**		
1	General Lighting and Receptacle Loads *220.12*				
	Do not include open porches, garages, or unused or unfinished spaces not adaptable for future use.	3 × 1050 (sq ft outside dimensions)	× 12 (number of units)	= 1	37,800

© Cengage Learning 2012

Line 2—Small-Appliance Branch Circuits

A Include small-appliance branch circuits when calculating a dwelling unit service 》*220.52(A)* 《.

B Each small-appliance branch circuit is calculated at no less than 1500 volt-amperes 》*220.52(A)* 《.

C If any unit(s) has a different number of small appliance branch circuits, calculate each separately.

D Each dwelling unit must have at least two small-appliance branch circuits 》*210.11(C)(1)* 《.

E These loads can be included with the general lighting load and are subject to Table 220.42 demand factors 》*220.52(A)* 《.

> ### N O T E
> A refrigeration equipment individual branch circuit (permitted by 210.52(B)(1) Exception No. 2) can be excluded from this calculation 》*220.52(A) Exception* 《.

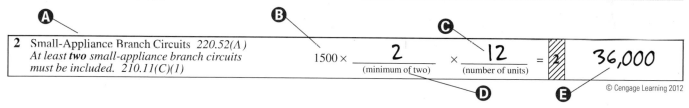

	A	**B**	**C**		
2	Small-Appliance Branch Circuits *220.52(A)*				
	*At least **two** small-appliance branch circuits must be included. 210.11(C)(1)*	1500 × 2 (minimum of two)	× 12 (number of units)	= 2	36,000

© Cengage Learning 2012

Line 3—Laundry Branch Circuit(s)

A Include laundry branch circuit(s) when calculating a dwelling unit service 》*220.52(B)* 《.

B Each laundry branch circuit is calculated at no less than 1500 volt-amperes 》*220.52(B)* 《.

C Include only the number of units having a laundry branch circuit. If only six of twelve units contain a laundry branch circuit, put six as the number of units. In this calculation, each unit has a laundry branch circuit.

D A receptacle for laundry equipment shall not be required in a dwelling unit of a multifamily building where laundry

facilities are provided on the premises for use by all building occupants 》*210.52(F) Exception No. 1* 《.

E A receptacle for laundry equipment shall not be required in other than one-family dwellings where laundry facilities are not to be installed or permitted 》*210.52(F) Exception No. 2* 《.

F Unless one of the exceptions is met, include at least one laundry branch circuit 》*210.11(C)(2)* 《.

G These loads can be included with the general lighting load and are subject to the Table 220.11 demand factors 》*220.52(B)* 《.

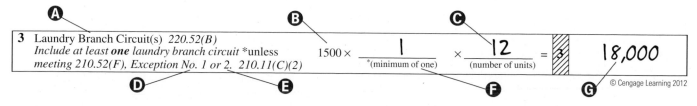

	A	**B**	**C**		
3	Laundry Branch Circuit(s) *220.52(B)*				
	*Include at least **one** laundry branch circuit *unless meeting 210.52(F), Exception No. 1 or 2. 210.11(C)(2)*	1500 × 1 *(minimum of one)	× 12 (number of units)	= 3	18,000

© Cengage Learning 2012

Lines 4 through 8—Applying Demand Factors Found in Table 220.11

(A) Line 4 represents the total of lines 1, 2, and 3.

(B) Lines 5 through 8 outline the derating procedure, using Table 220.42 demand factors.

(C) Line 5 is the result of subtracting 3000 from the number in Line 4.

(D) The first 3000 (or less) volt-amperes must remain at 100% »Table 220.42«.

(E) If line 5 is 117,000 or less, skip lines 6 and 7.

(F) If line 4 is 120,000 or less, then lines 6 and 7 remain empty.

(G) Insert the smaller of line 5 or 117,000.

(H) Multiply the next 117,000 (from 3001 to 120,000) by 35% »Table 220.42«.

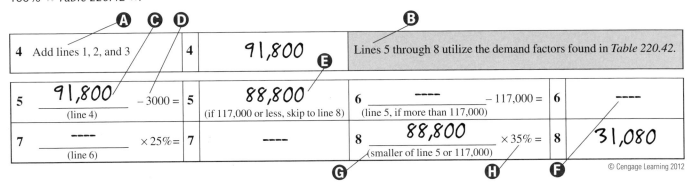

4 Add lines 1, 2, and 3	**4** 91,800	Lines 5 through 8 utilize the demand factors found in *Table 220.42*.

5	91,800 (line 4) − 3000 = **5**	88,800 (if 117,000 or less, skip to line 8)	6 ---- (line 5, if more than 117,000) − 117,000 = **6**	----
7	---- (line 6) × 25% = **7**	----	8 88,800 (smaller of line 5 or 117,000) × 35% = **8**	31,080

© Cengage Learning 2012

Line 9—Total General Lighting (and Receptacle) Load

(A) Line 9 is the total of general lighting, general-use receptacles, small-appliance branch circuits, and laundry branch circuit(s) after derating.

(B) If line 7 is blank, leave this blank also.

(C) Insert the result of line 8, not the number from line 5.

(D) This shaded block draws attention to the first of six lines (boxes) that will be added together resulting in the total volt-ampere demand load for the one-family dwelling.

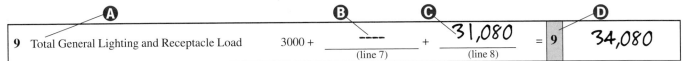

9 Total General Lighting and Receptacle Load	3000 + ---- (line 7) + 31,080 (line 8) = **9**	34,080

© Cengage Learning 2012

Line 10—Fastened-in-Place Appliances

(A) The term **appliance** designates utilization equipment (commonly built in standardized types and sizes) installed as a unit to perform specific function(s) »Article 100«.

(B) For the purpose of this load calculation, a kW rating is the same as a kVA rating.

(C) Additional space may be needed to calculate total appliance volt-amperes; (1200 × 12 = 14,400) and (900 × 12 = 10,800).

(D) Derating is allowed for a total of four (or more) fastened-in-place appliances. It is not necessary that the four fastened-in-place appliances be of different types. This derating factor can be used even if all units do not contain four appliances.

(E) If the calculation contains at least four fastened-in-place appliances, the combined volt-ampere rating of those appliances is multiplied by 75% and placed on Line 10 »220.53«.

> **NOTE**
>
> Do not include electric ranges, dryers, space-heating equipment, and air-conditioning equipment as fastened-in-place appliances »220.53«.
> Household cooking appliances individually rated in excess of 1750 watts are derated under 220.55 provisions.

10 Fastened-In-Place Appliances *220.53* Use the nameplate rating. Do not include electric ranges, clothes dryers, space-heating equipment, or air-conditioning equipment.	water heaters/ 54,000 microwave ovens / 14,400 /	dishwashers / 14,400 / /	disposers / 10,800 / /
(D) If fewer than four units, put total volt-amperes on line 10. If four or more units, multiply total volt-amperes by 75%.		93,600 (volt-amps of four or more) × 75% = **10**	70,200

© Cengage Learning 2012

Line 11—Clothes Dryers

A A clothes dryer is not a requirement for a load calculation. Skip this line if there is no clothes dryer.

B The dryer rating must be at least 5000 watts (volt-amperes) ≫ *220.54* ≪.

C Table 220.54 can be used to derate five (or more) clothes dryer loads. First, find the demand factor percentage across from the number of dryers. In this case the percentage is 46 (47 − 1 = 46%). Next, multiply the total dryer load by that percentage and put the result in line 11 (5000 × 12 − 60,000 × 46% = 27,600).

D A clothes dryer neutral load on a feeder is calculated at 70% ≫ *220.61(B)* ≪.

E If the nameplate rating exceeds 5000 watts (volt-amperes), use the larger number ≫ *220.54* ≪.

> **NOTE**
>
> For four (or fewer) clothes dryers, the demand is 100%.

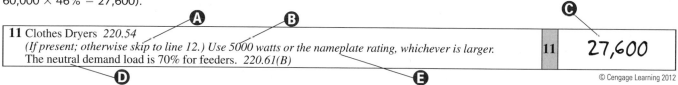

11 Clothes Dryers *220.54*
(If present; otherwise skip to line 12.) Use 5000 watts or the nameplate rating, whichever is larger.
The neutral demand load is 70% for feeders. 220.61(B)

| **11** | **27,600** |

© Cengage Learning 2012

Line 12—Household Cooking Appliances

A Household cooking appliances (ranges, wall-mounted ovens, counter-mounted cooking units, etc.) are not required in a load calculation. If no cooking appliances are rated over $1\frac{3}{4}$ kW, skip this line.

B Individual household cooking appliances rated more than 1750 watts (volt-amperes) can be derated using Table 220.55 demand factors (and all applicable notes).

C The demand for twelve 12 kW ranges comes from Table 220.55, Column C. First, look at the "Number of Appliances" column and find the number 12. Next, follow that row across to the last column (Column C) and find the kW demand load, which is 27. Finally, multiply this number by 1000 and place in line 12. Because each range is 12 kW, Columns A and B cannot be used.

D For household electric ranges, wall-mounted ovens, and counter-mounted cooking units, the neutral load on a feeder is calculated at 70% ≫ *220.61(B)* ≪.

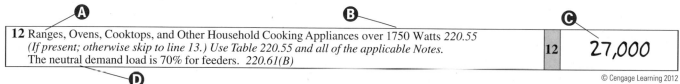

12 Ranges, Ovens, Cooktops, and Other Household Cooking Appliances over 1750 Watts *220.55*
(If present; otherwise skip to line 13.) Use Table 220.55 and all of the applicable Notes.
The neutral demand load is 70% for feeders. 220.61(B)

| **12** | **27,000** |

© Cengage Learning 2012

Line 13—Heating or Air-Conditioning System

A The smaller of two (or more) **noncoincident loads** can be omitted, as long as they are never energized simultaneously ≫ *220.60* ≪. By definition, noncoincident means not occurring at the same time. A dwelling's electric heating system and air-conditioning system may be noncoincident loads.

B The air handler (blower motor) has a rating of 345 volt-amperes (3 × 115 = 345). Because both loads are energized simultaneously, add the air handler to the 5 kW of electric heat.

C Fixed electric space-heating loads are calculated at 100% of the total connected load ≫ *220.51* ≪.

D Because the air handler (blower) works with either the heating or the air conditioning, add it to both calculations.

E Each unit contains an air-conditioner compressor (14 × 230 = 3220 VA), condenser fan motor (2 × 115 = 230 VA),

and blower motor (3 × 115 = 345 VA). The combined load is 3795 volt-amperes. Because this is less than the heating load, omit the air-conditioner compressor and the condenser fan motor. Multiply each unit's heat load by the number of units (5345 × 12 = 64,140).

> **NOTE**
>
> A heat pump with supplementary heat is not considered a noncoincident load. Add the compressor full-load current to the maximum amount of heat that can be energized while the compressor is running. Remember to include all associated motor(s), that is, blower, condenser, etc.

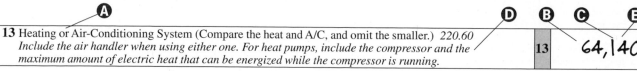

13 Heating or Air-Conditioning System (Compare the heat and A/C, and omit the smaller.) *220.60*
Include the air handler when using either one. For heat pumps, include the compressor and the
maximum amount of electric heat that can be energized while the compressor is running.

| **13** | **64,140** |

© Cengage Learning 2012

Line 14—Largest Motor

(A) Multiply the largest motor in the dwelling calculation by 25% 》*220.50* and *430.24* 《.

(B) This means a total of one motor, not one motor per unit.

(C) The largest motor is usually the air-conditioner compressor. If the air conditioner load has been omitted, use the second largest motor. In this case, use the 7.5-ampere garbage disposer motor (7.5 × 120 = 900).

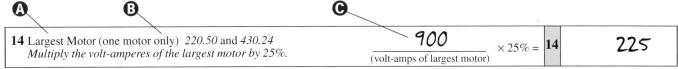

14 Largest Motor (one motor only) *220.50* and *430.24*
Multiply the volt-amperes of the largest motor by 25%.

$$\frac{900}{\text{(volt-amps of largest motor)}} \times 25\% = \boxed{14} \quad 225$$

© Cengage Learning 2012

Line 15—Total Demand Load

(A) Find the minimum volt-ampere load by adding lines 9 through 14.

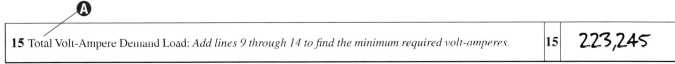

15 Total Volt-Ampere Demand Load: *Add lines 9 through 14 to find the minimum required volt-amperes.*   **15**   **223,245**

© Cengage Learning 2012

Lines 16 and 17—Minimum Service or Feeder

(A) Place the total volt-ampere figure found in line 15 here.

(B) Write down the source voltage supplying the service equipment.

(C) Fractions of an ampere 0.5 and higher are rounded up to the nearest whole ampere, and fractions less than 0.5 are dropped 》*220.5(B)* 《.

(D) The service (or feeder) overcurrent protection must be higher than the number in line 16. A list of standard ampere ratings for fuses and circuit breakers is provided in 240.6(A).

(E) Divide the volt-amperes by the voltage to determine the amperage.

(F) This is the minimum amperage rating required for the service or feeder being calculated.

(G) The next standard size fuse or breaker above 930 amperes is 1000 》*240.6(A)* 《.

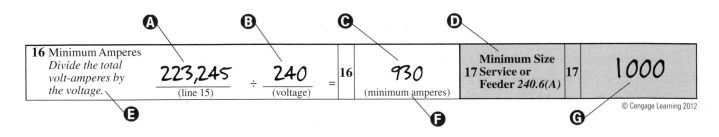

16 Minimum Amperes *Divide the total volt-amperes by the voltage.*

$$\frac{223,245}{\text{(line 15)}} \div \frac{240}{\text{(voltage)}} = \boxed{16} \quad \frac{930}{\text{(minimum amperes)}}$$

Minimum Size 17 Service or Feeder 240.6(A)   **17**   **1000**

© Cengage Learning 2012

Line 18—Minimum Size Conductors

Ⓐ For a service rated 100 through 400 A, the service conductors supplying the entire load associated with a one-family dwelling, or the service conductors supplying the entire load associated with an individual dwelling unit in a two-family or multifamily dwelling, shall be permitted to have an ampacity not less than 83 percent of the service rating ≫ *310.15(B)(7)(1)* ≪.

Ⓑ Section 310.15(B)(7) also applies to feeder conductors supplying the entire load associated with an individual dwelling ≫ *310.15(B)(7)(2)* ≪.

Ⓒ The maximum service or feeder rating permitted when using 310.15(B)(7) is 400 amperes.

Ⓓ When running **three** sets of paralleled copper conductors (for a 1000 ampere overcurrent device), the minimum size is 400 kcmil, unless the equipment is listed and identified for use with 90°C (194°F) conductors. Even if a 90°C (194°F) conductor (such as THHN) is installed, the ampacity rating shall not exceed the 75°C (167°F) ampacity, which is 335 ≫ *110.14(C)(2)* ≪. Because the overcurrent protection is rated over 800 amperes, the conductor ampacity must match (or exceed) the 1000 ampere rating (1000 ÷ 3 = 333.3 minimum amperes, each conductor) ≫ *240.4(C)* ≪.

Ⓔ Other parallel conductor combinations are possible, such as four sets of 250 kcmil copper (Cu), four sets of 350 kcmil aluminum (Al), five sets of 3/0 AWG Cu, five sets of 250 kcmil Al, etc. The smallest conductor that can be run in parallel is 1/0 AWG ≫ *310.10(H)(1)* ≪.

Ⓕ Since the rating is more than 400 amperes, use Table 310.15(B)(16) to find the minimum size conductors.

Ⓖ The number and type of paralleled conductor sets is a design consideration, not necessarily a *Code* issue. Exercise extreme care when designing a paralleled conductor installation, without violating provisions such as 110.14(A), 110.14(C), 240.4(C), 250.66, 250.122(F), 300.20(A), 310.10(H), etc.

Ⓗ 1/0 AWG and larger conductors can be connected in parallel (electrically joined at both ends to form a single conductor) ≫ *310.10(H)(1)* ≪.

> **CAUTION**
>
> Conductors carrying alternating current and installed in ferrous metal enclosures or **metal raceways** must be arranged to avoid heating of the surrounding metal by induction ≫*300.20(A)*≪.

> **N O T E**
>
> The paralleled conductors in each phase, polarity, neutral, grounded circuit conductor, equipment grounding conductor, or equipment bonding jumper must share the same characteristics ≫*310.10(H)(2)*≪.
> For example, all paralleled, Phase A conductors must:
> (1) be the same length,
> (2) consist of the same conductor material,
> (3) be the same size in circular mil area,
> (4) have the same insulation type, and
> (5) be terminated in the same manner.

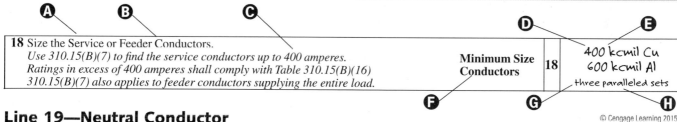

18 Size the Service or Feeder Conductors.					
Use 310.15(B)(7) to find the service conductors up to 400 amperes.			**Minimum Size**	**18**	**400 kcmil Cu**
Ratings in excess of 400 amperes shall comply with Table 310.15(B)(16)			**Conductors**		**600 kcmil Al**
310.15(B)(7) also applies to feeder conductors supplying the entire load.					*three paralleled sets*

Line 19—Neutral Conductor

Ⓐ For the purpose of this load calculation, the terms **neutral** and **grounded** are synonymous.

Ⓑ The feeder (or service) neutral load is the maximum unbalance of the load determined by Article 220 ≫ *220.61* ≪.

Ⓒ The neutral (grounded) conductor can be smaller than the ungrounded (hot) conductors provided certain requirements in 215.2, 220.61, and 230.42 are met ≫ *310.15(B)(7)* ≪.

Ⓓ Size the neutral by finding the maximum unbalanced load.

Ⓔ Although the neutral conductor requires only 152 amperes each (for three paralleled sets), it shall not be smaller than the appropriate grounding electrode conductor (see line 20).

> **N O T E**
>
> The smallest paralleled grounded (neutral) conductor is 1/0 AWG ≫*250.24(C)(2)*≪.

> **CAUTION**
>
> The grounded conductor must not be smaller than the required grounding electrode conductor specified in Table 250.102(C)(1) ≫*250.24(C)(1)*≪.

19 Size the Neutral Conductor. *220.61*					
310.15(B)(7) states that the neutral service or feeder conductor			**Minimum Size**		**3/0 Cu**
can be smaller than the ungrounded (hot) conductors, provided the			**Neutral**	**19**	**250 kcmil Al**
requirements of 215.2, 220.61, and 230.42 are met.			**Conductor**		*three paralleled sets*
250.24(C)(1) states that the grounded conductor shall not be smaller					
than specified in Table 250.102(C)(1).					

Line 19—Neutral Conductor (continued)

Ⓐ No neutral conductor is connected to the water heaters in this calculation.

Ⓑ All fastened-in-place appliances using a grounded conductor must be included.

Ⓒ The electric heat and air-conditioner compressor are 240-volt loads (no neutral).

Ⓓ Because all of these motors can be energized simultaneously, thereby contributing to the neutral load, include them in the neutral calculation.

Ⓔ To convert to amperes, simply divide volt-amperes by voltage.

Ⓕ The minimum conductor ampacity for three paralleled neutral conductors is 152 amperes. Although the minimum size copper conductor at 75°C (167°F) is 2/0 AWG, the neutral conductor cannot be smaller than the required grounding electrode conductor.

Ⓖ All numbers are shown as volt-amperes.

Ⓗ Since all of the general lighting and receptacle loads have a rating of 120 volts, the total volt-ampere rating (from line 9) must be included in the neutral calculation.

Ⓘ A demand factor of 75% can be applied to the nameplate rating load of four (or more) fastened-in-place appliances ⟫220.53⟪.

Ⓙ Multiply the clothes dryer load (line 11) by 70% to determine the neutral load ⟫220.61(B)⟪.

Ⓚ Multiplying line 12 by 70% provides the neutral cooking appliances load ⟫220.61(B)⟪.

Ⓛ Multiply the largest 120- (115-) volt motor (one motor only) by 25% and include in the calculation.

Ⓜ The minimum neutral load is 109,125 volt-amperes.

Ⓖ

General lighting and receptacles (line 9) . . . . 34,080 **Ⓗ**

Ⓐ Fastened-in-place appliances

Ⓑ
Water heaters 0
Dishwashers (1200 × 12) 14,400
Waste disposers (7.5 × 120 × 12) . . 10,800
Microwave ovens (1200 × 12) . . . . . 14,400 **Ⓘ**
Total 39,600

Ⓙ

Total fastened-in-place appliances (39,600 × 75%) 29,700 **Ⓚ**
Clothes dryers (line 11 × 70%) 19,320
Ranges (line 12 × 70%) 18,900

Ⓒ Electric heat 0
Air-conditioner compressors 0

Ⓓ Air handlers (3 × 115 × 12) 4140
Condenser fan motors (2 × 115 × 12) . . . . 2760 **Ⓛ**
Largest motor (900 × 25%) 225

TOTAL . . . . 109,125 **Ⓜ**

Ⓔ 109,125 ÷ 240 = 454.7 = 455 minimum neutral ampacity

455 ÷ 3 = 152 minimum amperes per neutral conductor

Ⓕ

© Cengage Learning 2012.

Line 20—Grounding Electrode Conductor

Ⓐ Table 250.66 has two headings—service conductor size and grounding electrode conductor size. Each heading contains two columns: one for copper and one for aluminum (or copper-clad aluminum). Because of the four columns, take extreme care in selecting the correct grounding electrode conductor.

Ⓑ The grounding electrode conductor is determined by the largest service-entrance conductor (or the equivalent area) if the service-entrance conductors are paralleled. The equivalent area for three parallel sets in this calculation is 1200 kcmil copper (3 × 400) or 1800 kcmil aluminum (3 × 600).

Ⓒ The minimum size grounding electrode conductor is 3/0 AWG copper (or 250 kcmil aluminum) ⟫Table 250.66⟪.

Ⓐ **Ⓑ**

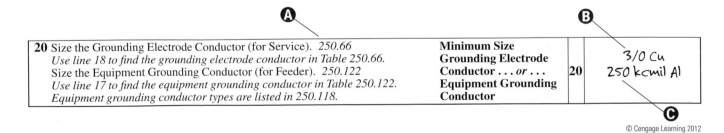

| 20 Size the Grounding Electrode Conductor (for Service). *250.66* *Use line 18 to find the grounding electrode conductor in Table 250.66.* Size the Equipment Grounding Conductor (for Feeder). *250.122* *Use line 17 to find the equipment grounding conductor in Table 250.122.* *Equipment grounding conductor types are listed in 250.118.* | **Minimum Size Grounding Electrode Conductor . . . or . . . Equipment Grounding Conductor** | **20** | 3/0 Cu 250 kcmil Al |

Ⓒ

© Cengage Learning 2012

The following page shows the completed standard method load calculation for a twelve-unit multifamily dwelling, having all units combined in a single service.

Standard Method Load Calculation for Multifamily Dwellings

1	General Lighting and Receptacle Loads *220.12* *Do not include open porches, garages, or unused or* *unfinished spaces not adaptable for future use.*	$3 \times$ __1050__ (sq ft outside dimensions)	$\times$ __12__ (number of units)	=	1	**37,800**
2	Small-Appliance Branch Circuits *220.52(A)* *At least two small-appliance branch circuits* *must be included. 210.11(C)(1)*	$1500 \times$ __2__ (minimum of two)	$\times$ __12__ (number of units)	=	2	**36,000**
3	Laundry Branch Circuit(s) *220.52(B)* *Include at least one laundry branch circuit *unless* *meeting 210.52(F), Exception No. 1 or 2. 210.11(C)(2)*	$1500 \times$ __1__ *(minimum of one)	$\times$ __12__ (number of units)	=	3	**18,000**

4	Add lines 1, 2, and 3	**4** **91,800**	Lines 5 through 8 utilize the demand factors found in *Table 220.42.*

5	__91,800__ $- 3000 =$ (line 4)	**5** **88,800** (if 117,000 or less, skip to line 8)	**6** __----__ $- 117,000 =$ (line 5, if more than 117,000)	**6**	**----**
7	__----__ $\times 25\% =$ (line 6)	**7** **----**	**8** __88,800__ $\times 35\% =$ (smaller of line 5 or 117,000)	**8**	**31,080**

9	Total General Lighting and Receptacle Load	$3000 +$ __----__ $+$ __31,080__ $=$ (line 7) (line 8)	**9** **34,080**

10	Fastened-In-Place Appliances *220.53* *Use the nameplate rating.* *Do not include electric ranges,* *clothes dryers, space-heating* *equipment, or air-conditioning* *equipment.*	water heaters/ **54,000** dishwashers / **14,400** disposers / **10,800** microwave ovens / **14,400** / / / / /
	If fewer than four units, put total volt-amperes on line 10. If four or more units, multiply total volt-amperes by 75%.	__93,600__ $\times 75\% =$ **10** **70,200** (volt-amps of four or more)

11	Clothes Dryers *220.54* *(If present; otherwise skip to line 12.) Use 5000 watts or the nameplate rating, whichever is larger.* The neutral demand load is 70% for feeders. *220.61(B)*	**11**	**27,600**
12	Ranges, Ovens, Cooktops, and Other Household Cooking Appliances over 1750 Watts *220.55* *(If present; otherwise skip to line 13.) Use Table 220.55 and all of the applicable Notes.* The neutral demand load is 70% for feeders. *220.61(B)*	**12**	**27,000**
13	Heating or Air-Conditioning System (Compare the heat and A/C, and omit the smaller.) *220.60* *Include the air handler when using either one. For heat pumps, include the compressor and the* *maximum amount of electric heat that can be energized while the compressor is running.*	**13**	**64,140**
14	Largest Motor (one motor only) *220.50 and 430.24* *Multiply the volt-amperes of the largest motor by 25%.*	__900__ $\times 25\% =$ **14** (volt-amps of largest motor)	**225**

15	Total Volt-Ampere Demand Load: *Add lines 9 through 14 to find the minimum required volt-amperes.*	**15**	**223,245**

16	Minimum Amperes *Divide the total* *volt-amperes by* *the voltage.*	__223,245__ $\div$ __240__ $=$ (line 15) (voltage)	**16** **930** (minimum amperes)	**Minimum Size** **17 Service or** **Feeder *240.6(A)*** **17** **1000**

18	Size the Service or Feeder Conductors. *Use 310.15(B)(7) to find the service conductors up to 400 amperes.* *Ratings in excess of 400 amperes shall comply with Table 310.15(B)(16)* *310.15(B)(7) also applies to feeder conductors supplying the entire load.*	**Minimum Size** **Conductors**	**18**	400 kcmil Cu 600 kcmil Al three paralleled sets
19	Size the Neutral Conductor. *220.61* *310.15(B)(7) states that the neutral service or feeder conductor* *can be smaller than the ungrounded (hot) conductors, provided the* *requirements of 215.2, 220.61, and 230.42 are met.* *250.24(C)(1) states that the grounded conductor shall not be smaller* *than specified in Table 250.102(C)(1).*	**Minimum Size** **Neutral** **Conductor**	**19**	3/0 Cu 250 kcmil Al three paralleled sets
20	Size the Grounding Electrode Conductor (for Service). *250.66* *Use line 18 to find the grounding electrode conductor in Table 250.66.* Size the Equipment Grounding Conductor (for Feeder). *250.122* *Use line 17 to find the equipment grounding conductor in Table 250.122.* *Equipment grounding conductor types are listed in 250.118.*	**Minimum Size** **Grounding Electrode** **Conductor . . . or . . .** **Equipment Grounding** **Conductor**	**20**	3/0 Cu 250 kcmil Al

SIX-UNIT MULTIFAMILY DWELLING CALCULATION

Two Meter Centers, Each Supplying Six Units

A Each service drop, set of overhead service conductors, set of underground service conductors, or service lateral supplies only one set of service-entrance conductors, unless two to six service disconnecting means (in separate enclosures) are grouped at one location and supply separate loads from one service drop ≫ *230.40 Exception No. 2* ≪ .

B Because service equipment installations can vary greatly, the arrangement must be known before a load calculation is performed. In this configuration, each service (conductors and equipment) must be able to carry the calculated load for six dwelling units.

C The following multifamily dwelling, standard method, load calculation is based on six units with identical configurations. Turn to page 233 for dwelling unit details.

D Service-entrance conductors must be able to carry the ampacity determined in accordance with Article 220 ≫ *230.42(A)* ≪ .

E No more than six disconnects per service can be grouped in one location ≫ *230.71(A)* ≪ .

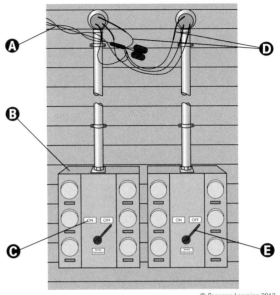

© Cengage Learning 2012

Lines 1 through 9—General Lighting and Receptacle Load

A Fill in the first three lines using six dwelling units instead of twelve.

B Insert the result of line 8, not the number from line 5.

C The general lighting and receptacle load for these six dwelling units is 18,015 volt-amperes. Compare this line 9 with the line 9 for twelve units (34,080). Obviously, this is not half of line 9 for twelve units; therefore, a calculation must be performed for each different application.

D Multiply the result of line 5 by 35% and place on line 8.

1	General Lighting and Receptacle Loads *220.12* *Do not include open porches, garages, or unused or unfinished spaces not adaptable for future use.*		$3 \times$ __1050__ (sq ft outside dimensions) $\times$ __6__ (number of units)	=	1	18,900
2	Small-Appliance Branch Circuits *220.52(A)* *At least **two** small-appliance branch circuits must be included. 210.11(C)(1)*		$1500 \times$ __2__ (minimum of two) $\times$ __6__ (number of units)	=	2	18,000
3	Laundry Branch Circuit(s) *220.52(B)* *Include at least **one** laundry branch circuit *unless meeting 210.52(F), Exception No. 1 or 2. 210.11(C)(2)*		$1500 \times$ __1__ *(minimum of one) $\times$ __6__ (number of units)	=	3	9000
4	Add lines 1, 2, and 3	4	45,900	Lines 5 through 8 utilize the demand factors found in *Table 220.42*.		
5	__45,900__ (line 4) $-3000 =$	5	42,900 (if 117,000 or less, skip to line 8)	6 __----__ (line 5, if more than 117,000) $- 117,000 =$	6	----
7	__----__ (line 6) $\times 25\% =$	7	----	8 __42,900__ (smaller of line 5 or 117,000) $\times 35\% =$	8	15,015
9	Total General Lighting and Receptacle Load	$3000 +$ __----__ (line 7) $+$ __15,015__ (line 8)		=	9	18,015

© Cengage Learning 2012

Lines 11 and 12—Clothes Dryers and Ranges

A Use Table 220.54 to derate five (or more) clothes dryer loads. First, find the demand factor percentage across from the number of dryers, in this case 75%. Next, multiply the total dryer load by that percentage and record the result on line 11 (5000 × 6 = 30,000 × 75% = 22,500).

B The demand for six 12-kW ranges comes from Table 220.55, Column C. First, look at the "Number of Appliances" column and find the number 6. Next, follow that row across to Column C and find the kW demand load, which is 21. Finally, multiply this number by 1000 and place in line 12. Since each range is 12 kW, Columns A and B cannot be used.

A

| 11 | Clothes Dryers 220.54 *(If present; otherwise skip to line 12.) Use 5000 watts or the nameplate rating, whichever is larger.* The neutral demand load is 70% for feeders. 220.61(B) | 11 | **22,500** |
| 12 | Ranges, Ovens, Cooktops, and Other Household Cooking Appliances over 1750 Watts 220.55 *(If present; otherwise skip to line 13.) Use Table 220.55 and all of the applicable Notes.* The neutral demand load is 70% for feeders. 220.61(B) | 12 | **21,000** |

B

© Cengage Learning 2012

Lines 15 through 17—Minimum Amperes

A The minimum volt-ampere load for these six dwelling units is 128,910.

B The service-entrance conductors and equipment must have a rating of at least 537 amperes.

C The next standard size fuse or circuit breaker above 537 is 600 amperes ≫ 240.6(A) ≪.

A

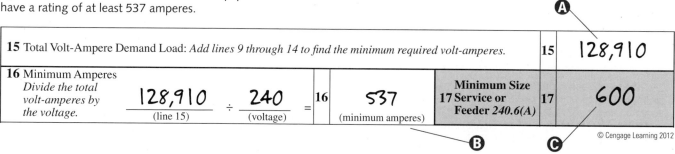

| 15 | Total Volt-Ampere Demand Load: *Add lines 9 through 14 to find the minimum required volt-amperes.* | | 15 | **128,910** |

| 16 | Minimum Amperes *Divide the total volt-amperes by the voltage.* $\dfrac{128,910}{\text{(line 15)}} \div \dfrac{240}{\text{(voltage)}} =$ | 16 | **537** (minimum amperes) | **Minimum Size** 17 **Service or Feeder** *240.6(A)* | 17 | **600** |

B   **C**

© Cengage Learning 2012

Line 18—Minimum Size Conductors

A At 75°C (167°F), the minimum size copper conductors for two parallel sets matching or exceeding 537 amperes are 300 kcmil. Because the overcurrent protection is less than 800 amperes, the conductor ampacity can be less than the 600-ampere rating. The next higher standard overcurrent device rating (above the ampacity of the conductors being protected) shall be permitted to be used, provided all of the following three conditions are met: (1) the conductors being protected are not part of a branch circuit supplying more than one receptacle for cord- and plug-connected portable loads, (2) the ampacity of the conductors does not correspond with the standard ampere rating of a fuse or a circuit breaker without overload trip adjustments above its rating (but that

shall be permitted to have other trip or rating adjustments), and (3) the next higher standard rating selected does not exceed 800 amperes ≫ 240.4(B) ≪. The combined rating of two 300-kcmil copper conductors is 570 amperes. Because the combined rating of 250-kcmil copper conductors is only 510 amperes, which is below the minimum required conductor ampacity of 531 amperes, 250-kcmil conductors are not permitted.

B Other parallel conductor combinations are possible, such as three sets of 3/0 AWG Cu and three sets of 4/0 AWG Al.

A

| 18 | Size the Service or Feeder Conductors. *Use 310.15(B)(7) to find the service conductors up to 400 amperes. Ratings in excess of 400 amperes shall comply with Table 310.15(B)(16) 310.15(B)(7) also applies to feeder conductors supplying the entire load.* | **Minimum Size Conductors** | 18 | 300 kcmil Cu 400 kcmil Al two paralleled sets |

B

© Cengage Learning 2015.

Line 19—Neutral Conductor

Ⓐ The minimum neutral load is 66,990 volt-amperes.

Ⓑ The minimum conductor ampacity for two paralleled neutral conductors is 140 amperes. At 75°C (167°F), the minimum size conductor is 1/0 AWG copper *or* 3/0 AWG aluminum.

> **N O T E**
>
> Even with a lower neutral ampacity, the conductors must remain this size because the neutral conductor shall not be smaller than specified in Table 250.102(C)(1) »*250.24(C)(1)*«.

General lighting and receptacles (line 9) 18,015
Fastened-in-place appliances
    Water heaters 0
    Dishwashers (1200 × 6) 7200
    Waste disposers (7.5 × 120 × 6) . 5400
    Microwave ovens (1200 × 6) 7200
    Total 19,800

Total fastened-in-place appliances (19,800 × 75%) . . 14,850
Clothes dryers (line 11 × 70%) 15,750
Ranges (line 12 × 70%) 14,700
Electric heat 0
Air-conditioner compressors 0
Air handlers (3 × 115 × 6) 2070
Condenser fan motors (2 × 115 × 6) 1380
Largest motor (900 × 25%) 225
TOTAL **66,990**

Ⓐ

66,990 ÷ 240 = 279.13 = 279 minimum neutral ampacity
279 ÷ 2 = 139.5 = 140 minimum amperes per neutral conductor

Ⓑ

19 Size the Neutral Conductor. *220.61*		
310.15(B)(7) states that the neutral service or feeder conductor can be smaller than the ungrounded (hot) conductors, provided the requirements of 215.2, 220.61, and 230.42 are met. 250.24(C)(1) states that the grounded conductor shall not be smaller than specified in Table 250.102(C)(1).	**Minimum Size Neutral Conductor** **19**	1/0 Cu 3/0 Al *two paralleled sets*

Line 20—Grounding Electrode Conductor

Ⓐ The equivalent area for two parallel sets of service entrance conductors, in this calculation, is 600 kcmil copper (2 × 300) . . . or . . . 800 kcmil aluminum (2 × 400).

Ⓑ The minimum size grounding electrode conductor is 1/0 AWG copper *or* 3/0 AWG aluminum »*Table 250.66*«.

Ⓐ

20 Size the Grounding Electrode Conductor (for Service). *250.66*		
Use line 18 to find the grounding electrode conductor in Table 250.66. Size the Equipment Grounding Conductor (for Feeder). *250.122* *Use line 17 to find the equipment grounding conductor in Table 250.122. Equipment grounding conductor types are listed in 250.118.*	**Minimum Size Grounding Electrode Conductor . . . or . . . Equipment Grounding Conductor** **20**	1/0 Cu 3/0 Al

Ⓑ

The following page shows the completed standard method load calculation for a six-unit multifamily dwelling.

Standard Method Load Calculation for Multifamily Dwellings

1	General Lighting and Receptacle Loads *220.12* *Do not include open porches, garages, or unused or* *unfinished spaces not adaptable for future use.*	$3 \times$ __1050__ $\times$ __6__ $=$ **1** (sq ft outside dimensions) (number of units)	**18,900**
2	Small-Appliance Branch Circuits *220.52(A)* At least **two** small-appliance branch circuits must be included. *210.11(C)(1)*	$1500 \times$ __2__ $\times$ __6__ $=$ **2** (minimum of two) (number of units)	**18,000**
3	Laundry Branch Circuit(s) *220.52(B)* Include at least **one** laundry branch circuit *unless* *meeting 210.52(F), Exception No. 1 or 2. 210.11(C)(2)*	$1500 \times$ __1__ $\times$ __6__ $=$ **3** *(minimum of one) (number of units)	**9000**

4	Add lines 1, 2, and 3	**4** **45,900**	Lines 5 through 8 utilize the demand factors found in *Table 220.42.*

5 __45,900__ $- 3000 =$ **5** **42,900** (line 4) (if 117,000 or less, skip to line 8)		**6** __----__ $- 117,000 =$ **6** (line 5, if more than 117,000)	**----**
7 __----__ $\times 25\% =$ **7** **----** (line 6)		**8** __42,900__ $\times 35\% =$ **8** (smaller of line 5 or 117,000)	**15,015**

9	Total General Lighting and Receptacle Load	$3000 +$ __----__ $+$ __15,015__ $=$ **9** (line 7) (line 8)	**18,015**

10	Fastened-In-Place Appliances *220.53* *Use the nameplate rating.* *Do not include electric ranges,* *clothes dryers, space-heating* *equipment, or air-conditioning* *equipment.*	water heaters/ **27,000** dishwashers / **7200** disposers / **5400** microwave ovens / **7200** / / / / /		
	If fewer than four units, put total volt-amperes on line 10. If four or more units, multiply total volt-amperes by 75%.	**46,800** $\times 75\% =$ **10** (volt-amps of four or more)		**35,100**
11	Clothes Dryers *220.54* *(If present; otherwise skip to line 12.) Use 5000 watts or the nameplate rating, whichever is larger.* The neutral demand load is 70% for feeders. *220.61(B)*		**11**	**22,500**
12	Ranges, Ovens, Cooktops, and Other Household Cooking Appliances over 1750 Watts *220.55* *(If present; otherwise skip to line 13.) Use Table 220.55 and all of the applicable Notes.* The neutral demand load is 70% for feeders. *220.61(B)*		**12**	**21,000**
13	Heating or Air-Conditioning System (Compare the heat and A/C, and omit the smaller.) *220.60* *Include the air handler when using either one. For heat pumps, include the compressor and the* *maximum amount of electric heat that can be energized while the compressor is running.*		**13**	**32,070**
14	Largest Motor (one motor only) *220.50 and 430.24* *Multiply the volt-amperes of the largest motor by 25%.*	**900** $\times 25\% =$ **14** (volt-amps of largest motor)		**225**
15	Total Volt-Ampere Demand Load: *Add lines 9 through 14 to find the minimum required volt-amperes.*		**15**	**128,910**

16	Minimum Amperes *Divide the total* *volt-amperes by* *the voltage.*	__128,910__ $\div$ __240__ $=$ **16** **537** (line 15) (voltage) (minimum amperes)	**Minimum Size** **17 Service or** **Feeder** *240.6(A)*	**17** **600**
18	Size the Service or Feeder Conductors. *Use 310.15(B)(7) to find the service conductors up to 400 amperes.* *Ratings in excess of 400 amperes shall comply with Table 310.15(B)(16)* *310.15(B)(7) also applies to feeder conductors supplying the entire load.*		**Minimum Size** **Conductors**	**18** 300 kcmil Cu 400 kcmil Al two paralleled sets
19	Size the Neutral Conductor. *220.61* *310.15(B)(7) states that the neutral service or feeder conductor* *can be smaller than the ungrounded (hot) conductors, provided the* *requirements of 215.2, 220.61, and 230.42 are met.* *250.24(C)(1) states that the grounded conductor shall not be smaller* *than specified in Table 250.102(C)(1).*		**Minimum Size** **Neutral** **Conductor**	**19** 1/0 Cu 3/0 Al two paralleled sets
20	Size the Grounding Electrode Conductor (for Service). *250.66* *Use line 18 to find the grounding electrode conductor in Table 250.66.* *Size the Equipment Grounding Conductor (for Feeder). 250.122* *Use line 17 to find the equipment grounding conductor in Table 250.122.* *Equipment grounding conductor types are listed in 250.118.*		**Minimum Size** **Grounding Electrode** **Conductor . . . or . . .** **Equipment Grounding** **Conductor**	**20** 1/0 Cu 3/0 Al

STANDARD LOAD CALCULATION FOR EACH UNIT OF A MULTIFAMILY DWELLING

Calculating Feeders for Individual Dwelling Units

A A load calculation on an individual unit in a multifamily dwelling must be calculated in accordance with one of the one-family dwelling load-calculation methods (standard or optional).

B The service disconnecting means must have a rating of at least the calculated load, determined in accordance with Article 220 ≫ 230.79 ≪.

C If the service disconnecting means consists of more than one switch or circuit breaker (permitted by 230.71), the combined ratings of all switches or circuit breakers used must not be less than that required by 230.79 ≫ 230.80 ≪.

D A load calculation must be performed on each unique unit, that is, different from any unit already calculated.

E A grounded circuit conductor (neutral) shall not be used for grounding non-current-carrying metal equipment on the load side of the service disconnecting means ≫ 250.142(B) ≪. In other words, the neutral must be isolated from the equipment ground in a remote panelboard (subpanel).

F For a feeder rated 100 through 400 A, the feeder conductors supplying the entire load associated with an individual dwelling unit in a two-family or multifamily dwelling, shall be permitted to have an ampacity not less than 83 percent of the feeder rating ≫ 310.15(B)(7)(2) ≪.

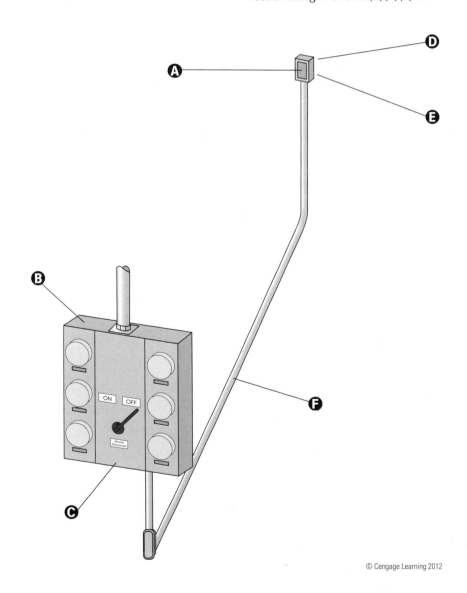

Standard Method—One Unit

Ⓐ Each identically configured unit—that is, square footage dimensions, small-appliance and laundry branch circuits, fastened-in-place appliances, etc.—results in the same size feeder.

Ⓑ Although located in a multifamily dwelling, this dwelling unit contains one feeder and one panelboard. It is, therefore, calculated in accordance with one of the one-family dwelling methods.

Ⓒ A load calculation must be performed on each dissimilar unit.

Ⓓ Using the standard method, the minimum required volt-amperes for this single unit is 29,048.

> **NOTE**
>
> Load calculations are calculated for each individual unit containing a panelboard (subpanel). Although located in a multifamily dwelling, calculate each unit by one of the methods for one-family dwellings **»Article 220«**. Performing a load calculation on multiple units and then dividing by the number of units will give an incorrect result.

Standard Method Load Calculation for One-Family Dwellings

#	Description		Calculation		Result
1	General Lighting and Receptacle Loads *220.12* *Do not include open porches, garages, or unused or unfinished spaces not adaptable for future use.*		$3 \times$ __1050__ (sq ft outside dimensions)	= 1	3150
2	Small-Appliance Branch-Circuits *220.52(A)* *Include at least **two** small-appliance branch circuits must be included. 210.11(C)(1)*		$1500 \times$ __2__ (minimum of two)	= 2	3000
3	Laundry Branch Circuit(s) *220.52(B)* *Include at least **one** laundry branch circuit *unless meeting 210.52(F), Exception No. 1 or 2. 210.11(C)(2)*		$1500 \times$ __1__ *(minimum of one)	= 3	1500
4	Add lines 1, 2, and 3	4	7650	Lines 5 through 8 utilize the demand factors found in *Table 220.42.*	
5	__7650__ (line 4) $- 3000 =$ 5		4650 (if 117,000 or less, skip to line 8)	6 ____ (line 5, if more than 117,000) $- 117,000 =$ 6	----
7	__----__ (line 6) $\times 25\% =$ 7		----	8 __4650__ (smaller of line 5 or 117,000) $\times 35\% =$ 8	1628
9	Total General Lighting and Receptacle Load $3000 +$ __----__ (line 7) $+$ __1628__ (line 8) $=$ 9				4628

10	Fastened-In-Place Appliances *220.53* *Use the nameplate rating. Do not include electric ranges, clothes dryers, space-heating equipment, or air-conditioning equipment.*	water heater / 4500	dishwasher / 1200	disposer / 900
		compactor /	microwave oven / 1200	/
		/	/	/

	If fewer than four units, put total volt-amperes on line 10. If four or more units, multiply total volt-amperes by 75%.	7800 (volt-amps of four or more) $\times 75\% =$ 10	5850

11	Clothes Dryers *220.54* *(If present; otherwise skip to line 12.) Use 5000 watts or the nameplate rating, whichever is larger. The neutral demand load is 70% for feeders. 220.61(B)*	11	5000
12	Ranges, Ovens, Cooktops, and Other Household Cooking Appliances over 1750 Watts *220.55* *(If present; otherwise skip to line 13.) Use Table 220.55 and all of the applicable Notes. The neutral demand load is 70% for feeders. 220.61(B)*	12	8000
13	Heating or Air-Conditioning System (Compare the heat and A/C, and omit the smaller.) *220.60* *Include the air handler when using either one. For heat pumps, include the compressor and the maximum amount of electric heat that can be energized while the compressor is running.*	13	5345
14	Largest Motor (one motor only) *220.50 and 430.24* *Multiply the volt-amperes of the largest motor by 25%.* __900__ (volt-amps of largest motor) $\times 25\% =$ 14		225
15	Total Volt-Ampere Demand Load: *Add lines 9 through 14 to find the minimum required volt-amperes.*	15	29,048

Minimum Size Feeder and Overcurrent Protection

A Feeder conductors shall have an ampacity not less than required to supply the load as calculated in Parts III, IV, and V of Article 220. Conductors shall be sized to carry not less than the larger of 215.2(A)(1)(a) or (b).

(a) Where a feeder supplies continuous loads or any combination of continuous and noncontinuous loads, the minimum feeder conductor size shall have an allowable ampacity not less than the noncontinuous load plus 125 percent of the continuous load.

(b) The minimum feeder conductor size shall have an allowable ampacity not less than the maximum load to be served

after the application of any adjustment or correction factors. » 215.2(A) «.

B Feeders must be protected against overcurrent in accordance with the Article 240, Part I provisions » 215.3 «.

C The minimum ampere rating for this one unit is 121.

D Use Table 310.15(B)(7) to determine the minimum size feeder conductors supplying the entire load associated with an individual dwelling » 310.15(B)(7)(2) «.

A   **B**

16 Minimum Amperes				
Divide the total volt-amperes by the voltage.	$\dfrac{29{,}048}{\text{(line 15)}} \div \dfrac{240}{\text{(voltage)}} = $ **16** $\boxed{121}$ (minimum amperes)		**Minimum Size** 17 **Service or Feeder** *240.6(A)* **17**	**125**
18 Size the Service or Feeder Conductors. *Use 310.15(B)(7) to find the service conductors up to 400 amperes. Ratings in excess of 400 amperes shall comply with Table 310.15(B)(16) 310.15(B)(7) also applies to feeder conductors supplying the entire load.*			**Minimum Size** **18** **Conductors**	2 copper 1/0 aluminum

© Cengage Learning 2015.

C   **D**

Minimum Size Neutral and Equipment Grounding Conductors

A The minimum neutral-conductor ampacity is 74.

B At 75°C (167°F), the minimum size conductor is 4 AWG copper *or* 3 AWG aluminum.

C If a feeder supplies branch circuits requiring equipment grounding conductors, it must include (or provide) a grounding means to which the equipment grounding conductors of the branch circuits shall be connected, in accordance with 250.134 provisions » 215.6 «.

D Copper, aluminum, or copper-clad aluminum equipment grounding conductors of the wire type must not be smaller than shown in Table 250.122, but do not have to be larger than the circuit conductors supplying the equipment » 250.122(A) «.

> **NOTE**
>
> A raceway or cable sheath acting as the equipment grounding conductor, as provided in 250.118 and 250.134(A), must comply with 250.4(A)(5) or (B)(4) **»250.122(A)«**.

General lighting and receptacles (line 9) 4628
Fastened-in-place appliances
      Water heater. 0
      Dishwasher. 1200
      In-sink waste disposer 900
      Microwave oven 1200
      Total 3300

Total fastened-in-place appliances (3 units = 100%) 3300
Clothes dryer (line 11 × 70%). 3500
Range (line 12 × 70%) 5600
Electric heat 0
Air-conditioner compressor 0
Air handler (3 × 115). 345
Condenser fan motor (2 × 115). 230
Largest motor (900 × 25%). 225

TOTAL 17,828

A 17,828 ÷ 240 = 74.28 minimum neutral-conductor ampacity

B

19 Size the Neutral Conductor. *220.61* *310.15(B)(7) states that the neutral service or feeder conductor can be smaller than the ungrounded (hot) conductors, provided the requirements of 215.2, 220.61, and 230.42 are met. 250.24(C)(1) states that the grounded conductor shall not be smaller than specified in Table 250.102(C)(1).*	**Minimum Size Neutral Conductor** **19**	4 copper 3 aluminum
20 Size the Grounding Electrode Conductor. *250.66* *Use line 18 to find the grounding electrode conductor in Table 250.66. Size the Equipment Grounding Conductor (for Feeder). 250.122 Use line 17 to find the equipment grounding conductor in Table 250.122. Equipment grounding conductor types are listed in 250.118.*	**Minimum Size Grounding Electrode Conductor . . . or . . . Equipment Grounding Conductor** **20**	6 copper 4 aluminum

C   **D**

© Cengage Learning 2015.

The following page shows the completed one-family dwelling load calculation for one multifamily dwelling-unit feeder and remote panelboard (subpanel), according to the standard method.

Standard Method Load Calculation for One-Family Dwellings

1	General Lighting and Receptacle Loads *220.12* *Do not include open porches, garages, or unused or* *unfinished spaces not adaptable for future use.*	$3 \times$ __1050__ (sq ft outside dimensions)	= 1	**3150**	

2 Small-Appliance Branch-Circuits *220.52(A)*
*Include at least **two** small-appliance branch circuits*
must be included. 210.11(C)(1) $1500 \times$ __2__ (minimum of two) = 2 **3000**

3 Laundry Branch Circuit(s) *220.52(B)*
*Include at least **one** laundry branch circuit *unless meeting*
210.52(F), Exception No. 1 or 2. 210.11(C)(2) $1500 \times$ __1__ *(minimum of one) = 3 **1500**

4 Add lines 1, 2, and 3	**4**	**7650**	Lines 5 through 8 utilize the demand factors found in *Table 220.42*.

5 __7650__ (line 4) $- 3000 =$ **5** __4650__ (if 117,000 or less, skip to line 8) **6** __----__ (line 5, if more than 117,000) $- 117,000 =$ **6** ----

7 __----__ (line 6) $\times 25\% =$ **7** ---- **8** __4650__ (smaller of line 5 or 117,000) $\times 35\% =$ **8** **1628**

9 Total General Lighting and Receptacle Load $3000 +$ __----__ (line 7) $+$ __1628__ (line 8) $=$ **9** **4628**

10 Fastened-In-Place Appliances *220.53*
Use the nameplate rating.
Do not include electric ranges,
clothes dryers, space-heating
equipment, or air-conditioning
equipment.

water heater / 4500   dishwasher / 1200   disposer / 900
compactor /   microwave oven / 1200   /
/   /   /

If fewer than four units, put total volt-amperes on line 10.
If four or more units, multiply total volt-amperes by 75%.   __7800__ (volt-amps of four or more) $\times 75\% =$ **10** **5850**

11 Clothes Dryers *220.54*
(If present; otherwise skip to line 12.) Use 5000 watts or the nameplate rating, whichever is larger.
The neutral demand load is 70% for feeders. 220.61(B)   **11** **5000**

12 Ranges, Ovens, Cooktops, and Other Household Cooking Appliances over 1750 Watts *220.55*
(If present; otherwise skip to line 13.) Use Table 220.55 and all of the applicable Notes.
The neutral demand load is 70% for feeders. 220.61(B)   **12** **8000**

13 Heating or Air-Conditioning System (Compare the heat and A/C, and omit the smaller.) *220.60*
Include the air handler when using either one. For heat pumps, include the compressor and the
maximum amount of electric heat that can be energized while the compressor is running.   **13** **5345**

14 Largest Motor (one motor only) *220.50 and 430.24*
Multiply the volt-amperes of the largest motor by 25%.   __900__ (volt-amps of largest motor) $\times 25\% =$ **14** **225**

15 Total Volt-Ampere Demand Load: *Add lines 9 through 14 to find the minimum required volt-amperes.*   **15** **29,048**

16 Minimum Amperes
Divide the total
volt-amperes by
the voltage.   __29,048__ (line 15) $\div$ __240__ (voltage) = **16** __121__ (minimum amperes) | **Minimum Size 17 Service or Feeder** *240.6(A)* | **17** | **125** |

18 Size the Service or Feeder Conductors.
Use 310.15(B)(7) to find the service conductors up to 400 amperes.
Ratings in excess of 400 amperes shall comply with Table 310.15(B)(16)
310.15(B)(7) also applies to feeder conductors supplying the entire load. | **Minimum Size Conductors** | **18** | 2 copper
1/0 aluminum |

19 Size the Neutral Conductor. *220.61*
310.15(B)(7) states that the neutral service or feeder conductor
can be smaller than the ungrounded (hot) conductors, provided the
requirements of 215.2, 220.61, and 230.42 are met.
250.24(C)(1) states that the grounded conductor shall not be smaller
than specified in Table 250.102(C)(1). | **Minimum Size Neutral Conductor** | **19** | 4 copper
3 aluminum |

20 Size the Grounding Electrode Conductor. *250.66*
Use line 18 to find the grounding electrode conductor in Table 250.66.
Size the Equipment Grounding Conductor (for Feeder). 250.122
Use line 17 to find the equipment grounding conductor in Table 250.122.
Equipment grounding conductor types are listed in 250.118. | **Minimum Size Grounding Electrode Conductor . . . or . . . Equipment Grounding Conductor** | **20** | 6 copper
4 aluminum |

OPTIONAL METHOD: MULTIFAMILY DWELLINGS

Optional Method Load Calculation for Multifamily Dwellings

1 General Lighting and Receptacle Loads *220.84(C)(1)* *Do not include open porches, garages, or unused or* *unfinished spaces not adaptable for future use.*	$3 \times$ _____ $\times$ _____ = (sq ft outside dimensions) (number of units)	**1**
2 Small-Appliance Branch Circuits *220.84(C)(2)* At least **two** small-appliance branch circuits must be included. *210.11(C)(1)*	$1500 \times$ _____ $\times$ _____ = (minimum of two) (number of units)	**2**
3 Laundry Branch Circuit(s) *220.84(C)(2)* *Include at least **one** laundry branch circuit *unless* *meeting 210.52(F), Exception No. 1 or 2. 210.11(C)(2)*	$1500 \times$ _____ $\times$ _____ = *(minimum of one) (number of units)	**3**
4 through 11 Appliances and Motors *220.84(C)(3) and (4)* *Use the nameplate rating of **all** appliances (fastened in place, permanently connected, or connected to a specific circuit), ranges, wall-mounted ovens, counter-mounted cooking units, motors, water heaters, and clothes dryers.* *Convert any nameplate rating given in amperes to volt-amperes by multiplying the amperes by the rated voltage.* **Do not include any heating or air-conditioning equipment in this section.**	water heater _____ / _____ $\times$ _____ = (volt-amperes each) (number) dishwashers _____ / _____ $\times$ _____ = (volt-amperes each) (number) disposers _____ / _____ $\times$ _____ = (volt-amperes each) (number) clothes dryers _____ / _____ $\times$ _____ = (volt-amperes each) (number) ranges _____ / _____ $\times$ _____ = (volt-amperes each) (number) microwave ovens _____ / _____ $\times$ _____ = (volt-amperes each) (number) _____ / _____ $\times$ _____ = (volt-amperes each) (number) _____ / _____ $\times$ _____ = (volt-amperes each) (number)	**4** **5** **6** **7** **8** **9** **10** **11**
12 Heating or Air-Conditioning System (Compare the heat and A/C, and omit the smaller.) *220.84(C)(5)* *Include the air handler when using either one.* *For heat pumps, include the compressor and the* *maximum amount of electric heat that can be* *energized while the compressor is running.* _____ $\times$ _____ = (volt-amperes each) (number)		**12**
13 Total Volt-Ampere Demand Load: *Multiply total VA by Table 220.84* *demand factor percentage.* _____ + _____ = (total volt-amperes from lines 1 through 12) (*Table 220.84* demand factor)		**13**
14 House Load (*If present; otherwise skip to line 15*) *Calculate in accordance with Article 220, Part III.* *Do not include Table 220.84 demand factors.*		**14**
15 Minimum Amperes *Divide the total* *volt-amperes* *by the voltage.* _____ ÷ _____ = **15** _____ (lines 13 and 14) (voltage) (minimum amperes)	**Minimum Size** **16 Service or** **Feeder** *240.6(A)*	**16**
17 Size the Service or Feeder Conductors. *Use 310.15(B)(7) to find the service conductors up to 400 amperes.* *Ratings in excess of 400 amperes shall comply with Table 310.15(B)(16)* *310.15(B)(7) also applies to feeder conductors supplying the entire load.*	**Minimum Size** **Conductors**	**17**
18 Size the Neutral Conductor. *220.61* **Note: There is no optional method for calculating the neutral conductor.** *310.15(B)(7) states that the neutral service or feeder conductor* *can be smaller than the ungrounded (hot) conductors, provided the* *requirements of 215.2, 220.61, and 230.42 are met.* *250.24(C)(1) states that the grounded conductor shall not be smaller* *than specified in Table 250.102(C)(1).*	**Minimum Size** **Neutral** **Conductor**	**18**
19 Size the Grounding Electrode Conductor. *250.66* *Use line 17 to find the grounding electrode conductor in Table 250.66.* Size the Equipment Grounding Conductor (for Feeder). *250.122* *Use line 16 to find the equipment grounding conductor in Table 250.122.* *Equipment grounding conductor types are listed in 250.118.*	**Minimum Size** **Grounding Electrode** **Conductor . . . *or* . . .** **Equipment Grounding** **Conductor**	**19**

Lines 1 through 3—General Lighting and Receptacle Load

Ⓐ In the **Optional Method,** lines 1, 2, and 3 are calculated exactly as described in the **Standard Method.** (See explanations on page 235.)

Ⓑ The feeder or service multifamily dwelling load can be calculated in accordance with Table 220.84 (instead of Article 220, Part III) where all of the following conditions are met: (1) no dwelling unit is supplied by more than one feeder; (2) each dwelling unit is equipped with electric cooking equipment; and (3) each dwelling unit is equipped with either electric space-heating or air-conditioning, or both ⟫*220.84(A)*⟪.

Ⓒ This optional load calculation is calculated on the same twelve-unit multifamily dwelling used for the standard method (located in the first part of this unit).

> **NOTE**
>
> When the calculated load for multifamily dwellings, without electric cooking, in Article 220, Part III, exceeds that calculated under Part IV for the identical load, plus electric cooking (based on 8 kW per unit), use the lesser of the two loads ⟫*220.84(A)(2) Exception*⟪.

Ⓐ **Ⓑ**

Optional Method Load Calculation for Multifamily Dwellings

1 General Lighting and Receptacle Loads *220.84(C)(1)* *Do not include open porches, garages, or unused or* *unfinished spaces not adaptable for future use.*	$3 \times \dfrac{1050}{\text{(sq ft outside dimensions)}}$	$\times \dfrac{12}{\text{(number of units)}} =$	**37,800**
2 Small-Appliance Branch Circuits *220.84(C)(2)* *At least **two** small-appliance branch circuits* *must be included. 210.11(C)(1)*	$1500 \times \dfrac{2}{\text{(minimum of two)}}$	$\times \dfrac{12}{\text{(number of units)}} =$	**36,000**
3 Laundry Branch Circuit(s) *220.84(C)(2)* *Include at least **one** laundry branch circuit *unless* *meeting 210.52(F), Exception No. 1 or 2. 210.11(C)(2)*	$1500 \times \dfrac{1}{\text{*(minimum of one)}}$	$\times \dfrac{12}{\text{(number of units)}} =$	**18,000**

© Cengage Learning 2012

Ⓒ

Lines 4 through 11—All Appliances

Ⓐ This section includes appliances (fastened in place, permanently connected, or connected to a specific circuit), ranges, wall-mounted ovens, counter-mounted cooking units, clothes dryers, and water heaters ⟫ *220.84(C)(3)* ⟪.

Ⓑ Low-power-factor loads and motors (except air-conditioning motors) are also included ⟫ *220.84(C)(4)* ⟪.

Ⓒ Use the exact clothes dryer nameplate rating, if any, even if less than 5000 volt-amperes. Do not apply the demand factors in Table 220.55 to household cooking equipment (ranges, cooktops, ovens, etc.); instead, list the nameplate ratings as they appear.

Ⓓ Insert the number of appliances (such as water heaters) having the same rating, which may not necessarily be the number of dwelling units. Here, all dwelling units contain identical equipment, so the number of each type of appliance is the same as the number of dwelling units. In another application, if only half of the units are equipped with laundry hookups, the number of units is not the same as the number of a given type of appliance.

Ⓔ Heating and air-conditioning systems are not included in this computation.

Ⓕ Additional lines have been provided for listing appliances.

Ⓖ Inserting numbers in all of the lines is not necessary. However, it is advisable to insert dashes, or some other indicator, to show that the line is intentionally blank.

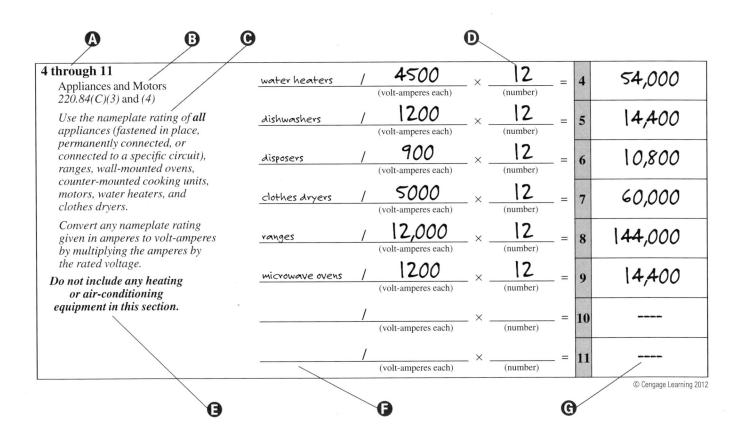

© Cengage Learning 2012

Line 12—Heating or Air-Conditioning System

Ⓐ Include only the larger of the heating or air-conditioning loads (except for heat pumps).

Ⓑ Electric heating loads (including associated motors) are calculated at 100% of the total connected load. Because the air handler (blower) works with either the heating or the air conditioning, add it to both calculations.

Ⓒ Air-conditioning loads (including associated motors) are calculated at 100% of the total connected load. No derating is allowed. Each unit contains an air-conditioner compressor (14 × 230 = 3220 VA), condenser fan motor (2 × 115 = 230 VA), and blower motor (3 × 115 = 345 VA). The combined load is 3795 volt-amperes. Because this is less than the heating load, omit the A/C compressor and the condenser fan motor.

Ⓓ The air handler (blower motor) has a rating of 345 volt-amperes (3 × 115 = 345). Because both loads are energized simultaneously, add the air handler to the 5 kW of electric heat (345 + 5000 = 5345 volt-amperes).

> ### N O T E
>
> A heat pump with supplementary heat is not considered a noncoincident load. Add the compressor full-load current to the maximum amount of heat that can be energized while the compressor is running. Remember to include all associated motor(s), that is, blower, condenser, etc.

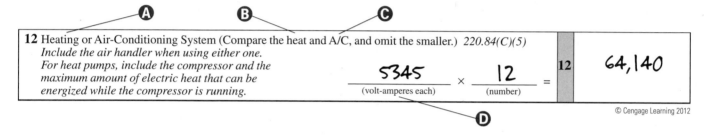

12 Heating or Air-Conditioning System (Compare the heat and A/C, and omit the smaller.) *220.84(C)(5)*
Include the air handler when using either one.
For heat pumps, include the compressor and the
maximum amount of electric heat that can be
energized while the compressor is running.

$$\underset{\text{(volt-amperes each)}}{5345} \times \underset{\text{(number)}}{12} = \boxed{12} \quad 64,140$$

© Cengage Learning 2012

Line 13—Total Volt-Ampere Demand Load

Ⓐ Adding lines 1 through 12 results in the volt-ampere load before derating.

Ⓑ Table 220.84 contains demand factors (percentages) for derating three (or more) multifamily dwelling units. Find the number of dwelling units in the first column and follow that

row across to the demand factor. The demand factors listed in the second column are percentages. The demand factor for a twelve-unit multifamily dwelling is 41 (41%).

Ⓒ This is the total calculated load excluding house loads (if any).

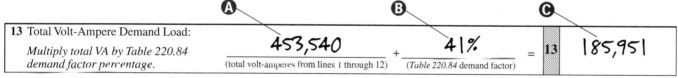

13 Total Volt-Ampere Demand Load:
Multiply total VA by Table 220.84
demand factor percentage.

$$\underset{\text{(total volt-amperes from lines 1 through 12)}}{453,540} + \underset{\text{(Table 220.84 demand factor)}}{41\%} = \boxed{13} \quad 185,951$$

© Cengage Learning 2012

Line 14—House Load

Ⓐ House loads must be calculated as stipulated in Article 220, Part III, and must be in addition to the dwelling-unit loads calculated in accordance with Table 220.84 »*220.84(B)* «.

Ⓑ House loads include areas such as clubhouse, office, hallway and stairway lighting, parking lot lighting, etc.

Ⓒ Insert a line (or dashes) in spaces left empty intentionally.

14 House Load *(If present; otherwise skip to line 15)*
Calculate in accordance with Article 220, Part III.
Do not include Table 220.84 demand factors.

$$\boxed{14} \quad \text{----}$$

© Cengage Learning 2012

Lines 15 and 16—Minimum Service or Feeder

Ⓐ Place the sum of lines 13 and 14 (total volt-amperes) here. If there is no house load, insert only line 13.

Ⓑ Fill in the source voltage supplying the service equipment.

Ⓒ Fractions of an ampere 0.5 and greater are rounded up, while fractions less than 0.5 are dropped (185,951 ÷ 240 = 774.8 = 775) ≫ *220.5(B)* ≪. The minimum ampere rating required for conductors and overcurrent protection, whether for the service or feeder, is 775.

Ⓓ The service (or feeder) overcurrent protection must be higher than the number found in line 15. Standard ampere ratings for fuses and circuit breakers are listed in 240.6(A).

Ⓔ The next standard size fuse or breaker above 775 amperes is 800 ≫ *240.6(A)* ≪.

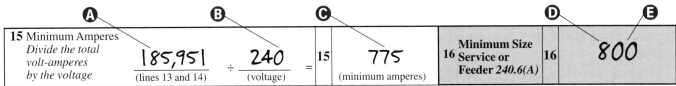

Line 17—Minimum Size Conductors

Ⓐ The total combined rating of paralleled conductors must meet or exceed the minimum calculated load.

Ⓑ At 75°C (167°F), the minimum size copper conductors for three parallel sets matching or exceeding 775 amperes are 300 kcmil. Although three paralleled sets of 250-kcmil copper

conductors (at 75°C [167°F]) are sufficient for an 800-ampere overcurrent protective device, they are not sufficient for the 775-ampere calculated load. (Three times 255 is only 765 amperes, short of the necessary 775 rating.)

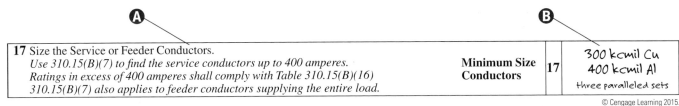

Line 18—Minimum Size Neutral Conductors

(A) There is no optional method for neutral conductor calculation. This neutral calculation is performed in accordance with Article 220, Parts II and III.

(B) The loads for clothes dryers and ranges have already been calculated and are in the "Standard Method Load Calculation for One-Family Dwellings" form, Lines 11 and 12. As permitted in 220.61(B)(1), apply a 70% demand factor to these loads.

(C) The minimum calculated volt-ampere load is 108,705.

(D) The minimum neutral ampacity is 453.

(E) The minimum rating for three paralleled sets of neutral conductors is 151 amperes each.

(F) The loads for general lighting and receptacles have already been calculated and are in the "Standard Method Load Calculation for One-Family Dwellings" form, Line 9.

(G) At 75°C (167°F), the minimum size copper conductor is 2/0 AWG, with an individual rating of 175 amperes ≫ *Table 310.15(B)(16)* ≪.

(H) At 75°C (167°F), the minimum size aluminum conductor, if determined by ampacity alone, would be 3/0 AWG, with a rating of 155 amperes each. However, 3/0 AWG would not be acceptable because the neutral shall not be smaller than specified in Table 250.102(C)(1), which in this example is 4/0 AWG ≫ *250.24(C)(1)* ≪.

General lighting and receptacles (line 9)	34,080
Fastened-in-place appliances	
Water heaters	0
Dishwashers (1200 × 12)	14,400
Waste disposers (7.5 × 120 × 12)	10,800
Microwave ovens (1200 × 12)	14,400
Total	39,600
Total fastened-in-place appliances (39,600 × 75%)	29,700
Clothes dryers (line 11 × 70%)	18,900
Ranges (line 12 × 70%)	18,900
Electric heat	0
Air-conditioner compressors	0
Air handlers (3 × 115 × 12)	4140
Condenser fan motors (2 × 115 × 12)	2760
Largest motor (900 × 25%)	225
TOTAL	**108,705**

108,705 ÷ 240 = 452.9 = 453 minimum neutral ampacity
453 ÷ 3 = 151 minimum amperes per neutral conductor

18 Size the Neutral Conductor. *220.61* **Note: There is no optional method for calculating the neutral conductor.** *310.15(B)(7) states that the neutral service or feeder conductor can be smaller than the ungrounded (hot) conductors, provided the requirements of 215.2, 220.61, and 230.42 are met. 250.24(C)(1) states that the grounded conductor shall not be smaller than specified in Table 250.102(C)(1).*	**Minimum Size Neutral Conductor**	**18**	2/0 Cu 4/0 Al three paralleled sets

© Cengage Learning 2015.

Line 19—Grounding Electrode Conductor

(A) The grounding electrode conductor size is determined by the largest service-entrance conductor (or the equivalent area) if the service-entrance conductors are paralleled. The equivalent area for three parallel sets in this calculation is 900-kcmil copper (3 × 300) *or* 1200-kcmil aluminum (3 × 400).

(B) The minimum size grounding electrode conductor is 2/0 AWG copper *or* 4/0 AWG aluminum ≫ *Table 250.66* ≪.

19 Size the Grounding Electrode Conductor. *250.66* *Use line 17 to find the grounding electrode conductor in Table 250.66.* Size the Equipment Grounding Conductor (for Feeder). *250.122* *Use line 16 to find the equipment grounding conductor in Table 250.122. Equipment grounding conductor types are listed in 250.118.*	**Minimum Size Grounding Electrode Conductor . . . or . . . Equipment Grounding Conductor**	**19**	2/0 Cu 4/0 Al

© Cengage Learning 2012

The following page shows the completed optional method for a twelve-unit multifamily dwelling, having all units combined into a single service.

Optional Method Load Calculation for Multifamily Dwellings

1	General Lighting and Receptacle Loads *220.84(C)(1)* *Do not include open porches, garages, or unused or* *unfinished spaces not adaptable for future use.*	$3 \times$ __1050__ (sq ft outside dimensions)	$\times$ __12__ (number of units)	$=$		**37,800**
2	Small-Appliance Branch Circuits *220.84(C)(2)* At least **two** small-appliance branch circuits must be included. *210.11(C)(1)*	$1500 \times$ __2__ (minimum of two)	$\times$ __12__ (number of units)	$=$		**36,000**
3	Laundry Branch Circuit(s) *220.84(C)(2)* Include at least **one** laundry branch circuit *unless* *meeting 210.52(F), Exception No. 1 or 2. 210.11(C)(2)*	$1500 \times$ __1__ *(minimum of one)	$\times$ __12__ (number of units)	$=$		**18,000**

4 through 11							
Appliances and Motors *220.84(C)(3) and (4)*	water heaters	/ __4500__ (volt-amperes each)	$\times$ __12__ (number)	$=$	**4**	**54,000**
*Use the nameplate rating of **all*** *appliances (fastened in place,*	dishwashers	/ __1200__ (volt-amperes each)	$\times$ __12__ (number)	$=$	**5**	**14,400**
permanently connected, or *connected to a specific circuit),* *ranges, wall-mounted ovens,*	disposers	/ __900__ (volt-amperes each)	$\times$ __12__ (number)	$=$	**6**	**10,800**
counter-mounted cooking units, *motors, water heaters, and* *clothes dryers.*	clothes dryers	/ __5000__ (volt-amperes each)	$\times$ __12__ (number)	$=$	**7**	**60,000**
Convert any nameplate rating *given in amperes to volt-amperes* *by multiplying the amperes by*	ranges	/ __12,000__ (volt-amperes each)	$\times$ __12__ (number)	$=$	**8**	**144,000**
the rated voltage.	microwave ovens	/ __1200__ (volt-amperes each)	$\times$ __12__ (number)	$=$	**9**	**14,400**
Do not include any heating **or air-conditioning** **equipment in this section.**		/ ____ (volt-amperes each)	$\times$ ____ (number)	$=$	**10**	----
		/ ____ (volt-amperes each)	$\times$ ____ (number)	$=$	**11**	----

12 Heating or Air-Conditioning System (Compare the heat and A/C, and omit the smaller.) *220.84(C)(5)* *Include the air handler when using either one.* *For heat pumps, include the compressor and the* *maximum amount of electric heat that can be* *energized while the compressor is running.*	__5345__ $\times$ __12__ $=$ (volt-amperes each) (number)	**12**	**64,140**

13 Total Volt-Ampere Demand Load: *Multiply total VA by Table 220.84* *demand factor percentage.*	__453,540__ $+$ __41%__ $=$ (total volt-amperes from lines 1 through 12) (*Table 220.84* demand factor)	**13**	**185,951**

14 House Load (*If present; otherwise skip to line 15*) *Calculate in accordance with Article 220, Part III.* *Do not include Table 220.84 demand factors.*	**14**	----

15 Minimum Amperes *Divide the total* *volt-amperes* *by the voltage*	__185,951__ $\div$ __240__ $=$ (lines 13 and 14) (voltage)	**15**	__775__ (minimum amperes)	**16** Minimum Size Service or Feeder *240.6(A)*	**16**	**800**

17 Size the Service or Feeder Conductors. *Use 310.15(B)(7) to find the service conductors up to 400 amperes.* *Ratings in excess of 400 amperes shall comply with Table 310.15(B)(16)* *310.15(B)(7) also applies to feeder conductors supplying the entire load.*	**Minimum Size Conductors**	**17**	300 kcmil Cu 400 kcmil Al three paralleled sets
18 Size the Neutral Conductor. *220.61* **Note: There is no optional method for calculating the neutral conductor.** *310.15(B)(7) states that the neutral service or feeder conductor* *can be smaller than the ungrounded (hot) conductors, provided the* *requirements of 215.2, 220.61, and 230.42 are met.* *250.24(C)(1) states that the grounded conductor shall not be smaller* *than specified in Table 250.102(C)(1).*	**Minimum Size Neutral Conductor**	**18**	2/0 Cu 4/0 Al three paralleled sets
19 Size the Grounding Electrode Conductor. *250.66* *Use line 17 to find the grounding electrode conductor in Table 250.66.* *Size the Equipment Grounding Conductor (for Feeder). 250.122* *Use line 16 to find the equipment grounding conductor in Table 250.122.* *Equipment grounding conductor types are listed in 250.118.*	**Minimum Size Grounding Electrode Conductor . . . or . . . Equipment Grounding Conductor**	**19**	2/0 Cu 4/0 Al

SIX-UNIT MULTIFAMILY DWELLING CALCULATION—OPTIONAL METHOD

Lines 1 through 12

A This optional method calculation is based on six units instead of twelve.

B Since everything is calculated at 100%, lines 1 through 12 are exactly half of that calculated for twelve units.

A

Optional Method Load Calculation for Multifamily Dwellings

1 General Lighting and Receptacle Loads *220.84(C)(1)* *Do not include open porches, garages, or unused or unfinished spaces not adaptable for future use.*	3 ×	**1050** (sq ft outside dimensions)	× **6** (number of units)	=	**1**	**18,900**
2 Small-Appliance Branch Circuits *220.84(C)(2)* *At least **two** small-appliance branch circuits must be included. 210.11(C)(1)*	1500 ×	**2** (minimum of two)	× **6** (number of units)	=	**2**	**18,000**
3 Laundry Branch Circuit(s) *220.84(C)(2)* *Include at least **one** laundry branch circuit *unless meeting 210.52(F), Exception No. 1 or 2. 210.11(C)(2)*	1500 ×	**1** *(minimum of one)	× **6** (number of units)	=	**3**	**9000**

4 through 11					
Appliances and Motors *220.84(C)(3) and (4)* water heaters /	**4500** (volt-amperes each)	× **6** (number)	=	**4**	**27,000**
*Use the nameplate rating of **all** appliances (fastened in place, permanently connected, or* dishwashers /	**1200** (volt-amperes each)	× **6** (number)	=	**5**	**7200**
connected to a specific circuit), ranges, wall-mounted ovens, disposers /	**900** (volt-amperes each)	× **6** (number)	=	**6**	**5400**
counter-mounted cooking units, motors, water heaters, and clothes dryers /	**5000** (volt-amperes each)	× **6** (number)	=	**7**	**30,000**
clothes dryers.					
Convert any nameplate rating given in amperes to volt-amperes ranges /	**12,000** (volt-amperes each)	× **6** (number)	=	**8**	**72,000**
by multiplying the amperes by the rated voltage. microwave ovens /	**1200** (volt-amperes each)	× **6** (number)	=	**9**	**7200**
Do not include any heating or air-conditioning equipment in this section. /	(volt-amperes each)	× (number)	=	**10**	----
	/ (volt-amperes each)	× (number)	=	**11**	----

12 Heating or Air-Conditioning System (Compare the heat and A/C, and omit the smaller.) *220.84(C)(5)* *Include the air handler when using either one. For heat pumps, include the compressor and the maximum amount of electric heat that can be energized while the compressor is running.*	**5345** (volt-amperes each)	× **6** (number)	=	**12**	**32,070**

© Cengage Learning 2012

Line 13—Total Volt-Ampere Demand Load

A Add lines 1 through 12 to find the volt-ampere load before derating.

B Find the demand factor, located in Table 220.84, and insert here. The demand factor for a six-unit multifamily dwelling is 44 (44%).

A   **B**

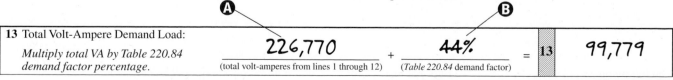

13 Total Volt-Ampere Demand Load: *Multiply total VA by Table 220.84 demand factor percentage.*	**226,770** (total volt-amperes from lines 1 through 12)	+	**44%** *(Table 220.84 demand factor)*	=	**13**	**99,779**

© Cengage Learning 2012

Lines 15 and 16—Minimum Service or Feeder

Ⓐ Place the sum of lines 13 and 14 (total volt-amperes) here. If there is no house load, insert only line 13.

Ⓑ Insert the source voltage supplying the service equipment.

Ⓒ The minimum amperage rating required for conductors and overcurrent protection is 416 (99,779 ÷ 240 = 415.75 = 416).

Ⓓ The minimum service (or feeder) determined by the optional method for these six units is 450 amperes.

Ⓔ The next standard size fuse or breaker above 416 amperes is 450 ≫ *240.6(A)* ≪ .

> **NOTE**
>
> The same six units calculated by the standard method require a rating of 600 amperes.

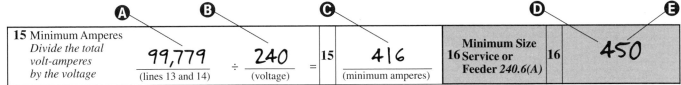

	Ⓐ	Ⓑ	Ⓒ	Ⓓ	Ⓔ
15 Minimum Amperes *Divide the total volt-amperes by the voltage*	**99,779** (lines 13 and 14)	÷ **240** (voltage)	= 15 **416** (minimum amperes)	**Minimum Size** **16** Service or Feeder *240.6(A)*	16 **450**

© Cengage Learning 2012

Line 17—Minimum Size Conductors

Ⓐ Ensure that the total combined rating of paralleled conductors meets or exceeds the minimum calculated load.

Ⓑ At 75°C (167°F), the minimum size conductors for two parallel sets matching or exceeding 416 amperes are 4/0 AWG copper or 300-kcmil aluminum. The minimum rating for each conductor is 208 amperes.

17 Size the Service or Feeder Conductors. *Use 310.15(B)(7) to find the service conductors up to 400 amperes. Ratings in excess of 400 amperes shall comply with Table 310.15(B)(16) 310.15(B)(7) also applies to feeder conductors supplying the entire load.*	**Minimum Size Conductors**	17	**4/0 Cu** **300 kcmil Al** *two paralleled sets*

© Cengage Learning 2015.

Lines 18 and 19—Minimum Size Neutral and Grounding Conductors

Ⓐ Since there is no optional method for the neutral conductor calculation, refer to page 244, which gives the neutral for six units. The result requires a minimum rating of 279 amperes, or 140 amperes each for two paralleled conductors.

Ⓑ At 75°C (167°F), the minimum size paralleled conductors are 1/0 AWG copper (with a rating of 150 amperes) *or* 3/0 AWG aluminum (with a rating of 155 amperes) ≫ *Table 310.15(B)(16)* ≪ .

Ⓒ The equivalent area for two paralleled conductors in this calculation is 423.2-kcmil copper (2 × 211,600 ÷ 1000) *or* 600-kcmil aluminum (2 × 300). The minimum size grounding electrode conductor is 1/0 AWG copper *or* 3/0 AWG aluminum ≫ *Table 250.66* ≪ .

18 Size the Neutral Conductor. *220.61* ***Note: There is no optional method for calculating the neutral conductor.*** *310.15(B)(7) states that the neutral service or feeder conductor can be smaller than the ungrounded (hot) conductors, provided the requirements of 215.2, 220.61, and 230.42 are met. 250.24(C)(1) states that the grounded conductor shall not be smaller than specified in Table 250.102(C)(1).*	**Minimum Size Neutral Conductor**	18	**1/0 Cu** **3/0 Al** *two paralleled sets*
19 Size the Grounding Electrode Conductor. *250.66* *Use line 17 to find the grounding electrode conductor in Table 250.66. Size the Equipment Grounding Conductor (for Feeder). 250.122 Use line 16 to find the equipment grounding conductor in Table 250.122. Equipment grounding conductor types are listed in 250.118.*	**Minimum Size Grounding Electrode Conductor . . . or . . . Equipment Grounding Conductor**	19	**1/0 Cu** **3/0 Al**

© Cengage Learning 2015.

The following page shows the completed optional method for a six-unit multifamily dwelling load calculation.

Optional Method Load Calculation for Multifamily Dwellings

1	General Lighting and Receptacle Loads *220.84(C)(1)* *Do not include open porches, garages, or unused or unfinished spaces not adaptable for future use.*	3 ×	**1050** (sq ft outside dimensions)	×	**6** (number of units)	= **1**	**18,900**
2	Small-Appliance Branch Circuits *220.84(C)(2)* *At least **two** small-appliance branch circuits must be included. 210.11(C)(1)*	1500 ×	**2** (minimum of two)	×	**6** (number of units)	= **2**	**18,000**
3	Laundry Branch Circuit(s) *220.84(C)(2)* *Include at least **one** laundry branch circuit *unless meeting 210.52(F), Exception No. 1 or 2. 210.11(C)(2)*	1500 ×	**1** *(minimum of one)	×	**6** (number of units)	= **3**	**9000**

4 through 11 Appliances and Motors *220.84(C)(3)* and *(4)*	water heaters /	**4500** (volt-amperes each)	×	**6** (number)	= **4**	**27,000**
*Use the nameplate rating of **all** appliances (fastened in place, permanently connected, or connected to a specific circuit),*	dishwashers /	**1200** (volt-amperes each)	×	**6** (number)	= **5**	**7200**
ranges, wall-mounted ovens, counter-mounted cooking units,	disposers /	**900** (volt-amperes each)	×	**6** (number)	= **6**	**5400**
motors, water heaters, and clothes dryers.	clothes dryers /	**5000** (volt-amperes each)	×	**6** (number)	= **7**	**30,000**
Convert any nameplate rating given in amperes to volt-amperes by multiplying the amperes by the rated voltage.	ranges /	**12,000** (volt-amperes each)	×	**6** (number)	= **8**	**72,000**
Do not include any heating or air-conditioning equipment in this section.	microwave ovens /	**1200** (volt-amperes each)	×	**6** (number)	= **9**	**7200**
	/	 (volt-amperes each)	×	 (number)	= **10**	-----
	/	 (volt-amperes each)	×	 (number)	= **11**	-----

12	Heating or Air-Conditioning System (Compare the heat and A/C, and omit the smaller.) *220.84(C)(5)* *Include the air handler when using either one. For heat pumps, include the compressor and the maximum amount of electric heat that can be energized while the compressor is running.*	**5345** (volt-amperes each) × **6** (number) =	**12**	**32,070**

13	Total Volt-Ampere Demand Load: *Multiply total VA by Table 220.84 demand factor percentage.*	**226,770** (total volt-amperes from lines 1 through 12)	+ **44%** (*Table 220.84* demand factor)	= **13**	**99,779**

14	House Load *(If present, otherwise skip to line 15)* *Calculate in accordance with Article 220, Part III. Do not include in Table 220.84 demand factors.*	**14**	-----

15	Minimum Amperes *Divide the total volt-amperes by the voltage*	**99,779** (lines 13 and 14)	÷ **240** (voltage)	= **15**	**416** (minimum amperes)	**16** Minimum Size Service or Feeder *240.6(A)* **16**	**450**

17	Size the Service or Feeder Conductors. *Use 310.15(B)(7) to find the service conductors up to 400 amperes. Ratings in excess of 400 amperes shall comply with Table 310.15(B)(16) 310.15(B)(7) also applies to feeder conductors supplying the entire load.*	Minimum Size Conductors **17**	**4/0 Cu** **300 kcmil Al** two paralleled sets
18	Size the Neutral Conductor. *220.61* **Note: There is no optional method for calculating the neutral conductor.** *310.15(B)(7) states that the neutral service or feeder conductor can be smaller than the ungrounded (hot) conductors, provided the requirements of 215.2, 220.61, and 230.42 are met. 250.24(C)(1) states that the grounded conductor shall not be smaller than specified in Table 250.102(C)(1).*	Minimum Size Neutral Conductor **18**	**1/0 Cu** **3/0 Al** two paralleled sets
19	Size the Grounding Electrode Conductor. *250.66* *Use line 17 to find the grounding electrode conductor in Table 250.66.* Size the Equipment Grounding Conductor (for Feeder). *250.122* *Use line 16 to find the equipment grounding conductor in Table 250.122. Equipment grounding conductor types are listed in 250.118.*	Minimum Size Grounding Electrode Conductor . . . *or* . . . Equipment Grounding Conductor **19**	**1/0 Cu** **3/0 Al**

OPTIONAL LOAD CALCULATION FOR EACH UNIT OF A MULTIFAMILY DWELLING

Optional Method—One Unit

Ⓐ The first three lines of both the optional and standard methods of one-family dwelling calculations are identical.

Ⓑ Although located in a multifamily dwelling, this unit contains one feeder and one panelboard. Therefore, it is calculated in accordance with a one-family dwelling method.

Ⓒ Under certain conditions, a laundry branch circuit is optional in multifamily dwellings ≫ *210.52(F) Exception No. 1 and 2* ≪.

Optional Method Load Calculation for One Multifamily Dwelling Unit

1 General Lighting and Receptacle Loads *220.82(B)(1)* *Do not include open porches, garages, or unused or unfinished spaces not adaptable for future use.*	$3 \times$ ___**1050**___ (sq ft outside dimensions)	= 1	**3150**
2 Small-Appliance Branch Circuits *220.82(B)(2)* *At least **two** small-appliance branch circuits must be included. 210.11(C)(1)*	$1500 \times$ ___**2**___ (minimum of two)	= 2	**3000**
3 Laundry Branch Circuit(s) *220.82(B)(2)* *Include at least **one** laundry branch circuit *unless meeting 210.52(F), Exception No. 1 or 2. 210.11(C)(2)*	$1500 \times$ ___**1**___ *(minimum of one)	= 3	**1500**

© Cengage Learning 2012

Line 4—Appliances

Ⓐ This section includes appliances (fastened in place, permanently connected, or connected to a specific circuit), ranges, wall-mounted ovens, counter-mounted cooking units, clothes dryers, and water heaters ≫ *220.82(B)(3)* ≪.

Ⓑ Motors (except air-conditioning motors) are also included in this section ≫ *220.82(B)(4)* ≪.

Ⓒ Heating and air-conditioning systems are **not** part of this calculation.

Ⓓ Use the exact nameplate rating of the clothes dryer, even if less than 5000 volt-amperes.

> **NOTE**
>
> Do not apply the demand factors in Table 220.55 to household cooking equipment (ranges, cook tops, and ovens). Instead, list the nameplate ratings as they appear.
>
> Do not increase motor loads by 25% ≫ *220.82(B)(4)* ≪.

4 Appliances *220.82(B)(3) and (4)* *Use the nameplate rating of **all** appliances (fastened in place, permanently connected, or connected to a specific circuit), ranges, ovens, cooktops, motors, and clothes dryers. Convert any nameplate rating given in amperes to volt-amperes by multiplying the amperes by the rated voltage.*	*Do not include any heating or air-conditioning equipment in this section.*		Total volt-amperes of all appliances LISTED BELOW	4	**24,800**
	water heater / **4500**	clothes dryer / **5500**	range / **12,000**		
	dishwasher / **1200**	disposer / **900**	microwave oven / **1200**		
	/	/	/		
	/	/	/		
	/	/	/		

© Cengage Learning 2012

Line 5—Applying Demand Factors to Lines 1 through 4

A This demand factor is found in 220.82(B).

B The newly calculated load includes everything except the heating and air-conditioning systems (3152 + 3000 + 1500 + 24,800).

C Because the first 10,000 volt-amperes are calculated at 100%, 10,000 is subtracted here.

D The remainder of the load is calculated at 40%.

E The result on line 5 will be part of line 7's calculation.

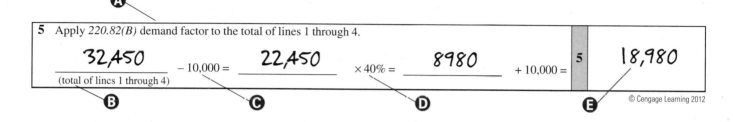

5	Apply *220.82(B)* demand factor to the total of lines 1 through 4.					
	32,450 (total of lines 1 through 4)	− 10,000 =	22,450 × 40% =	8980 + 10,000 =	**5**	18,980

© Cengage Learning 2012

Line 6—Heating or Air-Conditioning Systems

A The air-conditioner compressor (14 × 230 = 3220 VA), condenser fan motor (2 × 115 = 230 VA), and blower motor (3 × 115 = 345 VA) have a combined load of 3795 volt-amperes. Calculate the air-conditioning system at 100%.

B The electric heat (5000) added to the blower motor (3 × 115 = 345) is 5345 volt-amperes. The heating system is multiplied by 65% and the result placed on line 6d.

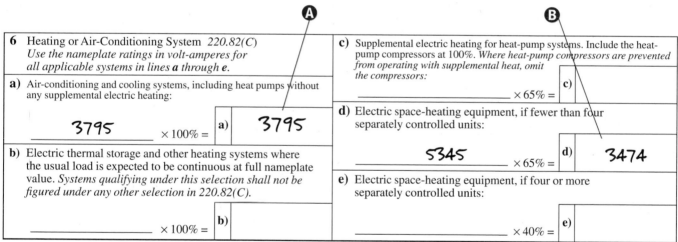

6	Heating or Air-Conditioning System *220.82(C)* *Use the nameplate ratings in volt-amperes for* *all applicable systems in lines **a** through **e**.*			
a)	Air-conditioning and cooling systems, including heat pumps without any supplemental electric heating:		**c)**	Supplemental electric heating for heat-pump systems. Include the heat-pump compressors at 100%. *Where heat-pump compressors are prevented from operating with supplemental heat, omit the compressors:*
	3795 _____ × 100% =	**a)** 3795		_____ × 65% = **c)**
b)	Electric thermal storage and other heating systems where the usual load is expected to be continuous at full nameplate value. *Systems qualifying under this selection shall not be figured under any other selection in 220.82(C).*		**d)**	Electric space-heating equipment, if fewer than four separately controlled units: 5345 _____ × 65% = **d)** 3474
	_____ × 100% =	**b)**	**e)**	Electric space-heating equipment, if four or more separately controlled units: _____ × 40% = **e)**

© Cengage Learning 2012

Line 7—Total Volt-Ampere Demand Load

A This is the largest volt-ampere rating of the heating or air-conditioning system(s), after application of demand factors.

B This is the volt-ampere demand load from line 5.

C This is the total volt-ampere load calculated by the optional method. (For comparison, this one-family dwelling calculated by the standard method is 29,048 volt-amperes.)

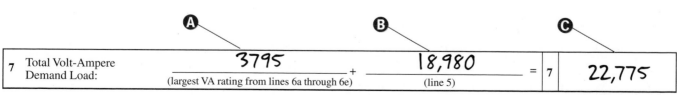

7	Total Volt-Ampere Demand Load:	3795 (largest VA rating from lines 6a through 6e)	+	18,980 (line 5)	=	7	22,775

© Cengage Learning 2012

Lines 8 and 9—Minimum Feeder

Ⓐ This is the volt-ampere load from line 7.

Ⓑ This is the source voltage.

Ⓒ Because the fraction is more than 0.5, the number 94.9 is rounded up to 95.

Ⓓ The overcurrent protection chosen for this feeder must be higher than the number found on line 8. Standard ampere ratings for fuses and circuit breakers are listed in 240.6(A).

Ⓔ This dwelling unit, calculated in accordance with the optional method, requires a 100-ampere feeder and panelboard.

Ⓕ A 100-ampere minimum restriction has been placed on this optional method for both services and feeders ⟫220.82(A)⟪. Feeders calculated in accordance with Article 220, Part III, are not restricted to a minimum rating of 100 amperes.

> **NOTE**
>
> For comparison, this same dwelling unit calculated by the standard method requires a minimum of 125 amperes.

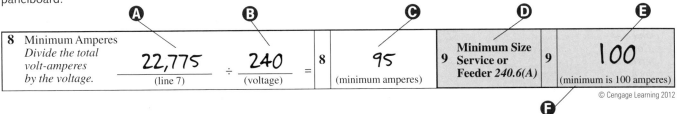

| 8 | Minimum Amperes *Divide the total volt-amperes by the voltage.* | 22,775 ÷ 240 = 8 (line 7)   (voltage) | 8 | 95 (minimum amperes) | 9 | Minimum Size Service or Feeder 240.6(A) | 9 | 100 (minimum is 100 amperes) |

© Cengage Learning 2012

Line 10—Minimum Size Conductors

Ⓐ Use 310.15(B)(7) to select feeder conductors supplying the entire load associated with an individual dwelling.

Ⓑ Multiply the minimum size feeder rating (Line 8) by 83 percent and then select a conductor from Table 310.15(B)(16).

100 × 83% = 83 amperes

In accordance with the 75°C column of Table 310.15(B)(16), the ampacity of a 4 AWG copper conductor is 85 amperes.

In accordance with the 75°C column of Table 310.15(B)(16), the ampacity of a 2 AWG aluminum conductor is 90 amperes. ⟫310.15(B)(7)(2)⟪.

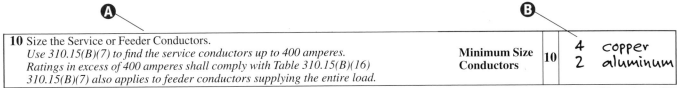

| 10 | Size the Service or Feeder Conductors. *Use 310.15(B)(7) to find the service conductors up to 400 amperes. Ratings in excess of 400 amperes shall comply with Table 310.15(B)(16) 310.15(B)(7) also applies to feeder conductors supplying the entire load.* | Minimum Size Conductors | 10 | 4 copper 2 aluminum |

© Cengage Learning 2015.

Lines 11 and 12—Minimum Size Neutral and Equipment Grounding Conductors

Ⓐ Since there is no optional method for neutral calculation, refer to page 248 for a detailed analysis.

Ⓑ At 75°C (167°F), the minimum size conductor is 4 AWG copper *or* 3 AWG aluminum.

Ⓒ Because these are feeder, not service, conductors, use Table 250.122 to size the equipment grounding conductor.

Ⓓ The minimum size equipment grounding conductor for a 100-ampere overcurrent protective device is 8 AWG copper or 6 AWG aluminum ⟫Table 250.122⟪.

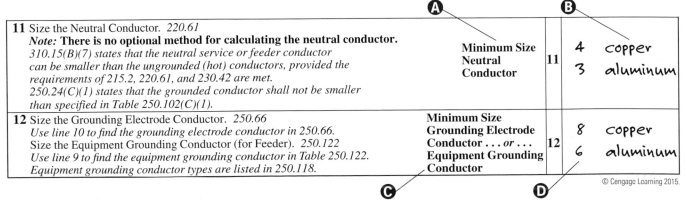

| 11 | Size the Neutral Conductor. *220.61* **Note: There is no optional method for calculating the neutral conductor.** *310.15(B)(7) states that the neutral service or feeder conductor can be smaller than the ungrounded (hot) conductors, provided the requirements of 215.2, 220.61, and 230.42 are met. 250.24(C)(1) states that the grounded conductor shall not be smaller than specified in Table 250.102(C)(1).* | Minimum Size Neutral Conductor | 11 | 4 copper 3 aluminum |
| 12 | Size the Grounding Electrode Conductor. *250.66* *Use line 10 to find the grounding electrode conductor in 250.66.* Size the Equipment Grounding Conductor (for Feeder). *250.122* *Use line 9 to find the equipment grounding conductor in Table 250.122. Equipment grounding conductor types are listed in 250.118.* | Minimum Size Grounding Electrode Conductor ... or ... Equipment Grounding Conductor | 12 | 8 copper 6 aluminum |

© Cengage Learning 2015.

The following page shows the completed optional method load calculation for a feeder and remote panelboard (subpanel) for one unit of a multifamily dwelling.

Optional Method Load Calculation for One Multifamily Dwelling Unit

1	General Lighting and Receptacle Loads *220.82(B)(1)* *Do not include open porches, garages, or unused or* *unfinished spaces not adaptable for future use.*	$3 \times$ ___**1050**___ (sq ft outside dimensions)	= 1	**3150**
2	Small-Appliance Branch Circuits *220.82(B)(2)* At least **two** small-appliance branch circuits must be included. *210.11(C)(1)*	$1500 \times$ ___**2**___ (minimum of two)	= 2	**3000**
3	Laundry Branch Circuit(s) *220.82(B)(2)* Include at least **one** laundry branch circuit *unless* meeting *210.52(F)*, Exception No. 1 or 2. *210.11(C)(2)*	$1500 \times$ ___**1**___ *(minimum of one)	= 3	**1500**

4 Appliances *220.82(B)(3)* and *(4)*
*Use the nameplate rating of **all***   *Do not include any heating*   Total volt-amperes
appliances (fastened in place,   *or air-conditioning*   of all appliances   **4**   **24,800**
permanently connected, or   *equipment in this section.*   LISTED BELOW
connected to a specific circuit),

water heater / **4500**	clothes dryer / **5500**	range / **12,000**
dishwasher / **1200**	disposer / **900**	microwave oven / **1200**
/	/	/
/	/	/
/	/	/

ranges, ovens, cooktops, motors,
and clothes dryers.
Convert any nameplate rating
given in amperes to volt-amperes
by multiplying the amperes
by the rated voltage.

5 Apply *220.82(B)* demand factor to the total of lines 1 through 4.

___**32,450**___ $- 10,000 =$ ___**22,450**___ $\times 40\% =$ ___**8980**___ $+ 10,000 =$ **5**   **18,980**
(total of lines 1 through 4)

6 Heating or Air-Conditioning System *220.82(C)*
Use the nameplate ratings in volt-amperes for
*all applicable systems in lines **a** through **e**.*

c) Supplemental electric heating for heat-pump systems. Include the heat-pump compressors at 100%. *Where heat-pump compressors are prevented from operating with supplemental heat, omit the compressors:*
_____ $\times 65\% =$ **c)**

a) Air-conditioning and cooling systems, including heat pumps without any supplemental electric heating:

___**3795**___ $\times 100\% =$ **a)**   **3795**

d) Electric space-heating equipment, if fewer than four separately controlled units:
___**5345**___ $\times 65\% =$ **d)**   **3474**

b) Electric thermal storage and other heating systems where the usual load is expected to be continuous at full nameplate value. *Systems qualifying under this selection shall not be figured under any other selection in 220.82(C).*

e) Electric space-heating equipment, if four or more separately controlled units:

_____ $\times 100\% =$ **b)**

_____ $\times 40\% =$ **e)**

7 Total Volt-Ampere Demand Load:
___**3795**___ $+$ ___**18,980**___ $=$ **7**   **22,775**
(largest VA rating from lines 6a through 6e)   (line 5)

8 Minimum Amperes
Divide the total volt-amperes by the voltage.
___**22,775**___ $\div$ ___**240**___ $=$ **8**   **95**
(line 7)   (voltage)   (minimum amperes)

9 Minimum Size Service or Feeder *240.6(A)*   **9**   **100**
(minimum is 100 amperes)

10 Size the Service or Feeder Conductors.
Use 310.15(B)(7) to find the service conductors up to 400 amperes.
Ratings in excess of 400 amperes shall comply with Table 310.15(B)(16)
310.15(B)(7) also applies to feeder conductors supplying the entire load.

Minimum Size Conductors   **10**   **4** copper
 2 aluminum

11 Size the Neutral Conductor. *220.61*
Note: There is no optional method for calculating the neutral conductor.
310.15(B)(7) states that the neutral service or feeder conductor
can be smaller than the ungrounded (hot) conductors, provided the
requirements of 215.2, 220.61, and 230.42 are met.
250.24(C)(1) states that the grounded conductor shall not be smaller
than specified in Table 250.102(C)(1).

Minimum Size Neutral Conductor   **11**   **4** copper
 3 aluminum

12 Size the Grounding Electrode Conductor. *250.66*
Use line 10 to find the grounding electrode conductor in 250.66.
Size the Equipment Grounding Conductor (for Feeder). 250.122
Use line 9 to find the equipment grounding conductor in Table 250.122.
Equipment grounding conductor types are listed in 250.118.

Minimum Size Grounding Electrode Conductor . . . or . . . Equipment Grounding Conductor   **12**   **8** copper
 6 aluminum

Summary

- Multifamily dwelling calculations are a progression of the one-family dwelling calculations (standard and optional).
- Some of the information needed for each individual dwelling unit includes total square-foot area, the number of small-appliance branch circuits and laundry branch circuits (if any), appliance ratings, dryer ratings (if any), cooking equipment ratings, and heating and air-conditioning loads.
- Identifying the number of dwelling units supplied by each service or disconnecting means is essential to the performance of multifamily load calculations.
- In certain instances, the laundry branch circuit may be omitted from both the dwelling unit and the calculation.
- Laundry (if installed) and small-appliance branch circuits are calculated using a rating of at least 1500 volt-amperes each.
- Table 220.42 demand factors are applied to general lighting, small-appliance, and laundry loads.
- The total volt-ampere rating of four (or more) fastened-in-place appliances is multiplied by 75% in the standard method calculation.
- Demand factors for household electric clothes dryers are found in Table 220.54.
- Table 220.55 lists the demand loads for household cooking appliances (over 1¾-kW rating).
- The heating and air-conditioning loads are compared, and the larger of the two is used.
- For heat pumps, include the compressor and the maximum amount of electric heat that can be energized while the compressor is running.
- Two multifamily load calculation methods are provided in Article 220: **Standard** and **Optional.**
- The neutral conductor can be smaller than the ungrounded (hot) conductors, provided the requirements of 215.2, 220.61, and 230.42 are met.
- Table 250.66 is used to size grounding electrode conductors.
- Remote panelboards (subpanels) require equipment grounding conductors, which can be one, more, or a combination of the types listed in 250.118.
- Equipment grounding conductors are sized in accordance with Table 250.122.

Unit 11 Competency Test

NEC Reference	Answer	
_____	_____	1. A feeder is supplying a 125-ampere, 240-volt panelboard in a multifamily dwelling. The conductors will be THHN aluminum conductors and all terminations will be rated 75°C (167°F). What is the minimum size aluminum equipment grounding conductor required?
_____	_____	2. What is the dryer demand load (in kW) for a twenty-five-unit multifamily dwelling with a 4.5-kW clothes dryer in each unit? (The service is 120/240-volt, single-phase.)
_____	_____	3. A four-unit apartment has the following ranges: a 15 kW, a 14 kW, a 10 kW, and a 9 kW. What is the kW demand load added to the service by these ranges?
_____	_____	4. An eighteen-unit multifamily dwelling contains the following appliances in each unit: a 1-kVA, 115-volt dishwasher; a ⅓-HP, 115-volt in-sink waste disposer; and a 4500-watt, 230-volt water heater. Using the standard method, what is the service neutral load contribution, in volt-amperes, for these appliances? (Assume water heater watt rating equivalent to volt-ampere.)
_____	_____	5. A multifamily dwelling's service-entrance conductors consist of four paralleled sets of 3/0 AWG THHN copper conductors. What size copper grounding electrode conductor is required?
_____	_____	6. A 5.4-kVA, 240-volt clothes dryer will contribute _____ amperes to the neutral load, when calculating the service by the standard method.
_____	_____	7. What is the minimum kW service demand load for twenty 6.5-kW ranges in a multifamily dwelling?

_____ _____ 8. An apartment has 3000 watts allocated for general lighting and receptacles. How many 15-ampere circuits are required? (The apartment contains laundry facilities.)

_____ _____ 9. Two paralleled sets of 500-kcmil copper service-entrance conductors supply a multifamily dwelling disconnecting means. What is the minimum size aluminum grounding electrode conductor required?

_____ _____ 10. A fifty-unit apartment building has a 4.5-kW clothes dryer in each apartment. By using the optional method, what is the service demand load?

Questions 11 through 18 are based on a 120/240-volt, 3-wire, single-phase, eight-unit multifamily dwelling. The eight identical units contain the following:

Floor area . . . . 1200 ft^2	Electric heat . . . . 4.2 kW, 240 volt	
Range. . . . . .8¾ kW, 240 volt	Air handler (blower motor) . . . . ¼ hp, 115 volt	
Water heater . . . . 4 kW, 240 volt	Air-conditioner compressor . . . . 3 hp, 230 volt	
Dishwasher . . . . 1 kW, 120 volt	Condenser fan motor. . . . . ⅙ hp, 115 volt	
Clothes dryer . . . . .5 kW, 240 volt		

Assume water heater, clothes dryer, range, and electric heat kW ratings equivalent to kVA.

Using Article 220, Parts II and III (the standard method):

_____ _____ 11. What is the service demand load for the clothes dryers?

_____ _____ 12. What is the service demand load for the cooking appliances?

_____ _____ 13. Of the heating or air-conditioning (and associated motors), which is omitted?

_____ _____ 14. What is the minimum rating (in amperes) for the service overcurrent device?

_____ _____ 15. A parallel run (two sets) of service-entrance conductors will be installed in two raceways. What is the minimum size THWN copper ungrounded conductors that can be used?

_____ _____ 16. What is the minimum rating (in amperes) for the neutral conductors?

_____ _____ 17. What is the minimum size THWN copper neutral (grounded) conductors that can be installed?

_____ _____ 18. What is the minimum size copper grounding electrode conductor?

Questions 19 through 22 are based on a 120/240-volt, 3-wire, single-phase, six-unit multifamily dwelling. The six identical units contain the following:

Floor area . . . . 1350 ft^2	Clothes dryer . . . . 4 kW, 240 volt	
Range. . . . . .10.6 kW, 240 volt	Electric heat (two banks at 3 kW each). . 6 kW, 240 volt	
Water heater . . . . .4.2 kW, 240 volt	Air handler (blower motor) . . . . ⅓ hp, 115 volt	
Dishwasher . . . . 1.2 kW, 120 volt	Air-conditioner compressor . . . . 3 hp, 230 volt	
Garbage disposer . . . . ⅓ hp, 115 volt	Condenser fan motor. . . . . ⅙ hp, 115 volt	

Assume water heater, clothes dryer, range, and electric heat kW ratings equivalent to kVA.

Using the Optional Method:

_____ _____ 19. What is the minimum rating (in amperes) for the service overcurrent device?

_____ _____ 20. What is the minimum size THWN copper ungrounded conductors that can be installed? (Do not parallel the conductors.)

_____ _____ 21. What is the minimum size THWN copper neutral (grounded) conductor that can be installed?

_____ _____ 22. What is the minimum acceptable size copper grounding electrode conductor?

COMMERCIAL LOCATIONS

section 4

UNIT 12

General Provisions

Objectives

After studying this unit, the student should:

▶ know that commercial occupancy receptacle placement differs from that of one-family and multifamily dwellings.

▶ understand that commercial bathrooms do not require receptacle outlets.

▶ be aware that each commercial building (or occupancy) accessible to pedestrians must have at least one sign outlet.

▶ be familiar with Article 430 motor provisions.

▶ know that air-conditioning and refrigeration provisions are found in Article 440.

▶ be able to determine receptacle volt-ampere ratings for single, duplex, quad, etc.

▶ know the maximum number of receptacles permitted on 15- and 20-ampere branch circuits.

▶ be aware that at least two receptacle outlets must be readily accessible in guest rooms of hotels, motels, and similar occupancies.

▶ thoroughly understand showcase and show window provisions.

▶ be able to determine an occupancy's general lighting load based on square-foot area.

▶ be familiar with fluorescent, HID, recessed, and track-lighting provisions.

▶ understand that while certain luminaires can be used as raceways, the number of branch circuits permitted therein is limited.

▶ know that a metal pole supporting a luminaire(s) requires a handhole, unless an exception is met.

Introduction

The broad topic of commercial wiring encompasses many types of structures—office buildings, restaurants, stores, schools, and warehouses, to name a few. In fact, most of the occupancies found in Table 220.12 are indeed commercial. Many of the electrical requirements that apply to one-family and multifamily dwellings also apply to commercial and industrial locations. Because this overlap of requirements is common, it is especially important to begin with a solid understanding of not only some but all of the preceding material. Of particular importance are Units 3 through 5 of Section 1, which contain fundamental principles applicable to all types of occupancies. In addition, Unit 9—Services and Electrical Equipment explains one-family dwelling provisions that extend into the realm of commercial wiring systems. Unit 12—General Provisions likewise covers overlapping topics, such as branch circuits, receptacles, and lighting. Because it would be very difficult, if not impossible, to cover such a large field as commercial wiring in the framework of this text, only a small but representative portion is presented. Unit 13—Nondwelling Load Calculations explains nondwelling (inclusive of commercial) load calculations, which are helpful in determining service and feeder loads. Complete subsequent units, and portions of others, apply to commercial wiring as well.

The inception of most commercial electrical projects includes multifaceted plans (blueprints) and detailed specifications. As a reference, illustrations for a retail store's electrical and lighting floor plans are provided. Various general provisions (branch circuits, receptacles, lighting, etc.) and numerous specific provisions (required receptacle outlets, show windows, showcases, etc.) are discussed. Proper receptacle placement is demonstrated through the use of a hotel/motel guest room floor plan, including furniture placement. Receptacle and lighting branch-circuit load calculations are also part of this unit.

While this unit is by no means a complete reference for commercial electrical wiring, taken together with the extensive information presented throughout this text, it does provide a substantial foundation toward understanding commercial electrical wiring systems.

BRANCH CIRCUITS

General Requirements

Article 210 covers branch circuits with the exception of branch circuits supplying only motor loads. These are covered in Article 430. Articles 210 and 430 provisions apply to branch circuits that have combination loads **》210.1《**.

Branch circuits recognized by Article 210 shall be rated in accordance with the maximum permitted ampere rating or setting of the overcurrent device. The rating for other than individual branch circuits shall be 15, 20, 30, 40, and 50 amperes. Where conductors of higher ampacity are used for any reason, the ampere rating or setting of the specified overcurrent device shall determine the circuit rating **》210.3《**.

Unless meeting one of the two exceptions, a single receptacle installed on an individual branch circuit shall have an ampere rating not less than that of the branch circuit **》210.21《**.

An individual branch circuit shall be permitted to supply any load for which it is rated, but in no case shall the load exceed the branch-circuit ampere rating **》210.22《**.

A Branch circuits recognized by Article 210 shall be permitted as multiwire circuits. A multiwire circuit shall be permitted to be considered as multiple circuits. All conductors of a multiwire branch circuit shall originate from the same panelboard or similar distribution equipment **》210.4(A)《**.

B Conductors shall be sized to carry not less than the larger of 210.19(A)(1)(a) or (b).

(a) Where a branch circuit supplies continuous loads or any combination of continuous and noncontinuous loads, the minimum branch-circuit conductor size shall have an allowable ampacity not less than the noncontinuous load plus 125 percent of the continuous load.

(b) The minimum branch-circuit conductor size shall have an allowable ampacity not less than the maximum load to be served after the application of any adjustment or correction factors **》210.19(A)(1)《**.

> ### N O T E
>
> Refer to 210.2 for a list of other articles (and sections) pertaining to branch-circuit requirements. The provisions for branch circuits supplying equipment in that list amend or supplement Article 210 provisions and apply to the referenced branch circuits **》210.2《**.

C The ungrounded and grounded circuit conductors of each multiwire branch circuit shall be grouped by cable ties or similar means in at least one location within the panelboard or other point of origination.

Exception: The requirement for grouping shall not apply if the circuit enters from a cable or raceway unique to the circuit that makes the grouping obvious or if the conductors are identified at their terminations with numbered wire markers corresponding to the appropriate circuit number **》210.4(D) and 210.4(D) Exception《**.

D Each multiwire branch circuit shall be provided with a means that will simultaneously disconnect all ungrounded conductors at the point where the branch circuit originates **》210.4(B)《**.

E Where a branch circuit supplies continuous loads, or any combination of continuous and noncontinuous loads, the rating of the overcurrent device must not be less than the noncontinuous load plus 125% of the continuous load **》210.20(A)《**.

F Multiwire branch circuits shall supply only line-to-neutral loads **》210.4(C)《**.

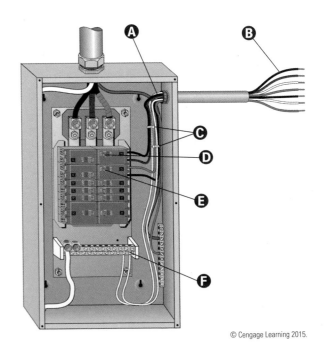

© Cengage Learning 2015.

Required Sign Outlet

The disconnecting means must be within sight of the sign or outline lighting system that it controls. If out of the line of sight from any section that is able to be energized, the disconnecting means must be capable of locking in the open position »*600.6(A)(2)*«.

Signs or outline lighting systems operated by external electronic or electromechanical controllers can have a disconnecting means located within sight of the controller or within the controller enclosure. The disconnecting means must disconnect both the sign (or outline lighting system) and the controller from all ungrounded supply conductors. It must be designed so that no pole operates independently and it shall be lockable in accordance with 110.25 »*600.6(A)(3)*«.

> ## N O T E
>
> Service hallways or corridors are not considered accessible to pedestrians »*600.5(A)*«.
>
> Switches, flashers, and similar devices controlling transformers and electronic power supplies must either be rated for controlling inductive loads or have a current rating not less than twice the transformer's current rating »*600.6(B)*«.

Ⓐ Each commercial building and occupancy open to pedestrians must have at least one accessible outlet at every entrance to each tenant space for sign or outline lighting systems use »*600.5(A)*«.

Ⓑ The outlet(s) must be supplied by a branch circuit rated 20 amperes or more and that supplies no other load »*600.5(A)*«.

Ⓒ Each sign and outline lighting system, feeder or branch circuit supplying a sign, outline lighting system, or skeleton tubing must be controlled by an externally operable switch (or circuit breaker) that opens all ungrounded conductors and controls no other load, unless (1) the sign is an exit directional sign located within a building or (2) the sign is cord-connected with an attachment plug. The switch or circuit breaker shall open all ungrounded conductors simultaneously on multiwire branch circuits in accordance with 210.4(B) »*600.6*«.

Ⓓ Sign and outline lighting outlets must be calculated at a minimum of 1200 volt-amperes per required branch circuit specified in 600.5(A) »*220.14(F)*«.

> ## N O T E
>
> The disconnecting means shall be capable of being locked in the open position. The provisions for locking shall remain in place with or without the lock installed »*110.25*«.

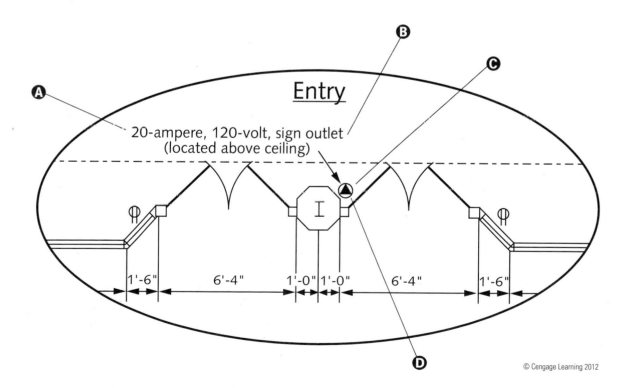

© Cengage Learning 2012

Motor-Operated and Combination Loads

A For circuits supplying loads consisting of motor-operated utilization equipment (fastened in place with a motor larger than 1/8 hp) in combination with other loads, the total calculated load must be based on 125% of the largest motor load plus the sum of the other loads ≫ *220.18(A)* ≪.

B Because equipment such as room air-conditioners, household refrigerators and freezers, drinking water coolers, and beverage dispensers are considered appliances, Article 422 provisions also apply ≫ *440.3(C)* ≪.

C For cord- and plug-connected appliances, an accessible, separable connector or an accessible plug and receptacle can serve as the disconnecting means. Where the separable connector or plug and receptacle are not accessible, cord- and plug-connected appliances must be provided with a disconnecting means in accordance with 422.31 ≫ *422.33(A)* ≪.

D The rating of a receptacle or a separable connector must not be less than the rating of any connected appliance ≫ *422.33(C)* ≪.

E This is a branch circuit, consisting of two duplex receptacles.

F If supplied by a 20-ampere circuit, this drinking fountain's maximum rating is 10 amperes (20 × 50%). If supplied by a 15-ampere circuit, the maximum rating is 7.5 (15 × 50%).

G This is a receptacle outlet for cord- and plug-connected, non-fastened-in-place utilization equipment.

H The receptacle outlet is fed from a dedicated branch circuit.

I If supplied by a 20-ampere circuit, the maximum rating for this copy machine is 16 amperes (20 × 80%). If supplied

by a 15-ampere circuit, the maximum rating is 12 amperes (15 × 80%).

J For cord-connected equipment such as room air-conditioners, household refrigerators (and freezers), drinking water coolers, and beverage dispensers, a separable connector or an attachment plug and receptacle can serve as the disconnecting means ≫ *440.13* ≪.

> ### NOTE
>
> A 15- or 20-ampere branch circuit can supply lighting units, other utilization equipment, or a combination of both. The rating of any one cord- and plug-connected utilization equipment that is not fastened in place must not exceed 80% of the branch-circuit ampere rating. The total rating of utilization equipment fastened in place (other than luminaries) cannot exceed 50% of the branch-circuit ampere rating in the event that lighting units, cord- and plug-connected utilization equipment not fastened in place, or both are also supplied ≫ *210.23(A)* ≪.

> ### CAUTION
>
> Electric drinking fountains shall be protected with ground-fault circuit-interrupter protection ≫ *422.52* ≪.
> The device providing GFCI protection required in Article 422 shall be readily accessible ≫ *422.5* ≪.

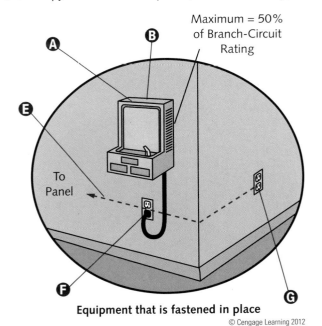

Maximum = 50% of Branch-Circuit Rating

To Panel

Equipment that is fastened in place

© Cengage Learning 2012

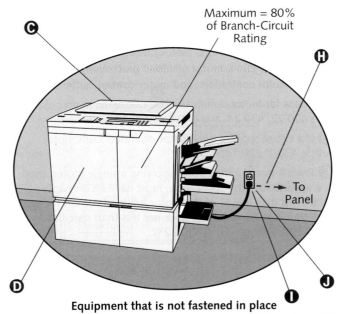

Maximum = 80% of Branch-Circuit Rating

To Panel

Equipment that is not fastened in place

© Cengage Learning 2012

Air-Conditioning and Refrigeration Equipment

Article 440, Part II, provisions require a means capable of disconnecting air-conditioning and refrigerating equipment (including motor compressors and controllers) from the circuit conductors ≫ *440.11* ≪.

The provisions of Article 440, Part III, specify devices intended to protect the branch-circuit conductors, control apparatus, and motors in circuits supplying hermetic refrigerant motor compressors against overcurrent due to short circuits and

Air-Conditioning and Refrigeration Equipment *(continued)*

ground faults. These provisions amend or supplement those in Article 240 》*440.21*《.

Article 310 and Article 440, Part IV, provisions specify conductor ampacities required to carry the motor current (without overheating) under the conditions specified, except as modified in 440.6(A), Exception No. 1 》*440.31*《.

Article 440, Part VI, specifies devices intended to protect the motor compressor, the motor-control apparatus, and the branch-circuit conductors against excessive heating due to motor overload and failure to start 》*440.51*《.

Ⓐ Loads for motor outlets shall be calculated in accordance with the requirements in 430.22, 430.24, and 440.6 》*220.14(C)*《.

Ⓑ Where a circuit supplies only air-conditioning equipment, refrigeration equipment, or both, apply Article 440 》*220.18(A)*《.

Ⓒ Disconnecting means must be located within sight and readily accessible from the air-conditioning or refrigerating equipment. The disconnecting means can be installed on or within the equipment 》*440.14*《.

Ⓓ For a hermetic refrigerant motor compressor, the rated-load current marked on the nameplate of the equipment in which the motor compressor is employed must be used in determining the rating or ampacity of the disconnecting means, the branch-circuit conductors, the controller, the branch-circuit short-circuit and ground-fault protection, and the separate motor overload protection. Where no rated-load current is shown on the equipment nameplate, use the rated-load current shown on the compressor nameplate 》*440.6(A)*《.

Ⓔ Article 440 provisions apply to electric motor-driven air-conditioning and refrigerating equipment, and to their branch circuits and controllers. It provides the special considerations necessary for circuits supplying hermetic refrigerant motor compressors and for any air-conditioning or refrigerating equipment supplied from a branch circuit that also supplies a hermetic refrigeration motor compressor 》*440.1*《.

CAUTION

The disconnecting means must not be located on panels that are designed to allow access to the air-conditioning or refrigeration equipment or to obscure the equipment name plate(s) 》*440.14*《.

NOTE

A 125-volt, single-phase, 15- or 20-ampere-rated receptacle outlet shall be installed at an accessible location for the servicing of heating, air-conditioning, and refrigeration equipment. The receptacle shall be located on the same level and within 25 ft (7.5 m) of the heating, air-conditioning, and refrigeration equipment. The receptacle outlet shall not be connected to the load side of the equipment disconnecting means 》*210.63*《.

NOTE

All 125-volt, single-phase, 15- and 20-ampere receptacles installed outdoors and on rooftops shall have ground-fault circuit-interrupter protection for personnel 》*210.8(B)*《. The ground-fault circuit-interrupter shall be installed in a readily accessible location 》*210.8(B)*《. Receptacles on rooftops shall not be required to be readily accessible other than from the rooftop 》*210.8(B) Exception No. 1*《.

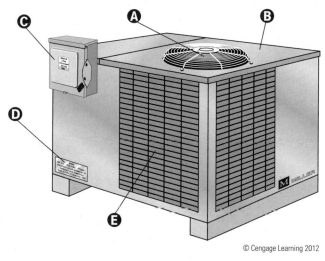

© Cengage Learning 2012

RECEPTACLES

Maximum Number of Receptacles on a Branch Circuit

Ⓐ Ten receptacles are permitted on a 15-ampere overcurrent protective device (15 ÷ 1.5 = 10).

Ⓑ Receptacles are calculated at 180 volt-amperes per strap 》*220.14(I)*《. Therefore, when the source is 120 volts, the rating of each receptacle is 1.5 amperes (180 ÷ 120 = 1.5).

Ⓒ Thirteen receptacles are permitted on a 20-ampere overcurrent protective device (20 ÷ 1.5 = 13.3 or 13).

Ⓓ A load having the maximum level of current sustained for three hours or more is referred to as a continuous load. A general-purpose receptacle is not a continuous load 》*Article 100*《.

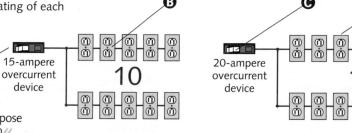

© Cengage Learning 2012

Nondwelling Receptacle Placement

Ⓐ Dwelling unit receptacles must be installed so that no point along the floor line in any wall space is more than 6 ft (1.8 m), measured horizontally, from a receptacle outlet in that space ≫ *210.52(A)(1)* ≪.

Ⓑ Receptacle outlets in hotels, motels, sleeping rooms in dormitories, and similar occupancy must be installed in accordance with 210.52(A) and (D). Guest rooms or guest suites provided with permanent provisions for cooking shall have receptacle outlets installed in accordance with all of the applicable rules in 210.52 ≫ *210.60(A)* ≪.

Ⓒ Although no provisions pertain to nondwelling general receptacle placement (except hotel and motel guest rooms and suites), certain requirements exist for specific occupancies or areas. Examples include but are not limited to show windows ≫ *210.62* ≪; rooftop, attic, and crawl space receptacle outlets for servicing heating, air-conditioning, and refrigeration equipment ≫ *210.63* ≪; and health care patient bed location receptacles ≫ *517.18(B)* ≪.

Ⓓ Generally speaking, there are no receptacle placement provisions for nondwelling occupancies.

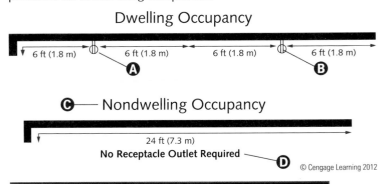

Dwelling Occupancy

6 ft (1.8 m) 6 ft (1.8 m) 6 ft (1.8 m) 6 ft (1.8 m)

Ⓐ   **Ⓑ**

Ⓒ— Nondwelling Occupancy

24 ft (7.3 m)

No Receptacle Outlet Required — **Ⓓ**

© Cengage Learning 2012

CAUTION

At least one 125-volt, single-phase, 15- or 20-ampere-rated receptacle outlet shall be installed within 50 ft (15 m) of the electrical service equipment ≫ *210.64* ≪.

Bathroom Receptacles

Ⓐ Nondwelling bathrooms do not require receptacles. A bathroom receptacle is required only in dwelling units, hotel or motel guest rooms, and sleeping rooms in dormitories ≫ *210.52(D)* and *210.60(A)* ≪.

Ⓑ Where installed, 125-volt, single-phase, 15- and 20-ampere bathroom receptacles must have GFCI protection for personnel ≫ *210.8(B)(1)* ≪.

NOTE

GFCIs are designed to trip when a 4- through 6-milliampere (0.004- through 0.006-ampere) difference between the ungrounded (hot) conductor and grounded conductor occurs.

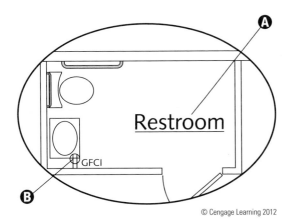

Restroom

© Cengage Learning 2012

Fixed Multioutlet Assemblies

Ⓐ Where appliances are unlikely to be used simultaneously, each 5 ft (1.5 m) or fraction thereof (of separate and continuous length) shall be considered one outlet of not less than 180 volt-amperes ≫ *220.14(H)(1)* ≪.

Ⓑ Because fixed multioutlet assemblies used in dwelling units or hotel and motel guest rooms and suites are included in the general lighting load calculation, no additional load calculation is required ≫ *220.14(H)* ≪.

Ⓒ Where appliances are likely to be used simultaneously, each 1 ft (300 mm) or fraction thereof shall be considered one outlet of not less than 180 volt-amperes ≫ *220.14(H)(2)* ≪.

Ⓓ It is not stipulated that each receptacle be rated 180 volt-amperes. This illustration shows one receptacle per foot; therefore, each receptacle is rated 180 volt-amperes. Multioutlet assemblies with two receptacles per ft (6 in. [150 mm] on center) have a rating of 180 volt-amperes for two receptacles.

Receptacles that are *unlikely* to be used simultaneously.

Ⓐ 5 ft (1.5 m)= **180 VA** **Ⓑ**

Receptacles that are *likely* to be used simultaneously.

Ⓒ   **Ⓓ**

1 ft (300 mm)= **180 VA** 1 ft (300 mm)= **180 VA** 1 ft (300 mm)= **180 VA** 1 ft (300 mm)= **180 VA** 1 ft (300 mm)= **180 VA**

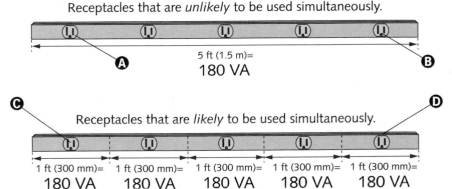

© Cengage Learning 2012

Receptacle Volt-Ampere Rating

A Except as covered in one-, two-, and multifamily dwellings, and in hotel and motel guest rooms and suites, receptacle outlets must be computed at not less than 180 volt-amperes for each single or multiple receptacle on one yoke ➤➤ *220.14(I)* ◀◀ .

B A single piece of equipment (consisting of four or more receptacles) must be calculated at not less than 90 volt-amperes per receptacle ➤➤ *220.14(I)* ◀◀ .

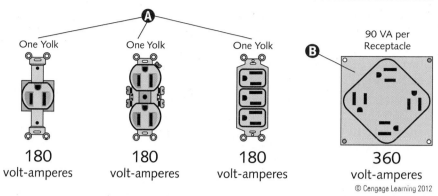

One Yoke — **180** volt-amperes
One Yoke — **180** volt-amperes
One Yoke — **180** volt-amperes
90 VA per Receptacle — **360** volt-amperes

© Cengage Learning 2012

Provisions for Hotel and Motel Guest Rooms and Suites

When performing service (or feeder) calculations, use Table 220.42 to derate hotel and motel (include apartment houses without cooking equipment) guest room lighting and receptacle loads. Do not apply Table 220.42 demand factors to areas where all lighting will probably be used at one time. Examples include but are not limited to office, lobby, restaurant, hallway, and parking areas.

A GFCI protection is required for every 125-volt bathroom receptacle ➤➤ *210.8(A)(1)* ◀◀ .

B Hallways of less than 10 ft (3.0 m) in length do not require a receptacle ➤➤ *210.52(H)* ◀◀ .

C At least one wall switch–controlled lighting outlet or receptacle must be installed in hotel, motel, or similar occupancy guest rooms or suites ➤➤ *210.70(B)* ◀◀ .

D Receptacles must be installed so that no point along the floor line of any wall space is more than 6 ft (1.8 m), measured horizontally, from an outlet in that space ➤➤ *210.52(A)(1)* ◀◀ . (See 210.60(B) provisions allowing a greater distance due to permanent luminaire layout.)

E Guest rooms or suites in hotels, motels, sleeping rooms in dormitories, and similar occupancies must have receptacle outlets installed in accordance with 210.52 (dwelling unit receptacle provisions) ➤➤ *210.60(A)* ◀◀ . (See Units 6 and 7 in this book for dwelling unit receptacle requirements.)

F Locate at least one wall receptacle within 36 in. (900 mm) of the outside edge of each basin (lavatory or sink). The receptacle outlet must be located on a wall or partition adjacent to the basin or basin countertop ➤➤ *210.52(D)* ◀◀ .

G Section 410.10(D) contains bathtub and shower area lighting provisions. (See the illustrated explanation in Unit 7 of this book.)

H Receptacles installed behind the bed must either be positioned so that the bed does not contact any installed attachment plug, or the receptacle must include a suitable guard ➤➤ *210.60(B)* ◀◀ .

I The total number of receptacle outlets must comply with the minimum number of receptacles provision of 210.52(A). These receptacle outlets can be located conveniently for permanent furniture layout ➤➤ *210.60(B)* ◀◀ . This receptacle could be located between the beds, even though the distance to the receptacle behind the desk exceeds 12 ft (3.7 m).

J At least two receptacle outlets must be readily accessible ➤➤ *210.60(B)* ◀◀ .

K Branch-circuit conductors supplying more than one receptacle for cord-and plug-connected portable loads must have an ampacity at least equal to the rating of the branch circuit ➤➤ *210.19(A)(2)* ◀◀ .

N O T E

Hotels, motels, and similar occupancies having guest kitchen facilities must meet 210.52(B) and (C) provisions. It is not required that occupants in hotel/motel guest rooms or suites that are intended for transient occupancy have ready access to all overcurrent devices protecting conductors supplying their room ➤➤ *240.24(B)(1)* ◀◀ .

CAUTION

In dwelling units, dormitories, and guest rooms or guest suites, overcurrent devices (other than supplementary overcurrent protection) shall not be located in bathrooms as defined in Article 100 ➤➤ *240.24(E)* ◀◀ .

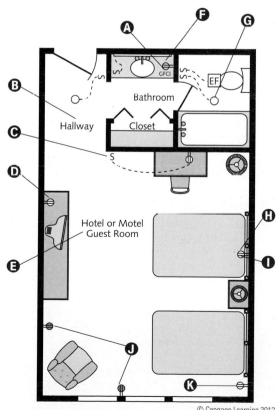

Hotel or Motel Guest Room

© Cengage Learning 2012

Show Windows

A *show window* is any window designed or used for the display of goods or advertising material, whether it is fully or partly enclosed or entirely open at the rear, and whether or not it has a platform raised higher than the street floor level »*Article 100* «.

Unless used in wiring of chain-supported luminaires, or as supply cords for portable lamps and other merchandise being displayed or exhibited, flexible cords used in show-cases and show windows must be Type S, SE, SEO, SEOO, SJ, SJE, SJEO, SJEOO, SJO, SJOO, SJT, SJTO, SJTOO, SO, SOO, ST, STO, STOO, SEW, SEOW, SEOOW, SJEW, SJEOW, SJEOOW, SJOW, SJOOW, SJTW, SJTOW, SJTOOW, SOW, SOOW, STW, STOW, or STOOW »*400.11* «.

A Show-window branch-circuit loads must be calculated either by (a) the unit load per outlet as required by other provisions of 220.14 or (b) at 200 volt-amperes per linear foot »*220.14(G)* «.

B At least one 125-volt, single-phase, 15- or 20-ampere-rated receptacle outlet shall be installed within 18 in. (450 mm) of the top of a show window for each 12 linear ft (3.7 m)—or major fraction thereof—of window area measured horizontally at its widest point »*210.62* «.

C The second method for calculating show-window branch-circuit loads is to multiply each show-window linear foot (or major fraction thereof) by 200 volt-amperes (2 + 11.5 + 2 = 15.5 = 16 linear ft . . . 16 × 200 = 3200 volt-amperes) »*220.14(G)* «.

D One of two methods for calculating show-window branch-circuit loads is to multiply each receptacle by 180 volt-amperes (180 × 2 = 360 volt-amperes) »*220.14(G)* and *(I)* «.

E Complete show-window feeder and/or service loads using a minimum rating of 200 volt-amperes for each linear ft of window, measured horizontally along its base (16 × 200 = 3200 volt-amperes) »*220.43(A)* «.

CAUTION

Where receptacles are installed in the floor, the boxes that house the receptacles must be listed specifically for the application »*314.27(B)*«.

NOTE

Where the AHJ judges them free from likely exposure to physical damage, moisture, and dirt, boxes located in **elevated floors of show windows** and similar locations can be other than those listed for floor applications. Receptacles and covers must be listed as an assembly for this type of location »*314.27(B) Exception*«.

Electric signs (including neon tubing) and associated wiring within show windows must comply with Article 600.

© Cengage Learning 2012

LIGHTING

Lampholder Installations

A Lampholders of the screw shell type shall be installed for use as lampholders only »*410.90* «.

B If the supply circuit has a grounded conductor, the grounded conductor must be connected to the screw shell »*410.90* «.

NOTE

Lampholders installed in wet locations shall be listed for use in wet locations. Lampholders installed in damp locations shall be listed for damp locations or shall be listed for wet locations »*410.96*«.

WARNING

Lampholders shall be constructed, installed, or equipped with shades or guards so that combustible material is not subjected to temperatures in excess of 90°C (194°F) »*410.97* «.

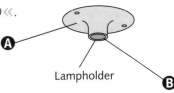

Lampholder

© Cengage Learning 2012

Reflective Ceiling/Lighting Plan

Branch circuits recognized by Article 210 must be rated in accordance with the maximum permitted overcurrent device ampere rating (or setting) »*210.3*«.

Branch circuits for lighting and appliances (including motor-operated appliances) shall be provided to supply the loads calculated in accordance with 220.10 »*210.11*«.

A 15- or 20-ampere branch circuit can supply lighting units or other utilization equipment, or a combination of both »*210.23(A)*«.

Wall switch–controlled lighting outlets are not required in commercial occupancies, except for attic and underfloor spaces containing equipment that requires servicing »*210.70(C)*«.

A unit load meeting Table 220.12 specifications for listed occupancies constitutes the minimum lighting load for each square foot of floor area. The area for each floor must be calculated using the outside dimensions of the building, dwelling unit, or other area involved »*220.12*«.

Supplementary overcurrent protection used for luminaires, appliances, and other equipment (including internal circuits and components) must not be used as a substitute for required branch-circuit overcurrent devices or in place of the branch-circuit protection. Ready accessibility is not required for supplementary overcurrent devices »*240.10*«.

A 30-, 40-, or 50-ampere branch circuit can supply fixed lighting units with heavy-duty lampholders in other than dwelling unit(s) »*210.23(B)* and *(C)*«.

Lighting outlet loads must not be supplied by branch circuits larger than 50 amperes »*210.23(D)*«.

A For circuits of over 250 volts to ground, the electrical continuity of metal raceways and cables with metal sheaths containing any conductor other than service conductors shall be ensured by one or more of the methods specified for services in 250.92(B) except for (B)(1) »*250.97*«.

B Circuit breakers used as switches in 120-volt and 277-volt fluorescent lighting circuits shall be listed and shall be marked SWD or HID »*240.83(D)*«. All switches and circuit breakers used as switches must be located so that they are operable from a readily accessible place. The center of the operating-handle grip of the switch or circuit breaker when in its highest position must not be more than 6 ft, 7 in. (2.0 m) above the floor or working platform »*404.8(A)*«.

CAUTION

Wiring located within the cavity of a floor–ceiling or roof–ceiling assembly shall not be secured to, or supported by, the ceiling assembly, including the ceiling support wires. An independent means of secure support shall be provided and shall be permitted to be attached to the assembly. Where independent support wires are used, they shall be distinguishable by color, tagging, or other effective means »*300.11(A)(1)* and *(2)*«.

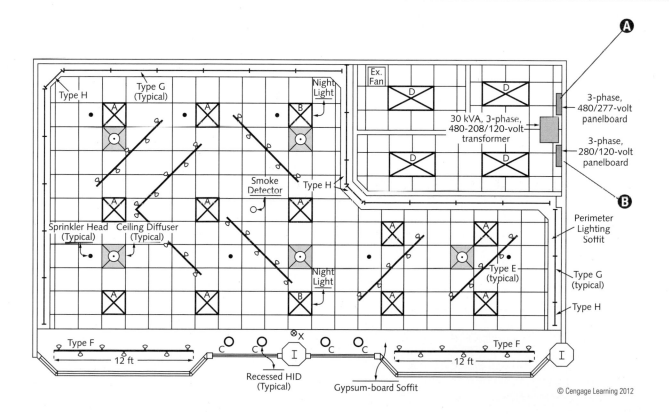

© Cengage Learning 2012

Luminaire Schedule

The minimum number of branch circuits is determined from the total calculated load and the size (or rating) of the circuits used. In all installations, the number of circuits must be sufficient to supply the load served. In no case shall the load on any circuit exceed the maximum specified by 220.18 ≫ *210.11(A)* ≪.

The load calculated on a volt-ampere per square-foot (or per square-meter) basis must be adequately served by the wiring system up to and including the branch-circuit panelboard(s). This load must be evenly proportioned among multioutlet branch circuits within the panelboard(s). Branch circuits and overcurrent devices must only be installed to serve the connected load ≫ *210.11(B)* ≪.

Ⓐ Conductors of alternating-current (ac) and direct-current (dc) circuits, rated 1000 volts, nominal or less, can occupy the same equipment wiring enclosure, cable, or raceway.

The conductors' insulation rating must equal or exceed the maximum circuit voltage applied to any conductor within the enclosure, cable, or raceway.

Secondary wiring to electric-discharge lamps of 1000 volts or less, if insulated for the secondary voltage involved, shall be permitted to occupy the same luminaire, sign, or outline lighting enclosure as the branch-circuit conductors ≫ *300.3(C)(1)* ≪.

Ⓑ For lighting units that have ballasts, transformers, or auto-transformers or LED drivers, the supply circuit's calculated load is based on the total ampere ratings of such units, not on the lamps' total wattage ≫ *220.18(B)* ≪. Obtain the ampere rating from the manufacturer or from the information listed on the ballast. If the luminaire contains more than one ballast, add the ampere ratings. Multiply the total ampere rating by the voltage to find the luminaire's volt-ampere rating.

LUMINAIRE SCHEDULE

Type	Manufacturer	Volts	Catalog Number	Comments
A	Lighthouse	277	P622-232U277	2′ × 2′ lay-in fluorescent deep cell parabolic
B	Lighthouse	277	P622-232U277-EM	emergency light
C	Lighthouse	120	10R-CM70MH120	10″ recessed, 70W metal halide
D	Lighthouse	277	RT24–432EB277	2′ × 4′ lay-in fluorescent
E	Lighthouse	120	8001BK/HT308	8′ lighting track with 5 heads
F	Lighthouse	120	8001BK/HT308	12′ lighting track with 7 heads
G	Lighthouse	120	SL48-32EB120	4′ single lamp fluorescent strip
H	Lighthouse	120	SL24-20EB120	2′ single lamp fluorescent strip
X	Lighthouse	120	EX7B1LEDR	Exit light with battery

LAMP SCHEDULE

Lamps per Luminaire	Watts per Lamp	Lamp Size
2	32	F32-T8-U
2	32	F32-T8-U
1	70	MH70/C/U/M
4	32	F32-T8
5	75	75W-PAR30FL
7	75	75W-PAR30FL
1	32	F32-T8
1	20	F20-T8

Lamps are included in the luminaire.

© Cengage Learning 2012

NOTE

Branch-circuit conductors shall have an ampacity not less than the maximum load to be served. Conductors shall be sized to carry not less than the larger of 210.19(A)(1)(a) or (b).
(a) Where a branch circuit supplies continuous loads or any combination of continuous and noncontinuous loads, the minimum branch-circuit conductor size shall have an allowable ampacity not less than the noncontinuous load plus 125 percent of the continuous load.
(b) The minimum branch-circuit conductor size shall have an allowable ampacity not less than the maximum load to be served after the application of any adjustment or correction factors ≫*210.19(A)(1)*≪.
Most commercial lighting is considered continuous.

CAUTION

Luminaires installed in exposed or concealed locations under metal-corrugated sheet roof decking shall be installed and supported so there is not less than 1-1/2 in. (38 mm) measured from the lowest surface of the roof decking to the top of the luminaire ≫*410.10(F)*≪.

Fluorescent Luminaires

Luminaires shall be equipped with shades (or guards) and constructed or installed so that combustible material will not be subjected to temperatures exceeding 90°C (194°F) »*410.11*«.

Securely support all luminaires, lampholders, and receptacles »*410.30(A)*«.

Exposed metal parts shall be connected to an equipment grounding conductor or insulated from the equipment grounding conductor and other conducting surfaces or be inaccessible to unqualified personnel. Grounding is not required for lamp tie wires, mounting screws, clips, and decorative bands on glass spaced at least 1½ in. (38 mm) from lamp terminals shall not be required to be grounded »*410.42*«.

Raceway fittings supporting a luminaire(s) shall be capable of supporting the combined weight of the luminaire assembly and lamp(s) »*410.36(E)*«.

A Suspended-ceiling system framing members used to support luminaires must be securely fastened to each other and to the building structure at appropriate intervals. Luminaires must be fastened securely to the ceiling framing member by mechanical means such as bolts, screws, or rivets. Listed clips identified for use with the type of ceiling framing member(s) and luminaire(s) are also permitted »*410.36(B)*«.

B Branch circuits recognized by Article 210 can be used as multiwire circuits. A multiwire branch circuit can be thought of as multiple circuits. All conductors must originate from the same panelboard or similar distribution equipment »*210.4(A)*«. Unless an exception is met, multiwire branch circuits can only supply line-to-neutral loads »*210.4(C)*«.

C A 3-phase, 4-wire, wye-connected power system supplying nonlinear loads may necessitate a design allowing for the possibility of high harmonic neutral currents on the neutral conductor »*210.4(A) Informational Note No. 1*«.

D Luminaires and equipment shall be mechanically connected to an equipment grounding conductor as specified in 250.118 and sized in accordance with 250.122 »*410.44*«.

E Wiring on or within luminaires must be neatly arranged and must not be exposed to physical damage. Avoid excess wiring. Arrange conductors so they are not subjected to temperatures greater than those for which they are rated »*410.48*«.

> **NOTE**
>
> Install luminaires so that the connections between luminaire conductors and circuit conductors can be inspected without having to disconnect any of the wiring, unless the luminaires are connected by attachment plugs and receptacles »*410.8*«.

> **CAUTION**
>
> Each multiwire branch circuit shall be provided with a means that will simultaneously disconnect all ungrounded conductors at the point where the branch circuit originates »*210.4(B)*«.

> **WARNING**
>
> The ungrounded and grounded circuit conductors of each multiwire branch circuit shall be grouped by cable ties or similar means in at least one location within the panelboard or other point of origin »*210.4(D)*«. The requirement for grouping shall not apply if the circuit enters from a cable or raceway unique to the circuit that makes the grouping obvious or if the conductors are identified at their terminations with numbered wire markers corresponding to the appropriate circuit number »*210.4(D) Exception*«.

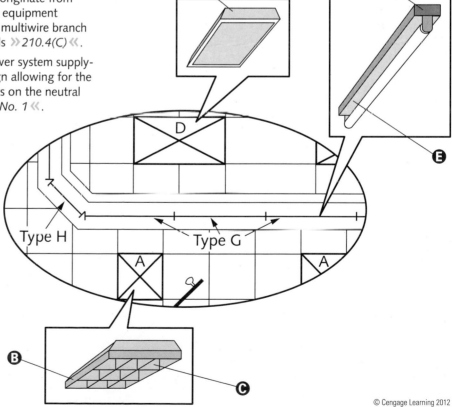

Show-Window and Track Lighting

Heavy-duty lighting track is identified for use exceeding 20 amperes. Each fitting attached to a heavy-duty lighting track shall have individual overcurrent protection » *410.153* «.

Ⓐ Lighting track (track lighting) is a manufactured assembly designed to support and energize luminaires that are capable of being readily repositioned along the track. Its length may be altered by the addition or subtraction of track sections » *Article 100* «.

Ⓑ Lighting track shall be permanently installed as well as permanently connected to a branch circuit. Install only lighting track fittings on the lighting track. Fittings equipped with general-purpose receptacles shall not be used on lighting track » *410.151(A)* «.

Ⓒ Lighting track shall be securely mounted so that each fastening suitably supports the maximum weight of luminaires that can be installed. Unless identified for greater support intervals, a single section 4 ft (1.2 m) or less in length shall have two supports. If installed in a continuous row, each individual section of not more than 4 ft (1.2 m) in length shall have one additional support » *410.154* «.

Ⓓ Lighting track shall be grounded in accordance with Article 250. Track sections shall be securely coupled to maintain continuity of the circuitry, polarization, and grounding throughout » *410.155(B)* «.

Ⓔ Track system ends shall be insulated and capped » *410.155(A)* «.

Ⓕ The load connected to a lighting track shall not exceed the track's rating. The rating of the branch circuit that supplies lighting track shall not exceed the track rating. The load calculation in 220.43(B) shall not be required to limit the length of track on a single branch circuit, and it shall not be required to limit the number of luminaires on a single track » *410.151(B)* «.

Ⓖ Fittings identified for use on lighting track shall be designed specifically for the track on which they are installed. They shall be securely fastened to the track, maintain polarization and connection to the equipment grounding conductor, and be designed to be suspended directly from the track » *410.151(D)* «.

Ⓗ When computing feeder or service track lighting in nondwelling units or hotel/motel guest rooms, include an additional load of 150 volt-amperes for every 2 ft (600 mm) of lighting track or fraction thereof » *220.43(B)* «.

Ⓘ No externally wired luminaire except chain-supported shall be used in show windows » *410.14* «.

Ⓙ When calculating feeder or service show-window lighting, include a unit load of not less than 200 volt-amperes per linear foot of show window, measured horizontally along the base » *220.43(A)* «
(2 + 11.5 + 2 = 15.5 = 16 × 200 = 3200 volt-amperes).

NOTE

Do not install lighting track in the following locations:
1. Where physical damage is likely
2. In wet or damp locations
3. Where subject to corrosive vapors
4. In storage battery rooms
5. In hazardous (classified) locations
6. Where concealed
7. Where extended through walls or partitions
8. Less than 5 ft (1.5 m) above the finished floor except where protected from physical damage or where the track operates at less than 30 volts root-mean–square (rms) open–circuit voltage
9. Where prohibited by 410.10(D) » **410.151(C)** «.

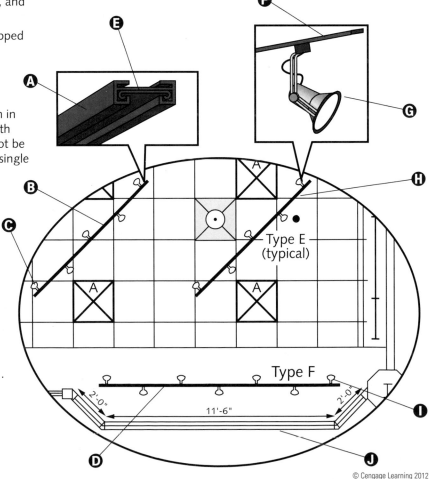

Type E (typical)

Type F

11'-6"

2'-0"   2'-0"

Recessed Luminaires

A recessed luminaire not identified for contact with insulation shall have all recessed parts spaced at least ½ in. (13 mm) from combustible materials. Support points and the trim finishing off the openings in the ceiling, wall, or other finished surface can be in contact with combustible materials ≫*410.116(A)(1)*≪.

A recessed luminaire identified for contact with insulation, Type IC, can contact combustible materials at recessed parts, points of support, and portions passing through or finishing off the structural opening ≫*410.116(A)(2)*≪.

A recessed HID luminaire identified for use and installed in poured concrete does not require thermal protection ≫*410.130(F)(3)*≪.

Ⓐ A luminaire can be recessed in fire-resistant material in a building of fire-resistant construction, subject to temperatures between 90°C (194°F) and 150°C (302°F), provided the luminaire is plainly marked as listed for that service ≫*410.115(B)*≪.

Ⓑ Install luminaires so that adjacent combustible material will not be subjected to temperatures greater than 90°C (194°F) ≫*410.115(A)*≪.

Ⓒ Luminaires installed in recessed cavities in walls or ceilings, including suspended ceilings, shall comply with 410.115 through 410.122 ≫*410.110*≪.

Ⓓ Thermal insulation shall not be installed above a recessed luminaire or within 3 in. (75 mm) of the recessed luminaire's enclosure, wiring compartment, ballast, transformer, LED driver, or power supply unless the luminaire is identified as Type IC for insulation contact ≫*410.116(B)*≪.

Ⓔ High-intensity discharge (HID) luminaires supply power to HID lamps. HID lamps produce light from gaseous discharge arc tubes and include mercury lamps, metal halide lamps, and high-pressure sodium lamps.

Ⓕ A recessed remote ballast for a HID luminaire shall have integral thermal protection and be identified as such ≫*410.130(F)(4)*≪.

Ⓖ Recessed high-intensity luminaires designed for installations in wall or ceiling cavities shall have thermal protection and be identified as such ≫*410.130(F)(1)*≪. Thermal protection is not required in a recessed high-intensity luminaire whose design, construction, and thermal performance characteristics are equivalent to a thermally protected luminaire and are identified as inherently protected ≫*410.130(F)(2)*≪.

> **N O T E**
>
> Luminaires that have exposed ballasts, transformers, LED drivers, or power supplies shall be installed such that ballasts, transformers, LED drivers, or power supplies shall not be in contact with combustible material unless listed for such condition ≫*410.136(A)*≪.

> **WARNING**
>
> Circuit breakers used as switches in 120-volt and 277-volt fluorescent lighting circuits shall be listed and shall be marked SWD or HID. Circuit breakers used as switches in high-intensity discharge lighting circuits shall be listed and shall be marked as HID ≫*240.83(D)*≪.

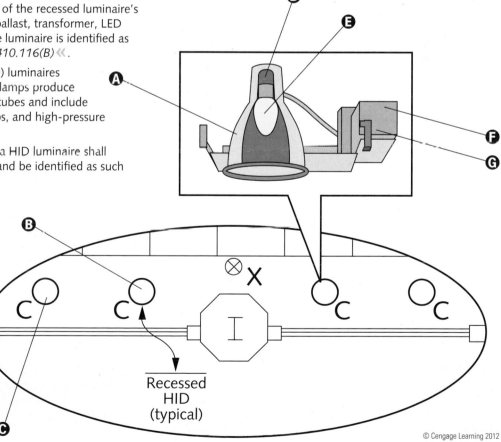

Recessed
HID
(typical)

Exit Luminaires and Egress Luminaires

A ballast in a fluorescent exit luminaire shall not have thermal protection ≫*410.130(E)(3)*≪.

A Framing members of suspended ceiling systems used to support luminaires shall be securely fastened to each other and to the building structure at appropriate intervals. Luminaires shall be securely fastened to the ceiling framing member by mechanical means such as bolts, screws, or rivets. Listed clips identified for use with the type of ceiling framing member(s) and luminaires are also permitted ≫*410.36(B)*≪.

B The ballast of a fluorescent luminaire installed indoors shall have integral thermal protection. Replacement ballasts shall also have similar thermal protection integral with the ballast ≫*410.130(E)(1)*≪.

C A fluorescent luminaire ballast used for egress lighting, energized only during a failure of the normal supply, shall not have thermal protection ≫*410.130(E)(4)*≪.

D An exit directional sign located within a building does not require a disconnecting means ≫*600.6 Exception No. 1*≪.

> **N O T E**
>
> Article 700 provisions apply to the electrical safety of emergency systems installation, operation, and maintenance. These systems consist of circuits and equipment intended to supply, distribute, and control electricity for facility illumination, power, or both whenever the normal electrical service is interrupted ≫**700.1**≪. Emergency systems are defined as those systems legally required and classed as emergency by municipal, state, federal, or other codes, or by any governmental agency having jurisdiction. The purpose of these systems is to automatically supply illumination, power, or both to designated areas and equipment in the event of failure of the normal supply, or in the event of accidental damage to systems elements intended to supply, distribute, and control power and illumination essential for safety to human life ≫**700.2**≪.

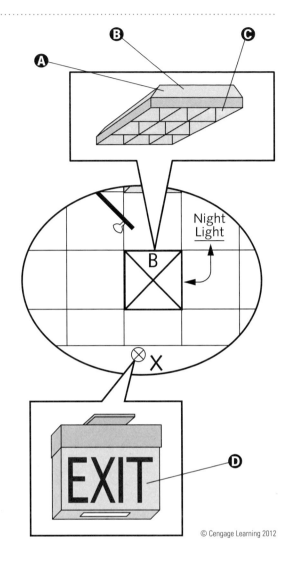

© Cengage Learning 2012

Adjustable Luminaires

A Two conduits can support an enclosure containing a device under the conditions found in 314.23(F): (1) the box shall not exceed 100 in.³ (1650 cm³); (2) the box has either threaded entries or identified hubs; (3) the box must be supported by two or more conduits threaded wrenchtight into the enclosure or hubs; and (4) each conduit must be secured within 18 in. (450 mm) of the enclosure, provided all entries are on the same side.

B Equipment grounding conductors that are part of flexible cords with the largest circuit conductor 10 AWG or smaller or used with luminaire wires in accordance with 240.5 shall not be smaller than 18 AWG copper and not smaller than the circuit conductors ≫*250.122(E)*≪.

C A wet location receptacle where the product plugged in is unattended while in use requires an enclosure that is weatherproof at all times (plug inserted or not) ≫*406.9(B)(2)(a)*≪.

D Luminaires that require adjusting or aiming after installation do not require an attachment plug or cord connector, provided the exposed cord is of the hard or extra-hard usage type and is not longer than that required for maximum adjustment. The cord shall not be subject to strain or physical damage ≫*410.62(B)*≪.

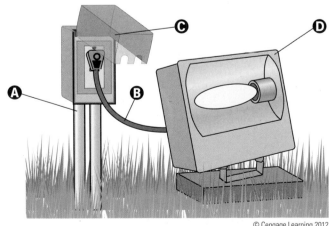

© Cengage Learning 2012

Switches Controlling Lighting Loads

The grounded circuit conductor for the controlled lighting circuit shall be provided at the location where switches control lighting loads that are supplied by a grounded general-purpose branch circuit for other than the following:

(1) Where conductors enter the box enclosing the switch through a raceway, provided that the raceway is large enough for all contained conductors, including a grounded conductor

(2) Where the box enclosing the switch is accessible for the installation of an additional or replacement cable without removing finish materials

(3) Where snap switches with integral enclosures comply with 300.15(E)

(4) Where a switch does not serve a habitable room or bathroom

(5) Where multiple switch locations control the same lighting load such that the entire floor area of the room or space is visible from the single or combined switch locations

(6) Where lighting in the area is controlled by automatic means

(7) Where a switch controls a receptacle load ≫ *404.2(C)* ≪.

A Because this switch is supplied by an MC cable, a grounded conductor is required.

B Because the conductors enter the box through a raceway and the raceway is large enough for all contained conductors, including a grounded conductor, a grounded conductor is not required.

C The provision for a (future) grounded conductor is to complete a circuit path for electronic lighting control devices ≫ *404.2(C) Informational Note* ≪.

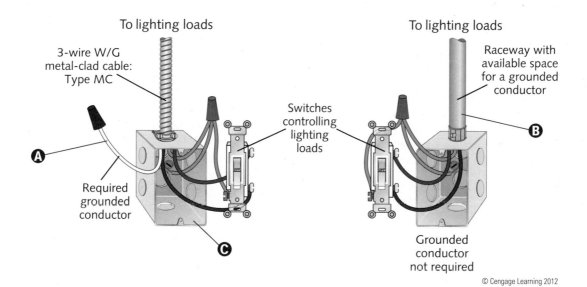

© Cengage Learning 2012

Luminaires Used as Raceways

In a completed installation, a cover shall be provided for each outlet box, unless covered by means of a luminaire canopy, lampholder, receptacle, or similar device ≫ *410.22* ≪.

Auxiliary equipment for electric-discharge lamps shall be enclosed in noncombustible cases and be treated as heat sources ≫ *410.104(A)* ≪.

A Luminaires designed for end-to-end connection, thereby forming a continuous assembly, or luminaires connected by recognized wiring methods, are permitted to contain the conductors of a 2-wire branch circuit or one multiwire branch

circuit supplying the connected luminaires and need not be listed as a raceway ≫ *410.64(C)* ≪.

B One additional 2-wire branch circuit separately supplying one or more of the connected luminaires is also permitted ≫ *410.64(C)* ≪.

C Feeder and branch-circuit conductors within 3 in. (75 mm) of a ballast, LED driver, power supply, or transformer shall have an insulation temperature rating no lower than 90°C (194°F) unless supplying a luminaire marked as suitable for a different insulation temperature ≫ *410.68* ≪.

Luminaires Used as Raceways *(continued)*

D Luminaires shall not be used as a raceway for circuit conductors unless listed and marked for use as a raceway ≫ *410.64(A)* ≪.

E Luminaires identified for through-wiring, as permitted by 410.21, shall be permitted to be used as a raceway ≫ *410.64(B)* ≪.

> **NOTE**
>
> Electric-discharge and LED luminaires supported independent of the outlet box shall be connected to the branch circuit through metal raceway, nonmetallic raceway, Type MC cable, Type AC cable, Type MI cable, or nonmetallic-sheathed cable, or by flexible cord as permitted in 410.62(B) or (C) ≫ *410.24(A)* ≪.

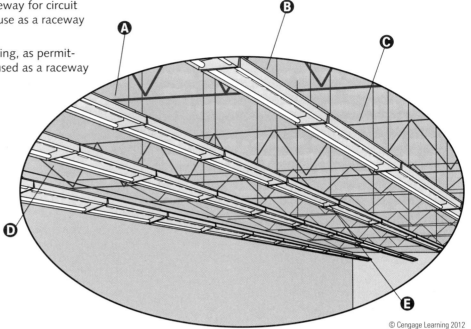

© Cengage Learning 2012

Cord-Connected Lampholders and Luminaires

The inlet of a metal lampholder attached to a flexible cord shall be equipped with an insulating bushing that, if threaded, is not smaller than nominal ³⁄₈-in. pipe size. The cord hole shall be of an appropriate size, and all burrs/fins must be removed, providing a smooth bearing surface for the cord ≫ *410.62(A)* ≪.

A Electric-discharge luminaires provided with mogul-base, screw shell lampholders can be connected to branch circuits of 50 amperes or less by cords that comply with 240.5. While receptacles and attachment plugs can be of a lower ampere rating than the branch circuit, they shall not be less than 125% of the luminaire full-load current ≫ *410.62(C)(2)* ≪.

B A luminaire or a listed assembly can be cord connected if the following conditions apply:

1. The luminaire is located directly below the outlet or busway.

2. The flexible cord meets all the following:

   a. Is visible for its entire length outside the luminaire.

   b. Is not subject to strain or physical damage.

   c. Is terminated in a grounding-type attachment plug cap or busway plug, or is a part of a listed assembly incorporating a manufactured wiring system connector in accordance with 604.6(C), or has a luminaire assembly with a strain relief and canopy having a maximum 6-in. (152-mm) long section of raceway for attachment to an outlet box above a suspended ceiling ≫ *410.62(C)(1)* ≪.

C For circuits that supply lighting units with ballasts, transformers, autotransformers, or LED drivers, the computed load shall be based on the total ampere rating of such units and not on the total lamp(s) wattage ≫ *220.18(B)* ≪.

D A load expected to continue at maximum current for 3 hours or more is a continuous load ≫ *Article 100* ≪.

E Electric-discharge luminaires having a flanged surface inlet can be supplied by cord pendants equipped with cord connectors. Inlets and connectors can be of a lower ampere rating than the branch circuit but not less than 125% of the luminaire load current ≫ *410.62(C)(3)* ≪.

F Cord-connected luminaires shall be located directly below the outlet box or busway ≫ *410.62(C)(1)(1)* ≪.

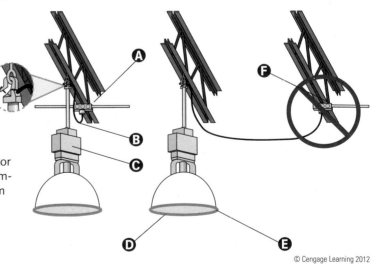

© Cengage Learning 2012

Portable Lamps

A Portable lamps must be wired with flexible cord, recognized by 400.4, and a polarized or grounding-type attachment plug ≫ *410.82(A)* ≪.

B Where used with Edison-base lampholders, the grounded conductor shall be identified and attached to the screw shell as well as the attachment plug's identified blade ≫ *410.82(A)* ≪.

C In addition to the provisions of 410.82(A), portable handlamps shall comply with 410.82(B)(1) through (5):

1. Metal shell, paper-lined lampholders must not be used.
2. Handlamps must be equipped with a handle of molded composition or other insulating material.
3. Handlamps must be equipped with a substantial guard attached to the lampholder or handle.
4. Metallic guards must be grounded by means of an equipment grounding conductor run with circuit conductors within the power-supply cord.
5. Portable handlamps supplied through an isolating transformer with an ungrounded secondary of not over 50 volts do not require grounding ≫ *410.82(B)* ≪.

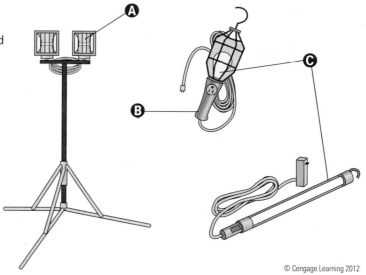

© Cengage Learning 2012

Required Handhole in a Metal Pole

A This metal pole is more than 8 ft (2.5 m) tall abovegrade.

B Metal raceways or other equipment grounding conductors shall be bonded to the pole by means of an equipment grounding conductor recognized by 250.118 and sized in accordance with 250.122 ≫ *410.30(B)(5)* ≪.

C Support conductors in vertical metal poles used as raceway, as provided in 300.19 ≫ *410.30(B)(6)* ≪.

D A pole shall have a handhole not less than 2 in. × 4 in. (50 mm × 100 mm) with a cover suitable for use in wet locations to provide access to the supply terminations within the pole or pole base ≫ *410.30(B)(1)* ≪.

E A metal pole with a handhole must have an equipment grounding terminal accessible from the handhole ≫ *410.30(B)(3)(a)* ≪.

F Direct-buried cable, conduit, or other raceways must meet Table 300.5 minimum cover requirements ≫ *300.5(A)* ≪.

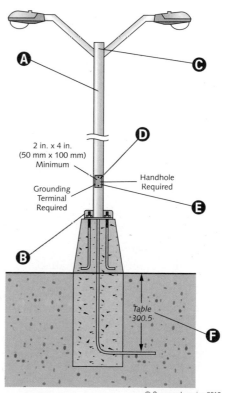

2 in. x 4 in.
(50 mm x 100 mm)
Minimum

Handhole
Required

Grounding
Terminal
Required

Table
300.5

© Cengage Learning 2012

NOTE

Where ungrounded conductors are increased in size from the minimum size that has sufficient ampacity for the intended installation, wire-type equipment grounding conductors, where installed, shall be increased in size proportionately according to the circular mil area of the ungrounded conductors ≫*250.122(B)*≪.

Hinged Metal Poles, 20 ft (6.0 m) or Less, Supporting Luminaires

Ⓐ A hinged-base pole longer than 20 ft (6.0 m) supporting a luminaire(s) requires a handhole ≫ *410.30(B)(1)* ≪.

Ⓑ A metal pole shall have an accessible grounding terminal within the base ≫ *410.30(B)(3)(b)* ≪.

Ⓒ A metal pole 20 ft (6.0 m) or less in height above grade that has a hinged base does not require a handhole ≫ *410.30(B)(1) Exception No. 2* ≪.

Ⓓ If equipped with a hinged base, the pole and base must be bonded together ≫ *410.30(B)(4)* ≪.

Ⓔ Metal raceways or other equipment grounding conductors shall be bonded to the pole by means of an equipment grounding conductor recognized by 250.118 and sized in accordance with 250.122 ≫ *410.30(B)(5)* ≪.

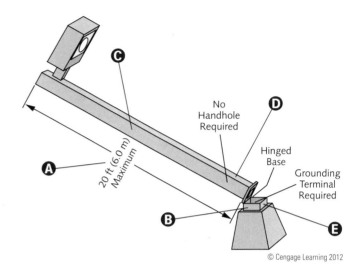

© Cengage Learning 2012

Metal Poles, 8 ft (2.5 m) or Less, Supporting Luminaires

Ⓐ No handhole is required in a pole no more than 8 ft (2.5 m) in height (abovegrade) where the supply wiring method continues without splice or pull point and where the pole interior and splices are accessible by removing the luminaire ≫ *410.30(B)(1) Exception No. 1* ≪.

Ⓑ Splices must remain accessible.

Ⓒ The wiring method must continue without splice to the luminaire termination point.

Ⓓ A pole longer than 8 ft (2.5 m) supporting a luminaire(s) requires a handhole ≫ *410.30(B)(1)* ≪.

Ⓔ No grounding terminal is required in a pole 8 ft (2.5 m) or less in height abovegrade where the supply wiring method continues without splice or pull point and where the pole interior and splices are accessible by luminaire removal ≫ *410.30(B)(3) Exception to (3)* ≪.

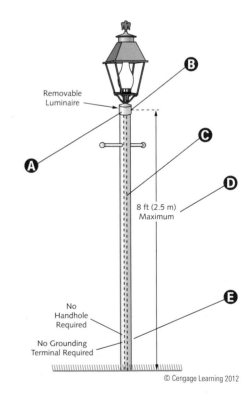

© Cengage Learning 2012

> ### NOTE
>
> Receptacles of 15 and 20 amperes installed in a wet location shall have an enclosure that is weatherproof whether or not the attachment plug cap is inserted. An outlet box hood installed for this purpose shall be listed and shall be identified as "extra duty." All 15- and 20-ampere, 125- and 250-volt nonlocking-type receptacles shall be listed weather-resistant type ≫**406.9(B)(1)**≪.

Supply Conductors Not Installed within the Pole

A Where cable or raceway risers are not installed within the pole, a threaded fitting (or nipple) shall be brazed (or welded) to the pole opposite the handhole for the supply connection ≫ *410.30(B)(2)* ≪.

B Installed raceways subject to physical damage may require additional protection.

C Raceways, fittings, boxes, etc. must be securely fastened in place ≫ *300.11(A)* ≪. Follow specific raceway support provisions for the type of raceway used.

D See Unit 5 for raceway provisions.

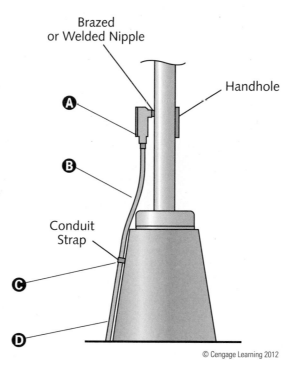

Brazed or Welded Nipple

Handhole

A

B

Conduit Strap

C

D

© Cengage Learning 2012

Summary

- Receptacle placement provisions for one-family and multi-family dwellings do not apply to commercial occupancies.
- Although a bathroom receptacle is not required in a commercial occupancy, it must be GFCI protected if installed.
- Each commercial building/occupancy open to pedestrians must be provided with at least one 20-ampere outlet in an accessible location at each entrance to every tenant space for sign or outline lighting systems use.
- One similarity in calculating branch-circuit conductors for continuous loads and motor loads is that each is multiplied by 125%.
- Article 440 contains air-conditioning and refrigeration equipment requirements.
- Commercial receptacles are computed at not less than 180 volt-amperes per strap.
- No more than ten receptacles are permitted on a 15-ampere branch circuit and thirteen on a 20-ampere branch circuit.

- At least two receptacle outlets must be readily accessible in guest rooms of hotels, motels, and similar occupancies.
- Table 220.12 gives the volt-ampere (per square foot) lighting load for listed occupancies.
- Calculate feeder or service track lighting at 150 volt-amperes for every 2 ft (600 mm) of lighting track or fraction thereof.
- Calculate feeder or service show-window lighting at not less than 200 volt-amperes per linear foot (300 mm) of show window.
- Certain luminaires, such as end-to-end connected fluorescent lights, have a limited use as raceways.
- A handhole is required in a metal pole taller than 8 ft (2.5 m) abovegrade, unless the pole is no more than 20 ft (6.0 m) with a hinged base.
- At least one 125-volt, single-phase, 15- or 20-ampere-rated receptacle outlet shall be installed within 50 ft (15 m) of the electrical service equipment.

Unit 12 Competency Test

NEC Reference   Answer

_____   _____   1. In calculating the load of lighting units that employ an autotransformer, the load shall be based on _____.

_____   _____   2. _____ luminaires used in a show window can be externally wired.

_____   _____   3. Luminaires shall be wired with conductors having insulation suitable for the _____ to which the conductors will be subjected.

_____   _____   4. Systems of illumination utilizing fluorescent lamps, high-intensity discharge (HID) lamps, or neon tubing are defined as _____.

_____   _____   5. The term _____ is defined as a complete lighting unit consisting of a light source such as a lamp(s), together with the parts designed to position the light source and connect it to the power supply.

_____   _____   6. The total rating of utilization equipment fastened in place, other than luminaires, shall not exceed _____% of the branch-circuit ampere rating where lighting units, cord- and plug-connected utilization equipment not fastened in place, or both, are also supplied.

_____   _____   7. A metal pole supporting a luminaire must have a handhole not less than _____ in. with a raintight cover to provide access to the supply terminations within the pole or pole base.

_____   _____   8. Branch circuits that supply neon tubing installations shall not be rated in excess of _____ amperes.

_____   _____   9. Multioutlet assembly shall not be permitted:

   I. to be run through hoistways.

   II. to be run within dry partitions.

   III. to be run through dry partitions.

      a) II only   b) III only   c) I and II only   d) I, II, and III

_____   _____   10. Where portable luminaires are used with Edison-base lampholders, the grounded conductor shall be identified and attached to the _____ of the attachment plug.

_____   _____   11. When calculating track lighting loads for a commercial service, an additional load of _____ volt-amperes shall be included for every _____ ft of lighting track or fraction thereof.

_____   _____   12. The terminal of an electric-discharge lamp shall be considered as a live part where any lamp terminal is connected to a circuit of over _____.

_____   _____   13. _____ luminaires surface mounted over concealed outlet, or pull or junction boxes and designed not to be supported solely by the outlet box shall be provided with suitable openings in back of the luminaire to provide access to the wiring in the boxes.

_____   _____   14. When calculating show-window lighting loads for a commercial service, a load of not less than _____ shall be included for each linear foot of show window, measured horizontally along its base.

_____   _____   15. Twenty-four ft (six 4-ft sections) of track lighting must be installed in a continuous row in a retail store. What is the minimum number of supports required?

_____   _____   16. No handhole shall be required in a pole supporting a luminaire that is _____ ft or less in height abovegrade where the supply wiring method continues without splice or pull point, and where the interior of the pole and any splices are accessible by removing the luminaire.

NEC Reference   **Answer**

17. A hermetic refrigerant motor compressor's branch-circuit short-circuit and ground-fault protective device shall be capable of carrying _____ of the motor's starting current.

    a) 100%   b) 125%   c) 150%   d) 175%

18. Heavy-duty lighting track is lighting track identified for use exceeding _____.

19. At least _____ receptacle outlet(s) shall be readily accessible in hotel and motel guest rooms.

20. Flexible cords shall be secured to the undersides of showcases so that a separation between cases not in excess of _____ in. nor more than _____ in. between the first case and the supply receptacle will be ensured.

21. Branch circuits that supply all other signs and outline lighting systems shall be rated not to exceed _____.

22. No handhole shall be required in a metal pole supporting a luminaire that is _____ ft or less in height abovegrade that is provided with a hinged base.

23. A single piece of equipment consisting of a multiple receptacle comprising four or more receptacles shall be computed at not less than _____ per receptacle.

24. Luminaires shall not be used as a raceway for circuit conductors unless they are _____ and marked for use as a raceway.

25. At least one 125-volt, single-phase, 15- or 20-ampere-rated receptacle outlet shall be installed within 18 in. of the top of a show window for each _____ linear foot or major fraction thereof of show-window area measured horizontally at its maximum width.

26. On hermetic refrigerant motor compressors, the value of branch-circuit selection current will always be _____ the marked rated-load current.

    I. less than

    II. equal to

    III. greater than

       a) I only   b) I or II   c) II or III   d) III only

27. Electric-discharge luminaires provided with mogul-base, screw shell lampholders shall be permitted to be connected to branch circuits of _____ amperes or less by cords complying with 240.5.

28. The load for the required sign branch circuit shall be calculated at a minimum of _____.

29. A(n) _____ branch circuit consists of two or more ungrounded conductors having a potential difference between them and a grounded conductor having equal potential difference between it and each ungrounded conductor of the circuit and that is connected to the neutral or grounded conductor of the system.

30. Lighting track shall be securely mounted so that each fastening will be suitable for supporting the maximum weight of luminaires that can be installed. Unless identified for supports at greater intervals, a single section _____ ft or shorter in length shall have _____ supports.

31. Switches, flashers, and similar devices controlling sign transformers and electronic power supplies shall be rated for controlling inductive loads or have a current rating not less than _____ of the transformer's current rating.

    a) 80%   b) 100%   c) 125%   d) 200%

NEC Reference	Answer
_____	_____

32. Conductors of circuits rated 1000 volts, nominal, or less, ac circuits, and dc circuits shall be permitted to occupy the same enclosure if:

    a) all conductors within the enclosure have an insulation rating equal to at least 1000 volts.

    b) all conductors operating at different voltage levels within the enclosure are separated by a permanent barrier.

    c) all conductors have an insulation rating equal to at least the maximum circuit voltage applied to any conductor within the enclosure.

    d) each conductor has an insulation rating equal to at least the maximum circuit voltage applied to each individual conductor within the enclosure.

33. Feeder and branch-circuit conductors within _____ in. of a ballast, LED driver, power supply, or transformer shall have an insulation temperature rating not lower than 90°C (194°F), unless supplying a luminaire listed and identified as suitable for a different insulation temperature.

34. The screw shell of all sign lampholders supplied by a grounded circuit shall be connected to the _____ conductor.

35. Where connected to a branch circuit having a rating in excess of _____, lampholders shall be of the heavy-duty type.

36. A recessed luminaire that is not identified for contact with insulation shall have all recessed parts spaced at least _____ in. from combustible materials.

37. At least one 125-volt, single-phase, 15- or 20-ampere-rated receptacle outlet shall be installed within _____ ft of the electrical service equipment in a commercial building.

UNIT 13

Nondwelling Load Calculations

Objectives

After studying this unit, the student should:

▶ have a good understanding of the elements required to perform a nondwelling load calculation.

▶ know that receptacle outlets are counted, unlike in dwelling-unit load calculations.

▶ understand how to apply Table 220.44 demand factors to receptacle loads in excess of 10 kVA.

▶ be able to calculate the receptacle load for banks and office buildings where the actual number of receptacle outlets is unknown.

▶ be familiar with volt-ampere unit loads for different types of occupancies, and even for different areas within certain occupancies.

▶ know when and how to apply Table 220.42 demand factors.

▶ be aware that track lighting is calculated in addition to the general lighting load.

▶ know when to include a sign or outline lighting outlet in a load calculation.

▶ understand the method for calculating show-window lighting loads.

▶ be able to determine whether a load is continuous or noncontinuous.

▶ understand that continuous loads require the inclusion of an additional 25% volt-ampere rating in the load calculation.

▶ know that the load calculation rating represents only a minimum rating.

Introduction

Unit 13 contains a nondwelling load calculation form, along with a detailed explanation of each line. Dwelling calculations are covered in Unit 8 (one family) and Unit 11 (multifamily) of this text. Unlike dwelling unit load calculations, receptacle outlets are counted.

Receptacle outlets (if known) and fixed multioutlet assemblies (if any) are entered into the calculation, and if the load is great enough, a demand factor is applied. As with dwelling-unit load calculations, general lighting is calculated using outside dimensions. Other items, such as a sign outlet (where required) and show window(s) (if present) are part of the nondwelling load calculation. The calculation for continuous loads (line 10) is very important. All continuous load ratings must be increased by 25%. While kitchen equipment (line 11) is not included in every type of calculation, be aware that kitchen equipment is not limited to restaurants. For instance, kitchen equipment could be a portion of a load calculation for a school. Line 14 (all other loads) is a catchall for any load not included otherwise in the calculation.

The load calculation form has little room for listing individual items, such as motors, equipment, etc. Depending on the size of the occupancy, the calculation could contain hundreds, if not thousands, of individual items. Space is also limited for certain calculations. When the need arises, simply attach additional sheets of paper containing extra items or calculations. Some procedures will not apply to certain load calculations. It is recommended that a line not be left completely empty. Some predetermined marking should fill the space, such as a dashed line or the letters NA. One of the procedures (line 4) is applicable only in banks and office buildings where the actual number of receptacle outlets is unknown.

The load calculation form results (overcurrent protection and conductors) represent only a *minimum* requirement. No consideration is given for the addition of future electrical loads. The size service or feeder is not restricted to the form's calculated size. For example, an electrician might install a 200-ampere service in an occupancy where the load calculation only required a 125-ampere rating.

Because certain cities, states, etc. require that some (if not all) electrical installations be designed by a licensed electrical engineer, caution is advised. Check with local authorities to determine these, as well as other, requirements.

NONDWELLING LOAD CALCULATIONS

Line 1—Receptacle Load

A load of 15,300 volt-amperes would be placed in line 1 for a commercial occupancy having seventy-five duplex and ten single receptacles (75 + 10 = 85 × 180 = 15,300).

A General-purpose receptacles are not continuous loads.

B Receptacle outlets are calculated at a minimum of 180 volt-amperes for each single (or multiple) receptacle on one yoke (or strap) ≫ *220.14(I)* ≪ (see Unit 12).

C A single piece of equipment (consisting of four or more receptacles) shall be calculated at no less than 90 volt-amperes per receptacle ≫ *220.14(I)* ≪ (see Unit 12).

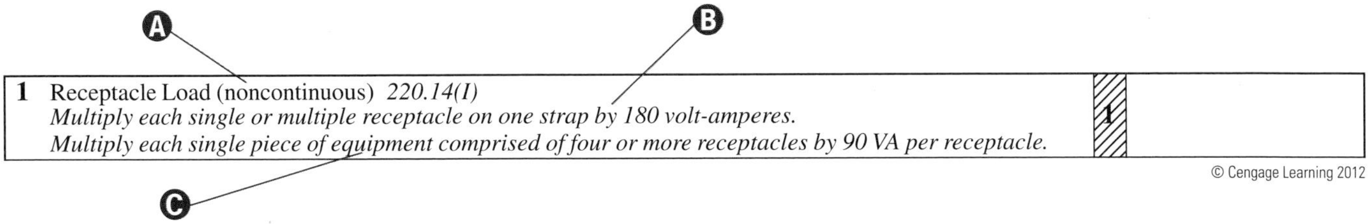

A   **B**

1 Receptacle Load (noncontinuous) *220.14(I)* *Multiply each single or multiple receptacle on one strap by 180 volt-amperes.* *Multiply each single piece of equipment comprised of four or more receptacles by 90 VA per receptacle.*	*1*

© Cengage Learning 2012

C

Line 2—Fixed Multioutlet Assembly Load

A commercial occupancy has 75 linear ft of fixed multioutlet assembly, with 15 ft of the assembly subject to simultaneous use. A load of 4860 volt-amperes is placed in line 2 because 75 − 15 = 60 ft (nonsimultaneous); 60 ÷ 5 = 12 × 180 = 2160 volt-amperes for nonsimultaneous use multioutlet assembly; 15 × 180 = 2700 volt-amperes for simultaneous use multioutlet assembly; and 2160 + 2700 = 4860 volt-amperes total for multioutlet assembly.

A Where simultaneous use of appliances is unlikely, each 5 ft (1.5 m) or fraction thereof (of separate and continuous lengths) are considered one outlet of no less than 180 volt-amperes ≫ *220.14(H)(1)* ≪. The number of receptacles within the 5-ft (1.5-m) measurement is irrelevant.

B Where simultaneous use of appliances is likely, each 1 ft (300 mm) or fraction thereof shall be considered one outlet of no less than 180 volt-amperes ≫ *220.14(H)(2)* ≪.

A

2 Fixed Multioutlet Assemblies (noncontinuous) *220.14(H)* *Where not likely to be used simultaneously, multiply each 5-ft section by 180 volt-amperes.* *Where likely to be used simultaneously, multiply each 1-ft section by 180 volt-amperes.*	*2*

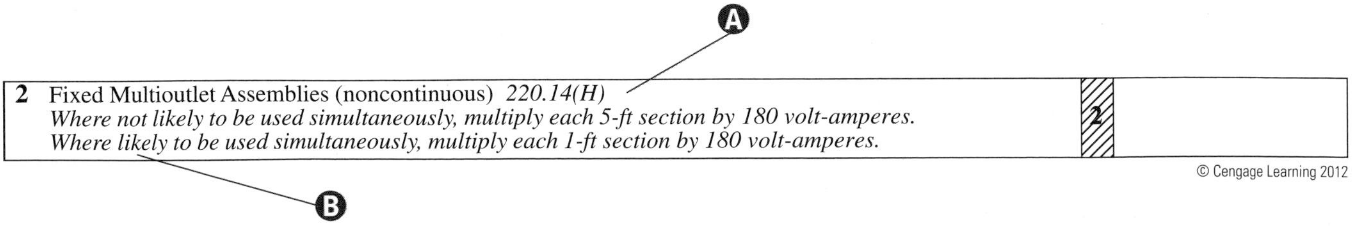

© Cengage Learning 2012

B

Nondwelling Feeder/Service Load Calculation

1 Receptacle Load (noncontinuous) *220.14(I)* *Multiply each single or multiple receptacle on one strap by 180 volt-amperes.* *Multiply each single piece of equipment comprised of four or more receptacles by 90 VA per receptacle.*	**1**	
2 Fixed Multioutlet Assemblies (noncontinuous) *220.14(H)* *Where not likely to be used simultaneously, multiply each 5-ft section by 180 volt-amperes.* *Where likely to be used simultaneously, multiply each 1-ft section by 180 volt-amperes.*	**2**	
3 Receptacle Load Demand Factor (for nondwelling receptacles) *220.44* *If the receptacle load is more than 10,000 volt-amperes, apply the demand factor from Table 220.44.* *Add lines 1 and 2. Multiply the first 10 kVA or less by 100%. Then multiply the remainder by 50%.*	**3**	
4 Receptacle Load (banks and office buildings only) *220.14(K)* *(1) Calculate the demand in line 3; (2) multiply 1 volt-ampere per square foot* $1 \times$ _____ = (sq-ft outside dimensions) *and add to line 4; and (3) compare lines 3 and 4 and omit the smaller.*	**4**	
5 General Lighting Load *Table 220.12* *Multiply the volt-ampere unit load (for the type of* _____ (VA unit load) $\times$ _____ (sq-ft outside dimensions) = *occupancy) by the square-foot outside dimensions.*	**5**	
6 Lighting Load Demand Factors *220.42.* *Apply Table 220.42 demand factors to certain portions of* *hospitals, hotels, motels, apartment houses (without provisions for cooking), and storage warehouses.* *Do not include areas in hospitals, hotels, and motels where the entire lighting will be used at one time.*	**6**	
7 Track Lighting (in addition to general lighting) *220.43(B)* *Include 150 volt-amperes for every 2 ft,* _____ (total linear ft) $\div 2$ _____ $\times 150 =$ *or fraction thereof, of lighting track.*	**7**	
8 Sign or Outline Lighting Outlet (where required) *220.14(F)* *Each commercial building (or occupancy) accessible to pedestrians must have at least one* *outlet per tenant space entrance. 600.5(A) Each outlet must be at least 1200 volt-amperes.*	**8**	
9 Show-Window Lighting *220.43(A)* *Include at least 200 volt-amperes for each linear foot,* _____ (total linear ft of show window) $\times 200 =$ *measured horizontally along the show window's base.*	**9**	
10 Continuous Loads *215.2(A), 215.3,* and *230.42(A)* *Multiply the continuous load volt-amperes (listed above) by 25%.* _____ (total continuous load volt-amperes) $\times 25\% =$ *(General-purpose receptacles are not considered continuous.)*	**10**	
11 Kitchen Equipment *220.56* *Multiply three or more pieces of equipment by the Table 220.56 demand factor (%).* Use *Table 220.55* for household cooking equipment used in instructional programs. *Table 220.55*, Note 5	**11**	
12 Noncoincident Loads *220.60.* *The smaller of two (or more) noncoincident loads can be omitted, as* *long as they will never be energized simultaneously (such as certain portions of heating and A/C systems).* *Calculate fixed electric space-heating loads at 100% of the total connected load. 220.51*	**12**	
13 Motor Loads *220.18(A), 430.24, 430.25, 430.26,* and *Article 440* *Motor-driven air-conditioning and refrigeration equipment is found in Article 440.* *Multiply the largest motor (one motor only) by 25% and add it to the load.*	**13**	
14 All Other Loads. *Add all other noncontinuous loads into the calculation at 100%.* *Multiply all other continuous loads (operating for 3 hours or more) by 125%.*	**14**	
15 Total Volt-Ampere Demand Load: *Add lines 3 through 14 to find the minimum required volt-amperes.*	**15**	

16 Minimum Amperes *Divide the total* *volt-amperes by* _____ (line 15) $\div$ _____ (voltage) $=$ *the voltage.*	**16**	_____ (minimum amperes)	**Minimum Size** **17 Service or** **Feeder** *240.6(A)*	**17**
18 Size the Service or Feeder Conductors. *Tables 310.15(B)(16)* through *310.15(B)(21)* *Use the tables along with 310.15(B)(1) through (6) to determine conductor size.* *If the overcurrent device is rated more than 800 amperes, the conductor ampacity* *must be equal to, or greater than, the rating of the overcurrent device. 240.4(C)*		**Minimum Size** **Conductors**	**18**	
19 Size the Neutral Conductor. *220.61* *The neutral service or feeder conductor can be smaller than the ungrounded* *(hot) conductors, but not smaller than the maximum unbalanced load determined* *by Article 220. 250.24(C)(1) states that the grounded conductor shall not be smaller* *than specified in Table 250.102(C)(1). A further demand factor is permitted for any* *neutral load over 200 amperes.*		**Minimum Size** **Neutral** **Conductor**	**19**	
20 Size the Grounding Electrode Conductor (for Service). *250.66* *Use line 18 to find the grounding electrode conductor in Table 250.66.* Size the Equipment Grounding Conductor (for Feeder). *250.122* *Use line 17 to find the equipment grounding conductor in Table 250.122.* *Equipment grounding conductor types are listed in 250.118.*		**Minimum Size** **Grounding Electrode** **Conductor . . . *or* . . .** **Equipment Grounding** **Conductor**	**20**	

Line 3—Receptacle Load Demand Factor

A commercial occupancy has a receptacle load of 15,300 volt-amperes and a fixed multioutlet assembly load of 4860. The total receptacle load is 20,160 volt-amperes. First, subtract 10,000 from the total receptacle load (20,160 − 10,000 = 10,160). Next, multiply the remainder by 50% (10,160 × 50% = 5080). Finally, add the result back to the original 10,000 volt-amperes (5080 + 10,000 = 15,080). The receptacle load, after demand, is then entered into line 3.

A Receptacle loads in nondwelling units are calculated at no more than 180 volt-amperes per outlet in accordance with 220.14(I), and fixed multioutlet assemblies are calculated per 220.14(H). Both can be added to the lighting load and made subject to the demand factors given in Table 220.42, or they can be made subject to Table 220.44 demand factors ≫ 220.44 ≪.

B If the total receptacle load is no more than 10,000 volt-amperes, insert the number directly in line 3.

C If the total receptacle load is more than 10,000 volt-amperes, subtract 10,000 from the total, and multiply the remainder by 50%. Finally, add that number to the original 10,000 and place the total in line 3.

> **N O T E**
>
> General-purpose receptacles are not considered continuous loads.

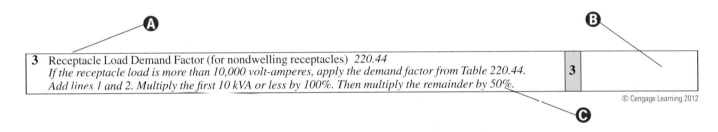

3   Receptacle Load Demand Factor (for nondwelling receptacles) *220.44*
If the receptacle load is more than 10,000 volt-amperes, apply the demand factor from Table 220.44.
Add lines 1 and 2. Multiply the first 10 kVA or less by 100%. Then multiply the remainder by 50%.   **3**

© Cengage Learning 2012

Line 4—Receptacle Load (Banks and Office Buildings Only)

It is not known how many receptacles are in an 8500-ft² bank. A load of 8500 volt-amperes should be placed in line 4 (8500 × 1 = 8500). Since the actual number of receptacle outlets is unknown, line 3 would not contain a number.

A In banks or office buildings, the receptacle load shall be calculated to be the larger of (1) the calculated load from 220.14(I), or (2) 1 volt-ampere per square foot ≫ 220.14(K) ≪.

B First, calculate the receptacle load in line 3. Next, multiply the square foot dimensions by 1 and insert in line 4. Finally, compare the two and omit the smaller.

C If the actual number of general-purpose receptacle outlets is unknown, multiply the square foot dimension by 1 and insert in line 4.

> **CAUTION**
>
> Line 4 applies only to bank and office building occupancies.

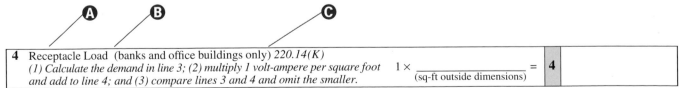

4   Receptacle Load (banks and office buildings only) *220.14(K)*
(1) Calculate the demand in line 3; (2) multiply 1 volt-ampere per square foot   1 × _____ = **4**
and add to line 4; and (3) compare lines 3 and 4 and omit the smaller.   (sq-ft outside dimensions)

© Cengage Learning 2012

Line 5—General Lighting Load

An 8500-ft² bank has a general lighting load of 29,750 volt-amperes (8500 × 3.5 = 29,750). The receptacle load, whether known or unknown, is not used here.

A 30,000-ft² store has a general lighting load of 90,000 volt-amperes (30,000 × 3 = 90,000).

A A unit load that meets or exceeds that specified in Table 220.12 for listed occupancies constitutes the minimum lighting load for each square foot of floor area ⟩⟩ *220.12* ⟨⟨ .

> **NOTE**
>
> The exception for using the minimum volt-ampere unit load in Table 220.12 states where the building is designed and constructed to comply with an energy code adopted by the local authority, the lighting load shall be permitted to be calculated at the values specified in the energy code where the following conditions are met:
> (1) A power monitoring system is installed that will provide continuous information regarding the total general lighting load of the building.
> (2) The power monitoring system will be set with alarm values to alert the building owner or manager if the lighting load exceeds the values set by the energy code.
> (3) The demand factors specified in 220.42 are not applied to the general lighting load.

B Find the correct volt-ampere unit load located across from the occupancy type, and insert it into the calculation ⟩⟩ *Table 220.12* ⟨⟨ .

C Each floor's area shall be calculated using the building's (or area's) outside dimensions ⟩⟩ *220.12* ⟨⟨ .

> **NOTE**
>
> In all Table 220.12 occupancies (except one-family dwellings and individual dwelling units of two-family and multifamily dwellings), specific areas can be separately multiplied by different volt-ampere unit loads. For example, assembly halls and auditoriums have a unit load of 1; halls, corridors, closets, and stairways have a unit load of 0.5; and storage spaces have a unit load of 0.25 volt-amperes per square foot.

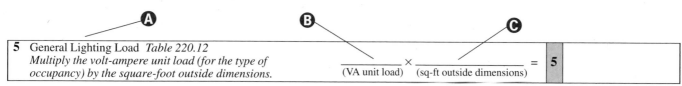

| 5 | General Lighting Load *Table 220.12*
 Multiply the volt-ampere unit load (for the type of
 occupancy) by the square-foot outside dimensions. | $\dfrac{}{\text{(VA unit load)}} \times \dfrac{}{\text{(sq-ft outside dimensions)}} =$ | **5** | |

© Cengage Learning 2012

Line 6—Lighting Load Demand Factors

After demand, the general lighting load for hospital patients' rooms where the room dimensions total 100,000 ft² is 50,000 volt-amperes. First, multiply 100,000 by the hospital volt-ampere unit load (2), found in Table 220.12 (100,000 × 2 = 200,000). Next, multiply the first 50,000 by 40% (50,000 × 40% = 20,000). Then, multiply the remainder by 20% (200,000 − 50,000 = 150,000 × 20% = 30,000). Finally, add the two figures (20,000 + 30,000 = 50,000).

A Demand factors apply to certain areas in hospitals, hotels, motels, apartment houses (without cooking provisions), and warehouses.

B Do not include areas in hospitals, hotels, and motels where all lighting is subject to simultaneous use. Primarily, such lighting is considered a continuous load. Continuous loads must not be derated by Table 220.42, but, instead, are increased by 25% (line 10).

> **NOTE**
>
> Do not use Table 220.42 to determine the total number of branch circuits ⟩⟩ *220.42* ⟨⟨ .

| 6 | Lighting Load Demand Factors *220.42. Apply Table 220.42 demand factors to certain portions of*
 hospitals, hotels, motels, apartment houses (without provisions for cooking), and storage warehouses.
 Do not include areas in hospitals, hotels, and motels where the entire lighting will be used at one time. | **6** | |

© Cengage Learning 2012

Line 7—Track Lighting

A store having 80 ft of lighting track has an additional lighting load of 6000 volt-amperes. First divide the total length of lighting track by two (80 ÷ 2 = 40). Then multiply the result by 150 volt-amperes (40 × 150 = 6000).

A Do not consider track lighting in dwelling unit (or hotel/motel guest room or suite) service or feeder load calculations 》220.43(B) 《.

B Track lighting is calculated in addition to the occupancy general lighting load, found in Table 220.12.

C Multiply each 2-ft section, or fraction thereof, by 150 volt-amperes 》220.43(B) 《.

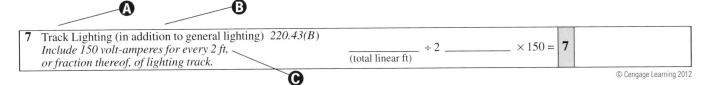

| 7 | Track Lighting (in addition to general lighting) 220.43(B) *Include 150 volt-amperes for every 2 ft, or fraction thereof, of lighting track.* | _____ ÷ 2 _____ × 150 = (total linear ft) | **7** | |

© Cengage Learning 2012

Line 8—Sign or Outline Lighting

A Each commercial building (or occupancy) open to pedestrians shall be provided with at least one outlet in a location accessible to each tenant space entrance for sign or outline lighting system use 》600.5(A) 《.

B Each sign and outline lighting outlet is computed at a minimum of 1200 volt-amperes 》220.14(F) 《.

> **NOTE**
>
> Because service hallways or corridors are not considered accessible to pedestrians, no outlet is required 》**600.5(A)《**.

> **CAUTION**
>
> If the rating is more than 1200 volt-amperes, use the actual rating of the sign or outline lighting.

| 8 | Sign or Outline Lighting Outlet (where required) 220.14(F) *Each commercial building (or occupancy) accessible to pedestrians must have at least one outlet per tenant space entrance. 600.5(A) Each outlet must be at least 1200 volt-amperes.* | **8** | |

© Cengage Learning 2012

Line 9—Show-Window Lighting

A store having 75 linear ft of show-window space has a feeder/service load of 15,000 volt-amperes (75 × 200 = 15,000).

A If an occupancy has show window(s), include at least 200 volt-amperes for each linear foot, measured horizontally along the show window's base 》220.43(A) 《.

> **NOTE**
>
> Calculate show-window branch circuits in accordance with 220.14(G).

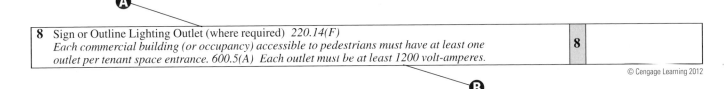

| 9 | Show-Window Lighting 220.43(A) *Include at least 200 volt-amperes for each linear foot, measured horizontally along the show window's base.* | _____ × 200 = (total linear ft of show window) | **9** | |

© Cengage Learning 2012

Line 10—Continuous Loads

Ⓐ A load where the maximum current is expected to continue for three hours or more is a continuous load ≫ *Article 100* ≪.

Ⓑ Feeder conductors and overcurrent protection shall be sized at 100% of noncontinuous loads *plus* 125% of continuous loads ≫ *215.2(A)* and *215.3* ≪. In other words, an additional 25% of the volt-ampere rating must be added to the continuous loads. This form simplifies that computation. Add all of the continuous loads (listed above line 10) and multiply by 25%. This rule also applies to service-entrance conductors ≫ *230.42(A)* ≪.

Ⓒ General-purpose receptacles and multioutlet assemblies are not included in this calculation.

Ⓓ Total all of the *continuous* loads listed in lines 5 through 9. These lines are not automatically considered continuous. Only if the load is expected to continue for three hours or more is it a continuous load. Examples may include, but are not limited to, general lighting, track lighting, show-window lighting, signs, and outline lighting.

> **N O T E**
>
> Feeder conductors shall have an ampacity not less than required to supply the load as calculated in Parts III, IV, and V of Article 220. Conductors shall be sized to carry not less than the larger of 215.2(A)(1)(a) or (b).
> (a) Where a feeder supplies continuous loads or any combination of continuous and noncontinuous loads, the minimum feeder conductor size shall have an allowable ampacity not less than the noncontinuous load plus 125 percent of the continuous load.
> (b) The minimum feeder conductor size shall have an allowable ampacity not less than the maximum load to be served after the application of any adjustment or correction factors ≫ *215.2(A)* ≪.

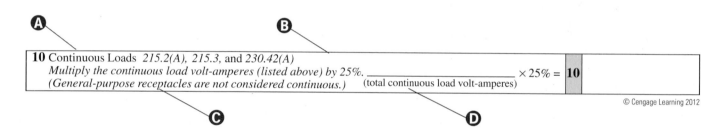

© Cengage Learning 2012

Line 11—Kitchen Equipment

Ⓐ Calculate the load for commercial electric cooking equipment, dishwasher booster heaters, water heaters, and other kitchen equipment in accordance with Table 220.56. Apply the demand factors to all equipment that (1) has thermostatic control or (2) is intermittently used as kitchen equipment ≫ *220.56* ≪.

Ⓑ Table 220.56 is relatively simple. No derating is allowed for only one or two pieces of equipment. The demand factor for three pieces of equipment is 90%; for four pieces, 80%; and for five pieces, 70%. The demand factor for more than five pieces of equipment is 65%.

Ⓒ For household cooking appliances rated over 1¾ kW and used in instructional programs, refer to Table 220.55 ≫ *Table 220.55, Note 5* ≪. For example, in a school home-economics classroom, a feeder supplying 20 12-kW household electric ranges requires a minimum rating of only 35 kW ≫ *Table 220.55, Column C* ≪.

> **N O T E**
>
> Table 220.56 does not apply to space-heating, ventilating, or air-conditioning equipment.

> **CAUTION**
>
> The feeder or service calculated load can never be smaller than the combined rating of the two largest kitchen equipment loads ≫ *220.56* ≪.

Ⓐ **Ⓑ**

11 Kitchen Equipment *220.56*		
Multiply three or more pieces of equipment by the Table 220.56 demand factor (%).	**11**	
Use *Table 220.55* for household cooking equipment used in instructional programs. *Table 220.55*, Note 5		

Ⓒ

© Cengage Learning 2012

Line 12—Noncoincident Loads

Ⓐ Where it is unlikely that multiple noncoincident loads will be used simultaneously, use only the largest load(s) that will be used at one time in calculating the total feeder/service load ≫ *220.60* ≪.

Ⓑ While fixed electric space-heating loads are calculated at 100% of the total connected load, in no case shall a feeder or service load current rating be less than the rating of the largest branch circuit supplied ≫ *220.51* ≪.

> **N O T E**
>
> Where reduced loading of the conductors results from units operating on duty cycle or intermittently, or from all units not operating at one time, the AHJ may grant permission for feeder and service conductors to have an ampacity less than 100%, provided the conductors have an ampacity for the load so determined ≫ *220.51 Exception* ≪.

12 Noncoincident Loads *220.60. The smaller of two (or more) noncoincident loads can be omitted, as long as they will never be energized simultaneously (such as certain portions of heating and A/C systems). Calculate fixed electric space-heating loads at 100% of the total connected load. 220.51* | **12** |

© Cengage Learning 2012

Line 13—Motor Loads

Where one or more of the motors of the group are used for short-time, intermittent, periodic, or varying duty, the ampere rating of such motors used in the summation shall be determined in accordance with 430.22(E). For the highest-rated motor, the greater of either the ampere rating from 430.22(E) or the largest continuous-duty motor full-load current multiplied by 1.25 shall be used in the summation ≫ *430.24 Exception No. 1* ≪.

Where interlocked circuitry prevents simultaneous operation of selected motors or other loads, the conductor ampacity shall be permitted to be based on the summation of the currents of the motors and other loads being operated simultaneously resulting in the highest total current ≫ *430.24 Exception No. 3* ≪.

Ⓐ The ampacity of the conductors supplying multimotor and combination-load equipment shall not be less than the minimum circuit ampacity marked on the equipment per 430.7(D). If the individual equipment nameplates are visible

(per 430.7[D][2]), but the equipment is not factory wired, use 430.24 to determine conductor ampacity ≫ *430.25* ≪.

Ⓑ Where conductor heating is reduced as a result of motors operating on duty cycle, intermittently, or from all motors not operating at one time, the AHJ may grant permission for feeder conductors to have an ampacity less than specified in 430.24, provided the conductors have sufficient ampacity for the number of motors supplied and the nature of their loads and duties ≫ *430.26* ≪.

Ⓒ Conductors supplying several motors or a motor(s) and other load(s) must have an ampacity not less than the sum of each of the following:
(1) 125% of the full-load current rating of the highest rated motor, as determined by 430.6(A)
(2) Sum of the full-load current ratings of all the other motors in the group, as determined by 430.6(A)
(3) 100% of the noncontinuous nonmotor load
(4) 125% of the continuous nonmotor load ≫ *430.24* ≪.

13 Motor Loads *220.18(A), 430.24, 430.25, 430.26, and Article 440*
Motor-driven air-conditioning and refrigeration equipment is found in Article 440.
Multiply the largest motor (one motor only) by 25% and add it to the load. | **13** |

© Cengage Learning 2012

Line 14—All Other Loads

Ⓐ This line is a catchall for loads not falling into one of the previous categories.

Ⓑ Calculate all noncontinuous loads (not yet input) at 100% of their volt-ampere rating.

Ⓒ Total all continuous loads (not yet calculated) and multiply by 125%. Add the result of the continuous loads to that for noncontinuous loads and put the total on line 14.

14 All Other Loads.
Add all other noncontinuous loads into the calculation at 100%.
Multiply all other continuous loads (operating for 3 hours or more) by 125%. | **14** |

© Cengage Learning 2012

Line 15—Total Volt-Ampere Demand Load

A Add all volt-ampere loads listed in lines 3 through 14 and place the result in line 15.

A

| **15** Total Volt-Ampere Demand Load: *Add lines 3 through 14 to find the minimum required volt-amperes.* | **15** | |

© Cengage Learning 2012

Lines 16 and 17—Minimum Size Service or Feeder

A Place the total volt-ampere amount found in line 15 here.

B Write down the source voltage that supplies the feeder/service.

C Fractions of an ampere 0.5 and higher are rounded up, while fractions less than 0.5 are dropped ≫*220.5(B)*≪.

D The service (or feeder) overcurrent protection must be higher than the number found in line 16. Refer to 240.6(A) for a list of fuse and circuit breaker standard ampere ratings.

E Divide the volt-amperes by the voltage to find the amperage.

F Three-phase voltage is found by multiplying the voltage (single-phase) by 1.732. For example, the source voltage for 208-volt, 3-phase is 360 (208 × 1.732).

G The result on line 16 is the minimum amperage rating required for the service or feeder being calculated.

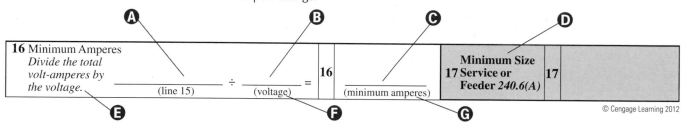

© Cengage Learning 2012

Line 18—Minimum Size Conductors

Conductors 1/0 AWG and larger can be connected in parallel (electrically joined at both ends to form a single conductor) ≫*310.10(H)(1)*≪.

The number and type of paralleled conductor sets is a design consideration, not necessarily a *Code* issue. Exercise extreme care when designing a paralleled conductor installation without violating provisions such as 110.14(A), 110.14(C), 240.4(C), 250.66, 250.122(F), 300.20(A), 310.10(H), etc.

A Using Tables 310.15(B)(16) through 310.15(B)(21), choose a conductor size that equals (or exceeds) line 16's minimum ampacity rating. The conductor's ampacity rating does not have to equal or exceed the overcurrent device rating unless its rating exceeds 800 amperes ≫*240.4(B)*≪.

B Ampacity adjustment (correction) factors for more than three current-carrying conductors are located in Table 310.15(B)(3)(a).

A

| **18** Size the Service or Feeder Conductors. *Tables 310.15(B)(16) through 310.15(B)(21)* *Use the tables along with 310.15(B)(1) through (6) to determine conductor size.* *If the overcurrent device is rated more than 800 amperes, the conductor ampacity* *must be equal to, or greater than, the rating of the overcurrent device. 240.4(C)* | **Minimum Size Conductors** | **18** | |

B

© Cengage Learning 2012

Line 19—Neutral Conductor

Include all lighting and receptacle loads having a 120-volt rating in the neutral calculation.

All appliances, equipment, motors, etc. utilizing a grounded conductor are included in the neutral calculation.

A For the purpose of this load calculation, the terms **neutral** and **grounded** are synonymous.

B The feeder (or service) neutral load is the maximum unbalance of the load determined by Article 220 ›› *220.61* ‹‹ .

C In addition to 220.61 demand factors, a 70% demand factor is permitted for that portion of the unbalanced load above 200 amperes ›› *220.61(B)* ‹‹ .

D The neutral is sized by finding the maximum unbalanced load.

> **N O T E**
>
> The grounded conductor shall not be smaller than specified in Table 250.102(C)(1) ››*250.24(C)(1)*‹‹ .

> **CAUTION**
>
> Reduction of neutral capacity is not permitted for that portion of the load consisting of nonlinear loads supplied from a 4-wire, wye-connected, 3-phase system nor the grounded conductor of a 3-wire circuit consisting of 2-phase wires and the neutral of a 4-wire, 3-phase, wye-connected system ››*220.61(C)*‹‹ .

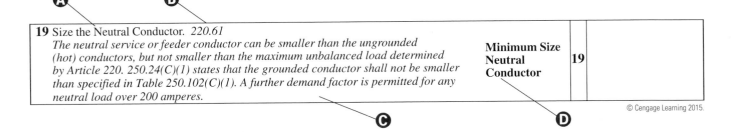

19 Size the Neutral Conductor. *220.61*
The neutral service or feeder conductor can be smaller than the ungrounded (hot) conductors, but not smaller than the maximum unbalanced load determined by Article 220. 250.24(C)(1) states that the grounded conductor shall not be smaller than specified in Table 250.102(C)(1). A further demand factor is permitted for any neutral load over 200 amperes.

Minimum Size Neutral Conductor | **19** |

© Cengage Learning 2015.

Line 20—Grounding Conductor

Where the grounding electrode conductor connects to a single or multiple rod, pipe, or plate electrode(s) or any combination thereof, and that conductor portion is the sole grounding electrode connection, the maximum size required is 6 AWG copper or 4 AWG aluminum ›› *250.66(A)* ‹‹ .

For a grounding electrode conductor connected to a single or multiple concrete-encased electrode(s) and serving as the only grounding electrode connection, the maximum size required is 4 AWG copper ›› *250.66(B)* ‹‹ .

If the grounding electrode conductor is connected to a ground ring and that conductor portion is the sole connection to the grounding electrode, the maximum size required is the size used for the ground ring ›› *250.66(C)* ‹‹ .

A Grounding electrode conductor size is based on the largest service-entrance conductor or equivalent area for parallel conductors ›› *Table 250.66* ‹‹ . (The minimum size service-entrance conductors were determined on line 18.)

B While Table 250.66 is used for service installations, it is also used in other wiring applications, such as separately derived systems ›› *250.30(A)(2)* ‹‹ and connections at separate buildings or structures ›› *250.32* ‹‹ .

C Equipment bonding jumpers on the load side of an overcurrent device(s) shall be sized in accordance with Table 250.122 ›› *250.102(D)* ‹‹ .

D The equipment grounding conductor size is based on the rating (or setting) of the circuit's automatic overcurrent device ahead of equipment, conduit, and so on ›› *Table 250.122* ‹‹ .

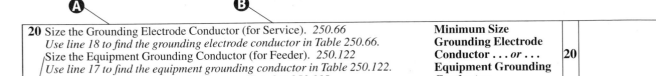

20 Size the Grounding Electrode Conductor (for Service). *250.66*
Use line 18 to find the grounding electrode conductor in Table 250.66.
Size the Equipment Grounding Conductor (for Feeder). *250.122*
Use line 17 to find the equipment grounding conductor in Table 250.122.
Equipment grounding conductor types are listed in 250.118.

Minimum Size Grounding Electrode Conductor . . . or . . . Equipment Grounding Conductor | **20** |

© Cengage Learning 2012

SAMPLE LOAD CALCULATION—STORE

Load Calculation Information

Ⓐ This is the load calculation data for a small retail store.

Ⓑ Calculate the floor area from the outside dimensions. In this example, the square-foot area is already provided.

Ⓒ A 3-phase transformer steps down the voltage from 480 to 208Y/120 volts.

Ⓐ   **Ⓑ**

Store's area (in ft²) . . . . .1200 ft²
Receptacles 10
Lighting track. . . . . .80 ft
Two sign circuits . . . . .14.5 amperes, 120 volt (each)
Show window . . . . .32 ft
Water heater . . . . .2 kW, 120 volt
Electric heat . . . . .15 kW, 480 volt, 3-phase
Air handler (blower motor). . . . . . .1.8 amperes, 480 volt, 3-phase
Air-conditioner compressor . . . . . .8.5 amperes, 480 volt, 3-phase
Condenser fan motor . . . . .0.8 amperes, 480 volt, 3-phase

Assume water heater and electric heat kW ratings equivalent to kVA.

The electric supply consists of a 480-volt, 3-phase, 4-wire service.

Ⓒ

© Cengage Learning 2012

Lines 1 and 2

Ⓐ The different style of box on line 1 (unlike lines 3 through 14) indicates that the load will become part (if not all) of line 3. Line 1 is not included in line 15's total volt-ampere demand load.

Ⓑ Since there are ten receptacles, the calculated load is 1800 volt-amperes (10 × 180).

Ⓒ As with line 1, line 2 has a different style box from lines 3 through 14 because the load (if any) becomes part of line 3. Line 2 is not included in line 15's total volt-ampere demand load.

Ⓓ Because this store has no fixed multioutlet assembly, a dashed line occupies line 2.

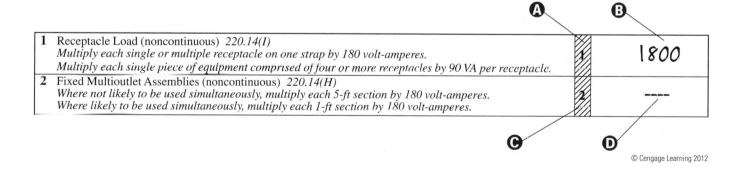

1	Receptacle Load (noncontinuous) *220.14(I)*	1	1800
	Multiply each single or multiple receptacle on one strap by 180 volt-amperes.		
	Multiply each single piece of equipment comprised of four or more receptacles by 90 VA per receptacle.		
2	Fixed Multioutlet Assemblies (noncontinuous) *220.14(H)*	2	----
	Where not likely to be used simultaneously, multiply each 5-ft section by 180 volt-amperes.		
	Where likely to be used simultaneously, multiply each 1-ft section by 180 volt-amperes.		

© Cengage Learning 2012

Lines 3 and 4

Ⓐ Line 3 is the first of 12 lines whose values are combined to determine the total volt-ampere demand load.

Ⓑ Because there is no multioutlet assembly and the receptacle load is less than 10 kVA, the 1800 volt-ampere load from line 1 is inserted in line 3.

Ⓒ The dashed line indicates line 4 is intentionally blank, because it only applies to banks and office buildings.

Ⓓ The dashed line indicates line 4 is intentionally blank, because it only applies to banks and office buildings.

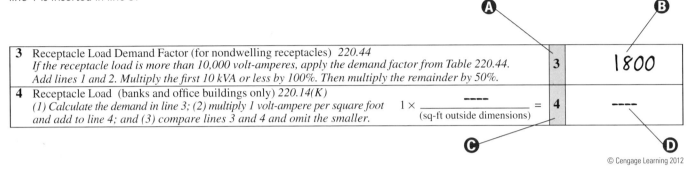

© Cengage Learning 2012

Lines 5 and 6

Ⓐ The volt-ampere unit load for a store is 3 ≫ *Table 220.12* ≪.

Ⓑ Since the store's square-foot area is provided, no calculation is required.

Ⓒ Store lighting loads are continuous. The result of line 5 becomes part of the continuous volt-ampere load calculation (line 10).

Ⓓ Use line 6 only for certain portions of hospitals, hotels, motels, storage warehouses, and apartment houses without provisions for cooking.

Ⓔ The marking in line 6 indicates that it has intentionally been left blank.

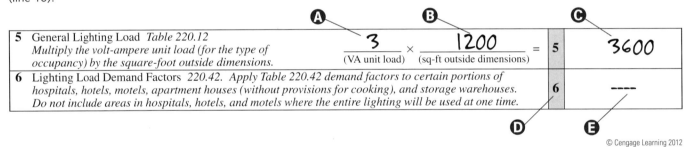

© Cengage Learning 2012

Lines 7 and 8

Ⓐ Except for dwelling units or hotel/motel guest rooms, track lighting loads are additional to Table 220.12's general lighting loads ≫ *220.43(B)* ≪.

Ⓑ This store contains 80 ft of lighting track.

Ⓒ Base the calculation on every 2 ft, or fraction thereof, of track.

Ⓓ A store's track lighting load is continuous. Line 7's result becomes part of the continuous volt-ampere load calculation (line 10).

Ⓔ At least one outlet having a minimum rating of 1200 volt-amperes is required.

Ⓕ This store's actual sign load is 3420 volt-amperes (2 × 14.5 × 120). This store has two sign circuits.

Ⓖ Since a store's sign load is continuous, the result of line 8 becomes part of the continuous volt-ampere load calculation (line 10).

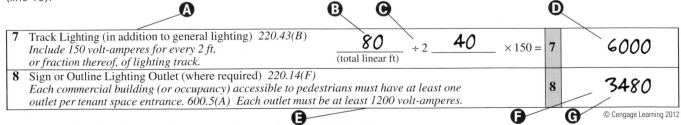

© Cengage Learning 2012

Line 9

(A) Not all commercial occupancies have show windows.

(B) The store in this example has 32 linear ft of show window.

(C) Because show-window load is continuous, the result of line 9 becomes part of the continuous volt-ampere load calculation (line 10).

9 Show-Window Lighting *220.43(A)* *Include at least 200 volt-amperes for each linear foot,* *measured horizontally along the show window's base.*	$\frac{32}{\text{(total linear ft of show window)}} \times 200 =$	**9**	**6400**

© Cengage Learning 2012

Lines 10 and 11

(A) Lines 5, 7, 8, and 9 represent the store's continuous loads.

(B) 3600 + 6000 + 3480 + 6400 = 19,480

(C) An additional 25% volt-ampere load has been added because of the continuous loads.

(D) The marking in line 11 indicates a lack of kitchen equipment.

10 Continuous Loads *215.2(A), 215.3,* and *230.42(A)* *Multiply the continuous load volt-amperes (listed above) by 25%.* *(General-purpose receptacles are not considered continuous.)*	$\frac{19,480}{\text{(total continuous load volt-amperes)}} \times 25\% =$	**10**	**4870**
11 Kitchen Equipment *220.56* *Multiply three or more pieces of equipment by the Table 220.56 demand factor (%).* *Use Table 220.55 for household cooking equipment used in instructional programs. Table 220.55, Note 5*		**11**	----

© Cengage Learning 2012

Lines 12 and 13

(A) Because the combined electric heat and air-handler load is larger than the combined compressor, condenser fan motor, and air handler, omit the compressor and condenser fan motor load.

(B) Find the largest motor. In this example, that would be the air-handler motor (1.8 × 480 × 1.732 = 1,496.45). Increase the largest motor by 25% (1,496.45 × 25% = 374).

12 Noncoincident Loads *220.60. The smaller of two (or more) noncoincident loads can be omitted, as* *long as they will never be energized simultaneously (such as certain portions of heating and A/C systems).* *Calculate fixed electric space-heating loads at 100% of the total connected load. 220.51*	**12**	**16,456**
13 Motor Loads *220.18(A), 430.24, 430.25, 430.26,* and *Article 440* *Motor-driven air-conditioning and refrigeration equipment is found in Article 440.* *Multiply the largest motor (one motor only) by 25% and add it to the load.*	**13**	**374**

© Cengage Learning 2012

Lines 14 and 15

(A) The only load not previously calculated is the water heater. A water heater is a noncontinuous load when calculating a feeder or service. Add noncontinuous loads to the calculation at 100%.

(B) Add lines 3 through 14 to find the minimum volt-ampere load (lines 4, 6, and 11 are blank).

> **WARNING**
>
> Although, under normal conditions, a water heater is not considered a continuous load when calculating a feeder or service, it must be considered a continuous load when calculating the branch circuit. A fixed storage-type water heater that has a capacity of 120 gal (450 L) or less shall be considered a continuous load for the purposes of sizing branch circuits » *422.13* «.

14 All Other Loads. *Add all other noncontinuous loads into the calculation at 100%.* *Multiply all other continuous loads (operating for 3 hours or more) by 125%.*	**14**	**2000**
15 Total Volt-Ampere Demand Load: *Add lines 3 through 14 to find the minimum required volt-amperes.*	**15**	**44,980**

© Cengage Learning 2012

Lines 16 and 17

Ⓐ Insert the volt-ampere load from line 15 on this line.

Ⓑ The total voltage for a 480-volt, 3-phase system is 831 (480 × 1.732).

Ⓒ The minimum ampacity for a 480-volt, 3-phase service is 54 amperes.

Ⓓ The next-higher standard size fuse (or breaker) above 54 amperes is 60 》 *240.6(A)* 《.

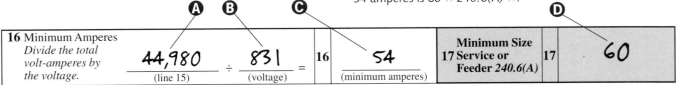

16 Minimum Amperes *Divide the total volt-amperes by the voltage.*	**44,980** (line 15)	÷	**831** (voltage)	=	16	**54** (minimum amperes)	**Minimum Size 17 Service or Feeder** *240.6(A)*	17	**60**

© Cengage Learning 2012

Line 18

Ⓐ The minimum 60/75°C (140/167°F) copper service conductor is 6 AWG 》 *Table 310.15(B)(16)* 《.

Ⓑ The minimum 60/75°C (140/167°F) aluminum service conductor is 4 AWG 》 *Table 310.15(B)(16)* 《.

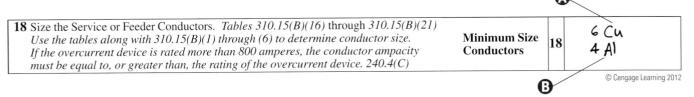

18 Size the Service or Feeder Conductors. *Tables 310.15(B)(16) through 310.15(B)(21)* *Use the tables along with 310.15(B)(1) through (6) to determine conductor size.* *If the overcurrent device is rated more than 800 amperes, the conductor ampacity must be equal to, or greater than, the rating of the overcurrent device. 240.4(C)*	**Minimum Size Conductors**	18	**6 Cu** **4 Al**

© Cengage Learning 2012

Line 19

Ⓐ Although there will be little (if any) load on the neutral, a neutral conductor is required. Where an ac system (operating at less than 1000 volts) is grounded at any point, the grounded conductor(s) shall be routed with the ungrounded conductors to each service disconnecting means and shall be connected to each disconnecting means grounded conductor(s) terminal or bus 》 *250.24(C)* 《.

Ⓑ The grounded conductor shall not be smaller than specified in Table 250.102(C)(1) 》 *250.24(C)(1)* 《.

Ⓒ In accordance with Table 250.102(C)(1), the minimum size grounded conductor for 2 AWG (or smaller) copper ungrounded conductors is 8 AWG copper.

Ⓓ In accordance with Table 250.102(C)(1), the minimum size grounded conductor for 1/0 AWG (or smaller) aluminum ungrounded conductors is 6 AWG aluminum.

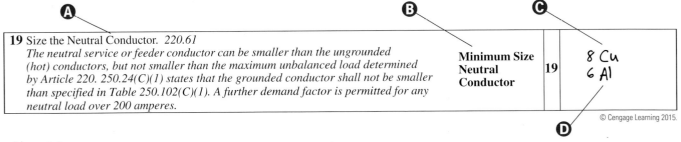

19 Size the Neutral Conductor. *220.61* *The neutral service or feeder conductor can be smaller than the ungrounded (hot) conductors, but not smaller than the maximum unbalanced load determined by Article 220. 250.24(C)(1) states that the grounded conductor shall not be smaller than specified in Table 250.102(C)(1). A further demand factor is permitted for any neutral load over 200 amperes.*	**Minimum Size Neutral Conductor**	19	**8 Cu** **6 Al**

© Cengage Learning 2015.

Line 20

Ⓐ The grounding electrode conductor's size is based on the service-entrance conductor 》 *Table 250.66* 《.

Ⓑ The minimum size copper grounding conductor is 8 AWG.

Ⓒ The minimum size aluminum grounding conductor is 6 AWG.

20 Size the Grounding Electrode Conductor (for Service). *250.66* *Use line 18 to find the grounding electrode conductor in Table 250.66.* Size the Equipment Grounding Conductor (for Feeder). *250.122* *Use line 17 to find the equipment grounding conductor in Table 250.122.* *Equipment grounding conductor types are listed in 250.118.*	**Minimum Size Grounding Electrode Conductor . . . or . . . Equipment Grounding Conductor**	20	**8 Cu** **6 Al**

© Cengage Learning 2012

The following page shows the store's completed load calculation.

Nondwelling Feeder/Service Load Calculation

1	Receptacle Load (noncontinuous) *220.14(I)* *Multiply each single or multiple receptacle on one strap by 180 volt-amperes.* *Multiply each single piece of equipment comprised of four or more receptacles by 90 VA per receptacle.*	**1**	1800
2	Fixed Multioutlet Assemblies (noncontinuous) *220.14(H)* *Where not likely to be used simultaneously, multiply each 5-ft section by 180 volt-amperes.* *Where likely to be used simultaneously, multiply each 1-ft section by 180 volt-amperes.*	**2**	----
3	Receptacle Load Demand Factor (for nondwelling receptacles) *220.44* *If the receptacle load is more than 10,000 volt-amperes, apply the demand factor from Table 220.44.* *Add lines 1 and 2. Multiply the first 10 kVA or less by 100%. Then multiply the remainder by 50%.*	**3**	1800
4	Receptacle Load (banks and office buildings only) *220.14(K)* *(1) Calculate the demand in line 3; (2) multiply 1 volt-ampere per square foot* $1 \times \underline{\quad ---- \quad}$ = *and add to line 4; and (3) compare lines 3 and 4 and omit the smaller.* (sq-ft outside dimensions)	**4**	----
5	General Lighting Load *Table 220.12* *Multiply the volt-ampere unit load (for the type of* $\underline{\quad 3 \quad} \times \underline{\quad 1200 \quad}$ = *occupancy) by the square foot outside dimensions.* (VA unit load) (sq-ft outside dimensions)	**5**	3600
6	Lighting Load Demand Factors *220.42.* *Apply Table 220.42 demand factors to certain portions of hospitals, hotels, motels, apartment houses (without provisions for cooking), and storage warehouses.* *Do not include areas in hospitals, hotels, and motels where the entire lighting will be used at one time.*	**6**	----
7	Track Lighting (in addition to general lighting) *220.43(B)* *Include 150 volt-amperes for every 2 ft,* $\underline{\quad 80 \quad} \div 2 \underline{\quad 40 \quad} \times 150 =$ *or fraction thereof, of lighting track.* (total linear ft)	**7**	6000
8	Sign or Outline Lighting Outlet (where required) *220.14(F)* *Each commercial building (or occupancy) accessible to pedestrians must have at least one outlet per tenant space entrance. 600.5(A) Each outlet must be at least 1200 volt-amperes.*	**8**	3480
9	Show-Window Lighting *220.43(A)* *Include at least 200 volt-amperes for each linear foot,* $\underline{\quad 32 \quad} \times 200 =$ *measured horizontally along the show window's base.* (total linear ft of show window)	**9**	6400
10	Continuous Loads *215.2(A), 215.3, and 230.42(A)* *Multiply the continuous load volt-amperes (listed above) by 25%.* $\underline{\quad 19,480 \quad} \times 25\% =$ *(General-purpose receptacles are not considered continuous.)* (total continuous load volt-amperes)	**10**	4870
11	Kitchen Equipment *220.56* *Multiply three or more pieces of equipment by the Table 220.56 demand factor (%).* *Use Table 220.55 for household cooking equipment used in instructional programs. Table 220.55, Note 5*	**11**	----
12	Noncoincident Loads *220.60.* *The smaller of two (or more) noncoincident loads can be omitted, as long as they will never be energized simultaneously (such as certain portions of heating and A/C systems). Calculate fixed electric space-heating loads at 100% of the total connected load. 220.51*	**12**	16,456
13	Motor Loads *220.18(A), 430.24, 430.25, 430.26, and Article 440* *Motor-driven air-conditioning and refrigeration equipment is found in Article 440.* *Multiply the largest motor (one motor only) by 25% and add it to the load.*	**13**	374
14	All Other Loads. *Add all other noncontinuous loads into the calculation at 100%.* *Multiply all other continuous loads (operating for 3 hours or more) by 125%.*	**14**	2000
15	Total Volt-Ampere Demand Load: *Add lines 3 through 14 to find the minimum required volt-amperes.*	**15**	44,980

16	Minimum Amperes *Divide the total* *volt-amperes by* $\dfrac{44,980}{\text{(line 15)}} \div \dfrac{831}{\text{(voltage)}} =$ *the voltage.*	**16**	$\dfrac{54}{\text{(minimum amperes)}}$	**Minimum Size** **17 Service or** **Feeder** *240.6(A)*	**17**	60

18	Size the Service or Feeder Conductors. *Tables 310.15(B)(16) through 310.15(B)(21)* *Use the tables along with 310.15(B)(1) through (6) to determine conductor size.* *If the overcurrent device is rated more than 800 amperes, the conductor ampacity must be equal to, or greater than, the rating of the overcurrent device. 240.4(C)*	**Minimum Size** **Conductors** **18**	6 Cu 4 Al
19	Size the Neutral Conductor. *220.61* *The neutral service or feeder conductor can be smaller than the ungrounded (hot) conductors, but not smaller than the maximum unbalanced load determined by Article 220. 250.24(C)(1) states that the grounded conductor shall not be smaller than specified in Table 250.102(C)(1). A further demand factor is permitted for any neutral load over 200 amperes.*	**Minimum Size** **Neutral** **Conductor** **19**	8 Cu 6 Al
20	Size the Grounding Electrode Conductor (for Service). *250.66* *Use line 18 to find the grounding electrode conductor in Table 250.66.* Size the Equipment Grounding Conductor (for Feeder). *250.122* *Use line 17 to find the equipment grounding conductor in Table 250.122.* *Equipment grounding conductor types are listed in 250.118.*	**Minimum Size** **Grounding Electrode** **Conductor . . . or . . .** **Equipment Grounding** **Conductor** **20**	8 Cu 6 Al

SAMPLE LOAD CALCULATION—BANK

Load Calculation Information

A Use the information below to calculate the service load of this bank.

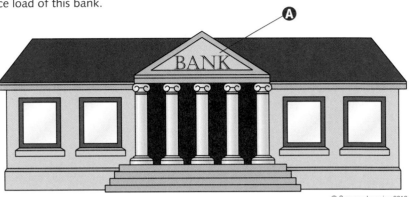

© Cengage Learning 2012

Bank's outside dimensions. . . . . 50′ × 100′
Actual inside connected lighting load . 16,200 volt-amperes
Receptacles . . . . unknown
Sign . . . . 9 amperes, 120 volt
Show window . . . . nonc
Parking lot lighting . . . . 57 amperes, 120 volt
Water heater . . . . 4 kW, 208 volt
Electric heating unit #1 . . . . 6 kW, 208 volt, 3-phase
Electric heating unit #2 . . . . 10 kW, 208 volt, 3-phase
Air handler (blower motor) #1 . . . . 4 amperes, 208 volt, 3-phase
Air handler (blower motor) #2 . . . . 4 amperes, 208 volt, 3-phase
Air-conditioner compressor #1. . . . . . . 22 amperes, 208 volt, 3-phase
Air-conditioner compressor #2. . . . . . . 22 amperes, 208 volt, 3-phase
Condenser fan motor #1 . . . . 2.5 amperes, 208 volt, 3-phase
Condenser fan motor #2 . . . . 2.5 amperes, 208 volt, 3-phase

Assume water heater and electric heat kW ratings equivalent to kVA.

The electric supply consists of a 208-volt, 3-phase, 4-wire service.

> **NOTE**
>
> Because the actual number of receptacles is unknown, line 1 does not contain a number.
>
> Line 2 will not contain a number since the bank has no fixed multioutlet assemblies.

Lines 3 and 4

A Since lines 1 and 2 are blank, line 3 also does not contain a number. A dashed line has been drawn in line 3 showing that it has been left blank intentionally.

B A bank (or office building) with an unknown number of general-purpose receptacle outlets requires a rating of 1 volt-ampere per square foot.

C Since the bank is 5000 ft² (50 × 100), the receptacle load is 5000 volt-amperes.

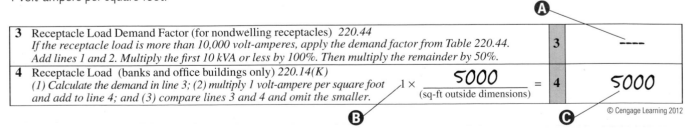

| 3 | Receptacle Load Demand Factor (for nondwelling receptacles) 220.44
If the receptacle load is more than 10,000 volt-amperes, apply the demand factor from Table 220.44.
Add lines 1 and 2. Multiply the first 10 kVA or less by 100%. Then multiply the remainder by 50%. | 3 | - - - - |
| 4 | Receptacle Load (banks and office buildings only) 220.14(K)
(1) Calculate the demand in line 3; (2) multiply 1 volt-ampere per square foot
and add to line 4; and (3) compare lines 3 and 4 and omit the smaller. $1 \times \underline{5000}$ =
(sq-ft outside dimensions) | 4 | 5000 |

© Cengage Learning 2012

Lines 5 and 6

Ⓐ The volt-ampere unit load for a bank is 3.5 »*Table 220.12*«.

Ⓑ This bank, with outside dimensions of 50 ft × 100 ft, has a total calculated area of 5000 ft².

Ⓒ A bank's lighting load is continuous. The result of line 5 becomes part of the continuous volt-ampere load calculation (line 10).

Ⓓ Because line 6 does not apply to this occupancy, a dashed line has been drawn showing that it has been left blank intentionally.

> **N O T E**
>
> Because this bank has no lighting track, calculating line 7 is not necessary.
> Although the actual connected lighting load, excluding exterior lighting, is 16,200 volt-amperes, Table 220.12 requires a minimum lighting load of 17,500 volt-amperes (3.5 × 50 × 100). The greater value of 17,500 must be used in this calculation. Use the actual connected lighting load (if any) only when it is greater than the value computed from Table 220.12. Actual versus computed lighting loads are both continuous and must be increased by 25%. This function is performed in line 10 as a separate calculation. Another example can be found in Informational Annex D, Example D3.

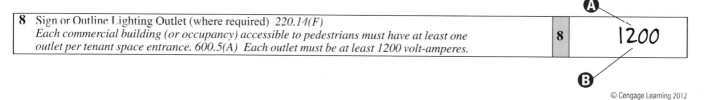

5	General Lighting Load *Table 220.12* *Multiply the volt-ampere unit load (for the type of occupancy) by the square-foot outside dimensions.*	$\dfrac{\textbf{3.5}}{\text{(VA unit load)}} \times \dfrac{\textbf{5000}}{\text{(sq-ft outside dimensions)}} =$ **5**	**17,500**
6	Lighting Load Demand Factors *220.42.* *Apply Table 220.42 demand factors to certain portions of hospitals, hotels, motels, apartment houses (without provisions for cooking), and storage warehouses. Do not include areas in hospitals, hotels, and motels where the entire lighting will be used at one time.*	**6**	----

© Cengage Learning 2012

Line 8

Ⓐ Although the bank's actual sign load is 1080 volt-amperes (9 × 120), each sign outlet must have a minimum rating of 1200 volt-amperes. Use the actual sign or outline load(s) only if the rating is greater than the required minimum.

Ⓑ This bank's sign load is continuous. Therefore, the rating is part of the continuous volt-ampere load calculation (line 10).

> **N O T E**
>
> Because this bank has no show window, a show-window lighting load (line 9) is not required.

8	Sign or Outline Lighting Outlet (where required) *220.14(F)* *Each commercial building (or occupancy) accessible to pedestrians must have at least one outlet per tenant space entrance. 600.5(A) Each outlet must be at least 1200 volt-amperes.*	**8**	**1200**

© Cengage Learning 2012

Line 10

Ⓐ This store's continuous loads are found on lines 5 and 8.

Ⓑ 17,500 + 1200 = 18,700

Ⓒ An additional 25% volt-ampere load is added to the calculation because of the continuous loads.

> **N O T E**
>
> Line 11 contains a dashed line because there is no kitchen equipment.

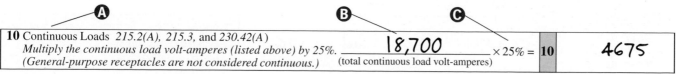

10	Continuous Loads *215.2(A), 215.3,* and *230.42(A)* *Multiply the continuous load volt-amperes (listed above) by 25%.* *(General-purpose receptacles are not considered continuous.)*	$\dfrac{\textbf{18,700}}{\text{(total continuous load volt-amperes)}} \times 25\% =$ **10**	**4675**

© Cengage Learning 2012

Lines 12 and 13

Each air-handler load is 1441 volt-amperes (4 × 208 × 1.732). Each condenser-fan motor load is 901 volt-amperes (2.5 × 208 × 1.732). Each compressor motor load is 7926 volt-amperes (22 × 208 × 1.732).

The total combined compressor, condenser-fan motor, and air-handler load is 20,536 volt-amperes (7926 + 7926 + 901 + 901 + 1441 + 1441).

The total combined electric heat and air-handler load is 18,882 (6000 + 10,000 + 1441 + 1441).

A Because the combined compressor, condenser-fan motor, and air-handler load is larger than the combined electric heat and air handler, omit the electric heat and air-handler load.

B Find the largest motor. In this example, that is the compressor motor (22 × 208 × 1.732 = 7925.63). Increase the largest motor by 25% (7925.63 × 25% = 1981).

12 Noncoincident Loads *220.60. The smaller of two (or more) noncoincident loads can be omitted, as long as they will never be energized simultaneously (such as certain portions of heating and A/C systems). Calculate fixed electric space-heating loads at 100% of the total connected load. 220.51*	**12**	**20,536**
13 Motor Loads *220.18(A), 430.24, 430.25, 430.26,* and *Article 440. Motor-driven air-conditioning and refrigeration equipment is found in Article 440. Multiply the largest motor (one motor only) by 25% and add it to the load.*	**13**	**1981**

© Cengage Learning 2012

Lines 14 and 15

A At this point, two loads have not been calculated: one continuous and one noncontinuous load. Because parking-lot lighting is continuous, multiply by 125% (57 × 120 × 125% = 8550 volt-amperes). The water heater is a noncontinuous load. These two loads have a combined total rating of 12,550 volt-amperes (8550 + 4000).

B Add lines 3 through 14 to determine the minimum volt-ampere load (lines 3, 6, 7, 9, and 11 are blank).

14 All Other Loads. *Add all other noncontinuous loads into the calculation at 100%. Multiply all other continuous loads (operating for 3 hours or more) by 125%.*	**14**	**12,550**
15 Total Volt-Ampere Demand Load: *Add lines 3 through 14 to find the minimum required volt-amperes.*	**15**	**63,442**

© Cengage Learning 2012

Lines 16 and 17

A Insert the total volt-ampere demand load from line 15.

B The total voltage for a 208-volt, 3-phase system is 360 (208 × 1.732).

C The minimum ampacity for a 208-volt, 3-phase service is 176 amperes.

D The next higher standard size fuse (or breaker) above 176 amperes is 200 》240.6(A) 《.

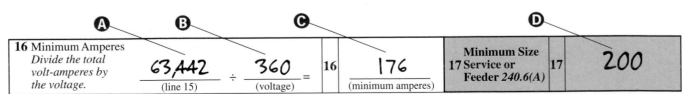

16 Minimum Amperes *Divide the total volt-amperes by the voltage.*	**63,442** (line 15) ÷ **360** (voltage) =	**16**	**176** (minimum amperes)	**Minimum Size** **17** Service or Feeder *240.6(A)*	**17**	**200**

© Cengage Learning 2012

Line 18

A The minimum 75°C (167°F) copper service conductor is 3/0 AWG ≫ *Table 310.15(B)(16)* ≪.

B The minimum 75°C (167°F) aluminum service conductor is 4/0 AWG ≫ *Table 310.15(B)(16)* ≪.

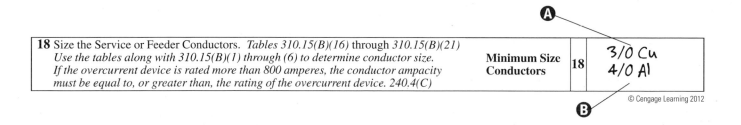

18 Size the Service or Feeder Conductors. *Tables 310.15(B)(16)* through *310.15(B)(21)*		
Use the tables along with *310.15(B)(1)* through *(6)* to determine conductor size.		
If the overcurrent device is rated more than 800 amperes, the conductor ampacity	**Minimum Size Conductors**	**18**	**3/0 Cu** **4/0 Al**
must be equal to, or greater than, the rating of the overcurrent device. *240.4(C)*		

© Cengage Learning 2012

Line 19

A The neutral conductor calculation requires only loads having a neutral or grounded conductor termination. In this example, include lines 4, 5, 8, 10, and part of line 14. Line 14's neutral load calculation is 8550 volt-amperes (57 × 120 × 125%). The total neutral load is 36,925 volt-amperes (5000 + 17,500 + 1200 + 4675 + 8550). The minimum required neutral rating is 103 amperes (36,925 ÷ 360).

B The minimum 75°C (167°F) copper neutral conductor is 2 AWG ≫ *Table 310.15(B)(16)* ≪.

C The minimum 75°C (167°F) aluminum neutral conductor is 1/0 AWG ≫ *Table 310.15(B)(16)* ≪.

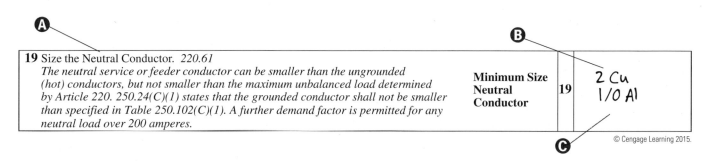

19 Size the Neutral Conductor. *220.61*		
The neutral service or feeder conductor can be smaller than the ungrounded		
(hot) conductors, but not smaller than the maximum unbalanced load determined	**Minimum Size Neutral Conductor**	**19**	**2 Cu** **1/0 Al**
by Article 220. 250.24(C)(1) states that the grounded conductor shall not be smaller		
than specified in Table 250.102(C)(1). A further demand factor is permitted for any		
neutral load over 200 amperes.		

© Cengage Learning 2015.

Line 20

A The grounding electrode conductor's size is based on the service-entrance conductor ≫ *Table 250.66* ≪.

B The minimum size copper grounding conductor is 4 AWG.

C The minimum size aluminum grounding conductor is 2 AWG.

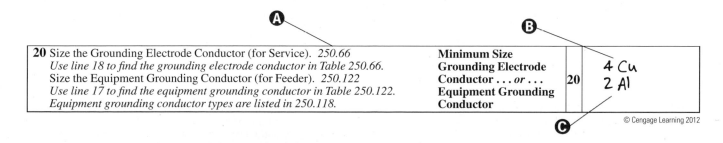

20 Size the Grounding Electrode Conductor (for Service). *250.66*	**Minimum Size**		
Use line 18 to find the grounding electrode conductor in Table 250.66.	**Grounding Electrode**		
Size the Equipment Grounding Conductor (for Feeder). *250.122*	**Conductor ... or ...**	**20**	**4 Cu** **2 Al**
Use line 17 to find the equipment grounding conductor in Table 250.122.	**Equipment Grounding**		
Equipment grounding conductor types are listed in 250.118.	**Conductor**		

© Cengage Learning 2012

The following page shows the bank's completed load calculation.

Nondwelling Feeder/Service Load Calculation

1 Receptacle Load (noncontinuous) *220.14(I)* *Multiply each single or multiple receptacle on one strap by 180 volt-amperes.* *Multiply each single piece of equipment comprised of four or more receptacles by 90 VA per receptacle.*	**1**	----
2 Fixed Multioutlet Assemblies (noncontinuous) *220.14(H)* *Where not likely to be used simultaneously, multiply each 5-ft section by 180 volt-amperes.* *Where likely to be used simultaneously, multiply each 1-ft section by 180 volt-amperes.*	**2**	----
3 Receptacle Load Demand Factor (for nondwelling receptacles) *220.44* *If the receptacle load is more than 10,000 volt-amperes, apply the demand factor from Table 220.44.* *Add lines 1 and 2. Multiply the first 10 kVA or less by 100%. Then multiply the remainder by 50%.*	**3**	----
4 Receptacle Load (banks and office buildings only) *220.14(K)* *(1) Calculate the demand in line 3; (2) multiply 1 volt-ampere per square foot* $1 \times$ __5000__ = *and add to line 4; and (3) compare lines 3 and 4 and omit the smaller.* (sq-ft outside dimensions)	**4**	**5000**
5 General Lighting Load *Table 220.12* *Multiply the volt-ampere unit load (for the type of* __3.5__ $\times$ __5000__ = *occupancy) by the square-foot outside dimensions.* (VA unit load) (sq-ft outside dimensions)	**5**	**17,500**
6 Lighting Load Demand Factors *220.42.* *Apply Table 220.42 demand factors to certain portions of* *hospitals, hotels, motels, apartment houses (without provisions for cooking), and storage warehouses.* *Do not include areas in hospitals, hotels, and motels where the entire lighting will be used at one time.*	**6**	----
7 Track Lighting (in addition to general lighting) *220.43(B)* *Include 150 volt-amperes for every 2 ft,* __----__ ÷ 2 __----__ $\times$ 150 = *or fraction thereof, of lighting track.* (total linear ft)	**7**	----
8 Sign or Outline Lighting Outlet (where required) *220.14(F)* *Each commercial building (or occupancy) accessible to pedestrians must have at least one* *outlet per tenant space entrance. 600.5(A) Each outlet must be at least 1200 volt-amperes.*	**8**	**1200**
9 Show-Window Lighting *220.43(A)* *Include at least 200 volt-amperes for each linear foot,* __----__ $\times$ 200 = *measured horizontally along the show window's base.* (total linear ft of show window)	**9**	----
10 Continuous Loads *215.2(A), 215.3, and 230.42(A)* *Multiply the continuous load volt-amperes (listed above) by 25%.* __18,700__ $\times 25\%$ = *(General-purpose receptacles are not considered continuous.)* (total continuous load volt-amperes)	**10**	**4675**
11 Kitchen Equipment *220.56* *Multiply three or more pieces of equipment by the Table 220.56 demand factor (%).* Use *Table 220.55* for household cooking equipment used in instructional programs. *Table 220.55,* Note 5	**11**	----
12 Noncoincident Loads *220.60.* *The smaller of two (or more) noncoincident loads can be omitted, as* *long as they will never be energized simultaneously (such as certain portions of heating and A/C systems).* *Calculate fixed electric space-heating loads at 100% of the total connected load. 220.51*	**12**	**20,536**
13 Motor Loads *220.18(A), 430.24, 430.25, 430.26, and Article 440* *Motor-driven air-conditioning and refrigeration equipment is found in Article 440.* *Multiply the largest motor (one motor only) by 25% and add it to the load.*	**13**	**1981**
14 All Other Loads. *Add all other noncontinuous loads into the calculation at 100%.* *Multiply all other continuous loads (operating for 3 hours or more) by 125%.*	**14**	**12,550**
15 Total Volt-Ampere Demand Load: *Add lines 3 through 14 to find the minimum required volt-amperes.*	**15**	**63,442**

16 Minimum Amperes *Divide the total* __63,442__ ÷ __360__ = *volt-amperes by* (line 15) (voltage) *the voltage.*	**16** __176__ (minimum amperes)	**Minimum Size** **17 Service or** **Feeder** *240.6(A)*	**17**	**200**
18 Size the Service or Feeder Conductors. *Tables 310.15(B)(16) through 310.15(B)(21)* *Use the tables along with 310.15(B)(1) through (6) to determine conductor size.* *If the overcurrent device is rated more than 800 amperes, the conductor ampacity* *must be equal to, or greater than, the rating of the overcurrent device. 240.4(C)*		**Minimum Size** **Conductors**	**18**	**3/0 Cu** **4/0 Al**
19 Size the Neutral Conductor. *220.61* *The neutral service or feeder conductor can be smaller than the ungrounded* *(hot) conductors, but not smaller than the maximum unbalanced load determined* *by Article 220. 250.24(C)(1) states that the grounded conductor shall not be smaller* *than specified in Table 250.102(C)(1). A further demand factor is permitted for any* *neutral load over 200 amperes.*		**Minimum Size** **Neutral** **Conductor**	**19**	**2 Cu** **1/0 Al**
20 Size the Grounding Electrode Conductor (for Service). *250.66* *Use line 18 to find the grounding electrode conductor in Table 250.66.* Size the Equipment Grounding Conductor (for Feeder). *250.122* *Use line 17 to find the equipment grounding conductor in Table 250.122.* *Equipment grounding conductor types are listed in 250.118.*		**Minimum Size** **Grounding Electrode** **Conductor . . . or . . .** **Equipment Grounding** **Conductor**	**20**	**4 Cu** **2 Al**

Summary

- Article 220 contains load calculation procedures for non-dwelling units.
- Unlike dwelling unit load calculations, receptacle outlets must be included.
- Receptacle loads, including fixed multioutlet assemblies, can be derated if the load is greater than 10,000 volt-amperes.
- If the actual number of receptacle outlets in a bank or office building is unknown, a load of one volt-ampere per square foot must be included.
- Various occupancy types have different volt-ampere unit loads per Table 220.12.
- One similarity between dwelling unit and nondwelling unit load calculations is that a volt-ampere unit load must be multiplied by the square footage using outside dimensions.

- Table 220.42 contains lighting load demand factors for certain portions of hospitals, hotels, motels, storage warehouses, and apartment houses without provisions for cooking.
- Track lighting loads must be included in addition to the general lighting load.
- Some occupancies require a sign or outline lighting outlet.
- Show-window minimum volt-ampere lighting loads are based on a linear footage measurement.
- Continuous load calculations must be increased by an additional 25% volt-ampere rating.
- The volt-ampere rating, calculated in accordance with Article 220, represents only the minimum rating.

Unit 13 Competency Test

NEC Reference	Answer	
_____	_____	1. What is the demand load contribution for a commercial laundry with ten 5-kVA clothes dryers and ten 1.5-kVA washing machines?
_____	_____	2. What is the volt-ampere unit load per square foot for a beauty salon?
_____	_____	3. Outlets for heavy-duty lampholders shall be calculated at a minimum of _____ volt-amperes.
_____	_____	4. What is the minimum general lighting load for an 11,750-ft² nightclub with an actual connected lighting load of 22,800 volt-amperes? (For this question, do not apply the continuous load demand factor.)
_____	_____	5. A restaurant contains the following commercial kitchen equipment: two 12-kW ovens, one 10-kW grill, two 8-kW deep fryers, one 1.2-kW disposer, one 1.5-kW dishwasher, one 10-kW booster heater, and one 4.5-kW water heater. What is the feeder calculated load (in kW) for this equipment?
_____	_____	6. For track lighting in a church, a load of not less than _____ shall be included for every _____ ft of lighting track or fraction thereof.
_____	_____	7. What is the receptacle demand load for a 9600-ft² bank with fifty-two receptacles?
_____	_____	8. Where it is unlikely that two or more _____ loads will be in use simultaneously, it shall be permissible to use only the largest load(s) that will be used at one time in calculating the total load of a feeder or service.
_____	_____	9. What is the receptacle load (after demand factors) for an office building with 250 receptacle outlets?
_____	_____	10. What is the minimum general lighting load for an 8000-ft² assembly hall? (Do not apply the continuous load demand factor.)
_____	_____	11. A manufacturing plant has 320 ft of multioutlet assemblies, of which 60 ft will contain equipment subject to simultaneous use. What is the volt-ampere calculated load contribution for these multioutlet assemblies?

NEC **Reference**   **Answer**

_____   _____   12. A school home economics class has twelve 8-kW household ranges. The kitchen equipment is supplied by a 230-volt, single-phase panelboard. What is the minimum size THW copper conductors required for this panelboard?

_____   _____   13. A motel has fifty guest rooms, each room measuring 18 ft × 30 ft. Each room contains eight duplex receptacles, of which one is GFCI protected. Exterior hallways have a total area of 3000 ft². What is the lighting and receptacle load (after demand factors) for this motel?

_____   _____   14. Fixed electric space-heating loads shall be calculated at _____% of the total connected load; however, in no case shall a feeder or service load current rating be less than the rating of the largest branch circuit supplied.

_____   _____   15. An outlet supplying recessed luminaire(s) shall be calculated based on _____.

_____   _____   16. For circuits supplying loads consisting of motor-operated utilization equipment that is fastened in place and that has a motor larger than ⅛ hp in combination with other loads, the total calculated load shall be based on _____.

_____   _____   17. In a commercial storage garage, duplex receptacle outlets shall be computed at not less than _____.

_____   _____   18. A restaurant contains the following commercial kitchen equipment: one 12-kW range, one 2.8-kW mixer, one 4.2-kW deep fryer, two 0.8-kW soup wells, and one 6-kW water heater. What is the feeder demand load (in kW) for this equipment?

_____   _____   19. A hospital has 150 patient's rooms, each room measuring 15 ft × 25 ft. What is the lighting load (after demand factors) for these rooms?

_____   _____   20. For circuits supplying lighting units that have ballasts, transformers, autotransformers, or LED drivers, the calculated load shall be based on _____.

Questions 21 through 30 are based on an office containing the following:

Office's total sq-ft area. . . . . 19,000 ft²
Receptacle outlets. 200
Multioutlet assemblies. . . . . 80 ft
Multioutlet assemblies. . . . . 25 ft
  (simultaneously used)
Sign . . . . 15 amperes, 120 volt
Track lighting. . . . . 48 ft
Water heater. . . . . 4 kW, 208 volt
Four electric heating units . . . . 10 kW, 208 volt (each)
Four air handlers (blower motors) . . . 4 amperes, 208 volt, 3-phase (each)
Four air-conditioner compressors. . . . 24.6 amperes, 208 volt, 3-phase (each)
Eight condenser fan motors . . . . 2.2 amperes, 208 volt, 3-phase (each)

Assume water heater and electric heat kW ratings equivalent to kVA.

The electric supply is a 208-volt, 3-phase, 4-wire service.

_____   _____   21. What is the total receptacle (including multioutlet assembly) volt-ampere service load after demand factors are applied?

_____   _____   22. What is the general lighting load for this 19,000-ft² office? (For this question, do not apply the continuous load demand factor.)

_____   _____   23. Without applying the continuous load demand factor, what track lighting load (if any) must be contributed to this office calculation?

NEC Reference	Answer

_____ _____ 24. What volt-ampere load shall be included for the sign? (For this question, do not apply the continuous load demand factor.)

_____ _____ 25. Which load(s), if any, can be omitted?

_____ _____ 26. What is the volt-ampere load contribution for the heating or air-conditioning system(s)? (Do not include 25% of the largest motor.)

_____ _____ 27. What is the minimum size service overcurrent protective device required for this office?

_____ _____ 28. A parallel run (two sets) of service-entrance conductors will be installed in two raceways. What is the minimum size THWN copper ungrounded conductors that can be used?

_____ _____ 29. What is the minimum size THWN copper neutral (grounded) conductors that can be installed? (The lighting does not consist of nonlinear loads.)

_____ _____ 30. What is the minimum size copper grounding electrode conductor?

UNIT 14

Services, Feeders, and Equipment

Objectives

After studying this unit, the student should:

▶ know the required minimum vertical wiring clearances for each electrical installation.

▶ be aware that other associated equipment, above or below the electrical installation, shall not extend more than 6 in. (150 mm) beyond the front of that electrical equipment.

▶ be familiar with the dedicated space above panelboards, switchboards, switchgear, and motor control centers.

▶ be aware that a space equal to the equipment's width and depth, above panelboards, switchboards, switchgear, and motor control centers, must be clear of foreign systems unless protection is provided.

▶ know that working space clearances vary depending on the conditions and voltages.

▶ understand the conditions that require two equipment working space entrances.

▶ know where to connect grounded and grounding conductors on both the service supply side and load side.

▶ understand the conditions that require ground-fault protection of equipment.

▶ have a good understanding of transformer and generator provisions.

▶ know that grounding and bonding connection points shall be made in the same place, at either the panel or the transformer.

▶ have an extensive understanding of busway provisions.

Introduction

Unit 14 contains electrical provisions relative to commercial type occupancies. Before beginning this unit, a review of Unit 9—Services and Electrical Equipment is recommended. Some of the same provisions that apply to one-family dwellings likewise apply to commercial installations. Vertical wiring clearances, found in Articles 225 and 230, apply to all electrical feeder and service installations. Also applicable to both dwelling and commercial areas is 110.26, which contains minimum working space provisions (600 volts or less) above, below, and in front of electrical equipment. Some requirements, such as for equipment rated 1200 amperes or more, over 6 ft (1.8 m) wide, and containing devices (overcurrent, switching, or control), are not included in Unit 9. This unit contains separately derived system provisions that pertain to certain battery, solar photovoltaic, generator, transformer, and converter winding-system installations. Because of space limitations, only transformer illustrations are used to explain these provisions. In addition to the separately derived system provisions, transformers have their own article (450). Although there are some exceptions, Article 450 covers the installation of all transformers see (450.1). Following the transformer material is an illustration briefly touching on generator provisions. Although generators have their own article (445), other generator provisions are found throughout the *Code*. Generators and their associated wiring and equipment must also adhere to applicable provisions of Articles 695, 700, 701, 702, and 705. Busway provisions close this unit. Busways can also be found in multiple occupancy types, that is, multifamily, commercial, and industrial.

Since state and local jurisdictions may modify *NEC* provisions, it is expedient to obtain a copy of local rules and regulations.

CLEARANCES AND WORKING SPACE

Vertical Clearances

A 10-ft (3.0-m) minimum vertical clearance is required above finished grade for sidewalks, or from any platform or projection from which overhead service conductors (of 150 volts, or less, to ground) might be reached 》*230.24(B)(1)* and *225.18(1)*《.

A minimum 12-ft (3.7-m) vertical clearance is required over commercial areas (not subject to truck traffic) provided conductor voltage does not exceed 300 volts to ground 》*230.24(B)(2)* and *225.18(2)*《.

The minimum vertical clearance increases from 12 ft (3.7 m) to 15 ft (4.5 m) for conductors whose voltage exceeds 300 volts to ground 》*230.24(B)(3)* and *225.18(3)*《.

A drip loop's lowest point has a minimum vertical clearance of 10 ft (3.0 m)》*230.24(B)(1)*《.

The minimum vertical clearance over track rails of railroads is 24.5 ft (7.5 m) 》*225.18(5)*《.

> **NOTE**
>
> Clearance from communication wires and cables shall be in accordance with 800.44(A)(4) 》*230.24(E)*《.

A The point of attachment of the service-drop conductors to a building (or other structure) shall meet 230.9 and 230.24 minimum clearance specifications. This point of attachment shall always be at least 10 ft (3.0 m) above finished grade 》*230.26*《.

B Public streets, alleys, roads, parking areas subject to truck traffic, driveways on nonresidential property, and other land traversed by vehicles (such as cultivated, grazing, forest, and orchard) require a minimum 18-ft (5.5-m) vertical clearance 》*230.24(B)(4)* and *225.18(4)*《.

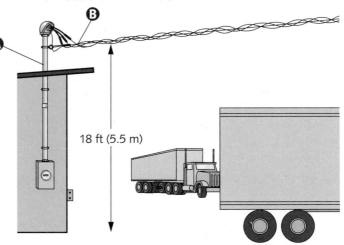

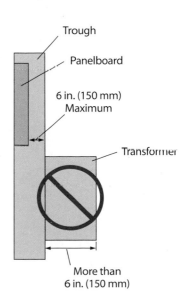

18 ft (5.5 m)

© Cengage Learning 2012

Maximum Depth of Associated Equipment

A The depth of the working space in the direction of access to live parts shall not be less than indicated by Table 110.26(A)(1) 》*110.26(A)(1)*《.

B All electrical equipment shall be constantly surrounded by sufficient access and work space permitting ready and safe equipment operation and maintenance 》*110.26*《.

C Wireways are troughs (sheet metal or nonmetallic) with hinged or removable covers used to house as well as protect electric wires and cables. A wireway shall be installed as a complete system before conductors are laid in place 》*376.2* and *378.2*《.

D Within the height requirements of 110.26, other equipment that is associated with the electrical installation and is located above or below the electrical equipment shall not extend more than 6 in. (150 mm) beyond the front of the electrical equipment 》*110.26(A)(3)*《.

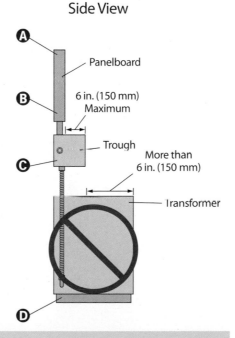

Side View

Panelboard

6 in. (150 mm) Maximum

Trough

More than 6 in. (150 mm)

Transformer

Top View

Trough

Panelboard

6 in. (150 mm) Maximum

Transformer

More than 6 in. (150 mm)

© Cengage Learning 2012

> **NOTE**
>
> Meters that are installed in meter sockets shall be permitted to extend beyond the other equipment. The meter socket must be installed in accordance with the provisions in 110.26 》*110.26(A)(3) Exception No. 2*《.

Dedicated Equipment Space

Sprinkler protection is permitted where piping complies with 110.26 ≫ *110.26(E)(1)(c)* ≪ .

Ⓐ The area above the dedicated space required by 110.26(E)(1)(a) can contain foreign systems, provided protection is installed to avoid damage to the electrical equipment from condensation, leaks, or breaks in such foreign systems ≫ *110.26(E)(1)(b)* ≪ .

Ⓑ The space equal to the equipment's width and depth extending from the floor to a height of 6 ft (1.8 m) above the equipment or to the structural ceiling (whichever is lower) is dedicated to the electrical installation. No piping, ducts, or foreign equipment shall be located in this zone ≫ *110.26(E)(1)(a)* ≪ .

Ⓒ Section 110.26(E) applies to motor control centers and equipment within the scope of Article 408.

Ⓓ Section 110.26(E)(1)(b) pertains to piping, ducts, etc. that could damage electrical equipment due to condensation, leaks, or breaks.

Ⓔ Foreign systems may enter the area above the dedicated electrical space if the electrical equipment is protected from condensation, leaks, or breaks in such foreign systems ≫ *110.26(E)(1)(b)* ≪ .

Ⓕ A dropped, suspended, or similar ceiling not adding strength to the building structure is not a structural ceiling ≫ *110.26(E)(1)(d)* ≪ .

Ⓖ The work space shall be clear and extend from the grade, floor, or platform to a height of 6½ ft (2.0 m) or the height of the equipment, whichever is greater ≫ *110.26(E)* ≪ .

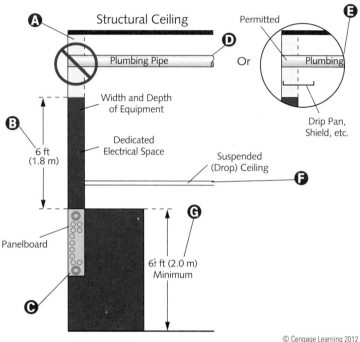

Structural Ceiling

Plumbing Pipe

Width and Depth of Equipment

6 ft (1.8 m)

Dedicated Electrical Space

Panelboard

Permitted

Or

Plumbing

Drip Pan, Shield, etc.

Suspended (Drop) Ceiling

6½ ft (2.0 m) Minimum

© Cengage Learning 2012

> **NOTE**
>
> Suspended ceilings with removable panels are permitted within the 6-ft (1.8-m) dedicated space ≫*110.26(E)(1)(a) Exception*≪ .

> **NOTE**
>
> Outdoor electrical equipment shall be installed in suitable enclosures and shall be protected from accidental contact by unauthorized personnel, or by vehicular traffic, or by accidental spillage or leakage from piping systems. The working clearance space shall include the zone described in 110.26(A). No architectural appurtenance or other equipment shall be located in this zone ≫*110.26(E)(2)(a)*≪.
>
> The space equal to the width and depth of the equipment, and extending from grade to a height of 6 ft (1.8 m) above the equipment, shall be dedicated to the electrical installation. No piping or other equipment foreign to the electrical installation shall be located in this zone ≫*110.26(E)(2)(b)*≪.

Working Space Access and Entrances

One access means is permitted where the location provides a continuous and unobstructed exit ›› *110.26(C)(2)(a)* ‹‹.

When normally enclosed live parts are exposed for inspection or servicing, any surrounding workspace, in a passageway or general open area, shall be suitably guarded ›› *110.26(B)* ‹‹.

Ⓐ Equipment rated 1200 amperes (or more) and over 6 ft (1.8 m) wide containing overcurrent devices, switching devices, or control devices, requires an entrance to and egress from the required working space at least 24 in. (610 mm) wide and 6½ ft (2.0 m) high at each end of the working space ›› *110.26(C)(2)* ‹‹.

Ⓑ If the workspace specified by 110.26(A) is doubled, only one entrance is required. The edge of the entrance nearest the equipment shall meet the minimum clear distance given in Table 110.26(A)(1) ›› *110.26(C)(2)(b)* ‹‹.

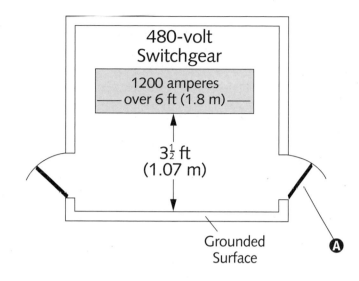

> ### N O T E
>
> Equipment operating over 600 volts, nominal, must comply with Article 110, Part III.

WARNING

Do not use working space required by 110.26 for storage ›› *110.26(B)* ‹‹.

CAUTION

Where equipment rated 800 amperes or more that contains overcurrent devices, switching devices, or control devices is installed and there is a personnel door(s) intended for entrance to and egress from the working space less than 25 ft (7.6 m) from the nearest edge of the working space, the door(s) shall open in the direction of egress and be equipped with listed panic hardware ›› *110.26(C)(3)* ‹‹.

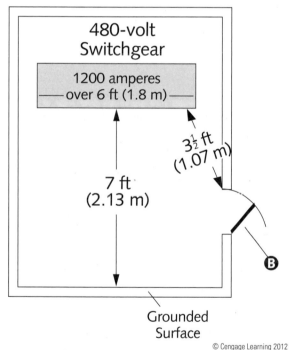

Working Space

Condition 1—**Exposed live parts on one side of the working space with no live or grounded parts on the other side of the working space, or exposed live parts on both sides of the working space that are effectively guarded by insulating materials** ≫ *Table 110.26(A)(1)* ≪.

Ⓐ The working space depth in the direction of access to live parts must meet Table 110.26(A)(1) specifications unless the requirements of 110.26(A)(1)(a), (A)(1)(b), or (A)(1)(c) are met. If exposed, measure the distance from the live part(s). Where enclosed, distance measurement begins from the enclosure front (or opening) ≫ *110.26(A)(1)* ≪.

Ⓑ Sufficient access and working space shall be provided and maintained about all electric equipment, permitting ready, safe equipment operation and maintenance ≫ *110.26* ≪.

> **NOTE**
>
> Electrical equipment rooms or enclosures housing electrical apparatus that are controlled by a lock(s) shall be considered accessible to qualified persons ≫ *110.26(F)* ≪.

Condition 2—**Exposed live parts on one side of the working space with grounded parts on the other side. Concrete, brick, or tile walls qualify as grounded** ≫ *Table 110.26(A)(1)* ≪.

Condition 3—**Exposed live parts on both sides of the working space** ≫ *Table 110.26(A)(1)* ≪.

Ⓔ In replacing equipment within existing buildings, Condition 2 working clearance is permitted between dead-front switchboards, switchgear, panelboards, or motor control centers located across the aisle from each other where (1) conditions of maintenance and supervision ensure that written procedures have been adopted to prohibit equipment on both sides of the aisle from being open at the same time and (2) authorized qualified personnel will service the installation ≫ *110.26(A)(1)(c)* ≪.

Ⓒ Working space is not required behind or beside assemblies, such as dead-front switchboards, switchgear, or motor control centers, having no renewable or adjustable parts (such as fuses or switches) on the back or sides, provided all connections are accessible from other locations ≫ *110.26(A)(1)(a)* ≪.

Ⓓ Where rear access is required to work on nonelectrical parts at the back of enclosed equipment, a minimum working space of 30 in. (762 mm) horizontally is required ≫ *110.26(A)(1)(a)* ≪.

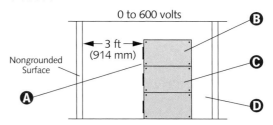

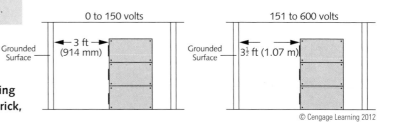

© Cengage Learning 2012

> **CAUTION**
>
> Working space for equipment operating at 600 volts, nominal, or less to ground and likely to require examination, adjustment, servicing, or maintenance while energized shall comply with 110.26(A)(1) through (3) dimensions, or as required/permitted elsewhere in the *Code* ≫ *110.26(A)* ≪.

> **NOTE**
>
> By special permission, smaller working spaces are allowed where all exposed live parts are no greater than 30 volts root-mean-square (rms), 42 peak, or 60 volts direct current (dc) ≫ *110.26(A)(1)(b)* ≪.
>
> Equipment operating over 600 volts, nominal, must comply with Article 110, Part III.

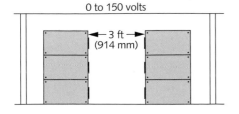

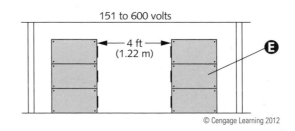

© Cengage Learning 2012

Guarding and Warning

Except as required (or permitted) elsewhere by the *NEC*, live parts of electric equipment operating at 50 volts or more shall be guarded against accidental contact by approved enclosures or by any of the following means listed in 110.27(A):

1. By location in a room, vault, or similar enclosure accessible only to qualified persons.

2. By suitable permanent, substantial partitions (or screens) arranged so that only qualified persons have access to the space within reach of live parts. Any openings in such partitions (barriers) shall be sized and located so that accidental human contact with the live parts or conducting objects is unlikely.

3. By locations on a suitable balcony, gallery, or elevated platform whose arrangement excludes unqualified persons.

4. By elevations above the floor or other working surface as shown in 110.27(A)(4)(a) or (b) below:

    a. A minimum of 8 ft (2.5 m) for 50 to 300 volts

    b. A minimum of 8-1⁄2 ft (2.6 m) for 301 to 600 volts

Where potential for physical damage to electric equipment exists, enclosures (or guards) shall be of sufficient strength and placement to prevent such damage 》*110.27(B)*《.

Ⓐ At least one entrance of sufficient area shall be provided, giving access to and egress from the electrical equipment's working space 》*110.26(C)(1)*《.

Ⓑ Entrances to rooms and other guarded locations containing exposed live parts shall be marked with conspicuous warning signs forbidding entrance of unqualified persons. The marking shall meet the requirements in 110.21(B) 》*110.27(C)*《.

Ⓒ When working on nonelectric parts on the back of enclosed equipment requires rear access, a minimum working space of 30 in. (762 mm) horizontally shall be provided 》*110.26(A)(1)(a)*《.

Ⓓ This is Condition 3 in Table 110.26(A)(1).

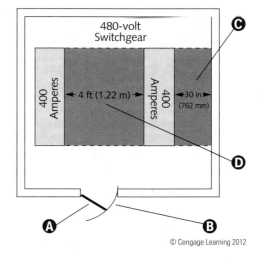

© Cengage Learning 2012

SWITCHBOARDS, SWITCHGEAR, AND PANELBOARDS

Phase Arrangement

The alternating current phase arrangement on 3-phase busses are A, B, C from front to back, top to bottom, or left to right, as viewed from the front of the switchboard, switchgear, or panelboard. The B phase must be that phase having the higher voltage to ground on 3-phase, 4-wire, delta-connected systems. Other clearly marked busbar arrangements are permitted for existing installation additions 》*408.3(E)(1)*《.

Equipment sharing the same single section or multisectional switchboard, switchgear, or panelboard as the meter (on 3-phase, 4-wire, delta-connected systems) is allowed to have the same phase configuration as the metering equipment 》*408.3(E)(1) Exception*《.

> **NOTE**
>
> A switchboard, switchgear, or panelboard supplied from a 4-wire, delta-connected system where the midpoint of one phase winding is grounded, requires the phase busbar or conductor having the higher voltage to ground to be durably and permanently marked by an outer orange finish or by other effective means. Such identification shall be placed at each point on the system where a connection is made if the grounded conductor is also present 》*110.15*《.

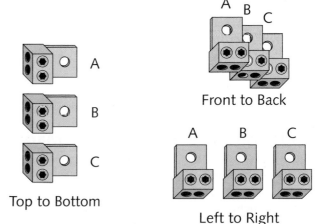

© Cengage Learning 2012

> **CAUTION**
>
> The area electric utility company may require a different phase arrangement on 3-phase, 4-wire, delta-connected systems. Obtain a copy of the utility's phase arrangement requirements.

Grounding—Supply Side of Service

Ⓐ A grounding electrode conductor shall be used to connect the equipment grounding conductors, the service-equipment enclosures, and, where the system is grounded, the grounded service conductor to the grounding electrode(s) required by Part III of Article 250. This conductor shall be sized in accordance with 250.66 ≫250.24(D)≪. See 250.64(A) through (F) for grounding electrode conductor installation requirements.

Ⓑ The connection between the grounded circuit conductor (neutral) and the equipment grounding conductor at the service is the main bonding jumper ≫Article 100≪. For a grounded system, an unspliced main bonding jumper shall be used to connect the equipment grounding conductor(s) and the service-disconnect enclosure to the grounded conductor within the enclosure for each service disconnect in accordance with 250.28 ≫250.24(B)≪.

Where a main bonding jumper or a system bonding jumper is a screw only, the screw shall be identified with a green finish that shall be visible with the screw installed≫250.28(B)≪.

Ⓒ In accordance with 250.92(A), the normally non-current-carrying metal parts of equipment indicated in (1) and (2) shall be bonded together:

1. All raceways, cable trays, cablebus framework, auxiliary gutters, or service cable armor/sheath that enclose, contain, or support service conductors, except as permitted in 250.80.

2. All enclosures containing service conductors, including meter fittings, boxes, or the like, interposed in the service raceway or armor.

Ⓓ Two to six circuit breakers (or sets of fuses) are permitted as the overcurrent device providing overload protection for service equipment. The sum of the circuit breakers (or fuses) ratings can exceed the service conductor's ampacity, provided the calculated load does not exceed the ampacity of the service conductors ≫230.90(A) Exception No. 3≪.

Ⓔ Each service-disconnecting means permitted by 230.2, or for each set of service-entrance conductors permitted by 230.40 Exception Nos. 1, 3, 4, or 5, shall consist of no more than six switches and sets of circuit breakers or a combination of not more than six switches and sets of circuit breakers mounted within a single enclosure, in a group of separate enclosures, or in (or on) a switchboard or in switchgear. The maximum number of disconnects per service grouped in any one location is six ≫230.71(A)≪.

> **NOTE**
>
> Metal water piping system(s) installed in or attached to a building or structure must be bonded to either the service equipment enclosure, the grounded conductor at the service, the grounding electrode conductor (if of sufficient size), or the grounding electrode(s) used ≫*250.104(A)(1)*≪.

Ⓕ A grounded circuit conductor can ground non-current-carrying metal equipment parts, raceways, and other enclosures at any of the following locations listed in 250.142(A):

1. On the supply side, or within the enclosure, of the alternating-current (ac) service disconnecting means.

2. On the supply side, or within the enclosure, of the main disconnecting means for separate buildings as provided in 250.32(B).

3. On the supply side, or within the enclosure, of the main disconnecting means or overcurrent devices of a separately derived system where 250.30(A)(1) permits.

Ⓖ The grounded conductor shall not be smaller than specified in Table 250.102(C)(1) ≫250.24(C)(1)≪.

> **WARNING**
>
> Where an ac system operating at 1000 volts or less is grounded at any point, the grounded conductor(s) shall be routed with the ungrounded conductors to each service disconnecting means and shall be connected to each disconnecting means grounded conductor(s) terminal or bus. A main bonding jumper shall connect the grounded conductor(s) to each service disconnecting means enclosure ≫250.24(C)≪.

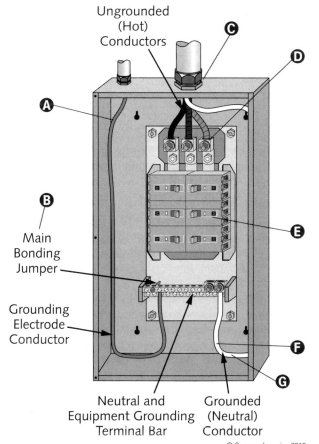

Ungrounded (Hot) Conductors — Ⓒ — Ⓓ — Ⓐ — Ⓑ — Main Bonding Jumper — Grounding Electrode Conductor — Ⓔ — Ⓕ — Ⓖ — Neutral and Equipment Grounding Terminal Bar — Grounded (Neutral) Conductor

Supply-Side Bonding Jumper

A supply-side bonding jumper is defined as a conductor installed on the supply side of a service or within a service equipment enclosure(s), or for a separately derived system, that ensures the required electrical conductivity between metal parts required to be electrically connected ≫ 250.2 *Bonding Jumper, Supply-Side* ≪.

Ⓐ The supply-side bonding jumper shall not be smaller than specified in Table 250.102(C)(1) ≫ *250.102(C)(1)* ≪. Where the ungrounded supply conductors are larger than 1100 kcmil copper or 1750 kcmil aluminum, the supply-side bonding jumper shall have an area not less than 12½% of the area of the largest set of ungrounded supply conductors or equivalent area for parallel supply conductors ≫ Note 1. under Table *250.102(C)(1)* ≪. Table 250.102(C)(1) is used for sizing

grounded conductors, main bonding jumpers, system bonding jumper, and supply-side bonding jumper for alternating-current systems.

Ⓑ Where the ungrounded supply conductors are paralleled in two or more raceways or cables, and an individual supply-side bonding jumper is used for bonding these raceways or cables, the size of the supply-side bonding jumper for each raceway or cable shall be selected from Table 250.102(C)(1), based on the size of the ungrounded supply conductors in each raceway or cable ≫ *250.102(C)(2)* ≪.

Ⓒ A single supply-side bonding jumper installed for bonding two or more raceways or cables shall be sized in accordance with 250.102(C)(1) ≫ *250.102(C)(2)* ≪.

NOTE

If the ungrounded supply conductors and the supply-side bonding jumper are of different materials (copper, aluminum, or copper-clad aluminum), the minimum size of the supply-side bonding jumper shall be based on the assumed use of ungrounded conductors of the same material as the supply-side bonding jumper and will have an ampacity equivalent to that of the installed ungrounded supply conductors ≫*Note 2. under Table 250.102(C)(1)*≪.

NOTE

The term supply conductors includes ungrounded conductors that do not have overcurrent protection on their supply side and terminate at service equipment or the first disconnecting means of a separately derived system ≫*250.102(C)(1) Informational Note*≪.

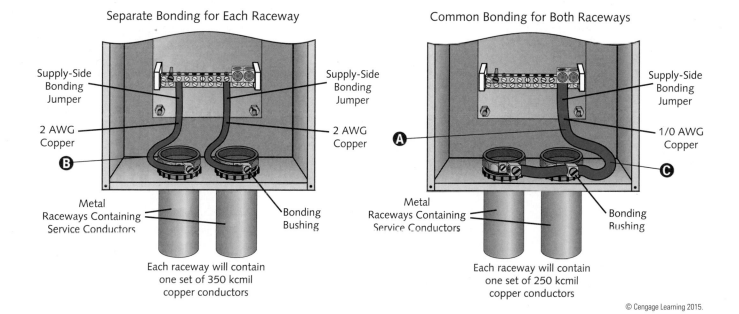

Separate Bonding for Each Raceway

Supply-Side Bonding Jumper

Supply-Side Bonding Jumper

2 AWG Copper

2 AWG Copper

Ⓑ

Metal Raceways Containing Service Conductors

Bonding Bushing

Each raceway will contain one set of 350 kcmil copper conductors

Common Bonding for Both Raceways

Supply-Side Bonding Jumper

Ⓐ

1/0 AWG Copper

Ⓒ

Metal Raceways Containing Service Conductors

Bonding Bushing

Each raceway will contain one set of 250 kcmil copper conductors

Grounding—Load Side of Service

Where ungrounded conductors are increased in size from the minimum size that has sufficient ampacity for the intended installation, wire-type equipment grounding conductors (where installed) shall be increased in size proportionately, according to the circular mil area of the ungrounded conductors »*250.122(B)*«.

A Equipment bonding jumpers on the load side of an overcurrent device(s) shall be sized in accordance with 250.122. A single common continuous equipment bonding jumper shall be permitted to connect two or more raceways or cables if the bonding jumper is sized in accordance with 250.122 for the largest overcurrent device supplying circuits therein »*250.102(D)*«.

B A grounded conductor shall not be connected to normally non-current-carrying metal parts of equipment, to equipment grounding conductor(s), or be reconnected to ground on the load side of the service disconnecting means except as otherwise permitted in Article 250 »*250.24(A)(5)*«.

C Equipment grounding conductors can be bare, covered, or insulated. Individually covered (or insulated) equipment grounding conductors shall have a continuous outer finish, either green or green with yellow stripe(s), unless otherwise permitted by Article 250, Part IV »*250.119*«.

D Metal panelboard cabinets and frames shall physically contact each other and shall be connected to an equipment grounding conductor. If the panelboard is used with nonmetallic raceway (or cable), or if separate equipment grounding conductors are provided, secure a terminal bar for the equipment grounding conductors within the cabinet. The terminal bar shall be bonded to the metal cabinet and panelboard frame. Otherwise, connect it to the equipment grounding conductor that is run with the conductors feeding the panelboard »*408.40*«.

E While equipment grounding conductors of wire type copper, aluminum, or copper-clad aluminum shall not be less than shown in Table 250.122, they are not required to be

larger than the circuit conductors supplying the equipment. A raceway, cable armor, or sheath used as the equipment grounding conductor (as provided in 250.118 and 250.134[A]), must comply with 250.4(A)(5) or 250.4(B)(4). Equipment grounding conductors shall be permitted to be sectioned within a multiconductor cable, provided the combined circular mil area complies with Table 250.122 »*250.122(A)*«.

F Unless grounded by connection to the grounded circuit conductor (permitted by 250.32, 250.140, and 250.142), non-current-carrying metal parts of equipment, raceways, and other enclosures, if grounded, shall be connected to an equipment grounding conductor by one of the following methods: (a) by connection to any of the equipment grounding conductors permitted by 250.118; or (b) by connection to an equipment grounding conductor contained within the same raceway, cable, or otherwise run with the circuit conductors »*250.134*«.

G Except as permitted in 250.30(A)(1) and 250.32(B) Exception, a grounded circuit conductor shall not be used for grounding non-current-carrying metal equipment parts on the load side of the service disconnecting means, on the load side of a separately derived system disconnecting means, or the overcurrent device for a separately derived system not having a main disconnecting means, unless an exception has been met »*250.142(B)*«.

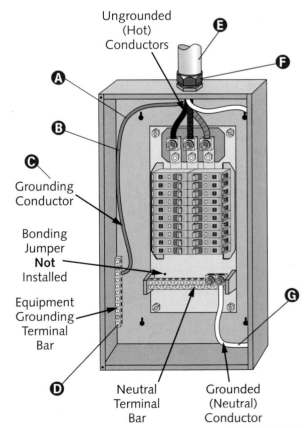

Ungrounded (Hot) Conductors

Grounding Conductor

Bonding Jumper **Not** Installed

Equipment Grounding Terminal Bar

Neutral Terminal Bar

Grounded (Neutral) Conductor

© Cengage Learning 2012

NOTE

Paralleled conductors within multiple raceways or cables, as permitted in 310.10(H), require that the equipment grounding conductors (where used) also be installed in parallel in each raceway or cable. Where conductors are installed in parallel in the same raceway, cable, or cable tray as permitted in 310.10(H), a single equipment grounding conductor shall be permitted. Equipment grounding conductors installed in cable tray shall meet the minimum requirements of 392.10(B)(1)(c). Each equipment grounding conductor shall be sized in compliance with 250.122 »**250.122(F)**«.

Ground-Fault Protection of Equipment

If a switch and fuse combination is used, the fuses employed shall be capable of interrupting any current higher than the interrupting capacity of the switch when the ground-fault protective system will not cause the switch to open 》*230.95(B)*《.

Ground-fault protection that functions to open the service disconnect will afford no protection from faults on the line side of the protective element. It serves to limit damage to conductors and equipment only on the load side of the protective element in the event of an arcing ground fault 》*230.95(C) Informational Note No. 1*《.

This added protective apparatus at the service equipment may necessitate a review of the overall wiring system for proper selective overcurrent protection coordination. Additional installations of ground-fault protective equipment may be needed on feeders and branch circuits requiring maximum continuity of electrical service 》*230.95(C) Informational Note No. 2*《.

Where ground-fault protection is provided for the service disconnect and another supply system is interconnected by a transfer device, a method may be needed to ensure proper sensing by the ground-fault protection equipment 》*230.95(C) Informational Note No. 3*《.

A Ground-fault protection of equipment shall be provided for solidly grounded wye electrical services of more than 150 volts to ground but not exceeding 1000 volts phase-to-phase for each service disconnect rated 1000 amperes or more. The grounded conductor for the solidly grounded wye system shall be connected directly to ground through a grounding electrode system, as specified in 250.50, without inserting any resistor or impedance device 》*230.95 and 240.13*《.

B The service disconnect rating is considered to be the rating of the largest fuse that can be installed or the highest continuous current trip setting for which the actual overcurrent device installed in a circuit breaker is rated (or can be adjusted) 》*230.95*《.

C Ground-fault protection systems, when activated, shall cause the service disconnect to open all ungrounded conductors of the faulted circuit. The maximum ground-fault setting is 1200 amperes 》*230.95(A)*《.

D One second is the maximum time delay for ground-fault currents of 3000 amperes or more 》*230.95(A)*《.

E **Ground-Fault Protection of Equipment** is a system intended to provide protection of equipment from damaging line-to-ground fault currents by causing the disconnecting means to open all ungrounded conductors of the faulted circuit. This protection is provided at current levels less than those required to protect conductors from damage through the operation of a supply circuit overcurrent device 》*Article 100*《.

> **N O T E**
>
> Section 230.95 ground-fault protection provisions do not apply to fire pumps 》*240.13(3)*《. The ground-fault protection provisions of 230.95 do not apply to a service disconnect for a continuous industrial process where a nonorderly shutdown introduces additional or increased hazards 》*230.95 Exception and 240.13(1)*《.

> **CAUTION**
>
> Bonding for circuits over 250 volts to ground must comply with 250.97.

> **N O T E**
>
> See 215.10 for ground-fault protection of equipment for feeders and 210.13 for ground-fault protection of equipment for branch circuits.

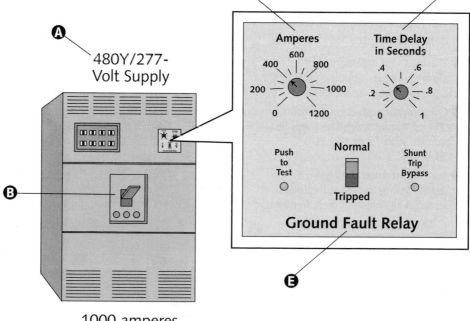

480Y/277-Volt Supply

1000 amperes or More

Amperes
600
400 800
200 1000
0 1200

Time Delay in Seconds
.4 .6
.2 .8
0 1

Push to Test   Normal   Shunt Trip Bypass

Tripped

Ground Fault Relay

Transformer—Primary and Secondary Protection 9 Amperes or More—1000 Volts or Less

Ⓐ Unlike Note 1 that permits the next standard rating above 125%, the **maximum** rating or setting is 250%.

Ⓑ A transformer equipped with coordinated thermal overload protection (by the manufacturer) and arranged to interrupt the primary current can have primary overcurrent protection rated (or set) at a current value not more than six times the rated current of transformers having not more than 6% impedance and not more than four times the rated current of transformers having more than 6%, but not more than 10%, impedance ≫ *Table 450.3(B) Note 3* ≪.

Ⓒ Overcurrent protection for transformers rated 1000 volts, nominal, or less must be provided according to Table 450.3(B), unless the transformer is installed as a motor-control circuit transformer per 430.72(C)(1) through (5) ≫ *450.3(B) and Exception* ≪.

Ⓓ Where 125% of this current does not correspond to a standard fuse or nonadjustable circuit breaker rating, the next higher rating described in 240.6 is permitted ≫ *Table 450.3(B) Note 1* ≪.

Ⓔ If secondary overcurrent protection is required, these devices are permitted to consist of no more than six circuit breakers (or six sets of fuses) grouped in one location. Where multiple overcurrent devices are implemented, the total of all the device ratings must not exceed the single overcurrent device allowed value ≫ *Table 450.3(B) Note 2* ≪.

Overcurrent Protection

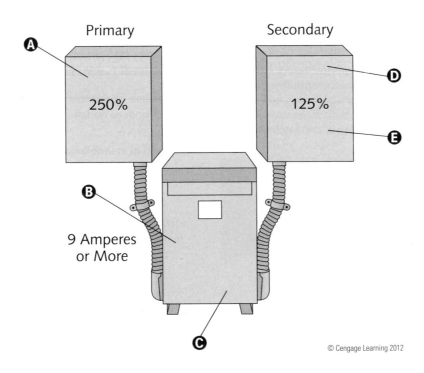

© Cengage Learning 2012

Grounding and Bonding—Terminating in the Panel

Premises wiring (system) is defined as interior and exterior wiring, including power, lighting, control, and signal circuit wiring together with all their associated hardware, fittings, and wiring devices, both permanently and temporarily installed. This includes (a) wiring from the service point or power source to the outlets or (b) wiring from and including the power source to the outlets where there is no service point. Such wiring does not include wiring internal to appliances, luminaires, motors, controllers, motor control centers, and similar equipment ≫ *Article 100* ≪.

A grounded circuit conductor can be used to ground non-current-carrying metal parts of equipment, raceways, and other enclosures on the supply side or within the enclosure of the main disconnecting means (or overcurrent devices) of a separately derived system as permitted by 250.30(A)(1) ≫ *250.142(A)(3)* ≪.

Grounding and Bonding—Terminating in the Panel *(continued)*

(A) An unspliced system bonding jumper shall comply with 250.28(A) through (D). This connection shall be made at any single point on the separately derived system from the source to the first system disconnecting means (or overcurrent devices) or shall be made at the source of a separately derived system that has no disconnecting means or overcurrent devices, in accordance with 250.30(A)(1)(a) or (b). The system bonding jumper shall remain within the enclosure where it originates. If the source is located outside the building or structure supplied, a system bonding jumper shall be installed at the grounding electrode connection in compliance with 250.30(C) ≫*250.30(A)(1)*≪.

(B) A grounding electrode conductor for a single separately derived system shall be sized in accordance with 250.66 for the derived ungrounded conductors. It shall be used to connect the grounded conductor of the derived system to the grounding electrode as specified in 250.30(A)(4). This connection shall be made at the same point on the separately derived system where the system bonding jumper is connected unless one of the exceptions is met ≫*250.30(A)(5)*≪.

(C) For vertically operated circuit-breaker handles, the "up" position of the handle shall be the "on" position ≫*240.81*≪.

(D) A grounded separately derived ac system shall comply with 250.30(A)(1) through (8), unless it is an impedance grounded neutral system installed as specified in 250.36 or 250.187, as applicable ≫*250.30(A)*≪.

(E) Size the equipment grounding conductor in accordance with 250.122. The equipment grounding conductor must be of the types listed in 250.118.

(F) If a building or structure is supplied by a feeder from an outdoor transformer, a system bonding jumper at both the source and the first disconnecting means is permitted, if doing so does not establish a parallel path for the grounded conductor. While grounded conductors used in this manner shall not be smaller than the size specified for the system bonding jumper, they do not have to be larger than the ungrounded conductor(s). In applying this exception, connection through the earth does not qualify as providing a parallel path ≫*250.30(A)(1) Exception No. 2*≪.

(G) Grounding (and bonding) terminations made at the system disconnecting means or overcurrent device require that the transformer's grounded (neutral) bus be insulated from the transformer enclosure. If the neutral bus is bonded to the enclosure and the raceway to the panelboard is metallic, then part of the neutral current flows on the raceway. Section 250.6(A) prohibits objectionable current flow over grounding paths or grounding conductors.

(H) Where separate equipment grounding conductors and supply-side bonding jumpers are installed, a terminal bar for all grounding and bonding conductor connections shall be secured inside the transformer enclosure. The terminal bar shall be bonded to the enclosure in accordance with 250.12 and shall not be installed on or over any vented portion of the enclosure ≫*450.10(A)*≪.

> **NOTE**
>
> Transformers, which are only one type of separately derived system, are used to illustrate and explain 250.30.

> **CAUTION**
>
> Except as otherwise permitted in Article 250, a grounded conductor shall not be connected to normally non–current-carrying metal parts of equipment, be connected to equipment grounding conductors, or be reconnected to ground on the load side of the system bonding jumper ≫*250.30(A)*≪.

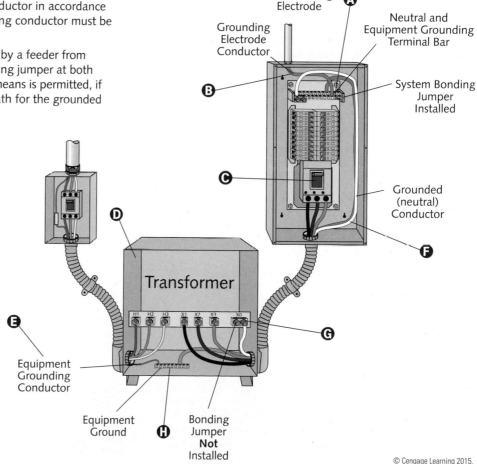

© Cengage Learning 2015.

Grounding and Bonding—Terminating in the Transformer

While a 3-phase separately derived system is used for this illustration, the same rules apply to single-phase systems.

A system supplying power to a Class 1, Class 2, or Class 3 circuit having the power derived from a transformer rated 1000 volt-amperes or less must have a system bonding jumper sized no smaller than the derived ungrounded conductors. The smallest size allowed is 14 AWG copper or 12 AWG aluminum ≫ *250.30(A)(1) Exception No. 3* ≪.

A system that supplies a Class 1, 2, or 3 circuit and is derived from a transformer rated no more than 1000 volt-amperes does not require a grounding electrode conductor, provided the grounded conductor is bonded to the transformer frame (or enclosure) by means of a jumper sized in accordance with 250.30(A)(1), Exception No. 3, and the transformer frame (or enclosure) is grounded by one of the means specified in 250.134 ≫ *250.30(A)(5) Exception No. 3* ≪.

A If the grounding (and bonding) terminations are made at the transformer, the system disconnecting means (or overcurrent device) grounded (neutral) terminal bar must be insulated from both the panelboard enclosure and the equipment grounding conductor. If the neutral terminal bar is bonded to the enclosure as well as the equipment grounding conductor, part of the neutral current will flow through the raceway (if metallic) and the equipment grounding conductor. Section 250.6(A) strictly prohibits such a condition.

B Except as permitted in 250.30(A)(1) and 250.32(B) Exception, a grounded circuit conductor shall not be used for grounding non-current-carrying metal equipment parts on the load side of a separately derived system disconnecting means (or the overcurrent devices) for a separately derived system having no main disconnecting means ≫ *250.142(B)* ≪.

C The system bonding jumper connection can be made at any point on the separately derived system, that is, in either the panel or transformer ≫ *250.30(A)(1)* ≪.

D A grounding electrode conductor for a single separately derived system shall be sized in accordance with 250.66 for the derived ungrounded conductors. It shall be used to connect the grounded conductor of the derived system to the grounding electrode as specified in 250.30(A)(41). This connection shall be made at the same point on the separately derived system where the system bonding jumper is connected ≫ *250.30(A)(5)* ≪.

E Use an unspliced system bonding jumper that complies with 250.28(A) through (D) to connect the equipment grounding conductors of the separately derived system to the grounded conductor ≫ *250.30(A)(1)* ≪.

F The grounding electrode conductor installation must comply with 250.64(A), (B), (C), and (E) ≫ *250.30(A)(7)* ≪.

G If the system bonding jumper specified in 250.30(A) (1) is a wire or busbar, it shall be permitted to connect the grounding electrode conductor to the equipment grounding terminal, bar, or bus, provided the equipment grounding terminal, bar, or bus is of sufficient size for the separately derived system ≫ *250.30(A)(5) Exception No. 1* ≪.

H Where separate equipment grounding conductors and supply-side bonding jumpers are installed, a terminal bar for all grounding and bonding conductor connections shall be secured inside the transformer enclosure. The terminal bar shall be bonded to the enclosure in accordance with 250.12 and shall not be installed on or over any vented portion of the enclosure ≫ *450.10(A)* ≪.

Where a dry-type transformer is equipped with wire-type connections (leads), the grounding and bonding connections shall be permitted to be connected together using any of the methods in 250.8 and shall be bonded to the enclosure if of metal ≫ *450.10(A) Exception* ≪.

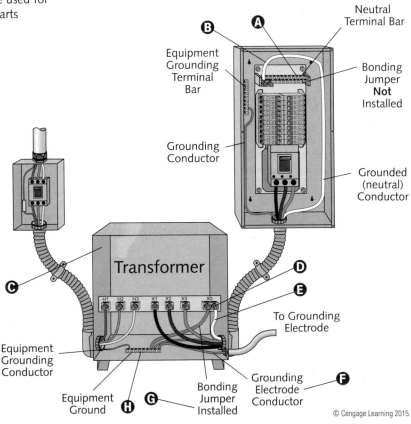

Grounding Electrode and Grounding Electrode Conductor

If a separately derived system originates in listed equipment suitable for use as service equipment, the service (or feeder) equipment's grounding electrode can serve as the separately derived system's grounding electrode ≫**250.30(A)(4) Exception No. 2**≪.

Ⓐ The grounding electrode shall be as near as practicable to, and preferably in the same area as, the grounding electrode conductor connection to the system. The grounding electrode shall be the nearest one of the following:

1. Metal water pipe grounding electrode as specified in 250.52(A)(1).

2. Structural metal grounding electrode as specified in 250.52(A)(2) ≫*250.30(A)(4)*≪.

Ⓑ Ferrous metal raceways and enclosures for grounding electrode conductors shall be electrically continuous from the point of attachment to cabinets or equipment to the grounding electrode and shall be securely fastened to the ground clamp or fitting. Ferrous metal raceways and enclosures shall be bonded at each end of the raceway or enclosure to the grounding electrode or grounding electrode conductor. Nonferrous metal raceways and enclosures shall not be required to be electrically continuous. Bonding shall be in compliance with 250.92(B) and ensured by one of the methods in 250.92(B)(2) through (B)(4). The bonding jumper for a grounding electrode conductor raceway or cable armor shall be the same size as, or larger than, the enclosed grounding electrode conductor. If a raceway is used as protection for a grounding electrode conductor, the installation shall comply with the requirements of the appropriate raceway article ≫*250.64(E)(1) through (4)*≪.

Ⓒ A grounding electrode conductor for a single separately derived system shall be sized according to 250.66 for the derived ungrounded conductors. It shall be used to connect the grounded conductor of the derived system to the grounding electrode as specified in 250.30(A)(4). This connection shall be made at the same point on the separately derived system where the system bonding jumper is connected ≫*250.30(A)(5)*≪.

Ⓓ All mechanical elements used to terminate a grounding electrode conductor or bonding jumper to a grounding electrode shall be accessible ≫*250.68(A)*≪.

Ⓔ Each separately derived system's grounded conductor shall be bonded to the nearest available point of interior metal water piping system(s) in the area served by each separately derived system. This connection shall be made at the same point on the separately derived system as the grounding electrode conductor connection. Size the bonding jumper per Table 250.66 based on the largest ungrounded conductor of the separately derived system ≫*250.104(D)(1)*≪.

Ⓕ Equipment requiring grounding and having a non-conductive coating (such as paint, lacquer, and enamel) shall have the coating removed from threads and other contact surfaces to ensure good electrical continuity (unless connected by fittings that are designed so as to make the removal unnecessary) ≫*250.12*≪.

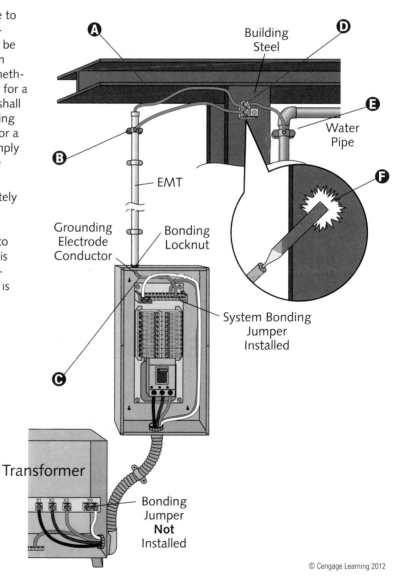

Building Steel

Water Pipe

EMT

Grounding Electrode Conductor

Bonding Locknut

System Bonding Jumper Installed

Transformer

Bonding Jumper **Not** Installed

Marking

A Equipment shall be marked to show the environment for which it has been evaluated. Unless otherwise specified in 500.8(C)(6), the markings shall include: (1) class, (2) division, (3) material classification group, (4) equipment temperature, and (5) ambient temperature range ≫ *500.8(C)* ≪ .

B The maximum safe operating temperature shall be marked on equipment for Class I and Class II locations, having been determined by simultaneous exposure to the combined conditions for both classes ≫ *500.8(C)(4)* ≪ .

C The temperature class, if provided, shall be indicated using the temperature class (T Codes) shown in Table 500.8(C) ≫ *500.8(C)(4)* ≪ .

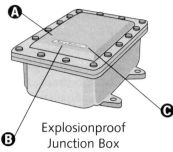

Explosionproof
Junction Box

© Cengage Learning 2012

> **N O T E**
>
> More than one marked temperature class or operating temperature, for gases and vapors, dusts, and different ambient temperatures, may appear ≫***500.8(C)(4)*** *Informational Note*≪ .

> **N O T E**
>
> For equipment installed in a Class II, Division 1 location, the temperature class or operating temperature shall be based on operation of the equipment when blanketed with the maximum amount of dust that can accumulate on the equipment ≫***500.8(C)(4)***≪ .

Protection Techniques

The purged and pressurized protection technique is permitted for equipment in any hazardous (classified) location for which it is identified ≫ *500.7(D)* ≪ .

Intrinsically safe apparatus and wiring is permitted in the specified hazardous (classified) location for which it is approved, provided the installation meets Article 504 requirements. Articles 501 through 503 and 510 through 516 are not applicable to such installations, unless required by Article 504 ≫ *500.7(E)* ≪ .

Dusttight: Constructed to prevent dust from entering the enclosed case, under stipulated test conditions ≫ *Article 100* ≪ .

A Acceptable protection techniques for electrical and electronic equipment in hazardous (classified) locations include explosionproof equipment; dust-ignitionproof; dusttight; purged and pressurized; intrinsically safe systems; nonincendive (circuit and equipment); oil immersion; hermetically sealed; combustible gas detection system; and other such techniques employing equipment specifically listed for a given location ≫ *500.7* ≪ .

B Dusttight-designated equipment is permitted for Class II, Division 2 or Class III, Division 1 or 2 locations ≫ *500.7(C)* ≪ .

C **Explosionproof equipment** is encased by constructive design to withstand a potential internal explosion of a specified gas/vapor, effectively preventing the ignition of surrounding gas/vapors by internal sparks, flashes, or explosions and operating at such an external temperature that any surrounding flammable atmosphere will not be ignited as a result ≫ *Article 100* ≪ .

D Explosionproof equipment is a protection technique that is permitted for equipment in Class I, Divisions 1 and 2 locations ≫ *500.7(A)* ≪ .

E Equipment of the dust-ignitionproof variety is acceptable for Class II, Divisions 1 or 2 locations ≫ *500.7(B)* ≪ .

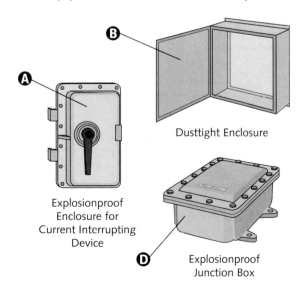

Explosionproof
Enclosure for
Current Interrupting
Device

Dusttight Enclosure

Explosionproof
Junction Box

Explosionproof Panelboard

Dust-Ignitionproof
Enclosure for Current
Interrupting Devices

© Cengage Learning 2012

Protection Techniques *(continued)*

Nonincendive Circuit: A circuit, other than field wiring, in which any arc or thermal effect produced under the equipment's intended operating conditions is incapable, under specified test conditions, of igniting the flammable gas–, vapor–, or dust–air mixture ≫*500.2*≪.

Nonincendive Field Wiring: Wiring that enters or leaves an equipment enclosure and, under the equipment's normal operating conditions, is not capable, due to arcing or thermal effects, of igniting the flammable gas–, vapor–, or dust–air mixture. Normal operation includes opening, shorting, or grounding the field wiring ≫*500.2*≪.

Under normal operating conditions, **nonincendive equipment's** electrical or electronic circuitry is incapable of igniting a given flammable substance due to arcing or thermal means ≫*500.2*≪.

A nonincendive component has a contacting mechanism capable of making or breaking an incendive circuit, and the contacting mechanism is constructed so that the component is incapable of igniting the specified flammable gas–air or vapor–air mixture. A nonincendive component's housing does not exclude the flammable atmosphere nor does it contain an explosion. This protection technique is allowed for equipment in Class I, Division 2; Class II, Division 2; or Class III, Division 1 or 2 locations ≫*500.2* and *500.7(H)*≪.

The oil-immersion protective technique is valid for current-interrupting contacts in Class I, Division 2 locations as described in 501.115(B)(1)(2) ≫*500.7(I)*≪.

Equipment can be **hermetically sealed** against the intrusion of an external atmosphere via a fusion seal, that is, soldering, brazing, welding, or the fusion of glass to metal ≫*500.2*≪.

Other protection techniques inherent to equipment listed for use in hazardous (classified) locations are allowed ≫*500.7(L)*≪.

Explosionproof
Incandescent
Light Fixture

Explosionproof
Vertical Seal

Explosionproof
Attachment Plug
and Receptacle

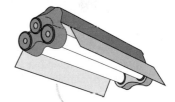

Explosionproof
Fluorescent
Light Fixture

Explosionproof
Outlet Box

Explosionproof
Vertical Seal with
Drain Fitting

Explosionproof Flexible Coupling

CLASS I LOCATIONS

Class I, Division 1 Locations

In addition to the specific requirements found within Article 500, general *Code* rules apply to the electric wiring and equipment in Class I locations.

In Class I, Division 1 locations, transformers and capacitors containing burnable liquid shall be installed only in approved vaults compliant with 450.41 through 450.48 in addition to all the following stipulations: (1) the vault shall not have any form of communicating opening into the Division 1 location; (2) ample ventilation shall continually remove flammable gases/vapors; (3) all vents/ducts lead to safe locations outside of buildings; and (4) vent ducts and openings shall be of sufficient area to relieve the vault's internal explosive pressures, having all portions of vent ducts (within the building) constructed of reinforced concrete »*501.100(A)(1)*«. Subsequently, transformers and capacitors not containing burnable liquids shall either be installed in 501.100(A)(1) compliant vaults or be identified for Class I locations »*501.100(A)(2)*«.

Identified enclosures shall be provided for Class I, Division 1 located meters, instruments, and relays. This category includes kilowatt-hour meters, instrument transformers, resistors, rectifiers, and thermionic tubes. Class I, Division 1 enclosures include explosionproof as well as purged and pressurized enclosures »*501.105(A)*«.

🅐 Class I, Division 1 classifications usually include the following locations:

1. Where volatile flammable liquids or liquefied flammable gases are transferred from one container to another
2. Spray booth interiors and areas in the vicinity of spraying/painting operations where flammable solvents are used
3. Locations containing open tanks/vats of volatile flammable liquids
4. Drying rooms or components for the evaporation of flammable solvents
5. Locations containing fat and oil extraction equipment using volatile flammable solvents
6. Portions of cleaning and dyeing plants where flammable liquids are used
7. Gas generator rooms and other portions of gas manufacturing plants where flammable gas may escape
8. Inadequately ventilated pump rooms for flammable gas or for volatile flammable liquids
9. Refrigerator/freezer interiors in which volatile flammable materials are stored in open, lightly stoppered, or easily ruptured containers
10. All other locations where ignitible concentrations of flammable vapors/gases are likely to occur in the course of normal operation »*500.5(B)(1) Informational Note No. 1*«

🅑 In some Division 1 locations, ignitible concentrations of flammable gases/vapors may be present continuously or for extended periods of time. Examples include the following:

1. The inside of inadequately vented enclosures containing instruments normally venting flammable gases/vapors within the enclosure
2. The inside of vented tanks containing volatile flammable liquids
3. The area between the inner and outer roof sections of a floating roof tank containing volatile flammable fluids
4. Inadequately ventilated areas within spraying/coating operations using volatile flammable fluids
5. The interior of an exhaust duct used to vent ignitible concentrations of gases/vapors

Avoiding the installation of electrical components in these areas altogether is strongly recommended. If relocation is unfeasible, use electric equipment or instrumentation approved specifically for the application, or consisting of what Article 504 describes as intrinsically safe systems »*500.5(B)(1) Informational Note No. 2*«.

🅒 Class I, Division 1 locations contain ignitible concentrations of flammable gases, flammable liquid-produced vapors, or combustible liquid-produced vapors that (1) can exist during normal operation; or (2) may often exist due to repair, maintenance, or leakage; or (3) might be released because operational failures cause simultaneous breakdowns of electrical equipment capable of direct ignition »*500.5(B)(1)*«.

🅓 Class I locations are those in which flammable gases, flammable liquid-produced vapors, or combustible liquid-produced vapors are or may be present in the air in quantities sufficient to produce explosive or ignitable mixtures »*500.5(B)*«.

NOTE

Bonding and grounding must comply with 501.30(A) and (B) provisions.

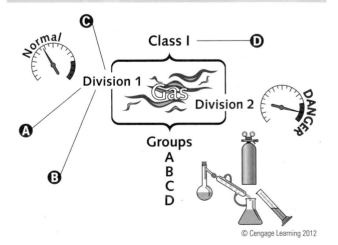

© Cengage Learning 2012

Class I, Division 2 Locations

Unless modified by this Article, 505.9(C)(2) approved equipment, specifically listed for use in Zone 0, 1, or 2 locations, is acceptable in Class I, Division 2 locations for the same gas and with a suitable temperature class ≫*501.5*≪.

In Class I, Division 2 locations, transformers shall comply with 450.21 through 450.27, and capacitors shall comply with 460.2 through 460.28 ≫*501.100(B)*≪.

Switches, circuit breakers, make-or-break push-button contacts, relays, and alarm bells and horns shall have enclosures identified for Class I, Division 1 locations per 501.105(A) ≫*501.105(B)(1)*≪. General-purpose enclosures are permitted if the current-interrupting contacts are (1) immersed in oil; (2) enclosed within a hermetically sealed chamber; (3) in nonincendive circuits; or (4) listed for Division 2 ≫*501.105(B)(1) Exception*≪.

Resistors, resistance devices, thermionic tubes, rectifiers, and similar equipment used in association with meters, instruments, and relays shall comply with 501.105(A) ≫*501.105(B)(2)*≪.

Transformer windings, impedance coils, solenoids, etc., without sliding/make-or-break contacts shall have enclosures, which may be of the general-purpose-type ≫*501.105(B)(3)*≪.

A single general-purpose enclosure is allowed for any assembly whose individual components qualify for general-purpose enclosures according to 501.105(B)(1) through (3). The exterior of an assembly enclosure containing resistors and similar equipment shall be clearly and permanently marked with the maximum obtainable surface temperature of any internal component. Alternatively, equipment can be marked with Table 500.8(B) identification numbers specifying its suitable temperature class (T Codes) ≫*501.105(B)(4)*≪.

Where general-purpose enclosures are allowed in 501.105(B)(1) through (4), they may contain overcurrent-protective fuses for instrument circuits (normally exempt from overloading), provided each fuse is preceded by a 501.105(B)(1) compliant switch ≫*501.105(B)(5)*≪.

Flexible cord, attachment plug, and receptacle connections are acceptable for process control instruments, provided all of the following conditions are met:

1. A 501.105(B)(1) compliant switch, rather than the attachment plug, provides for current interruption.

2. Maximum current is 3 amperes at 120 volts, nominal.

3. The power-supply cord is no more than 3 ft (900 mm) in length, is approved for extra-hard usage (where exposed) or for hard usage (if protected by location), and is supplied through a locking and grounding-type attachment plug and receptacle.

4. Only necessary receptacles are provided.

5. A receptacle label warns against unplugging under load ≫*501.105(B)(6)*≪.

Ⓐ Class I, Division 2 locations (1) are areas where volatile flammable gases, flammable liquid-produced vapors, or combustible liquid-produced vapors are handled, processed, or used, but the potentially volatile substance (liquid, gas, vapor) is normally enclosed (containers or systems) and is subject to escape only during accidental rupture (or breakdown) of such enclosures or in case of abnormal conditions; (2) are equipped with positive mechanical ventilation that normally prevents the buildup of gas/vapor concentrations to an ignition level; and (3) are adjacent to a Class I, Division 1 location that might communicate ignitible gaseous concentrations unless prevented by adequate positive-pressure ventilation from a clean air source, with adequate safeguards against ventilation failure ≫*500.5(B)(2)*≪.

Ⓑ Class I, Division 2 classification usually includes locations where volatile flammable liquids/gases, or vapors are used but that, in the judgement of the AHJ, would become hazardous only in case of an accident or unusual operating condition. The quantity of flammable material accidentally escaping, adequacy of ventilating equipment, total area involved, and the industry/business record with respect to explosions (or fires) are all factors that merit consideration in determining each location's classification and extent ≫*500.5(B)(2) Informational Note No. 1*≪.

Ⓒ Piping without valves, checks, meters, and similar devices would not ordinarily introduce a hazard even though used for flammable liquids or gases. Depending on factors such as quantity and size of containers and ventilation, locations used for the sealed-container storage of flammable liquids or liquefied or compressed gases may be considered either hazardous (classified) or unclassified locations ≫*500.(B)(2) Informational Note No. 2*≪.

> **NOTE**
>
> Bonding and grounding must comply with 501.30(A) and (B) provisions.

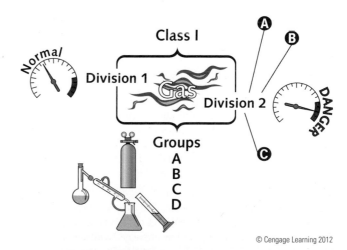

Wiring Methods—Class I, Division 1

Avoid tensile stress at Type MI cable termination fittings through proper installation and support methods ≫*501.10(A)(1)(b)*≪.

Type PVC conduit and Type RTRC conduit can be used when first surrounded by at least 2 in. (50 mm) of concrete, followed by a minimum cover of 24 in. (600 mm) (top of conduit to grade). In some cases, 514.8, Exception No. 2, and 515.8(A) permit the omission of the concrete encasement ≫*501.10(A)(1)(a) Exception*≪.

Threaded metal conduit (either rigid or steel intermediated) shall comprise the last 24 in. (600 mm) of a rigid nonmetallic underground conduit. This measurement ends at either the point of emergence or the point of aboveground raceway connection. An equipment grounding conductor is required for the dual purpose of providing raceway system electrical continuity, as well as grounding of non-current-carrying metal parts ≫*501.10(A)(1)(a) Exception*≪.

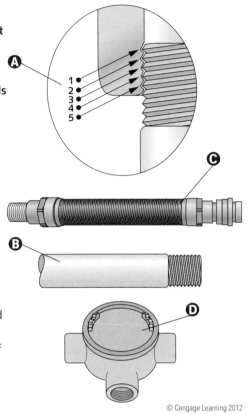

Ⓐ NPT threaded entries into explosionproof equipment shall fully engage at least five threads ≫*500.8(E)(1)*≪.

Ⓑ Fixed wiring in Class I, Division 1 locations shall consist of threaded metal conduit, either rigid metal or steel intermediate metal, or Type MI cable terminated with fittings listed for the location ≫*501.10(A)(1)*≪.

Ⓒ Where necessary to employ flexible connections, as at motor terminals, the following shall be permitted:

(1) Flexible fittings listed for the location, or

(2) Flexible cord in accordance with the provisions of 501.140, terminated with cord connectors listed for the location, or

(3) In industrial establishments with restricted public access, where the conditions of maintenance and supervision ensure that only qualified persons service the installation, for applications limited to 600 volts, nominal, or less, and where protected from damage by location or a suitable guard, listed Type TC-ER-HL cable with an overall jacket and a separate equipment grounding conductor(s) in accordance with 250.122 that is terminated with fittings listed for the location ≫*501.10(A)(2)*≪.

Ⓓ All boxes and fittings shall be approved for Class I, Division I ≫*501.10(A)(3)*≪.

© Cengage Learning 2012

Wiring Methods—Class I, Division 2

Explosionproof boxes and fittings are not mandatory unless required by 501.105(B)(1), 501.115(B)(1), and 501.150(B)(1) ≫*501.10(B)(4)*≪.

Nonincendive field wiring is permitted using any of the methods suitable for wiring in unclassified locations ≫*501.10(B)(3)*≪.

Ⓐ Where provisions must be made for flexibility, one of the following can be used: listed flexible metal fittings, flexible metal conduit with listed fittings, interlocked armor Type MC cable with listed fittings liquidtight flexible metal conduit with listed fittings, liquidtight flexible nonmetallic conduit with listed fittings, or flexible cord listed for extra-hard usage and terminated with listed fittings. Flexible cord shall include a conductor for use as an equipment grounding conductor. For elevator use, an identified elevator cable of Type EO, ETP, or ETT, shown under the "use" column in Table 400.4 for "hazardous (classified) locations" and terminated with listed fittings ≫*501.10(B)(2)*≪.

Ⓑ In Class I, Division 2 locations, the following wiring methods are permitted: (1) all wiring methods permitted in 501.10(A); (2) enclosed gasketed busways/wireways; (3) Type PLTC and Type PLTC-ER cable per Article 725, including installations in cable tray systems (the cable shall be terminated with

listed fittings); (4) Type ITC and Type ITC-ER cable as permitted in 727.4 and terminated with listed fittings; (5) Type MC, MV, or TC cable, including installations in cable tray systems (the cable shall be terminated with listed fittings); (6) in industrial establishments with restricted public access, where the

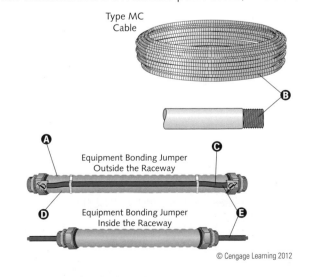

Type MC Cable

Equipment Bonding Jumper Outside the Raceway

Equipment Bonding Jumper Inside the Raceway

© Cengage Learning 2012

Wiring Methods—Class I, Division 2 *(continued)*

conditions of maintenance and supervision ensure that only qualified persons service the installation and where metallic conduit does not provide sufficient corrosion resistance, listed reinforced thermosetting resin conduit (RTRC), factory elbows, and associated fittings, all marked with the suffix -XW, and Schedule 80 PVC conduit, factory elbows, and associated fittings shall be permitted; and (7) Optical fiber cable Types OFNP, OFCP, OFNR, OFCR, OFNG, OFCG, OFN, and OFC shall be permitted to be installed in cable trays or any other raceway in accordance with 501.10(B). Optical fiber cables shall be sealed in accordance with 501.15 》》*501.10(B)(1)*《《.

C See 501.30(B) for grounding requirements where flexible conduit is used 》》*501.10(B)(2) Informational Note*《《.

D Flexible metal conduit and liquidtight flexible metal conduit shall include an equipment bonding jumper of

the wire type in compliance with 250.102 》》*501.30(B), 502.30(B), 503.30(B), 505.25(B), and 506.25(B)*《《.

E Bonding jumpers or conductors and equipment bonding jumpers shall be permitted to be installed inside or outside of a raceway or an enclosure.

(1) **Inside a Raceway or an Enclosure.** If installed inside a raceway, equipment bonding jumpers and bonding jumpers or conductors shall comply with the requirements of 250.119 and 250.148.

(2) **Outside a Raceway or an Enclosure.** If installed on the outside, the length of the bonding jumper or conductor or equipment bonding jumper shall not exceed 6 ft (1.8 m) and shall be routed with the raceway or enclosure 》》*250.102(E)*《《.

Conduit and Cable Seals—Class I, Divisions 1 and 2

A Seals in conduit and cable systems, within Class I, Divisions 1 and 2 shall comply with 501.15(A) through (F) 》》*501.15*《《.

B The cross-sectional area of the conductors or optical fiber tubes (metallic or nonmetallic) permitted in a seal shall not exceed 25% of the cross-sectional area of a rigid metal conduit of the same trade size unless the seal is specifically identified for a higher percentage of fill 》》*501.15(C)(6)*《《.

C If a probability exists that liquid or other condensed vapor may be trapped within control equipment enclosures, or at any point in the raceway system, install approved means to prevent accumulation or to permit periodic draining 》》*501.15(F)(1)*《《.

D Conduit/cable system seals minimize the passage of gas/vapors and prevent the passage of flames from one portion of the electrical installation to another through the conduit 》》*501.15 Informational Note No. 1*《《.

E The thickness of the sealing compound installed in completed seals, other than listed cable sealing fittings, shall at least equal the trade size of the sealing fitting; however, in no case shall the thickness of the compound be less than ⅝ in. (16 mm) 》》*501.15(C)(3)*《《.

F The compound shall provide a seal to minimize the passage of gas and/or vapors through the sealing fitting and shall not be affected by the surrounding atmosphere or liquids. The melting point of the compound shall not be less than 93°C (200°F) 》》*501.15(C)(2)*《《.

G A conduit seal shall be required in each conduit run leaving a Division 1 location. The sealing fitting shall be permitted to be installed on either side of the boundary within 10 ft (3.05 m) of the boundary, and it shall be designated and installed to minimize the amount of gas or vapor within the portion of the conduit installed in the Division 1 location that can be communicated beyond the seal 》》*501.15(A)(4) and 501.15(B)(2)*《《.

H The conduit run between the conduit seal and the point at which the conduit leaves the Division 1 location shall contain no union, coupling, box, or other fitting except for a listed explosionproof reducer installed at the conduit seal 》》*501.15(A)(4) and 501.15(B)(2)*《《.

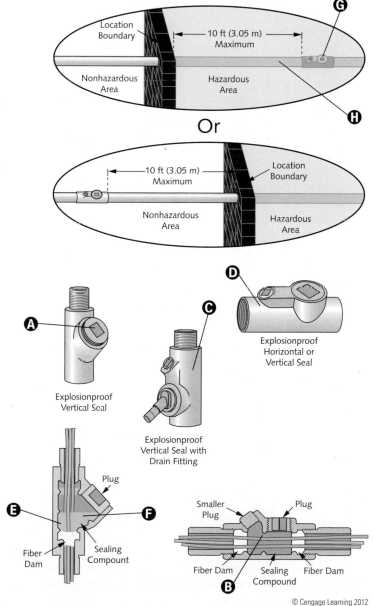

© Cengage Learning 2012

Conduit and Cable Seals—Class I, Divisions 1 and 2 *(continued)*

> **CAUTION**
>
> Sealing fittings shall be listed for use with one or more specific compounds and shall be accessible ≫ *501.15(C)(1)*◀◀.

> **N O T E**
>
> Study 501.15 carefully, because many exceptions and informational notes are contained therein.

Type MI cable termination fittings shall contain sealing compound that excludes moisture/fluids from the cable insulation ≫ *501.15* ◀◀.

In Class I, Division 1 locations, conduit seals shall be placed in each conduit entry into an explosionproof enclosure where either (1) the enclosure contains apparatus, such as switches, circuit breakers, fuses, relays, or resistors, that may produce arcs, sparks, or temperatures that exceed 80 percent of the autoignition temperature, in degrees Celsius, of the gas or vapor involved in normal operation or (2) the entry is trade size 2 or larger and the enclosure contains terminals, splices, or taps ≫ *501.15(A)(1)* ◀◀.

Class I, Division 2 connections to required explosionproof enclosures shall have a conduit seal in accordance with 501.15(A)(1)(1) and (A)(3). The entire conduit run or nipple between the seal and enclosure shall comply with 501.10(A) ≫ *501.15(B)(1)* ◀◀.

Splices/taps shall not be made inside fittings intended only for sealing with compound, nor can other fittings that contain splices/taps be filled with compound ≫ *501.15(C)(4)* ◀◀.

Commercial Garages, Repair, and Storage

Article 514 requirements apply to fuel-dispensing units (other than liquid petroleum gas, which is prohibited) located within buildings ≫ *511.4(B)(1)* ◀◀.

Within Class I locations as classified in 511.3, wiring shall conform to applicable Article 501 provisions ≫ *511.4(A)* ◀◀.

Seals conforming to 501.15 and 501.15(B)(2) provisions shall be provided and shall apply not only to horizontal but also to vertical boundaries of the defined Class I locations ≫ *511.9* ◀◀.

Equipment less than 12 ft (3.7 m) above the floor level that may produce arcs, sparks, or hot metal particles (such as cutouts, switches, charging panels, generators, motors, or other equipment and excluding receptacles, lamps, and lampholders) having make-or-break or sliding contacts, shall be of the totally enclosed type or constructed to prevent the escape of sparks or hot metal particles ≫ *511.7(B)(1)(a)* ◀◀.

Ⓐ Areas adjacent to classified locations in which flammable vapor release is unlikely (such as stock rooms, switchboard rooms, and similar locations), shall be unclassified where (1) mechanically ventilated at a rate of four (or more) air changes per hour; (2) designed with positive air pressure; or (3) effectively separated by walls or partitions ≫ *511.3(E)(1)* ◀◀.

Ⓑ A **major repair garage** is a building or portions of a building where major repairs, such as engine overhauls, painting, body and fender work, and repairs that require draining of the motor vehicle fuel tank are performed on motor vehicles, including associated floor space used for offices, parking, or showrooms ≫ *511.2* ◀◀.

Ⓒ Article 511 occupancies include those used for service and repair of self-propelled vehicles (including, but not limited to, passenger automobiles, buses, trucks, and tractors) that use volatile flammable liquids or flammable gases for fuel or power ≫ *511.1* ◀◀.

Ⓓ For pendants, flexible cord suitable for the type of service and listed for hard usage shall be used ≫ *511.7(A)(2)* ◀◀.

Ⓔ In major repair garages, unless there is mechanical ventilation meeting the specifications in 511.3(C)(1)(a), the entire floor area up to a level of 18 in. (450 mm) above the floor shall be classified as Class I, Division 2 ≫ *511.3(C)(1)(b)* ◀◀.

Ⓕ The pit area shall be a Class I, Division 2 location where there is mechanical ventilation providing a minimum of six air changes per hour ≫ *511.3(C)(3)(a)* ◀◀.

Ⓖ Fixed lighting lamps and lampholders that are located over lanes through which vehicles are commonly driven or that may otherwise be exposed to physical damage shall be

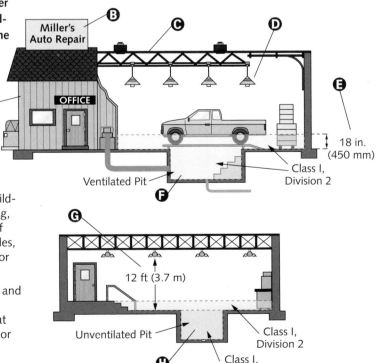

© Cengage Learning 2012

Commercial Garages, Repair, and Storage *(continued)*

located at least 12 ft (3.7 m) above floor level, unless totally enclosed or constructed to prevent the escape of sparks or hot metal particles ≫ *511.7(B)(1)(b)* ≪ .

H In major repair garages, unless there is mechanical ventilation meeting the specifications in 511.3(C)(3)(a), the pit or depression below floor level and extending up to the floor level shall be a Class I, Division 1 location ≫ *511.3(C)(3)(b)* ≪ .

> ### CAUTION
>
> All 125-volt, single-phase, 15- and 20-ampere receptacles in areas where electrical diagnostic equipment, electrical hand tools, or portable lighting equipment are used shall have GFCI protection for personnel ≫*511.12*≪ .

> ### NOTE
>
> The floor area shall be unclassified where there is mechanical ventilation providing a minimum of four air changes per hour or 1 cfm/ft² (0.3 m³/min/m²) of exchanged air for each square foot of floor area. Ventilation shall provide for air exchange across the entire floor area, and exhaust air shall be taken at a point within 12 in. (0.3 m) of the floor ≫*511.3(C)(1)(a)*≪ .

Any circuit in a Class I location supplying portables (or pendants) that include a grounded conductor (per Article 200) shall have receptacles, attachment plugs, connectors, and similar devices of the grounding type, and the flexible cord's grounded conductor shall either be connected to the screw shell of any lampholder or to the grounded terminal of any utilization equipment supplied. Equipment grounding conductor continuity (between the fixed wiring system and the non-current-carrying metal portions of pendant luminaires, portable lamps, and portable utilization equipment) shall be maintained via approved means ≫ *511.16(B)(1) and (2)* ≪ .

Battery chargers, their control equipment, and batteries being charged shall not be located within areas classified in 511.3 ≫ *511.10(A)* ≪ .

All metal raceways, the metal armor or metallic sheath on cables, and all non-current-carrying metal parts of fixed/portable electrical equipment, regardless of voltage, shall be grounded ≫ *511.16(A)* ≪ .

A All fixed wiring above Class I locations shall be placed within metal raceways, PVC, ENT, FMC, LFMC, or LFNC; or it shall be Type MC, AC, MI, manufactured wiring systems, PLTC cable in accordance with Article 725, or Type TC cable or Type ITC cable in accordance with Article 727 ≫ *511.7(A)(1)* ≪ .

B A **minor repair garage** is a building or portions of a building used for lubrication, inspection, and minor automotive maintenance work, such as engine tune-ups, replacement of parts, fluid changes (e.g., oil, antifreeze, transmission fluid, brake fluid, or air-conditioning refrigerants), brake system repairs, tire rotation, and similar routine maintenance work, including associated floor space used for offices, parking, or showrooms ≫ *511.2* ≪ .

C The floor area up to a level of 18 in. (450 mm) above any unventilated pit, belowgrade work area, or subfloor work area and extending a distance of 3 ft (900 mm) horizontally from the edge of any such pit, belowgrade work area, or subfloor work area, shall be classified as Class I, Division 2 ≫ *511.3(D)(1)(b)* ≪ .

D In minor repair garages, unless there is ventilation meeting the specifications in 511.3(D)(3)(a), any pit or depression below floor level and extending up to the floor level shall be a Class I, Division 2 location ≫ *511.3(D)(3)(b)* ≪ .

E Where lighter-than-air gaseous fuels (such as natural gas or hydrogen) will not be transferred, such locations shall be unclassified ≫ *511.3(D)(2)* ≪ .

> ### NOTE
>
> Where ventilation is provided to exhaust the pit area at a rate of not less than 1 cfm/ft² (0.3 m³/min/m²) of floor area at all times that the building is occupied or when vehicles are parked in or over this area and where exhaust air is taken from a point within 12 in. (300 mm) of the floor of the pit, belowgrade work area, or subfloor work area, the pit shall be unclassified ≫*511.3(D)(3)(a)*≪ .

> ### NOTE
>
> Floor areas in minor repair garages without pits, belowgrade work areas, or subfloor work areas shall be unclassified. Where floor areas include pits, belowgrade work areas, or subfloor work areas in lubrication or service rooms, the classification rules in (a) or (b) shall apply ≫*511.3(D)(1)*≪ .
> The entire floor area shall be unclassified where there is mechanical ventilation providing a minimum of four air changes per hour or 1 cfm/ft² (0.3 m³/min/m²) of exchanged air for each square foot of floor area. Ventilation shall provide for air exchange across the entire floor area, and exhaust air shall be taken at a point within 12 in. (0.3 m) of the floor ≫*511.3(D)(1)(a)*≪ .

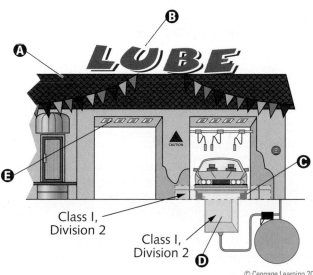

Class I, Division 2

Class I, Division 2

Parking Garages

For further information, see *Standard for Parking Structures*, NFPA 88A-2011, and *Code for Motor Fuel Dispensing Facilities and Repair Garages*, NFPA 30A-2012 》*511.3(A) Informational Note* 《.

A Garages used for parking or storage are permitted to be unclassified 》*511.3(A)* 《.

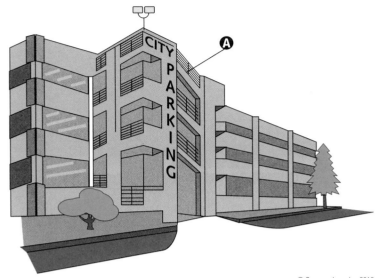

© Cengage Learning 2012

Motor Fuel Dispensing Facilities

Article 514 covers requirements applicable to motor fuel dispensing facilities, marine/motor fuel dispensing facilities, motor fuel dispensing facilities located inside buildings, and fleet vehicle motor fuel dispensing facilities 》*514.1* 《.

If the AHJ can satisfactorily determine that flammable liquids having a flash point below 38°C (100°F), such as gasoline, will not be handled, such location does not require a hazardous classification 》*514.3(A)* 《.

A Class I location does not extend beyond any unpierced wall, roof, or other solid partition 》*514.3(B)(1)* 《.

All metal raceways, the metal armor or metallic sheath on cables, and all non-current-carrying metal parts of fixed/portable electrical equipment, regardless of voltage, shall be grounded and bonded. Grounding and bonding in Class I locations shall comply with 501.30 》*514.16* 《.

A A **motor fuel dispensing facility** is that portion of a property where motor fuels are stored and dispensed from fixed equipment into fuel tanks of motor vehicles, marine craft, or into approved containers, including all equipment used in connection with motor fuel 》*514.2* 《.

B Wiring and equipment above Class I locations classified in 514.3 shall comply with 511.7 》*514.7* 《.

C All electrical equipment/wiring within Class I locations classified in 514.3 shall comply with the applicable provisions of Article 501 》*514.4* 《.

> **NOTE**
>
> Table 514.3(B)(1) shall be used where Class I liquids are stored, handled, or dispensed and to delineate and classify motor fuel dispensing facilities as well as any other location storing, handling, or dispensing Class I liquids 》*514.3(B)(1)* 《.

> **NOTE**
>
> Requirements for motor fuel dispensing stations in boatyards and marinas are in 514.3(C). Provisions for floating docks, piers, and wharfs are in 514.3(D) and (E).

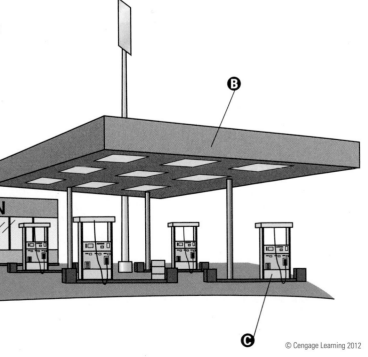

© Cengage Learning 2012

Dispensing Devices

Unattended self-service motor fuel dispensing facility emergency controls as specified in 514.11(A) shall be installed at a location acceptable to the AHJ, such location being more than 20 ft (6 m) but less than 100 ft (30 m) from dispensers. Additional emergency controls shall either be installed on each group of dispensers or on the outdoor equipment controlling the dispensers. Emergency controls shall completely shut off power to all of the station's dispensing equipment. Controls shall be manually reset only in a manner approved by the AHJ »514.11(C)«.

Attended self-service motor fuel dispensing facility emergency controls, as specified in 514.11(A), shall be installed at a location acceptable to the AHJ and not more than 100 ft (30 m) from dispensers »514.11(B)«.

Additional seals apply to horizontal as well as vertical boundaries of the Class I locations »514.9(B)«.

A Within 18 in. (450 mm) horizontally, extending in all directions from grade to the dispenser enclosure, or that portion of the dispenser enclosure containing liquid handling components, is a Class I, Division 2 location »Table 514.3(B)(1)«.

B The space inside the dispenser enclosure is classified as covered in *Power Operated Dispensing Devices for Petroleum Products,* ANSI/UL 87-1995 »Table 514.3(B)(1)«.

C Each circuit leading to (or through) dispensing equipment, including all associated power, communications, data, and video circuits, and equipment for remote pumping systems, shall have a clearly identified, readily accessible switch (or other approved means), located away from the dispensing devices, to simultaneously disconnect all circuit conductors (including the grounded conductor, if any) from the supply source. Single-pole breakers utilizing handle ties shall not be permited »514.11(A)«.

D From grade level up to a height of 18 in. (450 mm) within 20 ft (6.0 m) horizontally of any enclosure edge, is a Class I, Division 2 location »Table 514.3(B)(1)«.

E All or part of any pit, box, or space below grade level within a Division 1 or 2 (or Zone 1 or 2) classified location is designated Class I, Division 1 »Table 514.3(B)(1)«.

F A listed seal shall be provided in each conduit run entering/leaving a dispenser or any cavities (or enclosures) in direct communication with the dispensing area. The sealing fitting shall be the first fitting after the conduit emerges from the earth or concrete »514.9(A)«.

G Underground wiring shall be installed either in threaded RMC or threaded steel IMC. Any portion of electrical wiring below the surface of a Class I, Division 1 or 2 location shall be sealed within 10 ft (3.05 m) of the point of emergence above grade. Except for listed explosionproof reducers, at the conduit seal there shall be no union, coupling, box, or fitting between the conduit seal and the point of emergence above grade »514.8«.

> **NOTE**
>
> Each dispensing device shall be provided with a means to remove all external voltage sources, including power, communications, data, and video circuits and including feedback, during periods of maintenance and service of the dispensing equipment. The location of this means shall be permitted to be other than inside or adjacent to the dispensing device. The means shall be capable of being locked in the open position in accordance with 110.25 **»514.13«**.

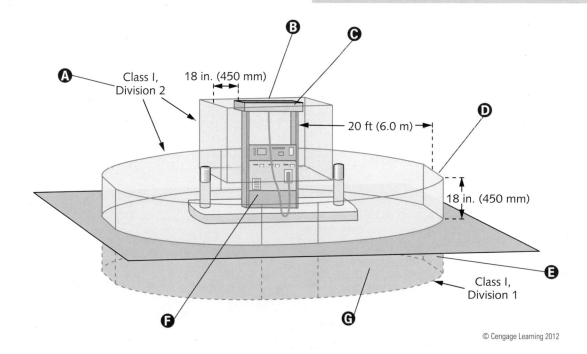

© Cengage Learning 2012

Aircraft Hangars

Mobile Equipment: Equipment having electric components designed for movement via mechanical aids, or having wheels to facilitate movement by person(s) or powered devices 》*513.2*《.

Portable Equipment (as applied to Article 513): Equipment with electric components that can be moved by a single person without using mechanical aids 》*513.2*《.

All wiring and equipment intended for installation that is installed or operated within any of the 513.3 defined Class I locations shall comply with the applicable provisions of Article 501 or Article 505 for the division or zone in which they are used 》*513.4(A)*《.

Attachment plugs and receptacles in Class I locations shall either be identified for Class I locations or designed so that they shall not be energized during connection/disconnection 》*513.4(A)*《.

All fixed wiring in a hangar not within a Class I location as defined in 513.3 shall be (1) installed in metal raceways or (2) Type MI, TC, or MC cable 》*513.7(A)*《.

Portable utilization equipment and luminaires require flexible cord suitable for the type of service and identified for extra-hard usage. Such cords shall include a separate equipment grounding conductor 》*513.10(E)(1) and (2)*《.

Adjacent areas in which flammable liquids or vapors are not likely to be released (stock rooms, electrical control rooms, etc.) shall be unclassified, provided such areas are adequately ventilated and effectively cut off from the hangar itself by walls or partitions 》*513.3(D)*《.

Portable utilization equipment that can be, or is, used within a hangar shall be of a type suitable for use in Class I, Division 2 or Zone 2 locations 》*513.10(E)(2)*《.

Seals shall be provided in accordance with 501.15 or 505.16, as applicable. Sealing requirements specified apply both to horizontal and vertical Class I location boundaries 》*513.9*《.

All metal raceways, the metal armor or metallic sheath on cables, and all non-current-carrying metal parts of fixed/portable electrical equipment, regardless of voltage, shall be grounded. Grounding in Class I locations shall comply with 501.30 or 505.25 》*513.16(A)*《.

Ⓐ Article 513 applies to buildings/structures or parts thereof inside which aircraft (containing Class I liquids or Class II liquids whose temperatures exceed their flash points) are housed, stored, serviced, repaired, or altered. It does not apply to locations used exclusively for aircraft that do not now, nor have ever, contained fuel 》*513.1*《.

Ⓑ The area within 5 ft (1.5 m) horizontally from aircraft power plants or fuel tanks is a Class I,

Division 2 or Zone 2 location extending upward from the floor to a level 5 ft (1.5 m) above the upper surface of wings and engine enclosures 》*513.3(C)(1)*《.

Ⓒ Any pit or depression below the hangar floor level is a Class I, Division 1 or Zone 1 location that extends up to floor level 》*513.3(A)*《.

Ⓓ The hangar's entire area, including any adjacent and communicating areas not suitably separated from the hangar, is designated as a Class I, Division 2 or Zone 2 location up to a level 18 in. (450 mm) above the floor 》*513.3(B)*《.

N O T E

In locations above those described in 513.3, equipment less than 10 ft (3.0 m) above aircraft wings and enclosures, which may produce arcs, sparks, or particles of hot metal (such as lamps and lampholders for fixed lighting, cutouts, switches, receptacles, charging panels, generators, motors, or other equipment having make-or-break or sliding contacts), shall be of the totally enclosed type or by construction shall prevent the escape of sparks or hot metal particles 》*513.7(C)*《.

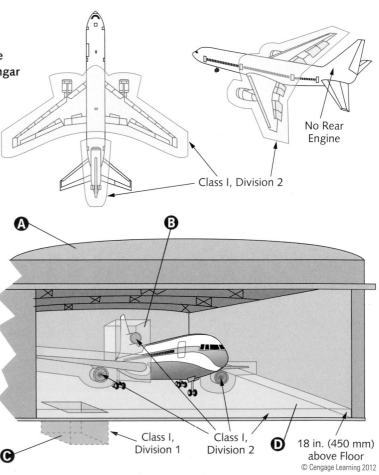

No Rear Engine

Class I, Division 2

Class I, Division 1   Class I, Division 2   18 in. (450 mm) above Floor

© Cengage Learning 2012

Bulk Storage Plants

Sealing requirements apply to horizontal as well as vertical boundaries of the defined Class I locations. Raceways buried under a defined Class I area are considered to be within a Class I, Division 1 or Zone 1 location ≫*515.9*≪.

Underground wiring shall be enclosed in either threaded RMC or threaded steel IMC; if buried under at least 2 ft (600 mm) of cover, Type PVC conduit, Type RTRC conduit (or a listed cable) can be used. If Type PVC conduit or Type RTRC conduit is used, threaded metal conduit (either rigid or steel intermediate) shall be used for not less than the last 2 ft (600 mm) of the conduit run to the conduit point of emergence from the underground location or to the point of connection to an aboveground raceway. Cable, if used, shall also be enclosed in threaded metal conduit (either rigid or steel intermediate) from the point of lowest buried cable to the point of aboveground raceway connection ≫*515.8(A)*≪.

Where Type PVC conduit, Type RTRC conduit, or cable with a nonmetallic sheath is used, include an equipment grounding conductor to provide electrical continuity of the raceway system and grounding of non-current-carrying metal parts ≫*515.8(C)*≪.

Where gasoline or other volatile flammable liquids or liquefied flammable gases are dispensed at bulk stations, the appropriate provisions of Article 514 apply ≫*515.10*≪.

All metal raceways, the metal armor or metallic sheath on cables, and all non-current-carrying metal parts of fixed/portable electrical equipment, regardless of voltage, shall be grounded and bonded as provided in Article 250 ≫*515.16*≪.

Ⓐ Article 515 covers a property or portion of a property where flammable liquids are received by tank vessel, pipelines, tank car, or tank vehicle and are stored or blended in bulk for the purpose of distributing such liquids by tank vessel, pipeline, tank car, tank vehicle, portable tank, or container ≫*515.1*≪.

Ⓑ Apply Table 515.3 where Class I liquids are stored, handled, or dispensed. Table 515.3 is also used to delineate and classify bulk storage plants. The class location does not extend beyond any floor, wall, roof, or other solid partition that has no communicating openings ≫*515.3*≪.

Ⓒ All electrical wiring and equipment within the Class I locations defined in 515.3 shall comply with the applicable provisions of Article 501 or Article 505 for the division or zone in which they are used ≫*515.4*≪.

Ⓓ All fixed wiring above Class I locations shall be in metal raceways, Schedule 80 PVC conduit, Type RTRC marked with suffix -XW, or Type MI, Type TC, or Type MC cable, or Type PLTC and Type PLTC-ER cable in accordance with the provisions of Article 725, including installation in cable tray systems or Type ITC and Type ITC-ER cable as permitted in 727.4. The cable shall be terminated with listed fittings. Fixed equipment that may produce arcs, sparks, or particles of hot metal (such as lamps and lampholders for fixed lighting, cutouts, switches, receptacles, motors, or other equipment having make-or-break or sliding contacts), shall either be of the totally enclosed type or otherwise be constructed to prevent the escape of sparks or hot metal particles. Portable luminaires or other utilization equipment and associated flexible cords shall comply with Article 501's or Article 505's requirements for the class of location above which they are connected or used ≫*515.7(A), (B), and (C)*≪.

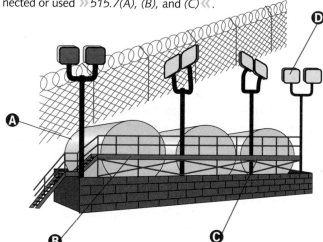

© Cengage Learning 2012

Spray Application, Dipping, Coating, and Printing Processes Using Flammable or Combustible Materials

Classification is based on quantities of flammable vapors, combustible mists, residues, dusts, or deposits that are present or might be present in quantities sufficient to produce ignitible or explosive mixtures with air ≫*516.3*≪.

The following spaces shall be considered Class I, Division I or Class I, Zone 0, as applicable:

1. The interior of any open or closed container of a flammable liquid
2. The interior of any dip tank or coating tank
3. The interior of any ink fountain, ink reservoir, or ink tank ≫*516.3(B)*≪

The following spaces shall be considered Class I, Division 1, or Class I, Zone 1, Class II, Division 1, or Zone 21 locations, as applicable:

1. Spray booth/room interiors except as specifically provided in 516.3(D)(7)
2. Exhaust duct interiors
3. Any area in the direct path of spray operations
4. For open dipping and coating operations, all space within a 5-ft (1.5-m) radius of the vapor source extending from the surface to the floor. The vapor source is liquid exposed during the process, which includes draining, and any dipped or coated object from which measurable vapor concentrations exceed 25% of the lower flammable limit within 1 ft (300 mm), in any direction, from the object.

Spray Application, Dipping, Coating, and Printing Processes Using Flammable or Combustible Materials (continued)

5. Sumps, pits, or belowgrade channels within 25 ft (7.5 m) horizontally of the vapor source. The sumps, pits, or belowgrade channels extending beyond 25 ft (7.5 m) must have a vapor stop, without which the entire pit is designated Class I, Division 1.

6. All space in all directions outside of but within 3 ft (900 mm) of open containers, supply containers, spray gun cleaners, and solvent distillation units containing flammable liquids.

7. For limited finishing workstations, the area inside the curtains or partitions »516.3(C)«

See 516.3(D)(1) through (D)(8) for Class I, Division 2; Class I, Zone 2; Class II, Division 2; and Zone 22 locations. The titles covered in this section include:

(1) Unenclosed Spray Processes.

(2) Closed-Top, Open-Face, and Open-Front Spray Booths and Spray Rooms.

(3) Open-Top Spray Booths.

(4) Enclosed Spray Booths and Spray Rooms.

(5) Limited Finishing Workstations.

(6) Areas Adjacent to Open Dipping and Coating Processes.

(7) Enclosed Coating and Dipping Operations.

(8) Open Containers.

The space adjacent to an enclosed dipping (or coating) process/apparatus is unclassified »516.3(D)(7)«.

Adjacent locations cut off from the defined Class I or Class II locations by tight partitions without communicating openings, where release of flammable vapors or combustible powders is unlikely, are unclassified »516.3(E)«.

Locations using drying, curing, or fusion apparatus having (1) positive mechanical ventilation adequate to prevent accumulation of flammable vapor concentrations and (2) effective interlocks to de-energize all electrical equipment (other than equipment approved for Class I locations) in case of inoperative ventilating equipment are unclassified where the AHJ deems appropriate »516.3(F)«.

While fixed electrostatic spraying and detearing equipment requirements are located in 516.10(A), electrostatic hand-spraying equipment provisions are found in 516.10(B).

All metal raceways, the metal armor or metallic sheath on cables, and all non-current-carrying metal parts of fixed/portable electrical equipment, regardless of voltage, shall be grounded and bonded. Grounding and bonding shall comply with 501.30, 502.30, or 505.25, as applicable »516.16«.

A Article 516 covers the regular (or frequent) spray application of flammable liquids, combustible liquids, and combustible powders, as well as the application of flammable/combustible liquids at temperatures above their flash point, by dipping, coating, or other means »516.1«.

B For open dipping and coating operations, all space within a 5-ft (1.5-m) radial distance from the vapor source, extending from these surfaces to the floor, is designated Class I, Division 1; Class I, Zone 1; Class II, Division 1; or Class I, Zone 21 locations »516.3(C)(4)«.

C A space 3 ft (915 mm) above the floor and extending 20 ft (6100 mm) horizontally in all directions from the Class I, Division 1 or Class I, Zone 1 location described in 6.4.3 shall be classified as Class I, Division 2 or Class I, Zone 2, and electrical wiring and electrical utilization equipment located within this space shall be suitable for Class I, Division 2 or Class I, Zone 2 locations, whichever is applicable »516.3(D)(4)«.

D A 3-ft (915-mm) space surrounding the dip tank and drain board's Class I, Division 1 or Class 1, Zone 1 location is considered Class I, Division 2; or Class I, Zone 2 »516.3(C)(5)«.

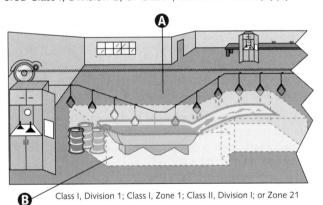

B Class I, Division 1; Class I, Zone 1; Class II, Division I; or Zone 21

Class I, Division 2 or Class I, Zone 2 (3 ft [915 mm] above floor, extending 20 ft [6100 mm] horizontally) **C**

D Class I, Division 2 or Class I, Zone 2 (outside area) | Class I, Division 1; Class I, Zone 1; Class II, Division 1; or Zone 21 (inside area) | Class I, Division 2 or Class I, Zone 2 (3 ft [915 mm] above floor, extending 20 ft [6100 mm] horizontally)

NOTE

Wiring and equipment above Class I and Class II locations must comply with 516.7 specifications.

CLASS II LOCATIONS

Class II, Division 1 Locations

In addition to the requirements found in Article 500, wiring and equipment in Class II locations shall meet Article 502's provisions.

Dust-ignitionproof means equipment enclosed in a manner that excludes dust, and, where installed and protected in accordance with this *Code,* which does not permit arcs, sparks, or heat otherwise generated or liberated inside the enclosure to ignite exterior accumulations or atmospheric suspensions of a specified dust on or near the enclosure »*500.2*«.

The temperature marking for Class II locations, per 500.8(C), shall be less than the ignition temperature of the specific dust to be encountered. For organic dusts that may dehydrate or carbonize, the temperature marking shall not exceed the lower of either the ignition temperature or 165°C (329°F) »*500.8(D)(2)*«.

The type of equipment and wiring defined as explosionproof (in Article 100) is not required and shall not be acceptable for Class II locations unless also identified for such locations »*502.5*«.

In Class II, Division 1 locations, transformers and capacitors containing burnable liquid shall be installed only in vaults compliant with 450.41 through 450.48, in addition to all the following stipulations: (1) openings (doors, etc.) communicating with the Division 1 locations shall have self-closing fire doors on both sides of the wall, carefully fitted and having suitable seals (such as weather stripping) to minimize the entrance of dust into the vault; (2) vent openings and ducts shall connect directly with outside air; and (3) suitable pressure-relief openings to the outside air shall be provided »*502.100(A)(1)*«. Subsequently, transformers and

capacitors not containing burnable liquids shall either be installed in vaults complying with 450.41 through 450.48, or be identified as a complete assembly, including terminal connections »*502.100(A)(2)*«.

No transformer or capacitor shall be installed in a Class II, Division 1, Group E location »*502.100(A)(3)*«.

Equipment listed and marked in accordance with 506.9(C)(2) for Zone 20 locations shall be permitted in Class II, Division 1 locations for the same dust atmosphere, and with a suitable temperature class »*502.6*«.

Ⓐ A Class II, Division 1 location is an area (1) in which combustible dust is airborne under normal operating conditions in quantities sufficient to produce explosive or ignitible mixtures; (2) where mechanical failure or abnormal operation of machinery/equipment might not only produce such explosive or ignitible mixtures, but also provide an ignition source through simultaneous failure of electric equipment, operation of protection devices, or by other causes; or (3) in which Group E combustible dusts may exist in hazardous quantities »*500.5(C)(1)*«.

Ⓑ Because dusts containing magnesium or aluminum are particularly hazardous, extreme precautions are necessary to avoid ignition and explosion »*500.5(C)(1) Informational Note*«.

Ⓒ Class II locations are hazardous because of the presence of combustible dust »*500.5(C)*«.

> **NOTE**
>
> Bonding and grounding must comply with 502.30(A) and (B) provisions.

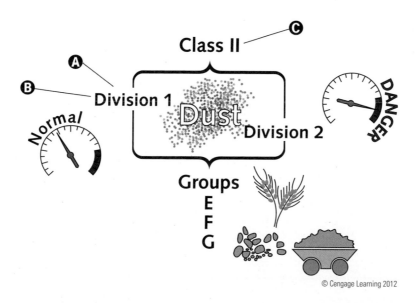

© Cengage Learning 2012

Class II, Division 2 Locations

In Class II, Division 2 locations, transformers and capacitors containing burnable liquid shall be installed within vaults compliant with 450.41 through 450.48 ≫*502.100(B)(1)*≪.

Askarel-containing transformers, rated in excess of 25 kVA, shall have (1) pressure-relief vents; (2) a means for absorbing any gases generated by arcing inside the case, or chimney/flue-connected pressure-relief vents that shall carry such gases outside the building; and (3) at least 6 in. (150 mm) of airspace between the transformer case and any adjacent combustible material ≫*502.100(B)(2)*≪.

Dry-type transformers, operating at not over 600 volts, nominal, shall either be installed in vaults or shall have their windings and terminal connections enclosed in tight metal housings without ventilation or other openings ≫*502.100(B)(3)*≪.

Equipment listed and marked in accordance with 506.9(C)(2) for Zone 20, 21, or 22 locations shall be permitted in Class II, Division 2 locations for the same dust atmosphere and with a suitable temperature class ≫*502.6*≪.

Ⓐ A Class II, Division 2 location is an area (1) where combustible dust due to abnormal operations may be present in the air in quantities sufficient to produce explosive or ignitible mixtures; (2) where combustible dust accumulations typically do not interfere with normal operation of electrical equipment or other apparatus, but may become airborne as a result of infrequent malfunctioning of processing/handling equipment; and (3) where combustible dust accumulations on, in, or near electrical equipment may be sufficient to interfere with safe heat dissipation from said equipment, or may be ignited by abnormal operation or failure of electrical equipment ≫*500.5(C)(2)*≪.

Ⓑ The quantity of combustible dust that may be present and the adequacy of dust removal systems are factors that merit consideration in determining the classification and may result in an unclassified area ≫*500.5(C)(2) Informational Note No. 1*≪.

Ⓒ Where the handling of substances such as seed produces low quantities of dust, the amount of dust deposited may not warrant classification ≫*500.5(C)(2) Informational Note No. 2*≪.

> **N O T E**
>
> Bonding and grounding must comply with 502.30(A) and (B) provisions.

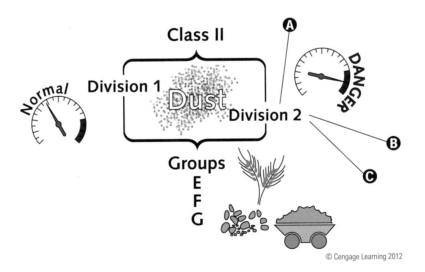

© Cengage Learning 2012

Wiring Methods—Class II, Divisions 1 and 2

Approved wiring methods in Class II, Division 1 locations include threaded RMC, threaded steel IMC, or Type MI cable with termination fittings listed for the location. Type MI cable shall be installed and supported in a manner to avoid tensile stress at the termination fittings ❯❯*502.10(A)(1)(1) and (2)*❮❮. In industrial establishments with limited public access, where maintenance and supervision conditions ensure that only qualified persons will service the installation, Class II, Division 1–listed Type MC-HL cable with a gas/vaportight continuous corrugated metallic sheath, an overall jacket of suitable polymeric material, separate equipment grounding conductors (per 250.122), and termination fittings listed for the application are permitted ❯❯*502.10(A)(1)(3)*❮❮.

Class II, Division 1 fittings and boxes shall be provided with threaded bosses for conduit connections or cable terminations and shall be dusttight. Fittings and boxes containing taps, joints, or terminal connections or those that are used in Group E locations shall be identified for Class II locations ❯❯*502.10(A)(1)(4)*❮❮.

Use only (1) dusttight flexible connectors; (2) LFMC with listed fittings; (3) LFNC with listed fittings; (4) interlocked armor Type MC cable having an overall jacket of suitable polymeric material and provided with termination fittings listed for Class II, Division 1 locations; (5) flexible cord listed for extra-hard usage and terminated with listed dusttight cord connectors. Any flexible cords used shall comply with 502.140; or (6) for elevator use, an identified elevator cable of Type EO, ETP, or ETT, shown under the "use" column in Table 400.4 for "hazardous (classified) locations" and terminated with listed dusttight fittings ❯❯*502.10(A)(2) and (B)(2)*❮❮.

Sealing fittings shall be accessible ❯❯*502.15*❮❮.

Unless an exception is met, approved wiring methods in Class II, Division 2 locations include (1) all wiring methods permitted in 502.10(A); (2) RMC, IMC, EMT, dusttight wireways; (3) Type MC or MI cable with listed termination fittings; (4) Type PLTC and Type PLTC-ER cable in accordance with the provisions of Article 725, including installation in cable tray systems (the cable shall be terminated with listed fittings); (5) Type ITC and Type ITC-ER cable as permitted in 727.4 and terminated with listed fittings; (6) Type MC, MI, or TC cable installed in cable trays (ladder, ventilated trough, or ventilated channel) in a single layer, with a space at least equal to the larger cable diameter between the two adjacent cables; or (7) in industrial establishments with restricted public access where the conditions of maintenance and supervision ensure that only qualified persons service the installation and where metallic conduit does not provide sufficient corrosion resistance, reinforced thermosetting resin conduit (RTRC) factory elbows, and associated fittings, all marked with suffix-XW, and Schedule 80 PVC Conduit, factory elbows, and associated fittings ❯❯*502.10(B)(1)*❮❮.

All Class II, Division 2 boxes and fittings shall be dusttight ❯❯*502.10(B)(4)*❮❮.

In Class II, Division 1 and 2 locations, where a raceway provides communication between an enclosure that is required to be dust-ignitionproof and one that is not, suitable means shall prevent the entrance of dust into the dust-ignitionproof enclosure via the raceway. The following means are permitted: (1) a permanent and effective seal; (2) a horizontal raceway at least 10 ft (3.05 m) long; (3) a vertical raceway not less than 5 ft (1.5 m) long that extends downward from the dust-ignitionproof enclosure; or (4) a raceway installed in a manner equivalent to (2) or (3) that extends only horizontally and downward from the dust-ignitionproof enclosures ❯❯*502.15*❮❮.

Seals are not required where a raceway provides communication between an enclosure that is required to be dust-ignitionproof and an enclosure in an unclassified location ❯❯*502.15*❮❮.

In Class II, Division 1 and 2 locations, explosionproof seals are not required ❯❯*502.15*❮❮.

Ⓐ Class II locations are hazardous because of the presence of combustible dust ❯❯*500.5(C)*❮❮.

CLASS III LOCATIONS

Class III, Division 1 Locations

In addition to Article 500 requirements, wiring and equipment in locations designated Class III must be installed in accordance with provisions in Article 503.

Class III located equipment shall be able to function at full rating without developing surface temperatures high enough to cause excessive dehydration or gradual carbonization of accumulated fibers or flyings. Carbonized or excessively dried organic material is highly susceptible to spontaneous ignition. During operation, the maximum surface temperature shall not exceed 165°C (329°F) for equipment not subject to overloading and 120°C (248°F) for equipment that may be overloaded (such as motors or power transformers) 》*503.5* 《.

Transformers and capacitors in Class III, Division 1 locations shall comply with 502.100(B) 》*503.100* 《.

Equipment listed and marked in accordance with 506.9(C)(2) for Zone 20 locations and with a temperature class of not greater than T120°C (for equipment that may be overloaded) or not greater than T165°C (for equipment not subject to overloading) shall be permitted in Class III, Division 1 locations 》*503.6* 《.

A A Class III, Division 1 location is an area where easily ignitible fibers/flyings are handled, manufactured, or used 》*500.5(D)(1)* 《.

B Easily ignitible fibers/flyings include rayon, cotton (including cotton linters and cotton waste), sisal or henequen, istle, jute, hemp, tow, cocoa fiber, oakum, baled waste kapok, Spanish moss, excelsior, and other materials of similar nature 》*500.5(D)(1) Informational Note No. 2* 《.

C Class III locations are hazardous because of the presence of easily ignitible fibers or where materials producing combustible flyings are handled, manufactured, or used, even though they are not likely to be airborne in quantities sufficient to produce ignitible mixtures 》*500.5(D)* 《.

> **NOTE**
>
> Bonding and grounding shall comply with 503.30(A) and (B) provisions.

> **CAUTION**
>
> In a Class III, Division 1 location, the operating temperature shall be the temperature of the equipment when blanketed with the maximum amount of dust (simulating fibers/flyings) that can accumulate on the equipment 》*503.5*《.

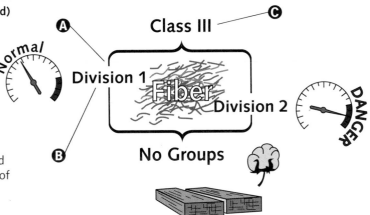

© Cengage Learning 2012

Class III, Division 2 Locations

In Class III, Division 2 locations, transformers and capacitors shall comply with 502.100(B) 》*503.100* 《.

Equipment listed and marked in accordance with 506.9(C)(2) for Zone 20, 21, or 22 locations and with a temperature class of not greater than T120°C (for equipment that may be overloaded) or not greater than T165°C (for equipment not subject to overloading) shall be permitted in Class III, Division 2 location 》*503.6* 《.

A A Class III, Division 2 location is an area in which easily ignitible fibers/flyings are stored or handled other than in the manufacturing process 》*500.5(D)(2)* 《.

> **NOTE**
>
> Bonding and grounding must comply with 503.30(A) and (B) provisions.

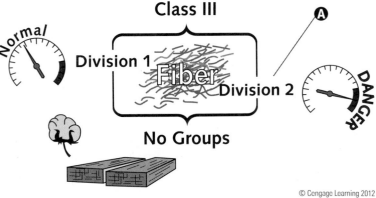

© Cengage Learning 2012

Wiring Methods—Class III, Division 1

In Class III, Division 1 locations, the following wiring methods are permitted: (1) RMC, PVC, RTRC, IMC, EMT, dusttight wireways, or Type MC or MI cable with listed termination fitting; (2) Type PLTC and Type PLTC-ER cable in accordance with the provisions of Article 725 including installation in cable tray systems (the cable shall be terminated with listed fittings); (3) Type ITC and Type ITC-ER cable as permitted in 727.4 and terminated with listed fittings; (4) Type MC, MI, or TC cable installed in ladder, ventilated trough, or ventilated channel cable trays in a single layer, with a space not less than the larger cable diameter between the two adjacent cables. The cable shall be terminated with listed fittings ⟫*503.10(A)*⟪. Note, see exception to (4).

All Class III, Division 1 boxes and fittings shall be dusttight ⟫*503.10(A)(2)*⟪.

Where necessary to employ flexible connections, one or more of the following shall be permitted: (1) dusttight flexible connectors, (2) liquidtight flexible metal conduit with listed fittings, (3) liquidtight flexible nonmetallic conduit with listed fittings, (4) interlocked armor Type MC cable having an overall jacket of suitable polymeric material and installed with listed dusttight termination fittings, (5) flexible cord in compliance with 503.140, and (6) for elevator use, an identified elevator cable of Type EO, ETP, or ETT, shown under the "use" column in Table 400.4 for "hazardous (classified) locations" and terminated with listed dusttight fittings ⟫*503.10(A)(3)*⟪. For grounding requirements where flexible conduit is used, see 503.30(B) ⟫*503.10(A)(3) Informational Note*⟪.

Class III, Division 1 and 2 switches, circuit breakers, motor controllers, and fuses (including push buttons, relays, and similar devices), shall have dusttight enclosures ⟫*503.115*⟪.

Class III, Division 1 and 2 flexible cords shall comply with the following: (1) be of a type listed for extra-hard usage; (2) contain, in addition to the conductors of the circuit, a 400.23 compliant equipment grounding conductor; (3) be supported by clamps or other suitable means so that there will be no tension on the terminal connections; and (4) be terminated with a listed dusttight cord and connector ⟫*503.140*⟪.

Class III, Division 1 and 2 receptacles and attachment plugs shall (1) be of the grounding type; (2) be designed to minimize the accumulation (or entry) of fibers/flyings; and (3) prevent the escape of sparks or molten particles ⟫*503.145*⟪. If, in the judgement of the AHJ, only moderate accumulations of lint or flyings are likely to collect in the vicinity of a receptacle, and where such receptacle is readily accessible for routine cleaning, general-purpose grounding-type receptacles (mounted to minimize the entry of fibers/flyings) are permitted ⟫*503.145 Exception*⟪.

Ⓐ Such locations usually include some parts of rayon, cotton, and other textile mills; combustible fibers/flyings manufacturing and processing plants; cotton gins and cotton-seed mills; flax-processing plants; clothing manufacturing plants; woodworking plants; and establishments and industries involving similar hazardous processes or conditions ⟫*500.5(D)(1) Informational Note No. 1*⟪.

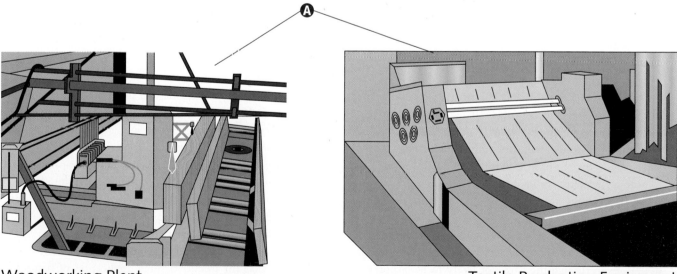

Woodworking Plant

Textile Production Equipment

Wiring Methods—Class III, Division 2

Unless an exception is met, Class III, Division 2 location wiring methods shall comply with 503.10(A) ❯❯*503.10(B)*❮❮.

All Class III, Division 2 boxes and fittings shall be dusttight ❯❯*503.10(A)(2)*❮❮.

Class III, Division 1 and 2 pendant luminaires shall be suspended by stems of threaded RMC, threaded IMC, threaded metal tubing of equivalent thickness, or by chains with approved fittings. Stems longer than 12 in. (300 mm) shall be permanently and effectively braced against lateral displacement at a level no more than 12 in. (300 mm) above the lower end of the stem, or flexibility in the form of an approved fitting or a flexible connector shall be provided no more than 12 in. (300 mm) from the point of attachment to the supporting box or fitting ❯❯*503.130(C)*❮❮.

Class III, Division 1 and 2 portable lighting equipment shall have both handles and substantial guards. Lampholders shall be of the unswitched type, incapable of receiving attachment plugs. There shall be no exposed current-carrying metal parts, and all exposed non-current-carrying metal parts shall be grounded. In all other respects, portable lighting equipment shall comply with 503.130(A) ❯❯*503.130(D)*❮❮.

Ⓐ Class III, Division 1 and 2 luminaires for fixed lighting shall provide enclosures for lamps and lampholders designed to minimize entrance of fibers/flyings and to prevent the escape of sparks, burning material, or hot metal. Each luminaire shall be clearly marked to show the maximum permitted lamp wattage without

exceeding an exposed surface temperature of 165°C (329°F) under normal conditions of use ❯❯*503.130(A)*❮❮.

Ⓑ A Class III, Division 2 location is a location in which easily ignitible fibers/flyings are stored or handled other than in the process of manufacture ❯❯*500.5(D)(2)*❮❮.

> **NOTE**
>
> Class III, Division 1 and 2 luminaires subject to physical damage shall be protected by a suitable guard ❯❯*503.130(B)*❮❮.

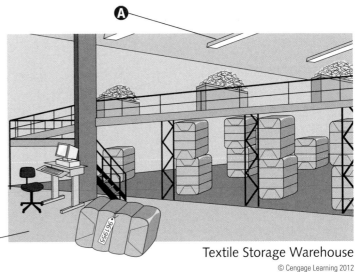

Textile Storage Warehouse

© Cengage Learning 2012

Summary

- Hazardous locations are divided into three classifications (Class I, II, and III); each class is subdivided into two categories (Division 1 and 2).

- While Article 500 covers general requirements for all classes, Articles 501, 502, and 503 provide specific provisions for each class individually.

- Class I locations are those in which flammable gases, flammable liquid-produced vapors, or combustible liquid-produced vapors are or may be present in the air in quantities sufficient to produce explosive or ignitable mixtures.

- Class II locations are hazardous due to the presence of combustible dust.

- Class III locations are hazardous because of the presence of easily ignitible fibers or on material producing combustible flyings, even though they are not likely to be airborne in quantities sufficient to produce ignitible mixtures.

- Under typical operating conditions, Division 1 locations contain ignitible or combustible elements (gases, dust, fibers, etc.) in quantities sufficient to produce explosive (or ignitible) mixtures.

- Locations where ordinarily present quantities of combustible or ignitible elements (gases, dust, fibers, etc.) in the air are insufficient to produce explosive (or ignitible) mixtures are classified as Division 2.

- Article 511 occupancies include those used for service and repair of self-propelled vehicles (including, but not limited to, passenger automobiles, buses, trucks, and tractors) that use volatile flammable liquids for fuel or power.

- Article 513 applies to buildings/structures or parts thereof inside which aircraft (containing Class I or II liquids whose temperatures exceed their flash points) are housed, stored, serviced, repaired, or altered.

- Motor fuel dispensing facility requirements are in Article 514.

- Article 515 provisions address bulk storage plants.

- The regular (or frequent) spray application of flammable liquids, combustible liquids, and combustible powders, as well as the application of flammable/combustible liquids at temperatures above their flash point, by dipping, coating, or other means, are covered by Article 516.

Unit 15 Competency Test

NEC Reference	Answer	
_____	_____	1. In aircraft hangars, any pit or depression below the level of the hangar floor shall be classified as a _____ location that shall extend up to said floor level.
_____	_____	2. Regardless of the classification of the location in which it is installed, equipment that depends on a single compression seal, diaphragm, or tube to prevent flammable or combustible fluids from entering the equipment, shall be identified for a _____ location.
_____	_____	3. A _____ location is a location in which volatile flammable gases, flammable liquid-produced vapors, or combustible liquid-produced vapors are handled, processed, or used, but in which the liquids, vapors, or gases will normally be confined within closed containers or closed systems from which they can escape only in case of accidental rupture or breakdown of such containers or systems or in case of abnormal operation of equipment.
_____	_____	4. All cut ends of rigid metal conduit (RMC) shall be reamed or otherwise finished to remove rough edges. Where conduit is threaded in the field, a standard cutting die with a _____ in. taper per foot shall be used.
_____	_____	5. In bulk storage plants, aboveground tank vents shall be classified as _____ within 5 ft (1.5 m) of the open end, extending in all directions.
_____	_____	6. In Class I, Division 1 locations, NPT threaded entries into explosionproof equipment shall be made up with at least _____ threads fully engaged.
_____	_____	7. In Class II, Division 1 locations, boxes and fittings shall be provided with _____ for connection to conduit or cable terminations, and shall be dusttight.
_____	_____	8. A 2-in. threaded RMC is terminating into a 2-in. sealing fitting. The minimum thickness of the sealing compound shall be _____ in.
_____	_____	9. In aircraft hangars, the area within _____ ft horizontally from aircraft power plants or aircraft fuel tanks shall be classified as a Class I, Division 2 or Zone 2 location that shall extend upward from the floor to a level _____ ft above the upper surface of wings and of the engine enclosures.
_____	_____	10. _____ is defined as equipment enclosed in a manner that will exclude dusts and does not permit arcs, sparks, or heat otherwise generated or liberated inside of the enclosure to cause ignition of exterior accumulations or atmospheric suspensions of a specified dust on or in the vicinity of the enclosure.
_____	_____	11. Class I locations are those in which _____ are or may be present in the air in quantities sufficient to produce explosive or ignitible mixtures.
_____	_____	12. In commercial garages, lamps and lampholders for fixed lighting that is located over lanes through which vehicles are commonly driven shall be located no less than _____ ft above the floor level.
_____	_____	13. A _____ location is a location in which easily ignitible fibers/flyings are handled, manufactured, or used.
_____	_____	14. _____, _____, _____, _____, and _____ are five acceptable hazardous (classified) location protection techniques for electrical and electronic equipment.
_____	_____	15. A(n) _____ is defined as a circuit other than field wiring in which any arc or thermal effect produced under intended operating conditions of the equipment is not capable, under specified test conditions, of igniting the flammable gas–, vapor–, or dust–air mixture.

GENERAL

Nursing Home and Limited-Care Facility

Requirements in 517.40(C) through 517.44 do not apply to freestanding buildings used as nursing homes and limited-care facilities, provided that:

(a) Admitting and discharge policies are maintained that preclude providing care to any patient (or resident) who may need to be sustained by electrical life-support equipment.

(b) No surgical treatment requiring general anesthesia is offered.

(c) Automatic battery-operated systems or equipment are provided, effective for a minimum of 1½ hours and otherwise in accordance with 700.12. Such a system shall capably supply lighting for exit lights, exit corridors, stairways, nursing stations, medical preparation areas, boiler rooms, and communications areas, as well as power to operate all alarm systems ≫517.40(A) Exception≪.

Nursing homes and limited-care facilities that are contiguous with or adjacent to a hospital can have essential electrical systems supplied by the hospital ≫517.40(C)≪.

Ⓐ **Nursing home:** A building (or part thereof) used for 24-hour housing and nursing care of four or more persons who, because of mental or physical incapacity, may be unable to safely provide for their own needs without assistance ≫517.2≪.

Ⓑ Nursing homes and limited-care facilities are included in the term **health care facilities.**

Ⓒ **Limited-care facility:** A building (or part thereof) used for 24-hour housing of four or more persons incapable of self-preservation because of age; physical limitation, whether due to accident or illness; or limitations, such as mental retardation/developmental disability, mental illness, or chemical dependency ≫517.2≪.

Ⓓ The requirements of Part III, 517.40(C) through 517.44, shall apply to nursing homes and limited-care facilities ≫517.40(A)≪.

Ⓔ Article 517, Part II (Wiring and Protection) requirements shall not apply to areas of nursing homes and limited-care facilities wired in accordance with Chapters 1 through 4 of the *NEC* where these areas are used exclusively as patient sleeping rooms ≫517.10(B)(2)≪.

> **N O T E**
>
> For those nursing home and limited-care facilities that admit patients who need to be sustained by electrical life-support equipment, the essential electrical system from the source to the portion of the facility where such patients are treated shall comply with the requirements of Part III, 517.30 through 517.35 ≫*517.40(B)*≪.

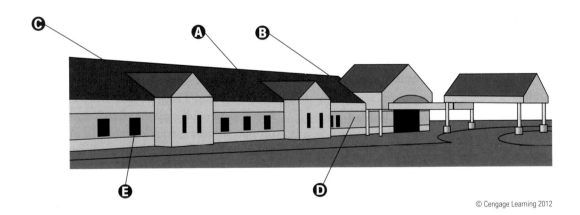

© Cengage Learning 2012

Hospital

The provisions of Article 517 shall apply to electrical construction and installation criteria in health care facilities that provide services to human beings ≫*517.1*≪.

Article 517 specifies the installation criteria and wiring methods that minimize electrical hazards by maintaining adequately low-potential differences between exposed conductive surfaces that are likely to become energized and are subject to patient contact ≫*517.11*≪.

Psychiatric hospital: A building used exclusively for 24-hour psychiatric care, with at least four inpatients ≫*517.2*≪.

Parts II and III requirements apply not only to single-function buildings, but also to their respective forms of occupancy within a multifunction building (for example, a doctor's examining room located within a limited-care facility shall meet 517.10 provisions) ≫*517.1*≪.

Ⓐ Hospital: A building (or part thereof) used for 24-hour medical, psychiatric, obstetrical, or surgical care of four or more inpatients ≫*517.2*≪.

Ⓑ Health care facilities: Buildings (or portions thereof) in which medical, dental, psychiatric, nursing, obstetrical, or surgical care is provided. Health care facilities include but are not limited to hospitals; nursing homes; limited-care facilities; clinics; medical and dental offices; and ambulatory care centers, whether permanent or movable ≫*517.2*≪.

Ⓒ Patient bed location: The location of a patient sleeping bed or the bed (or procedure table) of a critical care area ≫*517.2*≪.

Ⓓ Article 517, Part II applies to patient care space of all health care facilities ≫*517.10(A)*≪.

Ⓔ Patient care space: Space within a health care facility wherein patients are intended to be examined or treated ≫*517.2*≪. The four categories of patient care space are: basic care space, general care space, critical care space, and support space.

Ⓕ Critical Care Space: Space in which failure of equipment or a system is likely to cause major injury or death to patients or caregivers ≫*517.2 Critical Care Space under Patient Care Space*≪.

Critical care space includes special care units, intensive care units, coronary care units, angiography laboratories, cardiac catheterization laboratories, delivery rooms, operating rooms, and similar areas in which are patients are intended to be subjected to invasive procedures and are connected to line-operated, electromedical devices ≫*517.2 Patient Care Space, Informational Note No. 4*≪.

Ⓖ General Care Space: Space in which failure of equipment or a system is likely to cause minor injury to patients or caregivers ≫*517.2 General Care Space under Patient Care Space*≪.

General care space includes areas such as patient bedrooms, examining rooms, treatment rooms, clinics, and similar areas where the patient may come into contact with electromedical devices or ordinary appliances such as a nurse call system, electric beds, examining lamps, telephones, and entertainment devices ≫*517.2 Patient Care Space, Informational Note No. 3*≪.

> **N O T E**
>
> Article 517, Part II does not apply to the following: (1) business offices, corridors, waiting rooms, and the like in clinics, medical and dental offices, and outpatient facilities and (2) areas of nursing homes and limited-care facilities wired according to Chapters 1 through 4 of the *NEC*, where these areas are used exclusively as patient sleeping rooms ≫*517.10(B)*≪.
> See 517.2 for definitions pertaining to health care facilities.

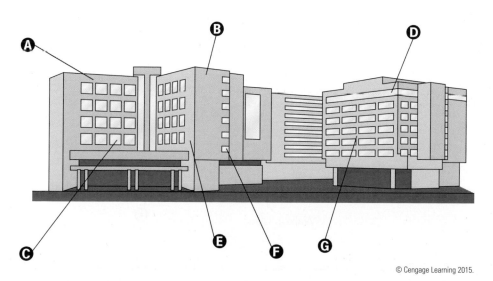

Patient Care Vicinity

Ⓐ Patient care vicinity: A space, within a location intended for the examination and treatment of patients, extending 6 ft (1.8 m) beyond the normal location of the patient bed, chair, table, treadmill, or other device that supports the patient during examination and treatment and extending vertically to 7 ft 6 in. (2.3 m) above the floor ≫ *517.2* ≪.

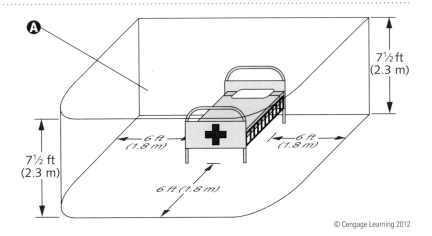

© Cengage Learning 2012

Ambulatory Health Care Occupancy

Where electrical life support equipment is required, the essential electrical distribution system shall comply with 517.30 through 517.35 ≫ *517.45(B)* ≪.

Where critical care areas are present, the essential electrical distribution system shall comply with 517.30 through 517.35 ≫ *517.45(C)* ≪.

Battery systems shall be installed per Article 700's requirements, and generator systems shall comply with 517.30 through 517.35 ≫ *517.45(D)* ≪.

Ⓐ Ambulatory health care occupancy: A building (or part thereof) used to provide simultaneous service (or treatment) to four or more patients that provides, on an outpatient basis, one or more of the following: (1) treatment for patients that renders the patients incapable of taking action for self-preservation under emergency conditions without assistance of others; (2) anesthesia that renders the patients incapable of taking action for self-preservation under

emergency conditions without the assistance of others; or (3) emergency or urgent care for patients who, due to the nature of their injury or illness, are incapable of taking action for self-preservation under emergency conditions without the assistance of others ≫ *517.2* ≪.

Ⓑ The essential electrical distribution system shall be a battery or generator system ≫ *517.45(A)* ≪.

Ⓒ Ambulatory health care occupancy is included in the term **health care facilities.**

N O T E

See NFPA 99-2012 *Health Care Facilities Code* ≫ *517.45 Informational Note* ≪.

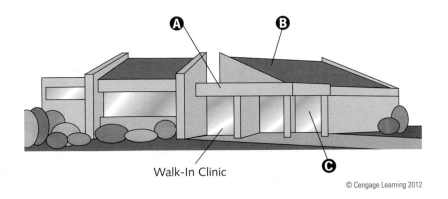

Walk-In Clinic

© Cengage Learning 2012

Electrical System Terminology

Electrical life-support equipment: Electrically powered equipment whose continuous operation is necessary to maintain a patient's life »517.2«.

Isolated power system: A system consisting of an isolating transformer (or equivalent) and a line isolation monitor with its ungrounded circuit conductors »517.2«. (See related definitions, which include: *hazard current, fault hazard current, monitor hazard current, total hazard current, and line isolation monitor.*)

Isolation transformer: A multiple-winding-type transformer, with physically separated primary and secondary windings which inductively couples its secondary winding(s) to circuit conductors connected to its primary winding(s) »517.2«.

Ⓐ Alternate power source: One or more generator sets (or battery systems, where permitted) to provide power during normal electrical service interruption, or the public utility electrical service intended to provide power during interruption of service normally provided by on-site generating facilities »517.2«.

Ⓑ Critical branch: A system of feeders and branch circuits supplying power for task illumination, fixed equipment, select receptacles, and select power circuits serving areas and functions related to patient care and that is automatically connected to alternate power sources by one or more transfer switches during interruption of normal power source »517.2«.

Ⓒ Life safety branch: A system of feeders and branch circuits supplying power for lighting, receptacles, and equipment essential for life safety that is automatically connected to alternate power sources by one or more transfer switches during interruption of the normal power source »517.2«.

Ⓓ Equipment branch: A system of feeders and branch circuits arranged for delayed, automatic, or manual connection to the alternate power source, primarily serving 3-phase power equipment »517.2«.

Ⓔ The number of transfer switches to be used shall be based on reliability, design, and load considerations. Each branch of the essential electrical system shall be served by one or more transfer switches. One transfer switch and downstream distribution system shall be permitted to serve one or more branches in a facility with a maximum demand on the essential electrical system of 150 kVA »517.30(B)(2)«.

Ⓕ Essential electrical system: A system comprising alternate power sources and all connected distribution systems (and ancillary equipment), designed both to ensure electrical power continuity to designated areas and functions of a health care facility during disruption of normal power sources and to minimize disruption within the internal wiring system »517.2«.

Ⓖ The essential electrical system for these facilities shall comprise a system capable of supplying a limited amount of lighting and power service, considered essential for life safety and orderly cessation of procedures, during normal electrical service interruption. This includes clinics, medical/dental offices, outpatient facilities, nursing homes, limited-care facilities, hospitals, and other patient-serving health care facilities »517.25«.

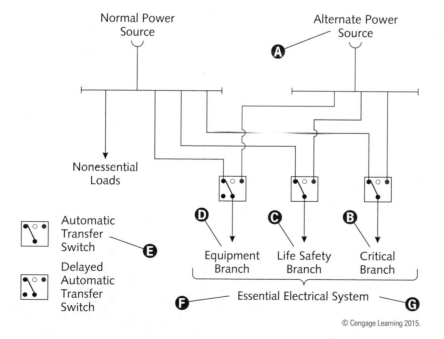

© Cengage Learning 2015.

Grounding and Bonding

An insulated equipment bonding jumper that directly connects to the equipment grounding conductor is permitted to connect the box and receptacle(s) to the equipment grounding conductor »517.13(B)(1) Exception «.

Metal faceplates can be connected to an equipment grounding conductor by means of metal mounting screw(s) securing the faceplate to a grounded outlet box (or grounded wiring device) »517.13(B) Exception No. 1 to (B)(3) «.

Luminaires more than 7½ ft (2.3 m) above the floor and switches located outside the patient care vicinity do not require grounding by an insulated grounding conductor »517.13(B) Exception No. 2 to (B)(3) «.

Patient equipment grounding point: A jack or terminal bus that serves as the collection point for electric appliance redundant grounding in a patient care vicinity, or for grounding other items to eliminate electromagnetic interference problems »517.2 «.

Reference grounding point: The ground bus of the panelboard or isolated power system panel supplying a patient care area »517.2 «.

Exposed conductive surfaces: Those unprotected, unenclosed, or unguarded surfaces capable of carrying electric current and permitting personal contact. Paint, anodizing, and similar coatings do not provide suitable insulation unless listed for such use »517.2 «.

A In general care and critical care areas, all patient bed location receptacles shall be listed "hospital grade" and shall be so identified »517.18(B) and 517.19(B)(2) «.

B An isolated ground receptacle shall not be installed within a patient care vicinity »517.16 «. For requirements pertaining to isolated ground receptacles, in other than patient care vicinities, see 250.146(D) and 408.40 Exception.

C Normal and essential branch-circuit panelboard equipment grounding terminal buses serving the same individual patient care vicinity shall be connected together with an insulated continuous copper conductor no smaller

than 10 AWG. Where two or more panelboards serving the same individual patient care vicinity are served from separate transfer switches on the essential electrical system, the equipment grounding terminal buses of those panelboards shall be connected together with an insulated continuous copper conductor not smaller than 10 AWG. This conductor can be broken in order to terminate on the equipment grounding terminal bus in each panelboard »517.14 «.

D In a patient care area, (1) all receptacle grounding terminals, (2) metal boxes and enclosures containing receptacles, and (3) all non-current-carrying conductive surfaces of fixed electric equipment likely to become energized that are subject to personal contact (operating at over 100 volts) shall be directly connected to an insulated copper equipment grounding conductor that is installed with the branch-circuit conductors in the wiring methods as provided in 517.13(A) »517.13(B) «.

E Equipment grounding conductors and equipment bonding jumpers shall be sized in accordance with 250.122 »517.13(B)(2) «.

F All branch circuits serving patient care areas shall be provided with an effective ground-fault current path by installation in a metal raceway system or a cable having a metallic armor or sheath assembly. The metal raceway system, or metallic cable armor or sheath assembly, shall qualify as an equipment grounding conductor according to 250.118 »517.13(A) «.

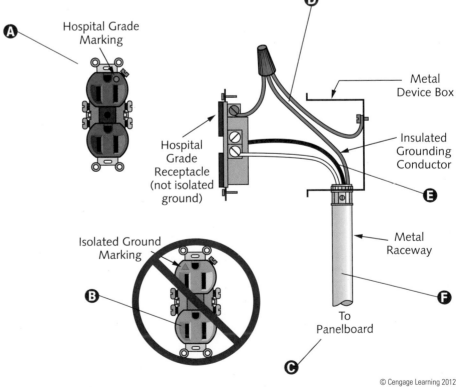

A Hospital Grade Marking

Hospital Grade Receptacle (not isolated ground)

Isolated Ground Marking

B

D

Metal Device Box

Insulated Grounding Conductor

E

Metal Raceway

To Panelboard

C

F

Ground-Fault Protection

Ground-fault protection for the service and feeder disconnecting means operation shall be fully selective so that the feeder device, and not the service device, opens for ground faults on the feeder device's load side. Separation of ground-fault protection time-current characteristics shall conform to manufacturer's recommendations and shall consider all required tolerances and disconnect operating time to achieve 100% selectivity »*517.17(C)*«.

On installation of equipment ground-fault protection, each level shall be performance tested to ensure compliance with 517.17(C) »*517.17(D)*«.

A If ground-fault protection is provided for service disconnecting means (or feeder disconnecting means) operation as specified by 230.95 or 215.10, an additional ground-fault protection step shall be provided in the next level of feeder disconnecting means downstream toward the load. Such protection shall consist of overcurrent devices and current transformers (or other equivalent protective equipment) that cause the feeder disconnecting means to open.

The additional levels of ground-fault protection shall not be installed on the load side of an essential electrical system transfer switch »*517.17(B)*«.

B Ground-fault protection of equipment is defined as a system intended to provide protection of equipment from damaging line-to-ground fault currents by operating to cause a disconnecting means to open all ungrounded conductors of the faulted circuit. This protection is provided at current levels less than those required to protect conductors from damage through the operation of a supply circuit overcurrent device »*Article 100*«.

C Ground-fault protection of equipment shall be provided for solidly grounded wye electric services of more than 150 volts to ground but not exceeding 1000 volts phase-to-phase for each service disconnect rated 1000 amperes or more. The grounded conductor for the solidly grounded wye system shall be connected directly to ground through a grounding electrode system, as specified in 250.50, without inserting any resistor or impedance device.

The rating of the service disconnect shall be considered to be the rating of the largest fuse that can be installed or the highest continuous current trip setting for which the actual overcurrent device installed in a circuit breaker is rated or can be adjusted »*230.95*«.

NOTE

The requirements of 517.17 apply to hospitals and other buildings (including multiple-occupancy buildings) with critical care space or utilizing electrical life-support equipment and buildings that provide the required essential utilities or service for the operation of critical care space of electrical life-support equipment »*517.17(A)*«.

NOTE

Each branch-circuit disconnect rated 1000 A or more and installed on solidly grounded wye electrical systems of more than 150 volts to ground, but not exceeding 1000 volts phase-to-phase, shall be provided with ground-fault protection of equipment in accordance with the provisions of 230.95 »*210.13*«.

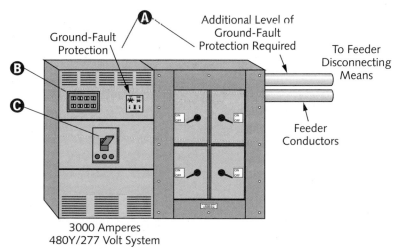

3000 Amperes 480Y/277 Volt System

© Cengage Learning 2012

Wet Procedure Locations

Spaces where a procedure is performed that subjects patients or staff to wet conditions are considered as wet procedure areas. It is the responsibility of the governing body of the health care facility to designate the wet procedure areas ≫ *517.2 Patient Care Space, Informational Note No. 5* ≪.

If an isolated power system is utilized, the isolated power equipment shall be listed as isolated power equipment and the isolated power system shall be designed and installed in accordance with 517.160 ≫ *517.20(B)* ≪.

A Wet procedure locations: The area in a patient care space where a procedure is performed that is normally subject to wet conditions while patients are present, including standing fluids on the floor or drenching of the work area, where either such condition is intimate to the patient or staff ≫ *517.2* ≪. Routine housekeeping procedures and incidental spillage of liquids do not define a wet procedure location ≫ *517.2 Wet Procedure Location, Informational Note* ≪

B Branch circuits supplying only listed, fixed, therapeutic, and diagnostic equipment can be supplied from a grounded service, single- or 3-phase system, provided that (a) wiring for grounded and isolated circuit wiring does not occupy the same raceway and (b) all conductive equipment surfaces are connected to an isolated copper equipment grounding conductor ≫ *517.20(A) Exception* ≪.

C For requirements for installation of therapeutic pools and tubs, see Part VI of Article 680.

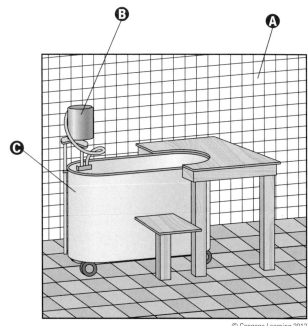

© Cengage Learning 2012

PATIENT CARE AREAS

Patient Bed Location Branch Circuits (General Care Areas)

Section 517.18(A) requirements do not apply to patient bed locations in clinics, medical/dental offices, and outpatient facilities; psychiatric, substance abuse, and rehabilitation hospitals; nursing home sleeping rooms; and limited-care facilities where 517.10(B)(2) requirements are met ≫ *517.18(A) Exception No. 2* ≪.

A Branch circuits serving only special-purpose outlets or receptacles (such as portable X-ray outlets) do not require service from the same distribution panel(s) ≫ *517.18(A) Exception No. 1* ≪.

B Each patient bed location shall be supplied by at least two branch circuits: one from the critical branch and one from the normal system. All normal system branch circuits must originate within the same panelboard ≫ *517.18(A)* ≪.

C The electrical receptacles or the cover plate for the electrical receptacles supplied from the critical branch shall have a distinctive color or marking so as to be readily identifiable and shall also indicate the panelboard and branch-circuit number supplying them ≫ *517.18(A)* ≪.

D A general care patient bed location served from two separate transfer switches on the critical branch shall not be required to have circuits from the normal system ≫ *517.18(A) Exception No. 3* ≪.

E Each patient bed location shall be provided with a minimum of eight receptacles ≫ *517.18(B)* ≪.

Patient Bed Location Branch Circuits (General Care Areas) *(continued)*

NOTE

General care space is defined as space in which failure of equipment or a system is likely to cause minor injury to patients or caregivers **»517.2«**.

General care space includes areas such as patient bedrooms, examining rooms, treatment rooms, clinics, and similar areas where the patient may come into contact with electromedical devices or ordinary appliances such as a nurse call system, electric beds, examining lamps, telephones, and entertainment devices **»517.2 General Care Space, Informational Note No. 3«**.

WARNING

Branch circuits serving patient bed locations shall not be part of a multiwire branch circuit **»517.18(A)«**.

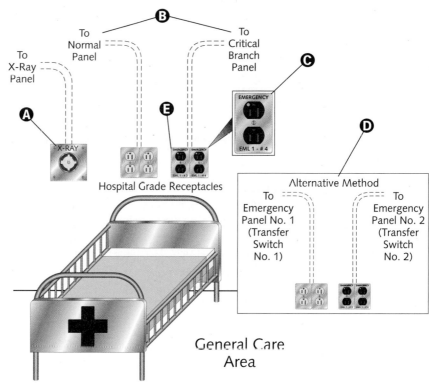

General Care Area

© Cengage Learning 2015.

Patient Bed Location Receptacles (General Care Areas)

A total, immediate replacement of existing non-hospital grade receptacles is not necessary. It is intended, however, that non-hospital grade receptacles be replaced with hospital grade receptacles on modification of use, or during renovation, or as receptacles need replacing »517.18(B) Informational Note«.

Ⓐ Special-purpose receptacles are not counted as required receptacles.

Ⓑ *Each* patient bed location shall be provided with a minimum of eight receptacles **»517.18(B)«**.

Ⓒ The eight receptacles can be either single, duplex, or quadruplex type, or any combination the three types **»517.18(B)«**.

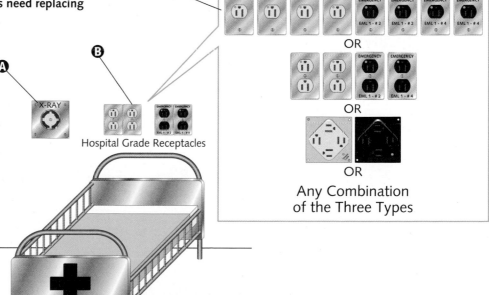

NOTE

All receptacles shall be listed "hospital grade" and shall be so identified. The grounding terminal of each receptacle shall be connected to an insulated copper equipment grounding conductor sized in accordance with Table 250.122 **»517.18(B)«**.

© Cengage Learning 2015.

Pediatric Location Receptacles (General Care Areas)

Ⓐ Receptacles that are located within the patient rooms, bathrooms, playrooms, and activity rooms of pediatric units, other than nurseries, shall be listed tamper-resistant or shall employ a listed tamper-resistant cover ≫ *517.18(C)* ≪.

Ⓑ Tamper-resistant receptacles are identified by the letters "TR" or the words *Tamper Resistant*. After the receptacle has been installed, the identification is only required to be visible with the cover plate removed. See UL (Underwriter Laboratories) White Book, category: Receptacles for Plugs and Attachment Plugs (RTRT).

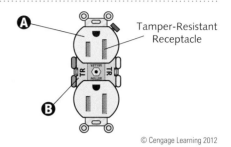

Tamper-Resistant Receptacle

© Cengage Learning 2012

Patient Bed Location Receptacles (Critical Care Areas)

All receptacles shall be listed "hospital grade" and shall be so identified. The grounding terminal of each receptacle shall be connected to the reference grounding point by means of an insulated copper equipment grounding conductor ≫ *517.19(B)(2)* ≪.

Although not required, a patient care vicinity shall be permitted to have a patient equipment grounding point. Where one is supplied, it can contain one (or more) listed grounding and bonding jacks. An equipment bonding jumper no smaller than 10 AWG shall connect the grounding terminal of all grounding-type receptacles to the patient equipment grounding point. The bonding conductor can be arranged centrically or looped as convenient ≫ *517.19(D)* ≪.

Where there is no patient equipment grounding point, it is important that the distance between the reference grounding point and the patient care vicinity be as short as possible to minimize any potential differences ≫ *517.19(D) Informational Note* ≪.

Ⓐ The equipment grounding conductor for special-purpose receptacles (such as the operation of mobile X-ray equipment) shall extend to the branch-circuit reference grounding points for all locations potentially served from such receptacles. Where an isolated ungrounded system serves such a circuit, the grounding conductor does not have to run with the power conductors; however, the special-purpose receptacle's equipment grounding terminal shall connect to the reference grounding point ≫ *517.19(H)* ≪.

Ⓑ Each patient bed location shall have a minimum of fourteen receptacles. At least one of the receptacles must be connected to (1) the normal system branch circuit as required by 517.19(A) or (2) a critical branch circuit supplied by a different transfer switch not associated with other receptacles at the same patient bed location ≫ *517.19(B)(1)* ≪.

Ⓒ The fourteen receptacles can be either single, duplex, or quadruplex type, or any combination the three ≫ *517.19(B)(2)* ≪.

NOTE

Where a grounded electrical distribution system is used and metal feeder raceway or Type MC or MI cable that qualifies as an equipment grounding conductor in accordance with 250.118 is installed, grounding of enclosures and equipment, such as panelboards, switchboards, and switchgear, shall be ensured by one of the following bonding means at each termination or junction point of the metal raceway or Type MC or MI cable:

(1) A grounding bushing and a continuous copper bonding jumper, sized according to 250.122, with the bonding jumper connected to the junction enclosure or the panel ground bus.

(2) Connection of feeder raceways or Type MC or MI cables to threaded hubs or bosses on terminating enclosures.

(3) Other approved devices such as bonding-type locknuts or bushings ≫ *517.19(E)* ≪.

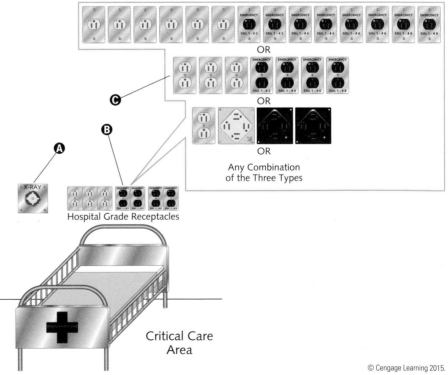

Hospital Grade Receptacles

Critical Care Area

© Cengage Learning 2015.

Patient Bed Location Branch Circuits (Critical Care Areas)

A Branch circuits serving only special-purpose receptacles or equipment in critical care spaces can be served by other panelboards ≫ *517.19(A) Exception No. 1* ≪.

B Each patient bed location shall be supplied by at least two branch circuits: one (or more) from the critical branch and one (or more) from the normal system. At least one branch circuit from the critical branch shall supply outlet(s) at that bed location only ≫ *517.19(A)* ≪.

C Critical branch receptacles shall be identified and shall also indicate the supplying panelboard and circuit number ≫ *517.19(A)* ≪.

D The cover plates for the electrical receptacles or the electrical receptacles themselves supplied from the essential electrical system shall have a distinctive color or marking so as to be readily identifiable ≫ *517.30(E)* ≪.

E Critical care space served from two separate critical branch transfer switches shall not be required to have circuits from the normal system ≫ *517.19(A) Exception No. 2* ≪.

F Each patient bed location shall be provided with a minimum of fourteen receptacles ≫ *517.19(B)('1)* ≪.

> ### NOTE
>
> **Critical care space** is the space in which failure of equipment or a system is likely to cause major injury or death to patients or caregivers. Critical care space is one of the four spaces within the patient care space ≫ *517.2 Patient Care Space* ≪.

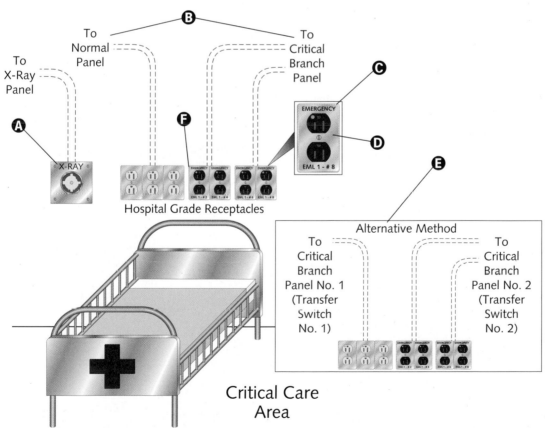

Hospital Grade Receptacles

Alternative Method

To Critical Branch Panel No. 1 (Transfer Switch No. 1)

To Critical Branch Panel No. 2 (Transfer Switch No. 2)

Critical Care Area

© Cengage Learning 2015.

Operating Room Receptacles

Ⓐ Each operating room shall be provided with a minimum of thirty-six receptacles, at least twelve of which shall be connected to either of the following:

(1) The normal system branch circuit required in 517.19(A)

(2) A critical branch circuit supplied by a different transfer switch than the other receptacles at the same location ≫ 517.19(C)(1) ≪.

Ⓑ The receptacles required in 517.19(C)(1) shall be permitted to be of the single or duplex types or a combination of both ≫ 517.19(C)(2) ≪.

Ⓒ All receptacles shall be listed hospital grade and so identified ≫ 517.19(C)(2) ≪.

> **NOTE**
>
> The grounding terminal of each receptacle shall be connected to the reference grounding point by means of an insulated copper equipment grounding conductor ≫**517.19(C)(2)**≪.
>
> The reference grounding point is defined as the ground bus of the panelboard or isolated power system panel supplying the patient care area ≫**517.2**≪.

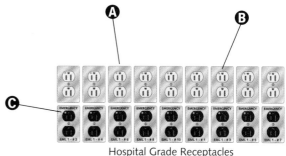

Hospital Grade Receptacles

© Cengage Learning 2015.

HOSPITALS

Life Safety Branch

Ⓐ The life safety branch of the essential electrical system shall supply power for lighting, receptacles, and equipment as listed in 517.32(A) through (H) ≫ 517.32 ≪.

Ⓑ Illumination of egress means is required, such as lighting for corridors, passageways, stairways, landings at exit doors, and all paths of approach to exits shall be connected to the life safety branch. Switching arrangement to transfer patient corridor lighting from general illumination circuits is permitted, provided only one of two circuits can be selected and both circuits cannot be simultaneously extinguished ≫ 517.32(A) ≪.

Ⓒ Task illumination battery charger for battery-powered lighting unit(s) and selected receptacles at the generator set and essential transfer switch locations shall be connected to the life safety branch ≫ 517.32(E) ≪. Generator set accessories as required for generator performance shall be connected to the life safety branch. Loads dedicated to a specific generator, including the fuel transfer pump(s), ventilation fans, electrically operated louvers, controls, cooling system, and other generator accessories essential for generator operation, shall be connected to the life safety branch or to the output terminals of the generator with overcurrent protective devices ≫ 517.32(F) ≪.

Ⓓ Exit and exit directional signs shall be connected to the life safety branch ≫ 517.32(B) ≪.

Ⓔ Alarm and alerting systems, including (1) fire alarms, (2) alarms required by systems used for piping nonflammable medical gases, and (3) mechanical, control, and other accessories required for effective life safety systems operation shall be permitted to be connected to the life safety branch ≫ 517.32(C) ≪.

Ⓕ Automatically operated doors used for building egress shall be connected to the life safety branch ≫ 517.32(H) ≪.

Ⓖ Only functions listed in 517.32(A) through (H) shall be connected to the life safety branch of the essential electrical system ≫ 517.32 ≪.

Ⓗ Hospital communication systems used for issuing instructions during emergency conditions shall be connected to the life safety branch ≫ 517.32(D) ≪.

Ⓘ Elevator cab lighting, control, communications, and signal systems shall be connected to the life safety branch ≫ 517.32(G) ≪.

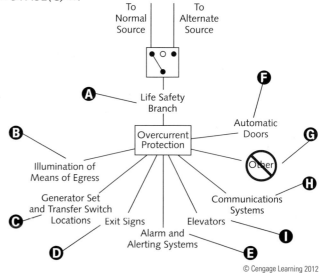

© Cengage Learning 2012

Essential Electrical Systems (Greater than 150 kVA)

(A) Hospital power sources and alternate power sources shall be permitted to serve the essential electrical systems of contiguous or same site facilities ≫ *517.30(B)(4)* ≪.

(B) Demand calculations for sizing of the generator set(s) shall be based on any of the following:

(1) Prudent demand factors and historical data

(2) Connected load

(3) Feeder calculation procedures described in Article 220

(4) Any combination of the above ≫ *517.30(D)* ≪.

(C) Loads served by generating equipment not specifically named in Article 517 shall be served by their own transfer switches so that these loads (1) are not transferred if the transfer will overload the generating equipment and (2) are automatically shed on generating equipment overloading ≫ *517.30(B)(3)* ≪.

(D) The number of transfer switches to be used shall be based on reliability, design, and load considerations. Each branch of the essential electrical system shall have one or more transfer switches ≫ *517.30(B)(2)* ≪.

(E) Those functions of patient care depending on lighting or appliances that are connected to the essential electrical system shall be divided into the life safety branch and the critical branch, as described in 517.32 and 517.33. The life safety and critical branches shall be installed and connected to the alternate power source so that all functions supplied by these branches specified here shall be automatically restored to operation within 10 seconds after interruption of the normal source ≫ *517.31* ≪.

(F) Essential electrical systems for hospitals shall be comprised of three separate branches capable of supplying a limited amount of lighting and power service that is considered essential for life safety and effective hospital operation during the time the normal electrical service is interrupted for any reason. The three branches are life safety, critical, and equipment ≫ *517.30(B)(1)* ≪.

(G) The essential electrical system shall have the capacity and rating to meet the maximum actual demand likely to be produced by the connected load. Feeders shall be sized in accordance with 215.2 and Part III of Article 220. The generator

set(s) shall have the capacity and rating to meet the demand produced by the load at any given time ≫ *517.30(D)* ≪.

(H) The wiring of the life safety and critical branches shall be mechanically protected. The permitted wiring methods are listed in 517.30(C)(3)(1) through (5) ≫ *517.30(C)(3)* ≪.

(I) Wiring of the life safety branch and the critical branch shall be permitted to occupy the same raceways, boxes, or cabinets of other circuits not part of the branch where such wiring complies with one of the following:

1. Is in transfer equipment enclosures

2. Is in exit or emergency luminaires supplied from two sources

3. Is in a common junction box attached to exit or emergency luminaires supplied from two sources

4. Is for two or more circuits supplied from the same branch and same transfer switch ≫ *517.30(C)(1)* ≪.

> **N O T E**
>
> Where isolated power systems are installed in any of the areas in 517.33(A)(1) and (A)(2), each system shall be supplied by an individual circuit serving no other load ≫*517.30(C)(2)*≪.

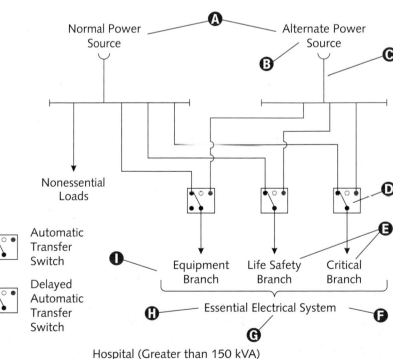

Hospital (Greater than 150 kVA)

Essential Electrical Systems (150 kVA or Less)

Ⓐ One transfer switch and downstream distribution system shall be permitted to serve one or more branches in a facility with a maximum demand on the essential electrical system of 150 kVA ≫ *517.30(B)(2)* ≪ .

> **NOTE**
>
> For essential electrical systems under 150 kVA, deletion of the time-lag intervals feature for delayed automatic connection to the equipment branch shall be permitted ≫ *517.34 Exception* ≪ .

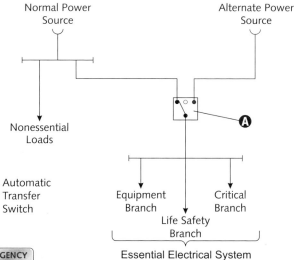

Hospital (150 kVA or Less)

© Cengage Learning 2015.

Receptacle Identification

Ⓐ The cover plates for the electrical receptacles or the electrical receptacles themselves supplied from the essential electrical system shall have a distinctive color or marking so as to be readily identifiable ≫ *517.30(E)* ≪ .

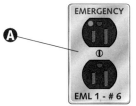

© Cengage Learning 2012

Critical Branch

Ⓐ The critical branch of the essential electrical system shall supply power for task illumination, fixed equipment, selected receptacles, and special power circuits serving patient care areas and function as listed in 517.33(A)(1) through (9) ≫ *517.33(A)* ≪ .

Ⓑ Critical care areas that utilize anesthetizing gases, including task illumination, selected receptacles, and fixed equipment shall be connected to the critical branch ≫ *517.33(A)(1)* ≪ .

Ⓒ The isolating power systems in special environments shall be connected to the critical branch ≫ *517.33(A)(2)* ≪ .

Ⓓ Additional specialized patient care task illumination and receptacles, where needed, shall be connected to the critical branch ≫ *517.33(A)(4)* ≪ .

Ⓔ Patient care areas, including task illumination and selected receptacles, in the following: (a) infant nurseries; (b) medication preparation areas; (c) pharmacy dispensing areas; (d) selected acute nursing areas; (e) psychiatric bed areas (omit receptacles); (f) ward treatment rooms; (g) nurses' stations (unless adequately lighted by corridor luminaires) shall be connected to the critical branch ≫ *517.33(A)(3)* ≪ .

Ⓕ Subdividing the critical branch into two or more branches is permitted ≫ *517.33(B)* ≪ .

Ⓖ Additional task illumination, receptacles, and selected power circuits needed for effective hospital operation shall be connected to the critical branch. Single-phase fractional horsepower motors can be connected to the critical branch ≫ *517.33(A)(9)* ≪ .

Ⓗ Task illumination, selected receptacles, and selected power circuits for the following shall be connected to the critical branch: (a) general care beds (at least one duplex receptacle in each patient bedroom); (b) angiographic labs; (c) cardiac catheterization labs; (d) coronary care units; (e) hemodialysis rooms or areas; (f) emergency room treatment areas (selected); (g) human physiology labs; (h) intensive care units; and (i) postoperative recovery rooms (selected) ≫ *517.33(A)(8)* ≪ .

Ⓘ Telephone and data equipment rooms and closets shall be connected to the critical branch ≫ *517.33(A)(7)* ≪ .

Ⓙ Blood, bone, and tissue banks shall be connected to the critical branch ≫ *517.33(A)(6)* ≪ .

Ⓚ Nurse call systems shall be connected to the critical branch ≫ *517.33(A)(5)* ≪ .

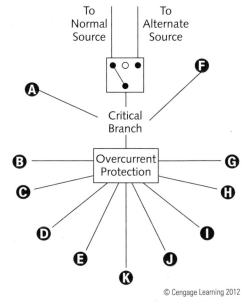

© Cengage Learning 2012

Power Sources

Facilities whose normal power source is supplied by multiple separate central station-fed services experience greater than normal electrical service reliability than those having only a single feed. Such a dual source of normal power consists of multiple electrical services fed from separate generator sets, or a utility distribution network having multiple power input sources. These dual services are arranged to provide mechanical and electrical separation when a fault between the facility and the generating sources is unlikely to cause an interruption of more than one of the facility service feeders ≫517.35(C) Informational Note ≪.

> **NOTE**
>
> Carefully consider the location of the essential electrical system components (housing and spaces) to minimize interruptions caused by natural forces common to the area (such as storms, floods, earthquakes, or hazards created by adjoining structures or activities). Consideration must also be given to the potential normal electrical service interruptions resulting from similar causes, as well as possible service disruption due to internal wiring and equipment failures. Consideration shall be given to the physical separation of the main feeders of the alternate source from the main feeders of the normal electrical source to prevent possible simultaneous interruption ≫517.35(C)≪.

> **NOTE**
>
> The alternate power source can also be a battery system located on the premises ≫517.35(B)(4)≪.

Ⓐ Essential electrical systems shall have a minimum of two independent sources of power: a normal source generally supplying the entire electrical system and one or more alternate sources for use when the normal source is interrupted ≫517.35(A)≪.

Ⓑ The alternate power source can be on-site generator(s) driven by some form of prime mover(s) ≫517.35(B)(1)≪.

Ⓒ The alternate power source can also be another generating unit(s) provided the normal source consists of on-site generating unit(s) ≫517.35(B)(2)≪.

Ⓓ An external utility service can act as the alternate power source when the normal source consists of on-site generating unit(s) ≫517.35(B)(3)≪.

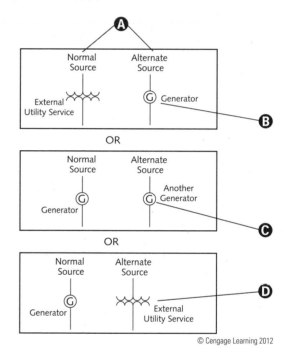

© Cengage Learning 2012

Equipment Branch

Generator accessories, including but not limited to, the transfer fuel pump, electrically operated louvers, and other generator accessories essential for generator operation, shall be arranged for automatic connection to the alternate power source ≫517.34(C)≪.

Deletion of the time-lag intervals feature for delayed automatic connection to the equipment system shall be permitted for essential electrical systems under 150 kVA ≫517.34 Exception≪.

Ⓐ The following equipment shall be permitted to be arranged for either delayed automatic or manual connection to the alternate power source:

1. Heating equipment to provide heat for operating, delivery, labor, recovery, intensive care, coronary care, nurseries, infection/isolation rooms, emergency treatment spaces, general patient rooms, and pressure

maintenance (jockey or make-up) pump(s) for water-based fire protection systems.
2. Elevator(s) providing service to patient, surgical, obstetrical, and ground floors during normal power interruption. (Where interruption of normal power would result in other elevators stopping between floors, throw-over facilities shall be provided to allow temporary operation of any elevator for patient/other person release.)
3. Hyperbaric facilities.
4. Hypobaric facilities.
5. Automatically operated doors.
6. Minimal electrically heated autoclaving equipment can be arranged for either automatic or manual connection to the alternate source.
7. Controls for equipment listed in 517.34.
8. Other selected equipment can be served by the equipment system ≫517.34(B)≪.

Equipment Branch *(continued)*

B The equipment branch shall be installed and connected to the alternate power source such that the equipment described in 517.34(A) is automatically restored to operation at appropriate time-lag intervals following the energizing of the essential electrical system. Its arrangement shall also provide for the subsequent connection of equipment described in 517.34(B) ≫ *517.34* ≪ .

C Delayed automatic connection to the alternate power source shall be permitted to be arranged for the following equipment:

1. Central suction systems serving medical/surgical functions, including controls. (Such suction systems are permitted on the critical branch.)

2. Sump pumps and other equipment required for the safe operation of major apparatus, including associated control systems and alarms.

3. Compressed air systems serving medical/surgical functions, including controls. (Such air systems are permitted on the critical branch.)

4. Smoke control, stair pressurization systems, or both.

> **NOTE**
>
> Sequential delayed automatic connection to the alternate power source, thereby preventing generator overload, is permitted where engineering studies indicate it is necessary ≫ *517.34(A) Exception* ≪ .

5. Kitchen hood supply, exhaust systems, or both, if operation is required during a fire in (or under) the hood.

6. Supply return, and exhaust ventilating systems for airborne infectious/isolation rooms, protective environment rooms, exhaust fans for laboratory fume hoods, nuclear medicine areas where radioactive material is used, ethylene oxide evacuation, and anesthesia evacuation. Where delayed automatic connection is not appropriate, such ventilation systems shall be permitted to be placed on the critical branch.

7. Supply, return, and exhaust ventilating systems for operating and delivery rooms.

8. Supply, return, exhaust ventilating systems and/or air-conditioning systems serving telephone equipment rooms and closets and data equipment rooms and closets ≫ *517.34(A)* ≪ .

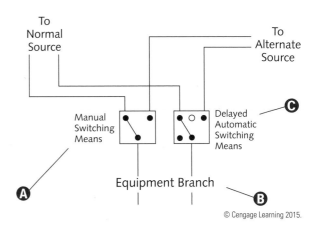

© Cengage Learning 2015.

NURSING HOMES AND LIMITED-CARE FACILITIES

Essential Electrical Systems (Greater than 150 kVA)

Where the normal source consists of on-site generating units, the alternate source shall be either another generator set, or an external utility service ≫ *517.44(B) Exception No. 1* ≪ .

Nursing homes or limited-care facilities meeting the requirements of 517.40(A) can use a battery system or a self-contained battery integral with the equipment ≫ *517.44(B) Exception No. 2* ≪ .

A Essential electrical systems shall have at least two independent power sources: a normal source generally supplying the entire system and one (or more) alternate sources for use during normal source interruption ≫ *517.44(A)* ≪ .

B The alternate power source shall be on-site generator(s) driven by some form of prime mover(s), unless an exception is met ≫ *517.44(B)* ≪ .

C The number of transfer switches used is based on reliability, design, and load considerations. Each essential electrical system branch (life safety and critical) shall be served by one (or more) transfer switches ≫ *517.41(B)* ≪ .

D Life safety branch installation and alternate power source connection shall provide that all functions (specified herein) be automatically restored to operation within 10 seconds after interruption occurs ≫ *517.42* ≪ .

E The essential electrical system shall be capable of meeting the demand for operation of all functions and equipment being served simultaneously by each branch ≫ *517.41(C)* ≪ .

Essential Electrical Systems (Greater than 150 kVA) *(continued)*

F Essential electrical systems for nursing homes and limited-care facilities shall comprise two separate branches, each capable of supplying a limited amount of lighting and power service for those functions considered essential for life safety and effective institution operation during normal service interruption, regardless of cause. These are the *life safety branch* and the *critical branch* ≫*517.41(A)*≪.

G The critical branch shall be installed and connected to the alternate power source so that the equipment listed in 517.43(A) shall be automatically restored to operation at appropriate time-lag intervals following the restoration of the life safety branch to operation. Its arrangement shall also provide for the additional connection of equipment listed in 517.43(B) by either delayed automatic or manual operation ≫*517.43*≪.

> **NOTE**
>
> Give careful consideration to the essential electrical system components location to minimize interruptions caused by natural forces common to the area (such as storms, floods, earthquakes, or hazards created by adjoining structures or activities). Consideration shall also be given to possible normal services interruption resulting from similar causes, as well as potential service disruption due to internal wiring and equipment failures ≫*517.44(C)*≪.

> **WARNING**
>
> The life safety branch shall be entirely independent of all other wiring (and equipment) and shall not enter the same raceways, boxes, or cabinets with other wiring except as follows:
> 1. In transfer switches,
> 2. In exit or emergency luminaires supplied from two sources, or
> 3. In a common junction box attached to exit or emergency luminaires supplied from two sources
>
> The critical branch wiring can occupy the same raceways, boxes, or cabinets of other circuits that are not part of the life safety branch ≫*517.41(D)*≪.

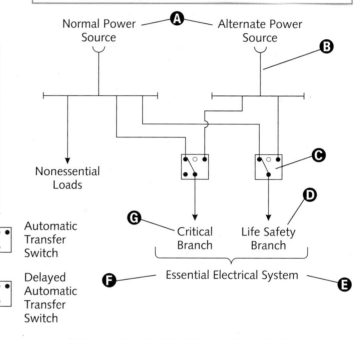

Nursing Home and Limited-Care Facility (Greater than 150 kVA)

© Cengage Learning 2015.

Essential Electrical Systems (150 kVA or Less)

A One transfer switch shall be permitted to serve one or more branches or systems in a facility with a maximum demand on the essential electrical system of 150 kVA ≫*517.41(B)*≪.

> **NOTE**
>
> For essential electrical systems under 150 kVA, deletion of the time-lag intervals feature for delayed automatic connection to the equipment branch shall be permitted ≫*517.43 Exception*≪.

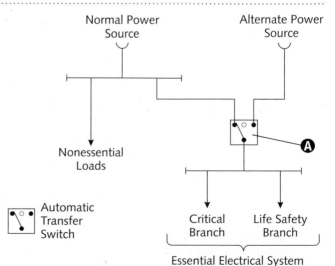

Nursing Home and Limited-Care Facility (150 kVA or Less)

© Cengage Learning 2015.

Life Safety Branch

Ⓐ Life safety branch installation and alternate power source connection shall provide that all functions specified in 517.42(A) through (G) are automatically restored to operation within 10 seconds after interruption of the normal source. The life safety branch shall supply power for the following lighting, receptacles, and equipment ≫ *517.42* ≪.

Ⓑ Illumination of means of egress as is necessary for corridors, passageways, stairways, landings, and exit doors and all ways of approach to exits shall be connected to the critical branch. Patient corridor lighting can be transferred via switching arrangement from general illumination circuits, provided only one circuit can be selected and both circuits cannot be simultaneously extinguished ≫ *517.42(A)* ≪.

Ⓒ Task illumination and selected receptacles in the generator set location shall be connected to the critical branch ≫ *517.42(F)* ≪.

Ⓓ Sufficient lighting in dining and recreation areas to provide exit way illumination shall be connected to the critical branch ≫ *517.42(E)* ≪.

Ⓔ Exit and exit directional signs shall be connected to the critical branch ≫ *517.42(B)* ≪.

Ⓕ Elevator cab lighting, control, communications, and signal systems shall be connected to the critical branch ≫ *517.42(G)* ≪.

Ⓖ Connect only those functions listed in 517.42(A) through (G) to the life safety branch ≫ *517.42* ≪.

Ⓗ Communications systems used for issuing emergency condition instructions shall be connected to the critical branch ≫ *517.42(D)* ≪.

Ⓘ Alarm and alerting systems, including the following shall be connected to the critical branch: (1) fire alarms and (2) alarms required for systems piping nonflammable medical gases ≫ *517.42(C)* ≪.

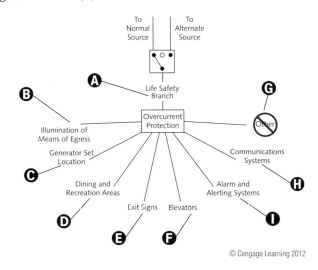

© Cengage Learning 2012

Critical Branch

Ⓐ The following equipment shall be permitted to be connected to the critical branch and shall be arranged for either delayed automatic or manual connection to the alternate power source:

1. Heating equipment providing patient room heating.

2. Elevator service where power disruption would result in between-floor stops. Throw-over facilities shall be provided to allow temporary operation of any elevator passenger release. See 517.42(G) for elevator cab lighting, control, and signal system requirements.

3. Additional illumination, receptacles, and equipment can be connected only to the critical branch ≫ *517.43(B)* ≪.

Ⓑ The critical branch shall be installed and connected to the alternate power source so that equipment listed in 517.43(A) is automatically restored to operation at appropriate time-lag intervals following the restoration of life safety branch operation. Arrangement shall also provide for additional connection of equipment listed in 517.43(B), by either delayed automatic or manual operation ≫ *517.43* ≪.

Ⓒ The following equipment shall be permitted to be connected to the critical branch and shall be arranged for delayed automatic connection to the alternate power source:

1. Patient care areas, including task illumination and selected receptacles in (a) medication prep areas; (b) pharmacy dispensing areas; and (c) nurses' stations (unless corridor luminaires provide adequate light)

2. Sump pumps and other equipment required for major apparatus safe operation, as well as associated control systems/alarms

3. Smoke control and stair pressurization systems

4. Kitchen hood supply or exhaust systems, if operation is required during a fire in (or under) the hood

5. Supply, return, and exhaust ventilating systems for airborne infectious isolation rooms ≫ *517.43(A)* ≪

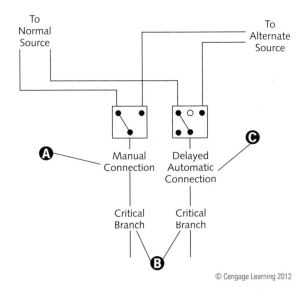

© Cengage Learning 2012

INHALATION ANESTHETIZING LOCATIONS

Hazardous (Classified) Anesthetizing Locations

Anesthetizing location: Any facility area designated for use in administering any flammable inhalation anesthetic agent (during examination or treatment), including the use of such agents for relative analgesia 》*517.2*《.

Flammable Anesthetics: Gases or vapors, such as fluroxene, cyclopropane, divinyl ether, ethyl chloride, ethyl ether, and ethylene, which may form flammable or explosive mixtures with air, oxygen, or reducing gases such as nitrous oxide 》*517.2*《.

Flammable Anesthetizing Location: Any area of the facility that has been designated to be used for the administration of any flammable inhalation anesthetic agents in the normal course of examination or treatment 》*517.2*《.

Unless otherwise allowed by 517.160, each power circuit within (or partially within) a flammable anesthetizing location (517.60) shall be isolated from any distribution system by means of an isolated power system 》*517.61(A)(1)*《.

Any inhalation anesthetizing location designated for the exclusive use of nonflammable anesthetizing agents is considered to be an other-than-hazardous (classified) location 》*517.60(B)*《. Installations in other-than-hazardous (classified) locations must comply with 517.61(C).

Isolated power system equipment shall be listed as isolated power equipment; the system shall be designed and installed in accordance with 517.160 》*517.61(A)(2)*《.

In hazardous (classified) locations referenced in 517.60, all fixed wiring/equipment and all portable equipment (including lamps and other utilization equipment) operating at more than 10 volts between conductors shall comply with the requirements of 501.1 through 501.25, 501.100 through 501.150, and 501.30(A) and (B) for Class I, Division 1 locations. All such equipment shall be approved for the specific hazardous atmosphere involved 》*517.61(A)(3)*《.

If a box, fitting, or enclosure is only partially within a hazardous (classified) location(s), the hazardous area extends to include the entire box, fitting, or enclosure 》*517.61(A)(4)*《.

Receptacles and attachment plugs in a hazardous (classified) location(s) shall be listed for use in Class I, Group C hazardous (classified) locations and shall provide a grounding conductor connection 》*517.61(A)(5)*《.

Flexible cords used in hazardous (classified) locations for connection of portable utilization equipment (including lamps operating at more than 8 volts between conductors) shall be of a type approved for extra-hard usage per Table 400.4 and shall include an additional conductor for grounding 》*517.61(A)(6)*《.

The flexible cord shall include a storage device that does not subject the cord to less than a 3-in. (75-mm) bending radius 》*517.61(A)(7)*《.

A Any room (or location) storing flammable anesthetics or volatile flammable disinfecting agents is a Class I, Division 1 location from floor to ceiling 》*517.60(A)(2)*《.

B Wherever flammable anesthetics are employed, the entire area is a Class I, Division 1 location, extending from the floor upward to a level of 5 ft (1.52 m). The remaining space (up to the structural ceiling) is considered to be above a hazardous (classified) location 》*517.60(A)(1)*《.

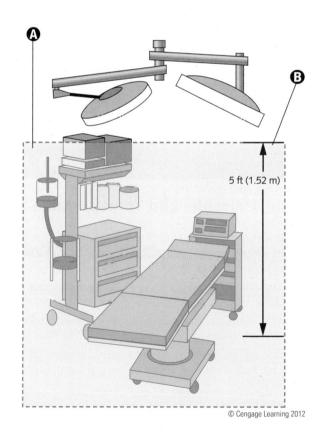

5 ft (1.52 m)

Above Hazardous (Classified) Anesthetizing Locations

Wiring above a hazardous (classified) location (referenced in 517.60) shall be installed in RMC, EMT, IMC, Type MI cable, or Type MC cable employing a continuous gas/vaportight metal sheath ≫*517.61(B)(1)*≪.

Installed equipment that may produce arcs, sparks, or particles of hot metal (such as lamps and lampholders for fixed lighting, cutouts, switches, generators, motors, or other equipment having make-or-break or sliding contacts) shall be of the totally enclosed type, or by design shall prevent the escape of sparks or hot metal particles ≫*517.61(B)(2)*≪. Wall-mounted receptacles installed above the hazardous (classified) location in flammable anesthetizing locations require neither total enclosure nor guarded or screened openings to prevent particle dispersion ≫*517.61(B)(2) Exception*≪.

Listed seals shall be provided in conformance with 501.15. Section 501.15(A)(4) applies to both horizontal and to vertical boundaries of the defined hazardous (classified) locations ≫*517.61(B)(4)*≪.

Receptacles and attachment plugs located above hazardous (classified) anesthetizing locations shall be listed for hospital use for services of prescribed voltage, frequency, rating, and number of conductors with provision for the connection of the grounding conductor. This requirement applies to attachment plugs and receptacles of the 2-pole, 3-wire grounding type for single-phase, 120-volt, nominal, ac service ≫*517.61(B)(5)*≪.

Plugs and receptacles rated 250 volts for connection of 50- and 60-ampere ac medical equipment above hazardous (classified) locations shall be arranged so that the 60-ampere receptacle accepts either the 50- or 60-ampere plug. Fifty-ampere

receptacles shall *not* accept a 60-ampere attachment plug. The plugs shall be on the 2-pole, 3-wire design, with a third contact connecting to the insulated (green or green with yellow stripe) electrical system equipment grounding conductor ≫*517.61(B)(6)*≪.

A Wiring above a hazardous (classified) location must meet 517.61(B)(1) through (6) provisions.

B Surgical and other luminaires shall conform to 501.130(B) ≫*517.61(B)(3)*≪.

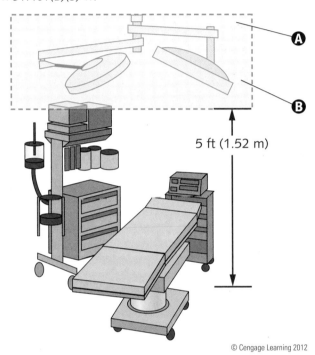

5 ft (1.52 m)

X-RAY INSTALLATIONS

Mobile, Portable, and Transportable X-Ray Equipment

A X-ray installations (mobile): X-ray equipment mounted on a permanent base with wheels, casters, or a combination of both, facilitating movement of equipment while fully assembled ≫517.2≪

X-ray installations (portable): X-ray equipment designed to be hand-carried ≫517.2≪

X-ray installations (transportable): X-ray equipment to be conveyed by a vehicle or that is readily disassembled for transport by a vehicle ≫517.2≪

B Individual branch circuits are not required for portable, mobile, and transportable medical X-ray equipment requiring a capacity of 60 amperes or less ≫517.71(B)≪.

C For portable X-ray equipment connected to a 120-volt branch circuit of 30 amperes (or less), a properly rated grounding-type attachment plug and receptacle can serve as a disconnecting means ≫517.72(C)≪.

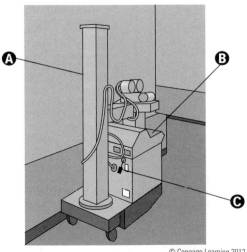

Fixed and Stationary X-Ray Equipment

Fixed and stationary X-ray equipment shall be connected to the power supply by means of a wiring method complying with applicable requirement, of Chapters 1 through 4 of the *NEC*, as modified by Article 517 》*517.71(A)*《. Equipment properly supplied by a branch circuit rated at 30 amperes (or less) can be supplied through a suitable attachment plug and hard-service cable (or cord) 》*517.71(A) Exception*《.

A supply circuit disconnecting means of adequate capacity for at least 50% of the input (for momentary rating) or 100% of the input (for long-time rating) of the X-ray equipment, whichever is greater, shall be provided 》*517.72(A)*《.

Location and operation of the disconnecting means shall be readily accessible from the X-ray control 》*517.72(B)*《.

X-ray installations, momentary rating: a rating based on an operating interval of 5 seconds or less 》*517.2*《.

X-ray installations, long-time rating: a rating based on an operating interval of 5 minutes or longer 》*517.2*《.

Supply branch-circuit conductor ampacity and the current rating of overcurrent protective devices shall not be less than 50% of the momentary rating or 100% of the long-time rating, whichever is greater 》*517.73(A)(1)*《.

Supply feeder ampacity and the current rating of overcurrent protective devices supplying multiple branch circuits feeding X-ray units shall not be less than 50% of the largest unit's momentary demand rating, plus 25% of the momentary demand rating of the next largest unit, plus 10% of the momentary demand rating of each additional unit. If the X-ray unit performs simultaneous biplane examinations,

the supply conductors and overcurrent protective devices shall be 100% of each unit's momentary demand rating 》*517.73(A)(2)*《. The minimum conductor size for branch and feeder circuits is also governed by voltage regulation requirements. For a specific installation, the manufacturer typically specifies minimum distribution transformer and conductor sizes, disconnecting means rating, and overcurrent protection 》*517.73(A)(2) Informational Note*《.

Conductor ampacity and overcurrent protective device ratings shall not be less than 100% of the current rating of medical X-ray *therapy* equipment 》*517.73(B)*《.

Circuits and equipment operated on a supply circuit of over 1000 volts shall comply with Article 490 》*517.71(C)*《.

N O T E

Non-current-carrying metal parts of X-ray and associated equipment (controls, tables, X-ray tube supports, transformer tanks, shielded cables, X-ray tube heads, etc.) shall be connected to an equipment grounding conductor in the manner specified in Part VII of Article 250, as modified by 517.13(A) and (B) 》*517.78(C)*《.

WARNING

Do not construe anything in Article 517, Part V as specifying safeguards against the useful beam or stray X-ray radiation 》*517.70*《. (See Informational Note Nos. 1 and 2 for information pertaining to safety and performance.)

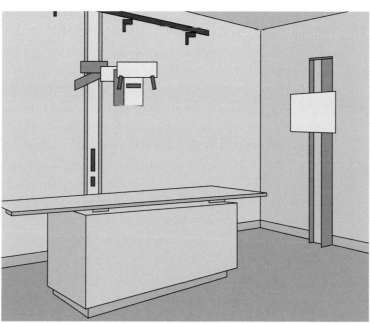

Summary

- Article 517 provisions apply to electrical construction and installation criteria in health care facilities that provide services to human beings.
- Many commonly used health care facility terms are defined in 517.2.
- Article 517's installation criteria and wiring methods serve to minimize electrical hazards.
- Nursing homes and limited-care facilities share the same provisions, found in 517.40 through 517.44.
- Each patient bed location, in general care areas, shall have at least eight receptacles.
- Within critical care areas, each patient bed location requires a minimum of fourteen receptacles.
- The required receptacles in both general and critical care areas can be of the single, duplex, or quadruplex types or any combination of the three.
- An isolated ground receptacle shall not be installed within a patient care vicinity.
- Receptacles that are located within the patient rooms, bathrooms, playrooms, and activity rooms of pediatric units, other than nurseries, shall be listed tamper-resistant or shall employ a listed tamper-resistant cover.

- Essential electrical systems for hospitals shall be comprised of three separate branches: life safety branch, critical branch, and equipment branch.
- An alternate power source for a hospital can be (1) on-site generator(s) driven by some form of prime mover(s); (2) another generating unit(s) where generating unit(s) located on the premises is the normal source; (3) an external utility service when the normal source consists of on-site generating unit(s); or (4) a battery system located on the premises.
- Essential electrical systems for nursing homes and limited-care facilities are composed of two separate branches: the life safety branch and the critical branch.
- The entire area surrounding any location employing flammable anesthetics is a Class I, Division 1 location, including the area extending upward from the floor to a level of 5 ft (1.52 m).
- Wiring and equipment within anesthetizing locations must comply with 517.61 requirements.
- X-ray installations must comply with 517, Part V requirements.
- Each operating room shall be provided with a minimum of thirty-six receptacles.

Unit 16 Competency Test

NEC Reference	Answer	
_____	_____	1. X-ray low-voltage cables connecting to oil-filled units that are not completely sealed, such as transformers, condensers, oil coolers, and high-voltage switches, shall have insulation of the _____ type.
_____	_____	2. The life safety and critical branches shall be installed and connected to the alternate power source so that all functions supplied by these branches specified here shall be automatically restored to operation within _____ after interruption of the normal source.
_____	_____	3. The equipment-grounding terminal buses of the normal and essential branch-circuit panelboards serving the same individual patient care vicinity shall be connected together with an insulated continuous copper conductor not smaller than _____.
_____	_____	4. The _____ is a system of feeders and branch circuits supplying power for task illumination, fixed equipment, select receptacles, and select power circuits serving areas and functions related to patient care and that is automatically connected to alternate power sources by one or more transfer switches during interruption of normal power source.
_____	_____	5. The receptacles or the coverplates for receptacles supplied by the hospital's essential electrical system shall have a distinctive _____ so as to be readily identifiable.
_____	_____	6. _____ is a jack or terminal that serves as the collection point for redundant grounding of electric appliances serving a patient care vicinity or for grounding other items in order to eliminate electromagnetic interference problems.

_____ _____ 7. In any anesthetizing area, all metal raceways and metal-sheathed cables, and all normally non-current-carrying conductive portions of fixed electric equipment, shall be _____.

_____ _____ 8. A disconnecting means of adequate capacity for at least _____% of the input required for the long-time rating or _____% of the input required for the momentary rating of the X-ray equipment, whichever is greater, shall be provided in the supply circuit.

_____ _____ 9. Each patient bed location in a critical care area shall be supplied by at least _____ branch circuit(s).

_____ _____ 10. A patient care vicinity is defined as a space, within a location intended for the examination and treatment of patients, extending _____ feet beyond the normal location of the patient bed, chair, table, treadmill, or other device that supports the patient during examination and treatment and extending vertically to _____ feet above the floor.

_____ _____ 11. The line isolation monitor shall not alarm for a fault hazard of less than _____ mA or for a total hazard current of less than _____ mA.

_____ _____ 12. Receptacles that are located within the patient rooms, bathrooms, playrooms, and activity rooms of pediatric units, other than nurseries, shall be _____ or shall employ a _____ cover.

_____ _____ 13. _____ is space in which failure of equipment or a system is likely to cause minor injury to patients or caregivers.

_____ _____ 14. In locations where flammable anesthetics are employed, the entire area shall be considered to be a Class I, Division 1 location that shall extend upward to a level _____ ft above the floor.

_____ _____ 15. Separation of ground-fault protection time-current characteristics shall conform to the manufacturer's recommendations and shall consider all required tolerances and disconnect operating time to achieve _____ selectivity.

_____ _____ 16. For portable X-ray equipment connected to a 120-volt branch circuit of _____ or less, a grounding-type attachment plug and receptacle of proper rating shall be permitted to serve as a disconnecting means.

_____ _____ 17. The _____ is a system of feeders and branch circuits supplying power for lighting, receptacles, and equipment essential for life safety that is automatically connected to alternate power sources by one or more transfer switches during interruption of the normal power source.

_____ _____ 18. The ampacity of conductors and the rating of overcurrent protective devices shall not be less than _____ of the current rating of medical X-ray therapy equipment.

_____ _____ 19. As a general rule, the wiring of a hospital's life safety and critical branches shall be mechanically protected by installation in _____ metal raceways, Type _____ cable, Type RTRC marked with the suffix -XW or Schedule 80 PVC conduit.

_____ _____ 20. A(n) _____ is any area of the health care facility that has been designated to be used for the administration of any flammable inhalation anesthetic agents in the normal course of examination or treatment.

_____ _____ 21. Each patient bed location in a critical care area shall be provided with a minimum of _____ receptacle(s).

_____ _____ 22. In areas used for patient care, all non-current-carrying conductive surfaces of fixed electrical equipment likely to become energized that are subject to personal contact, operating at over _____, shall be directly connected to an insulated copper equipment grounding conductor.

NEC Reference   Answer

_____ _____ 23. Wall-mounted remote-control stations for remote-control switches operating at _____ or less shall be permitted to be installed in any anesthetizing location.

_____ _____ 24. A(n) _____ is a system comprised of alternate sources of power and all connected distribution systems and ancillary equipment, designed to ensure continuity of electrical power to designated areas and functions of a health care facility during disruption of normal power sources, and also designed to minimize disruption within the internal wiring system.

_____ _____ 25. In critical care areas, a patient care vicinity shall be permitted to have a patient equipment grounding point. An equipment-bonding jumper, not smaller than _____, shall be used to connect the grounding terminal of all grounding-type receptacles to the patient equipment grounding point.

_____ _____ 26. Individual branch circuits shall not be required for portable, mobile, and transportable medical X-ray equipment requiring a capacity of not over _____.

_____ _____ 27. Each general care area patient bed location shall be provided with a minimum of _____ receptacle(s).

_____ _____ 28. _____ is space in which failure of equipment or a system is likely to cause major injury or death to patients or caregivers.

_____ _____ 29. One transfer switch shall be permitted to serve one or more branches or systems in a limited care facility with a maximum demand on the essential electrical system of _____ kVA.

_____ _____ 30. Size 18 or 16 AWG fixture wires as specified in 725.49 and flexible cords shall be permitted for the control and operating circuits of X-ray and auxiliary equipment where protected by not larger than _____-ampere overcurrent devices.

_____ _____ 31. A flexible cord storage device, within a hazardous (classified) anesthetizing location, shall not subject the cord to bending at a radius of less than _____ in.

_____ _____ 32. All branch circuits serving patient care areas shall be provided with an effective _____ path by installation in a metal raceway system or a cable having a metallic armor or sheath assembly.

_____ _____ 33. Essential electrical systems for nursing homes and limited-care facilities shall comprise two separate branches capable of supplying a limited amount of lighting and power service, which is considered essential for the protection of life safety and effective operation of the institution during the time normal electrical service is interrupted for any reason. These two separate branches shall be the _____ branch and the _____ branch.

_____ _____ 34. The ampacity of supply branch-circuit conductors and the current rating of overcurrent protective devices for X-ray equipment shall not be less than _____% of the long-time rating or _____% of the momentary rating, whichever is greater.

_____ _____ 35. In critical care areas, the equipment grounding conductor for special-purpose receptacles, such as the operation of mobile X-ray equipment, shall be extended to the _____ of branch circuits for all locations likely to be served from such receptacles.

_____ _____ 36. _____ is the area in a patient care space where a procedure is performed that is normally subject to wet conditions while patients are present, including standing fluids on the floor or drenching of the work area, where either such condition is intimate to the patient or staff.

_____ _____ 37. Each operating room shall be provided with a minimum of _____ receptacles.

Industrial Locations

Objectives

After studying this unit, the student should:

▶ be familiar with service entrance provisions with or without a single main disconnecting means.

▶ know that a raceway (or cable) shall not contain both service conductors and nonservice conductors.

▶ have a good understanding of feeder tap rules.

▶ understand the provisions pertaining to conductors supplying a transformer.

▶ be familiar with transformer vault provisions.

▶ know how to correctly bond service raceways and equipment.

▶ be able to accurately connect taps to grounding electrode conductors.

▶ understand the different methods for sizing bonding conductors in paralleled service raceways.

▶ be familiar with the provision requiring a full size equipment grounding conductor (when used) in each paralleled raceway.

▶ have a thorough understanding of cable tray provisions.

▶ be familiar with both general and specific motor provisions, that is, motor and branch-circuit overload protection, branch-circuit conductors, branch-circuit protection, feeder conductors, feeder protection, etc.

▶ understand specific equipment provisions for cranes and hoists, electric welders, electroplating equipment, industrial machinery, and capacitors.

Introduction

Because industrial wiring covers such a broad spectrum of electrical applications, one unit simply cannot contain all of the related provisions. As with other topics, this text's building-block approach has laid the foundation for industrial locations, and it has steadily built upon that foundation with layers of additional information. For instance, previous units address items pertinent to industrial projects, such as raceways, services, clearances, busways, transformers, etc. Provisions for services and service equipment can be found in Units 9, 10, and 14, as well as here (as they apply more specifically to industrial applications).

In particular, this unit explains various feeder tap installation provisions, as found in 240.21. These include, among others, the 10-ft (3.0-m) and 25-ft (7.5-m) tap rules, plus transformer tap rules. While general transformer provisions were introduced in Unit 14, this unit expands that topic by discussing transformer vault requirements. Grounding and bonding provisions make up a part of this unit. Included are grounding electrode conductor taps, bonding service raceways, and equipment grounding conductors in paralleled raceways. Cable tray requirements are also addressed in this unit. Cable tray installations, of course, are not limited to industrial environments. They are often part of commercial projects and are used in many other areas as well.

Article 430 (Motors, Motor Circuits, and Controllers) applies a great number of requirements to the installation of motors, motor branch-circuit and feeder conductors and their protection, motor overload protection, motor-control circuits, motor controllers, and motor-control centers. The most commonly used of these provisions are illustrated and explained within this unit. Finally, specific equipment typically found in industrial-type locations is presented. Such specific equipment includes cranes and hoists, electric welders, electroplating equipment, industrial machinery, and capacitors.

The aim of this unit is to introduce, in a broad manner, the more common components of a very complex area. See Delmar Cengage Learning's extensive list of offerings for a more in-depth study of industrial electrical wiring.

GENERAL

Service Entrance: One Main Disconnect

Where an orderly shutdown is required to minimize the personnel/equipment hazard(s), a system of coordination based on the following two conditions is permitted: (1) coordinated short-circuit protection and (2) overload indication based on monitoring systems or devices »240.12«.

Knife switches rated more than 1200 amperes at 250 volts or less, and at over 1000 amperes at 251 to 1000 volts, shall be used only as isolating switches and shall not be opened under load »404.13(A)«.

Use a circuit breaker or a switch of special design listed for such purpose to interrupt currents over 1200 amperes at 250 volts, nominal, or less; or, over 600 amperes at 251 to 600 volts, nominal »404.13(B)«.

A The next higher standard overcurrent device rating (above the ampacity of the conductors being protected) is permitted, provided all the following conditions are met.

1. The protected conductors are not part of a branch circuit supplying more than one receptacle for cord- and plug-connected portable loads.

2. The conductors' ampacity does not correspond with a fuse/circuit breaker's standard ampere rating without overload trip adjustments above its rating (but it can have other trip or rating adjustments).

3. The next higher standard rating does not exceed 800 amperes »240.4(B)«.

B Service equipment rated no more than 1000 volts shall be marked to identify it as being suitable for use as service equipment »230.66«.

C Each ungrounded circuit conductor shall have overcurrent protection located at the point where the conductors receive their supply, except as specified in 240.21(A) through (H) »240.21«.

D The number of disconnects on the load side of the service disconnecting means is not limited to six.

E Cartridge fuse and fuseholder overcurrent protection provisions are located in 240.60 and 240.61.

F All switches, and circuit breakers used as switches, shall be located in a readily accessible and operable place. The center of the switch (or circuit breaker) operating handle's grip in its highest position shall not be more than 6 ft 7 in. (2.0 m) above the floor/working platform »404.8(A)«.

G Where insulated conductors are deflected within a metallic wireway, either at the ends or where conduits, fittings, or other raceways or cables enter or leave the metallic wireway, or where the direction of the metallic wireway is deflected greater than 30 degrees, dimensions corresponding to one wire per terminal in Table 312.6(A) shall apply »376.23(A)«.

> **NOTE**
>
> See this book's Units 8 and 14 for electrical equipment working space clearances as specified in 110.26.

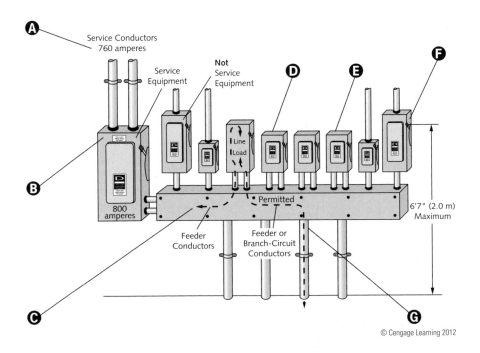

© Cengage Learning 2012

Service Entrance: Six Disconnects

Means shall be provided to disconnect all conductors in a building (or other structure) from the service-entrance conductors »*230.70*«.

Install the service disconnecting means at a readily accessible location, either outside of a building/structure or inside nearest the service conductors' entry point »*230.70(A)(1)*«.

A set of fuses is considered all the fuses required to protect all of the circuit's ungrounded conductors »*230.90(A)*«.

A The service disconnecting means for each service (permitted by 230.2), or for each set of service-entrance conductors (permitted by 230.40, Exception Nos. 1, 3, 4, or 5), shall consist of no more than six switches/circuit breakers in a single enclosure, in a group of separate enclosures, or in (or on) a switchboard or in switchgear »*230.71(A)*«.

B Two to six disconnects, permitted by 230.71, shall be grouped »*230.72(A)*«.

C Two to six circuit breakers (or sets of fuses) can serve as the overcurrent device providing overload protection. The sum of the circuit breakers' or fuses' ratings can exceed the service conductor's ampacity, provided the calculated load (according to Article 220) does not exceed the ampacity of the service conductors »*230.90(A) Exception No. 3*«.

D Each service disconnect shall be permanently marked to identify it as a service disconnect »*230.70(B)*«.

E Where two to six separately enclosed service disconnecting means (supplying separate loads from one service drop, set of overhead service conductors, set of underground service conductors, or service lateral) are grouped at one location, one set of service-entrance conductors can supply each, or several such, service equipment enclosures »*230.40 Exception No. 2*«.

F Article 376 contains metal wireway provisions.

G Service-entrance conductors can be spliced (or tapped) in accordance with 110.14, 300.5(E), 300.13, and 300.15 »*230.46*«.

H Conductors other than service conductors shall not be installed in the same service raceway or service cable »*230.7*«. This provision does not apply to grounding electrode conductors and equipment bonding jumpers or conductors. This provision also does not apply to load management control conductors having overcurrent protection »*230.7 Exception No. 1* and *No. 2*«.

N O T E

Each service disconnect shall simultaneously disconnect all ungrounded service conductors under its control from the premises wiring system »***230.74***«.

Section 230.2 permits one (or more) additional service disconnecting means for fire pumps, emergency systems, and legally required or optional standby services. These additional service disconnecting means shall be installed at a sufficient distance from the normal (one to six) service disconnecting means to minimize the possibility of simultaneous interruption of supply »***230.72(B)***«.

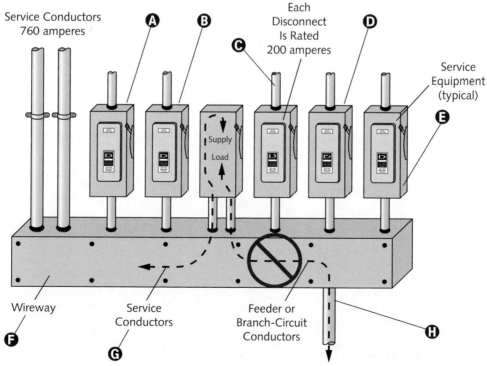

Service Conductors 760 amperes — **A** — **B** — Each Disconnect Is Rated 200 amperes — **C** — **D** — Service Equipment (typical) — **E**

Supply / Load

Wireway **F** — Service Conductors **G** — Feeder or Branch-Circuit Conductors — **H**

Feeder Taps

Because of dissimilar metals' characteristics, devices such as pressure terminal/splicing connectors and soldering lugs shall be identified for the conductor's material, shall be properly installed, and shall be properly used. Conductors of dissimilar metals shall not be intermixed in a terminal (or splicing) connector where physical contact occurs between dissimilar conductors (such as copper and aluminum, copper and copper-clad aluminum, etc.), unless the device is identified for the purpose and conditions of use »110.14«.

While single-phase system illustrations are used, be aware that the same rules also apply to 3-phase systems.

Ⓐ Tap conductors can be protected against overcurrent in accordance with 210.19(A)(3) and (4), 240.5(B)(2), 240.21, 368.17(B) and (C), 368.17(C), and 430.53(D) »240.4(E)«.

Ⓑ Conductors can be tapped to a feeder without having overcurrent protection at the tap as specified in 240.21(B)(1) through (5). The provisions of 240.4(B) shall not be permitted for tap conductors »240.21(B)«.

Ⓒ Per Article 240, a **tap conductor** is a conductor other than a service conductor having overcurrent protection ahead

of its point of supply that exceeds the value permitted for similar conductors protected as otherwise described in 240.4 »240.2«.

Ⓓ All conductor splices, joints, and free ends shall be covered either with an insulation equivalent to that of the conductors or with an identified insulating device »110.14(B)«.

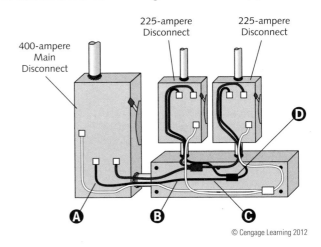

© Cengage Learning 2012

10-ft (3.0-m) Feeder Tap Rule

Ⓐ A 60-ampere rated conductor is the minimum size allowed if the overcurrent protection on the tap conductor's line side is 600 amperes (600 ÷ 10 = 60) »240.21(B)(1)(4)«.

Ⓑ The tap conductor's ampacity shall not be less than the rating of the equipment containing an overcurrent device(s) supplied by the tap conductors or less than the overcurrent-protective device rating at the tap conductor's termination »240.21(B)(1)(1)(b)«.

Ⓒ In addition to the requirement of 408.30, a panelboard shall be protected by an overcurrent protective device having a rating not greater than that of the panelboard. This overcurrent protective device shall be located within or at any point on the supply side of the panelboard »408.36«.

Ⓓ The tap conductor's ampacity shall not be less than the combined calculated loads on the circuits it supplies »240.21(B)(1)(1)(a)«.

Ⓔ The tap conductors shall not extend beyond the switchboard, switchgear, panelboard, disconnecting means, or control devices they supply »240.21(B)(1)(2)«.

Ⓕ Except where connected to the feeder, tap conductors shall be enclosed in a raceway extending from the tap to the enclosure of an enclosed switchboard, switchgear, panelboard, or control device, or to the back of an open switchboard »240.21(B)(1)(3)«.

Ⓖ If the tap conductor's length does not exceed 10 ft (3 m) and all of 240.21(B)(1)(1) through (4)

stipulations are satisfied, overcurrent protection at the tap to the feeder is not required »240.21(B)(1)«.

Ⓗ Field installations with the tap conductors exiting the enclosure (or vault) where the tap is made require that the ampacity of the tap conductors be not less than one-tenth of the rating of the overcurrent device protecting the feeder conductors »240.21(B)(1)(4)«.

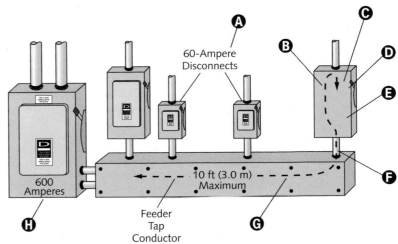

© Cengage Learning 2012

25-ft (7.5-m) Feeder Tap Rule

A A 200-ampere rated conductor is the minimum size allowed if the overcurrent protection on the tap conductor's line side is 600 amperes (600 ÷ 3 = 200) ≫ *240.21(B)(2)(1)* ≪ .

B The tap conductors shall terminate in a single circuit breaker (or single set of fuses) that will limit the load of the tap conductor's ampacity. This device can supply unlimited additional load side overcurrent devices ≫ *240.21(B)(2)(2)* ≪ .

C The tap conductors shall be protected from physical damage in an approved manner, such as enclosed in a raceway ≫ *240.21(B)(2)(3)* ≪ .

D If the tap conductor's length does not exceed 25 ft (7.5 m) and all of 240.21(B)(2)(1) through (3) stipulations are satisfied, over-current protection at the tap to the feeder is not required ≫ *240.21(B)(2)* ≪ .

E The tap conductors' ampacity shall not be less than one-third of the feeder conductor's over-current device rating ≫ *240.21(B)(2)(1)* ≪ .

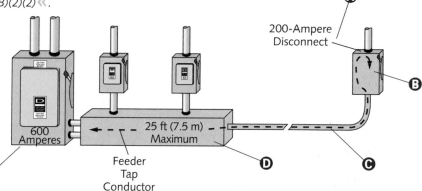

> **N O T E**
> Tap conductors longer than 25 ft (7.5 m) shall comply with 240.21(B)(4) specifications.

© Cengage Learning 2012

Bonding Service Raceways

A The supply-side bonding jumper shall not be smaller than specified in Table 250.102(C)(1) ≫ *250.102(C)(1)* ≪ . If the ungrounded supply conductors are larger than 1100 kcmil copper or 1750 kcmil aluminum, the supply-side bonding jumper shall have an area not less than 12-1/2 percent of the area of the largest ungrounded supply conductor or equivalent area for parallel supply conductors. If the ungrounded supply conductors and the bonding jumper are of different materials (copper, aluminum, or copper-clad aluminum), the minimum size of the supply-side bonding jumper shall be based on the assumed use of ungrounded supply conductors of the same material as the supply-side bonding jumper and will have an ampacity equivalent to that of the installed ungrounded supply conductors ≫ *Notes 1* and *2 under Table 250.102(C)(1)* ≪ .

B Here, one supply-side bonding jumper is used to bond *all* service raceways. This supply-side bonding jumper's size is based on the total combined ungrounded supply conductors' area.

C Here, a supply-side bonding jumper bonds *each individual* service raceway. Each supply-side bonding jumper is sized according to each individual raceway's ungrounded supply conductors.

D Where the ungrounded supply conductors are paralleled in two or more raceways or cables, and an individual supply-side bonding jumper is used for bonding these raceways or cables, the size of the supply-side bonding jumper for each raceway

or cable shall be selected from Table 250.102(C)(1) based on the size of the ungrounded supply conductors in each raceway or cable. A single supply-side bonding jumper installed for bonding two or more raceways or cables shall be sized in accordance with 250.102(C)(1) ≫ *250.102(C)(1)* ≪ .

E The total ungrounded supply conductors' area is 1400 kcmil (350 × 4). The supply-side bonding jumper shall be at least 12½% of that area (1400 × 12.5% = 175 kcmil or 175,000 cmil). In accordance with Chapter 9, Table 8, a 3/0 AWG copper conductor (with a circular mil area of 167,800) is insufficient. Therefore, a 4/0 AWG copper conductor (with a circular mil area of 211,600) meets the minimum size requirement.

F The following normally non-current-carrying metal parts of equipment shall be bonded together: (1) all raceways, cable trays, cablebus framework, auxiliary gutters, or service cable armor or sheath that enclose, contain, or support service conductors, except as permitted in 250.80; and (2) all enclosures containing service conductors, including meter fittings, boxes, or the like, interposed in the service raceway or armor ≫ *250.92(A)(1)* and *(2)* ≪ .

G A supply-side bonding jumper is defined as a conductor installed on the supply side of a service or within a service equipment enclosure(s), for a separately derived system, that ensures the required electrical conductivity between metal parts required to be electrically connected ≫ *250.2* ≪ .

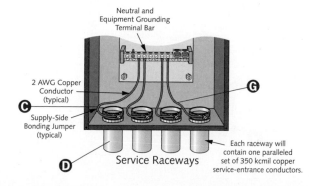

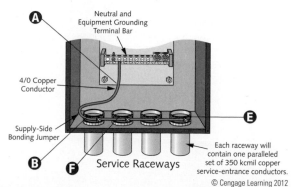

© Cengage Learning 2012

Tap Conductors Supplying a Transformer

Ⓐ A transformer's primary conductors shall have an ampacity at least one-third of the feeder conductors' overcurrent device rating ≫ *240.21(B)(3)(1)* ≪ .

Ⓑ A 100-ampere rated conductor is the minimum size allowed if the overcurrent protection on the tap conductor's line side is 300 amperes (300 ÷ 3 = 100) ≫ *240.21(B)(3)(1)* ≪ .

Ⓒ Primary and secondary conductors shall be protected from physical damage in an approved manner, such as enclosed in a raceway ≫ *240.21(B)(3)(4)* ≪ .

Ⓓ Secondary conductors shall terminate in a single circuit breaker or set of fuses that will limit the load current to no more than the conductor ampacity permitted by 310.15 ≫ *240.21(B)(3)(5)* ≪ .

Ⓔ Conductors supplied by the transformer's secondary shall have an ampacity that is not less than the value of the primary-to-secondary voltage ratio multiplied by one-third of the feeder conductors' overcurrent device rating ≫ *240.21(B)(3)(2)* ≪ .

Ⓕ The combined length of one primary plus one secondary conductor (excluding any portion of the primary conductor protected at its ampacity) shall not exceed 25 ft (7.5 m) ≫ *240.21(B)(3)(3)* ≪ .

Ⓖ Conductors supplied from a set of fuses (or circuit breaker) and feeding a transformer are feeder conductors, not feeder tap conductors.

Ⓗ Refer to this book's Unit 14 for transformer provisions (overcurrent protection, grounding, etc.).

Ⓘ Transformer secondary conductor provisions are found in 240.21(C).

WARNING
The provisions of 240.4(B) shall not be permitted for tap conductors ≫ *240.21(B)* ≪ .

CAUTION
Transformers, other than Class 2 or Class 3 transformers, shall have a disconnecting means located either in sight of the transformer or in a remote location. Where located in a remote location, the disconnecting means shall be lockable in accordance with 110.25, and its location shall be field-marked on the transformer ≫ *450.14* ≪ .

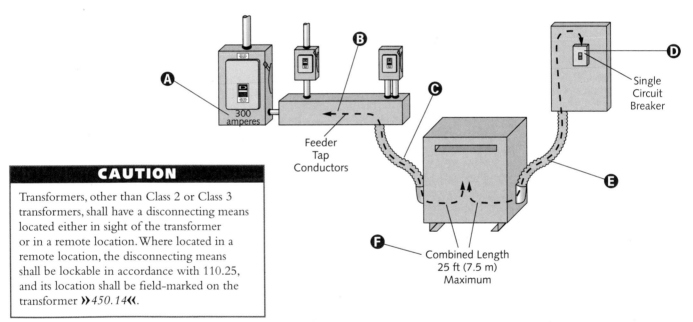

Single Circuit Breaker

Feeder Tap Conductors

Combined Length 25 ft (7.5 m) Maximum

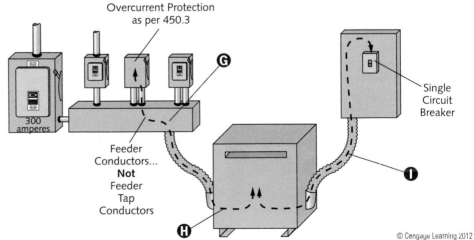

Overcurrent Protection as per 450.3

Feeder Conductors... **Not** Feeder Tap Conductors

Single Circuit Breaker

© Cengage Learning 2012

Transformer Vaults

Wherever such an arrangement is practicable, transformer vaults shall be located so that ventilation to outside air does not rely on flues or ducts ⟫ *450.41* ⟪.

Where required by 450.9, openings for ventilation shall also comply with 450.45(A) through (F) ⟫ *450.45* ⟪. That section includes provisions pertaining to location, arrangement, size, covering, dampers, and ducts.

Each doorway from the building interior into the vault shall have a tight-fitting door with a minimum fire rating of three hours, unless the exception is met. The AHJ may require such a door for exterior wall openings, should conditions warrant ⟫ *450.43(A)* ⟪.

Doors shall be equipped with fully engaged locks, thereby limiting access to qualified persons only ⟫ *450.43(C)* ⟪.

Personnel doors shall swing out and be equipped with panic bars, pressure plates, or other devices that are normally latched but that open under simple pressure ⟫ *450.43(C)* ⟪.

Vaults containing more than 100-kVA transformer capacity shall have a drain or other means to carry off any oil or water accumulation, unless local conditions make this impracticable. The vault floor must pitch toward the drain, where provided ⟫ *450.46* ⟪.

Nonelectrical pipe or duct systems shall neither enter nor pass through a transformer vault. Piping or other facilities providing vault fire protection, or transformer cooling, are allowed ⟫ *450.47* ⟪.

Ⓐ Vault walls and roofs shall be constructed of materials having approved structural strength for the conditions and a minimum fire resistance of three hours. For this section's purpose, studs and wallboard construction shall not be permitted ⟫ *450.42* ⟪.

Ⓑ Construction of one hour rating is permitted where transformers are protected with automatic sprinkler, water sprays, carbon dioxide, or halon ⟫ *450.42 Exception and 450.43(A) Exception* ⟪.

Ⓒ Vault floors in contact with the earth shall be of concrete, no less than 4 in. (100 mm) thick. Where the vault is constructed above a vacant space (or other stories), the floor shall have approved structural strength for the load imposed on it and shall have a minimum fire resistance of three hours ⟫ *450.42* ⟪.

Ⓓ A door sill (or curb) of an approved height, at least 4 in. (100 mm), that will confine the oil from the largest transformer within the vault shall be provided ⟫ *450.43(B)* ⟪.

NOTE

Entrances to all buildings, rooms, or enclosures containing exposed live parts or exposed conductors operating at over 600 volts, nominal, shall be kept locked, unless such entrances are constantly observed by qualified personnel. Permanent and conspicuous danger signs shall be provided. The danger sign shall meet the requirements in 110.21(B) and shall read as follows: *DANGER—HIGH VOLTAGE—KEEP OUT* ⟫*110.34(C)*⟪.

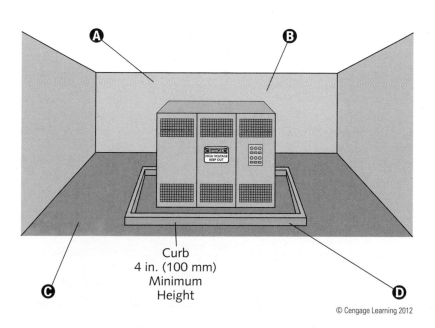

Curb
4 in. (100 mm)
Minimum
Height

Grounding Electrode Conductor Taps

The supply-side bonding jumper shall not be smaller than specified in Table 250.102(C)(1) *»250.102(C)(1)«*.

Where the ungrounded supply conductors are paralleled in two or more raceways or cables, and an individual supply-side bonding jumper is used for bonding these raceways or cables, the size of the supply-side bonding jumper for each raceway or cable shall be selected from Table 250.102(C)(1) based on the size of the ungrounded supply conductors in each raceway or cable. A single supply-side bonding jumper installed for bonding two or more raceways or cables shall be sized in accordance with (C)(1) *»250.102(C)(2)«*.

Ⓐ For a service or feeder with two or more disconnecting means in separate enclosures supplying a building or structure, the grounding electrode connections shall be made in accordance with 250.64(D)(1), (D)(2), or (D)(3) *»250.64(D)«*. A common grounding electrode conductor and grounding electrode conductor taps shall be installed. The common grounding electrode conductor shall be sized in accordance with 250.66, based on the sum of the circular mil area of the largest ungrounded conductor(s) of each set of conductors that supplies the disconnecting means. If the service-entrance conductors connect directly to the overhead service conductors, service drop, underground service conductors, or service lateral, the common grounding electrode conductor shall be sized in accordance with Table 250.66, Note 1. A grounding electrode conductor tap shall extend to the inside of each service disconnecting means enclosure. The grounding electrode conductor taps shall be sized in accordance with 250.66 for the largest service-entrance or feeder conductor serving the individual enclosure. The tap conductors shall be connected to the common grounding electrode conductor by one of the following methods in such a manner that the common grounding electrode conductor remains without a splice or joint; (1) exothermic welding, (2) connectors listed as grounding and bonding equipment, (3) connections to an aluminum or copper busbar not less than ¼ in. thick × 2 in. wide (6 mm thick × 50 mm wide) and of sufficient length to accommodate the number of terminations necessary for the installation. The busbar shall be securely fastened and shall be installed in an accessible location. Connections shall be made by a listed connector or by the exothermic welding process. If aluminum busbars are used, the installation shall comply with 250.64(A) *»250.64(D)(1)«*.

Ⓑ The following normally non-current-carrying metal parts of equipment shall be bonded together: (1) all raceways, cable trays, cablebus framework, auxiliary gutters, or service cable armor or sheath that enclose, contain, or support service conductors, except as permitted in 250.80; and (2) all enclosures containing service conductors, including meter fittings, boxes, or the like, interposed in the service raceway or armor *»250.92(A)(1) and (2)«*.

Ⓒ Bonding shall be provided where necessary to ensure electrical continuity and the capacity to conduct safely any fault current likely to be imposed *»250.90«*.

Ⓓ The grounding electrode conductor shall be of one continuous length (without splice or joint), except as permitted in 250.30(A)(5) and (A)(6), 250.30(B)(1), and 250.68(C).

If necessary, splices or connections shall be made as permitted in 250.64(C)(1) through (4) *»250.64(C)«*.

Ⓔ Ferrous metal raceways and enclosures for grounding electrode conductors shall be electrically continuous from the point of attachment to cabinets or equipment to the grounding electrode and shall be securely fastened to the ground clamp or fitting. Ferrous metal raceways and enclosures shall be bonded at each end of the raceway or enclosure to the grounding electrode or grounding electrode conductor. Bonding shall be in compliance with 250.92(B) and ensured by one of the methods in 250.92(B)(2) through (B)(4) *»250.64(E)(1) and (E)(2)«*.

Ⓕ A grounding electrode conductor shall be connected in a wireway or other accessible enclosure on the supply side of the disconnecting means to one or more of the following, as applicable:

(1) Grounded service conductor(s)

(2) Equipment grounding conductor installed with the feeder

(3) Supply-side bonding jumper.

The connection shall be made with exothermic welding or a connector listed as grounding and bonding equipment. The grounding electrode conductor shall be sized in accordance with 250.66 based on the service-entrance or feeder conductor(s) at the common location where the connection is made *»250.64(D)(3)«*.

> ### NOTE
>
> A grounding electrode conductor shall be connected between the grounding electrode system and one or more of the following, as applicable:
> (1) Grounded conductor in each service equipment disconnecting means enclosure
> (2) Equipment grounding conductor installed with the feeder
> (3) Supply-side bonding jumper.
> Each grounding electrode conductor shall be sized in accordance with 250.66 based on the service-entrance or feeder conductor(s) supplying the individual disconnecting means *»250.64(D)(2)«*.

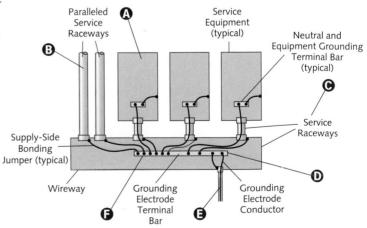

Paralleled Service Raceways **Ⓐ** Service Equipment (typical) Neutral and Equipment Grounding Terminal Bar (typical) **Ⓒ**

Ⓑ

Supply-Side Bonding Jumper (typical)

Service Raceways

Wireway Grounding Electrode Terminal Bar **Ⓕ** **Ⓔ** Grounding Electrode Conductor **Ⓓ**

© Cengage Learning 2012

Equipment Grounding Conductors

Where a single equipment grounding conductor is run with multiple circuits in the same raceway, cable, or cable tray, it shall be sized for the largest overcurrent device protecting conductors in the raceway, cable, or cable tray. Equipment grounding conductors installed in cable trays shall meet the minimum requirements of 392.10(B)(1)(c) ≫*250.122(C)*≪.

While copper, aluminum, or copper-clad aluminum equipment grounding conductors of the wire type shall not be smaller than shown in Table 250.122, in no case, are they required to be larger than the circuit conductors supplying the equipment ≫*250.122(A)*≪.

Where a raceway, cable tray, or cable armor (or sheath) serves as the equipment grounding conductor, as provided in 250.118 and 250.134(A), it shall comply with 250.4(A)(5) or 250.4(B)(4). Equipment grounding conductors shall be permitted to be sectioned within a multiconductor cable, provided the combined circular mil area complies with Table 250.122 ≫*250.122(A)*≪.

Ⓐ Since the overcurrent protection has a rating of 1200 amperes, the minimum size equipment grounding conductor is 3/0 AWG copper (or 250-kcmil aluminum) ≫*Table 250.122*≪.

WARNING

An equipment grounding conductor shall not be used as a grounding electrode conductor ≫*250.121*≪. A wire-type equipment grounding conductor installed in compliance with 250.6(A) and the applicable requirements for both the equipment grounding conductor and the grounding electrode conductor in Parts II, III, and VI of this article shall be permitted to serve as both an equipment grounding conductor and a grounding electrode conductor ≫*250.121 Exception*≪.

Ⓑ Each equipment grounding conductor shall be sized in compliance with 250.122 ≫*250.122(F)*≪.

Ⓒ Equipment grounding conductor provisions are not limited to feeders. For example, an alternative drawing could have shown one piece of equipment supplied by a paralleled set of branch-circuit conductors. Equipment grounding conductor provisions also apply to branch circuits.

Ⓓ Unlike service-equipment bonding jumpers, which can be sized according to each paralleled raceway's service-entrance conductor size, a **full-size** equipment grounding conductor (where used) must be installed in **each** raceway.

Ⓔ For paralleled conductors in multiple raceways (or cables) as permitted by 310.10(H), any equipment grounding conductors installed shall also run in parallel within **each** raceway ≫*250.122(F)*≪.

Ⓕ If a single raceway supplies the equipment, only one 3/0 AWG copper (or 250-kcmil aluminum) equipment grounding conductor is required.

N O T E

Where ungrounded conductors are increased in size from the minimum size that has sufficient ampacity for the intended installation, wire-type equipment grounding conductors, where installed, shall be increased in size proportionately according to the circular mil area of the ungrounded conductors ≫***250.122(B)***≪.

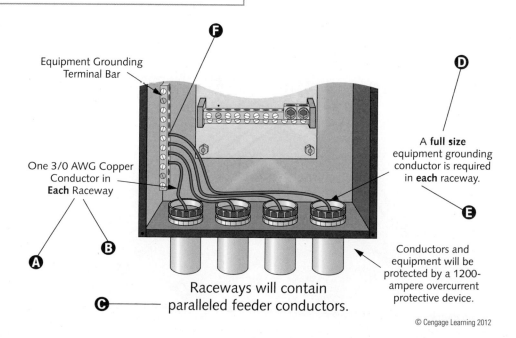

Equipment Grounding Terminal Bar

One 3/0 AWG Copper Conductor in **Each** Raceway

A **full size** equipment grounding conductor is required in **each** raceway.

Conductors and equipment will be protected by a 1200-ampere overcurrent protective device.

Raceways will contain paralleled feeder conductors.

© Cengage Learning 2012

Cable Trays

Cable tray installations are not limited to industrial establishments. Wiring methods in Table 392.10(A) can be installed in cable tray systems under the conditions described in their respective articles and sections ≫*392.10 and 10(A)*≪.

Only in industrial establishments, where maintenance and supervision conditions restrict cable tray system service to qualified personnel, can any of the cables in 392.10(B)(1) and (2) be installed in ladder, ventilated troughs, solid-bottom, or ventilated-channel cable trays ≫*392.10(B)*≪.

Cable tray systems shall be corrosion resistant. If made of ferrous material, the system shall be protected from corrosion per 300.6 ≫*392.100(C)*≪.

Steel or aluminum cable tray systems can be used as equipment grounding conductors provided all 392.60(B)(1) through (4) conditions are met.

While provisions pertaining to the number of multiconductor cables (rated 2000 volts or less) can be found in 392.22(A), 392.22(B) contains single conductor cable provisions.

A Cable trays can extend transversely through partitions and walls, or vertically through platforms and floors, in either wet or dry locations where the completed installation meets 300.21 requirements ≫*392.18(D)*≪.

B Sufficient space shall be provided and maintained around cable trays to permit adequate access for cable installation and maintenance ≫*392.18(F)*≪.

C Cable trays shall be of adequate strength and rigidity to provide support for all contained wiring ≫*392.100(A)*≪.

D A cable tray system is a unit (or assembly of units or sections) and associated fittings that together form a structural system used to securely fasten/support cables and raceways ≫*392.2*≪.

E Cable trays shall be exposed and accessible except as permitted by 392.10(D) ≫*392.18(E)*≪.

F Cable trays shall include fittings or other suitable means to facilitate changes in direction and elevation ≫*392.100(E)*≪.

G Cable trays shall have side rails or similar structural members ≫*392.100(D)*≪.

H Metallic cable trays can serve as equipment grounding conductors where (1) continuous maintenance and supervision ensures that qualified persons service the installed cable tray system and (2) the cable tray complies with 392.60 provisions ≫*392.60(A)*≪.

I Each cable tray run shall be completed before cables are installed ≫*392.18(B)*≪.

CAUTION

Cable trays in hazardous (classified) locations shall contain only the cable types and raceways permitted by other articles in the *NEC* ≫*392.10(C)*≪.

NOTE

Cable tray systems shall not be used in hoistways or where subject to severe physical damage ≫*392.12*≪.

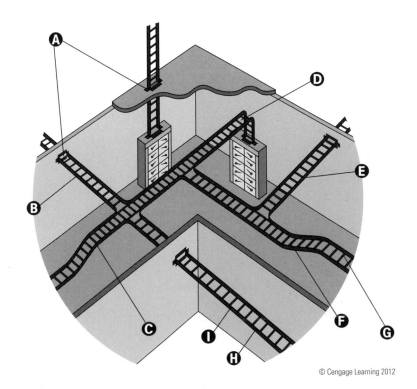

© Cengage Learning 2012

MOTORS

General Motor Provisions

Article 430 covers motors and all of the following as they relate specifically to motors: branch-circuit and feeder conductors and their protection, overload protection, control circuits, controllers, and motor-control centers ≫*430.1*≪.

Wires passing through an opening in an enclosure, conduit box, or barrier require a bushing to protect the conductors from any sharp opening edges ≫*430.13*≪.

Ⓐ Article 430, Part V specifies protective devices designed to protect feeder conductors supplying motors against overcurrents due to short circuits or grounds ≫*430.61*≪.

Ⓑ Sections 430.24 through 430.26 contain provisions for conductors supplying multiple motors, or a motor(s) and other load(s). These provisions are applicable to more than feeder conductors.

Ⓒ Article 430, Part IV specifies devices that are used to protect motor branch-circuit conductors, motor-control apparatus, and motors against overcurrent due to short circuits or ground faults. The rules add to or amend Article 240 provisions ≫*430.51*≪.

Ⓓ Article 430, Part II specifies ampacities of conductors capable of carrying the motor current without overheating, under specified conditions ≫*430.21*≪.

Ⓔ Article 430, Part III specifies overload devices intended to protect motors, motor-control apparatus, and motor branch-circuit conductors against excessive heating due to motor overloads and failure to start ≫*430.31*≪.

Overload is defined as operation of equipment in excess of normal, full-load rating, or of a conductor in excess of rated ampacity that, when it persists for a sufficient length of time, would cause damage or dangerous overheating. A fault, such as a short circuit or ground fault, is not an overload ≫*Article 100*≪.

Ⓕ General motor provisions are found in 430.1 through 430.18 (Part I).

Ⓖ Part XIII of Article 430 addresses the grounding of exposed non-current-carrying metal parts of motor and controller frames that are likely to become energized to prevent a voltage aboveground should accidental contact between energized parts and frames occur. Insulating, isolating, or guarding are suitable alternatives to motor grounding under certain conditions ≫*430.241*≪.

Ⓗ For general motor applications, base current ratings on 430.6(A)(1) and (2): (1) Other than for motors built for low speeds (less than 1200 rpm) or high torques, and for multi-speed motors, use the values given in Tables 430.247 through 430.250 (including notes) to determine conductor ampacity or ampere ratings of switches, branch-circuit short-circuit, and ground-fault protection. Do not use the actual current ratings marked on the motor nameplate unless an exception applies, and (2) Separate motor overload protection is based on the motor nameplate current rating ≫*430.6(A)*≪.

Ⓘ Motor disconnecting means provisions are found in 430.101 through 430.113 (Part IX).

> **N O T E**
>
> Motor tables are located in Part XIV of Article 430.

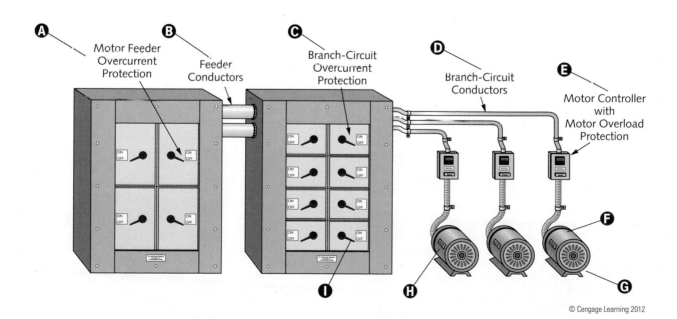

Ⓐ Motor Feeder Overcurrent Protection
Ⓑ Feeder Conductors
Ⓒ Branch-Circuit Overcurrent Protection
Ⓓ Branch-Circuit Conductors
Ⓔ Motor Controller with Motor Overload Protection
Ⓘ
Ⓗ
Ⓕ
Ⓖ

Motor and Branch-Circuit Overload Protection

Each continuous-duty motor rated 1 hp or less, *not* permanently installed, nonautomatically started, and within sight of the controller can be protected against overloads by the branch-circuit short-circuit and ground-fault device. Comply with branch-circuit overcurrent protection size specifications found in Article 430, Part IV ≫ *430.32(D)* ≪ .

Any automatically started motor, 1 hp or less, must be protected against overload by one of the methods found in 430.32(B)(1) through (4).

A motor controller can serve as an overload device if the number of overload units complies with Table 430.37, and if these units are operative in both the starting and running position (dc motor), or in the running position (ac motor) ≫ *430.39* ≪ .

Overload relays and other devices for motor overload protection incapable of opening short circuits or ground faults shall be protected by fuses (or circuit breakers) rated or set according to 430.52, or by a motor short-circuit protector per 430.52 ≫ *430.40* ≪ .

Overload protection for motors on general-purpose branch circuits (permitted by Article 210) is required and must meet 430.42(A), (B), (C), or (D) specifications.

A As an approved method of protecting a motor against overload, 430.32(A)(1) lists a separate overload device responsive to motor current. This device must be selected to trip or shall be rated at no more than the percentage of the motor nameplate full-load current rating shown here:

Motors with a marked service
factor not less than 1.15 . . . . .125%

Motors with a marked temperature
rise not over 40°C 125%

All other motors 115%

B Controllers must be marked according to 430.8 provisions.

C Article 430, Part III lists overload devices intended to protect motors, motor-control apparatus, and motor branch-circuit conductors against excessive heating due to motor overloads and failure to start ≫ *430.31* ≪ .

D Separate motor overload protection shall be based on the actual motor nameplate current rating, not on the ratings listed in Tables 430.247 through 430.250 ≫ *430.6(A)(2)* ≪ .

E Motors (in usual applications) must be marked with the information listed in 430.7(A)(1) through (15).

F Each motor used in a continuous-duty application rated more than 1 hp must be protected against overload by a means listed in 430.32(A)(1) through (4).

G **Overload** is defined as operation of equipment in excess of normal, full-load rating, or of a conductor in excess of rated ampacity that, when it persists for a sufficient length of time, would cause damage or dangerous overheating. A fault, such as a short circuit or ground fault, is not an overload ≫ *Article 100* ≪ .

CAUTION

A motor overload device, which restarts a motor automatically after overload tripping, shall not be installed if automatic restarting can result in injury to persons ≫ *430.43* ≪ .

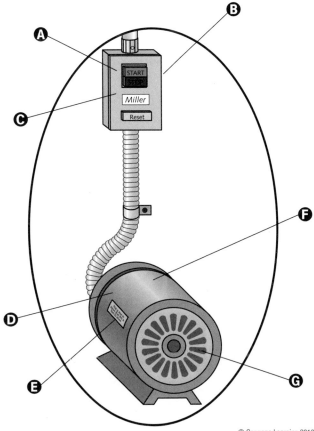

N O T E

Any motor application shall be considered continuous duty unless the driven apparatus, by nature, means the motor shall not, under any conditions, operate continuously with the load ≫ *430.33* ≪ .

WARNING

Section 430, Part III provisions *do not* require overload protection where a power loss would cause a hazard, such as with fire pumps ≫ *430.31* ≪ . For fire pump supply conductor protection, see 695.6.

Motors on General-Purpose Branch Circuits

Where a motor without individual overload protection (as provided in 430.42[A]) is connected to a branch circuit by means of an attachment plug and a receptacle or cord connector, the attachment plug and receptacle or cord connector rating shall not exceed 15 amperes at either 125 or 250 volts. Where individual overload protection is required per 430.42(B) for a motor or motor-operated appliance attached to a branch circuit through an attachment plug and a receptacle or a cord connector, the overload device shall be an integral part of the motor/appliance. The rating of the attachment plug and receptacle or cord connector determines the rating of the circuit to which the motor may be connected, as provided in 210.21(B) ›› *430.42(C)* ‹‹.

The branch-circuit short-circuit and ground-fault protective device of a circuit to which a motor (or motor-operated appliance) is connected shall have a time delay sufficient to start the motor and allow it to accelerate ›› *430.42(D)* ‹‹.

Ⓐ One or more motors without individual overload protection can be connected to a general-purpose branch circuit only if the installation complies with the limiting conditions of 430.32(B) and (D) and 430.53(A)(1) and (A)(2) ›› *430.42(A)* ‹‹.

Ⓑ Motors used on general-purpose branch circuits (as permitted in Article 210) require overload protection per 430.42(A), (B), (C), or (D) specifications ›› *430.42* ‹‹.

Motor Branch-Circuit Conductors

Conductors for motors used in short-time, intermittent, periodic, or varying duty applications shall have an ampacity of no less than the percentage of the motor nameplate current rating shown in Table 430.22(E), unless the AHJ grants special permission for conductors of lower ampacity ›› *430.22(E)* ‹‹.

Where motor circuits contain capacitors, conductors shall comply with 460.8 and 460.9 ›› *430.27* ‹‹.

For a multispeed motor, the selection of branch-circuit conductors on the line side of the controller shall be based on the highest of the full-load current ratings shown on the motor nameplate. The ampacity of the branch-circuit conductors between the controller and the motor shall not be less than 125 percent of the current rating of the winding(s) that the conductors energize ›› *430.22(B)* ‹‹.

Ⓐ Branch-circuit conductors supplying a single motor used in a continuous-duty application shall have an ampacity of at least 125% of the motor's full load current rating as determined by 430.6(A)(1), or not less than specified in 430.22(A) through (G) ›› *430.22* ‹‹.

NOTE

Conductors supplying several motors, or a motor(s) and other load(s), must comply with 430.24.

Ⓒ Motors of larger ratings than specified in 430.53(A) can be connected to general-purpose branch circuits only if each motor has overload protection specifically listed for that motor as specified in 430.32. In the case of more than one motor, both the controller and the motor overload device shall be approved for group installation, and the short-circuit and ground-fault protective device shall be selected according to 430.53 ›› *430.42(B)* ‹‹.

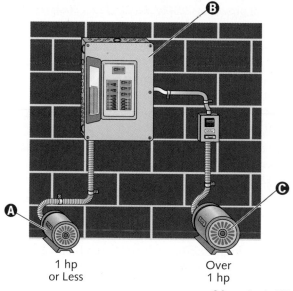

1 hp
or Less

Over
1 hp

© Cengage Learning 2012

Ⓑ Annex D contains a motor application example (Example No. D8).

Ⓒ Placement of motors shall allow adequate ventilation and facilitate maintenance, such as bearing lubrication and brush replacement ›› *430.14(A)* ‹‹.

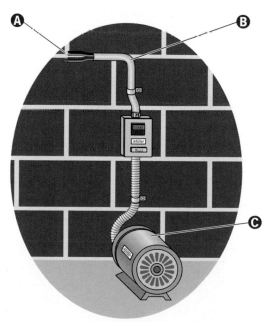

© Cengage Learning 2012

Motor Branch-Circuit Short-Circuit and Ground-Fault Protection

Part IV specifies devices intended to protect the motor branch-circuit conductors, motor-control apparatus, and the motors against overcurrent due to short circuits or ground faults »*430.51*«.

Where the rating specified in Table 430.52, or the rating modified by Exception No. 1, is not sufficient for the motor's starting current, apply 430.52(C)(1) Exception No. 2 (a) through (d) provisions »*430.52(C)(1) Exception No. 2*«.

Semiconductor fuses intended for the protection of electronic devices shall be permitted in lieu of devices listed in Table 430.52 for power electronic devices, associated electromechanical devices (such as bypass contactors and isolation contactors), and conductors in a solid-state motor controller system, provided replacement fuse markings are adjacent to the fuses »*430.52(C)(5)*«.

A listed self-protected combination controller may be substituted for the devices specified in Table 430.52 »*430.52(C)(6)*«.

Torque motor branch circuits shall be protected at the motor nameplate current rating according to 240.4(B) »*430.52(D)*«.

The branch-circuit short-circuit and ground-fault protective device rating for multimotor and combination-load equipment shall not exceed the equipment's marked rating in accordance with 430.7(D) »*430.54*«.

Where fuses are used for motor branch-circuit short-circuit and ground-fault protection, the fuseholders shall be of adequate size to accommodate the fuses specified by Table 430.52 »*430.57*«. (An exception is provided for fuses having a time delay appropriate for the motor's starting characteristics.)

A circuit breaker for motor branch-circuit short-circuit and ground-fault protection shall have a current rating as determined by 430.52 and 430.110 »*430.58*«.

Ⓐ Should the values for branch-circuit short-circuit and ground-fault protective devices (determined by Table 430.52) not correspond to a standard size or rating of fuses, nonadjustable circuit breakers, thermal

protective devices, or possible settings of adjustable circuit breakers, then the next higher standard size, rating, or possible setting is permitted. This rating or setting shall not exceed the next higher standard ampere rating »*430.52(C)(1) Exception No. 1*«.

Ⓑ Use a protective device whose rating (or setting) does not exceed the value calculated according to Table 430.52 »*430.52(C)(1)*«.

Ⓒ The motor branch-circuit short-circuit and ground-fault protective device shall comply with 430.52(B) and either (C) or (D), as applicable »*430.52(A)*«.

Ⓓ Branch-circuit protective devices shall comply with 240.15 provisions »*430.56*«.

NOTE

The motor branch-circuit short-circuit and ground-fault protective device shall be able to carry the motor's starting current »*430.52(B)*«.

CAUTION

If the maximum branch-circuit short-circuit and ground-fault protective device ratings are shown in the manufacturer's overload relay table (for use with a motor controller), or are otherwise marked on the equipment, they shall not be exceeded even if 430.52(C)(1) allows higher values »*430.52(C)(2)*«.

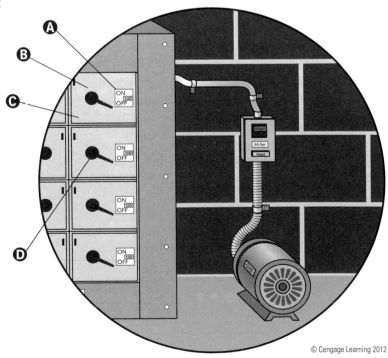

Motor Disconnecting Means

The disconnecting means shall open all ungrounded supply conductors and shall be designed so that no pole operates independently. The disconnecting means can occupy the same enclosure as the controller. The disconnecting means shall be designed so that it shall not be closed automatically ≫*430.103*≪.

The disconnecting means shall plainly indicate whether it is in the open (off) or closed (on) position ≫*430.104*≪.

Motor-control circuits' arrangement shall accommodate disconnection from all supply sources when the disconnecting means is in the open position ≫*430.75(A)*≪.

Where two or more motors are used together or where one or more motors are used in combination with other loads, such as resistance heaters, and where the combined load may be simultaneous on a single disconnecting means, the ampere and horsepower ratings of the combined load shall be determined in accordance with provisions in 430.110(C)(1) though (3) ≫*430.110(C)*≪.

A switch (or circuit breaker) can serve as both controller and disconnecting means if it complies with 430.111(A) and is one of a type listed in 430.111(B) ≫*430.111*≪.

Every motor-circuit disconnecting means between the feeder's point of attachment and the motor's connection point shall comply with the requirements of 430.109 and 430.110 ≫*430.108*≪.

The disconnecting means shall be a type specified in 430.109(A), unless otherwise permitted in (B) through (G), under the conditions given ≫*430.109*≪.

Ⓐ Motor circuits rated 1000 volts, nominal, or less, shall have a disconnecting means with an ampere rating of at least 115% of the motor's full-load current rating ≫*430.110(A)*≪.

Ⓑ According to Article 430, a controller is any switch (or device) normally used to start and stop a motor by making and breaking motor circuit current ≫*430.2*≪.

Ⓒ An individual, fully functional disconnecting means shall be provided for each controller. The disconnecting means shall be located in sight from the controller, unless an exception is met ≫*430.102(A)*≪. The controller disconnecting means required in accordance with 430.102(A) shall be permitted to serve as the disconnecting means for the motor if it is in sight from the motor location and the driven machinery location ≫*430.102(B)(2)*≪.

Ⓓ A disconnecting means shall be located in sight from the motor and driven machinery location ≫*430.102(B)(1)*≪.

Ⓔ Motor branch-circuit and ground-fault protection and motor overload protection can be combined into a single device, where the device's rating (or setting) provides the overload protection required by 430.32 ≫*430.55*≪.

Ⓕ Article 430, Part IX requires a disconnecting means capable of breaking the connection between the motors/controllers and the circuit ≫*430.101*≪.

Ⓖ The phrase "in sight from" indicates that specified items of equipment are visible and are no more than 50 ft (15 m) apart ≫*Article 100*≪.

> ### NOTE
> At least one of the disconnecting means shall be **readily** accessible ≫*430.107*≪.

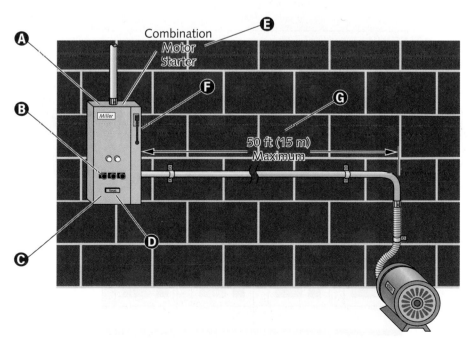

Combination Motor Starter

Miller

50 ft (15 m) Maximum

© Cengage Learning 2012

Motor Feeder Short-Circuit and Ground-Fault Protection

Where instantaneous trip circuit breaker(s) or motor short-circuit protector(s) are used for motor branch-circuit short-circuit and ground-fault protection, per 430.52(C), the preceding procedure for determining the maximum feeder protective device rating applies with the following provision: For calculation purposes, each instantaneous trip circuit breaker or motor short-circuit protector is assumed to have a rating not exceeding the maximum percentage of motor FLC permitted by Table 430.52 for the type of feeder protective device used ≫ *430.62(A) Exception No. 1* ≪ .

Where the feeder overcurrent protective device also provides overcurrent protection for a motor control center, the provisions of 430.94 shall apply ≫ *430.62 Exception No. 2* ≪ .

Where feeder conductor ampacity is greater than 430.24 requires, the rating (or setting) of the feeder overcurrent protective device can be based on the feeder conductor's ampacity ≫ *430.62(B)* ≪ .

If the same rating (or setting) of the branch-circuit short-circuit and ground-fault protective device is used on multiple branch circuits supplied by the feeder, one of the protective devices is considered the largest for the above calculation ≫ *430.62(A)* ≪ .

A A feeder consisting of conductor sizes based on 430.24 that supplies a specific fixed motor load(s) shall have a protective device with a rating (or setting) no greater than the largest rating (or setting) of the branch-circuit short-circuit and ground-fault protective device for any motor supplied by the feeder (based on the maximum permitted value for the specific protective device in accordance with 430.52 or 440.22[A] for hermetic refrigerant motor compressors), plus the sum of the FLCs of the other motors in the group ≫ *430.62(A)* ≪ .

B Article 430, Part V specifies devices for protection of feeder conductors supplying motors against overcurrents due to short circuits or grounds ≫ *430.61* ≪ .

> **NOTE**
>
> Where a feeder supplies a motor load and other load(s), the feeder protective device shall have a rating not less than that required for the sum of the other load(s) plus the following:
> (1) For a single motor, the rating permitted by 430.52,
> (2) For a single hermetic refrigerant motor compressor, the rating permitted by 440.22,
> (3) For two or more motors, the rating permitted by 430.62 ≫ *430.63* ≪ .

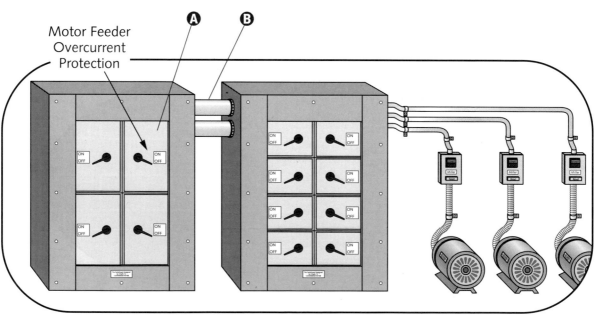

Motor Feeder Overcurrent Protection

SPECIFIC EQUIPMENT

Cranes and Hoists

Conductors shall either be enclosed in raceways or be Type AC cable with insulated grounding conductor, Type MC cable, or Type MI cable, unless 610.11(A) through (E) otherwise permits ≫*610.11*≪.

If a crane, hoist, or monorail hoist operates above readily combustible material, the resistors shall be located as outlined in 610.3(B)(1) and (2) ≫*610.3(B)*≪.

Conductors shall comply with Table 310.104(A) unless otherwise permitted in 610.13(A) through (D) ≫*610.13*≪.

Conductors exiting raceways/cables shall comply with 610.12(A) or (B) ≫*610.12*≪.

Table 610.14(A) dictates the allowable ampacities of conductors ≫*610.14(A)*≪.

Where the secondary resistor and the controller are separate, calculate the minimum size conductors (between controller and resistor) by multiplying the motor secondary current by the appropriate Table 610.14(B) factor, and select a wire from Table 610.14(A) ≫*610.14(B)*≪.

Motor and control external conductors shall not be smaller than 16 AWG, unless otherwise permitted in 610.14(C)(1) and (2) ≫*610.14(C)*≪.

A disconnecting means with a continuous ampere rating not less than that calculated in 610.14(E) and (F) shall be provided between the runway contact conductors and the power supply. The disconnecting means shall comply with 430.109. This disconnecting means shall meet 610.31(1) through (4) provisions ≫*610.31*≪.

A disconnecting means in compliance with 430.109 shall be provided in the leads from the runway contact conductors or other power supply on all cranes and monorail hoists. The disconnecting means shall be lockable open in accordance with 110.25 ≫*610.32*≪.

Section 610.32 requires that the continuous ampere rating of the switch or circuit breaker not be less than 50% of the combined short-time ampere rating of the motors, nor less than 75% of the sum of the short-time ampere rating of the motors required for any single motion ≫*610.33*≪.

Each motor shall have an individual controller unless 610.51(A) or (B) allows otherwise ≫*610.51*≪.

Ⓐ Article 610 covers electrical equipment and wiring installation for use with cranes, monorail hoist, hoists, and all runways ≫*610.1*≪.

Ⓑ All exposed non-current-carrying metal parts of cranes, monorail hoists, hoists, and accessories, including pendant controls, shall be bonded either by mechanical connections or bonding jumpers, where applicable, so that the entire crane or hoist is a ground-fault current path as required or permitted by Article 250, Parts V and VII. Moving parts, other than removable accessories, or attachments that have metal-to-metal bearing surfaces, shall be considered to be electrically bonded to each other through bearing surfaces for grounding purposes. The trolley frame and bridge frame shall not be considered as electrically grounded through the bridge and trolley wheels and its respective tracks. A separate bonding conductor shall be provided ≫*610.61*≪.

Ⓒ Crane, hoist, and monorail hoist overload protection must comply with 610.43.

Ⓓ Crane, hoist, and monorail hoist motor branch circuits shall be protected by fuses or inverse-time circuit breakers that have a rating in accordance with Table 430.52. Where two or more motors operate a single motion, the sum of their nameplate current ratings shall be considered as that of a single motor ≫*610.42(A)*≪.

> **NOTE**
>
> All equipment operating in a hazardous (classified) location shall comply with Article 500 ≫*610.3(A)*≪.

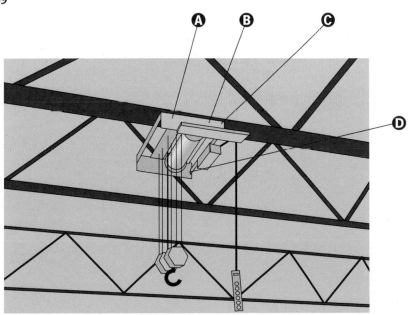

Electric Welders

Each arc welder shall have overcurrent protection rated (or set) at no more than 200% of $I_{1\,max}$. If the $I_{1\,max}$ is not given, the overcurrent protection shall be rated (or set) at no more than 200% of the welder's rated primary current »630.12(A)«.

An overcurrent device is not required for an arc welder having supply conductors protected by an overcurrent device rated (or set) at no more than 200% of $I_{1\,max}$, or the welder's rated primary current »*630.12(A)*«.

Supply conductors for an arc welder protected by an overcurrent device rated (or set) at no more than 200% of $I_{1\,max}$ or rated primary welder current do not require a separate overcurrent device »*630.12(A)*«.

Ⓐ Minimum ampacity of arc welder conductors supplying a group of welders shall be based on individual currents determined in 630.11(A) as the sum of 100% of the two largest welders, plus 85% of the third largest welder, plus 70% of the fourth largest, plus 60% of any other welder »630.11(B)«.

Ⓑ Section 630.12(A) and (B) contain requirements for arc welder overcurrent protection. Where the resulting values do not correspond with 240.6(A) standard ampere ratings, or the rating or setting specified results in unnecessary opening of the overcurrent device, the next higher standard rating or setting is acceptable »630.12«.

Ⓒ Lower percentages are permitted in cases where the work is such that a high-operating duty cycle for individual welders is impossible »*630.11(B) Exception*«.

Ⓓ Supply conductor ampacity, for a single arc welder shall not be less than the $I_{1\,eff}$ value on the rating plate. If the $I_{1\,eff}$ is not given, the supply conductor ampacity shall meet 630.11 requirements »630.11(A)«.

Ⓔ Conductors supplying one or more arc welders shall have an overcurrent protective device rated (or set) not more than 200% of the conductor ampacity »630.12(B)«. (See 630.12[B] Informational Note for calculation explanation.)

Ⓕ Article 630 covers apparatus for electric arc welding, resistance welding, plasma cutting, and similar welding and cutting process equipment connected to an electric supply system »630.1«.

Ⓖ Conductors used in the secondary circuit of electric welders shall have flame-retardant insulation »630.41«.

Ⓗ Arc welders must have a rating plate with the information listed in 630.14.

> **NOTE**
>
> Resistance welders must comply with Article 630, Part III (630.31 through 630.34) provisions.

> **NOTE**
>
> A disconnecting means shall be provided in the supply circuit for each arc welder that is not equipped with a disconnect mounted as an integral part of the welder. The disconnecting means identity shall be marked in accordance with 110.22(A).
>
> The disconnecting means shall be a switch or circuit breaker, and its rating shall be not less than that necessary to accommodate overcurrent protection as specified under 630.12 »*630.13*«.

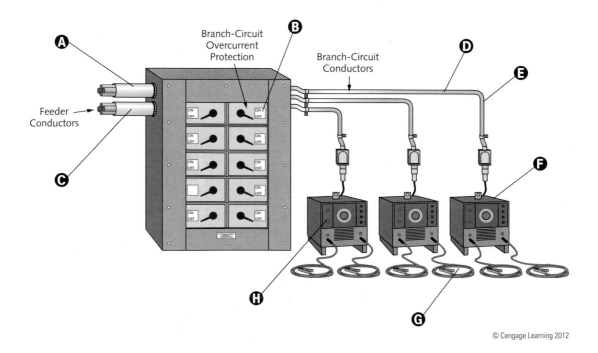

© Cengage Learning 2012

Electroplating

Equipment used in the electroplating process must be specifically identified for such service »*669.3*«.

Branch-circuit conductors supplying one or more units of equipment shall have an ampacity of at least 125% of the total connected load. Busbar ampacities shall meet 366.23 requirements »*669.5*«.

The following stipulations apply to conductors connecting the electrolyte tank equipment to the conversion equipment: (A) Insulated conductors in systems not exceeding 50 volts direct current (dc) can be run without insulated support, provided they are protected from physical damage. Bare copper (or aluminum) conductors are permitted where supported on insulators. (B) Insulated conductors in systems exceeding 50 volts direct current can be run on insulated supports, provided they are protected from physical damage. Bare copper (or aluminum) conductors are permitted where (1) supported on insulators and (2) guarded against accidental contact up to the termination point, per 110.27 »*669.6*«.

Where multiple power supplies serve the same dc system, the dc side of each power supply shall have a disconnecting means »*669.8(A)*«.

Removable links/conductors can serve as the disconnecting means »*669.8(B)*«.

Direct-current conductors shall be protected from overcurrent by at least one of the following: (1) fuses or circuit breakers, (2) a current-sensing device that operates a disconnecting means, or (3) other approved means »*669.9*«.

A Article 669 provisions apply to the installation of the electrical components and accessory equipment that supply the power and controls for electroplating, anodizing, electropolishing, and electrostripping, herein referred to simply as *electroplating* »*669.1*«.

CAUTION

Warning signs shall be posted to indicate the presence of bare conductors. The warning sign(s) or label(s) shall comply with 110.21(B) »*669.7*«.

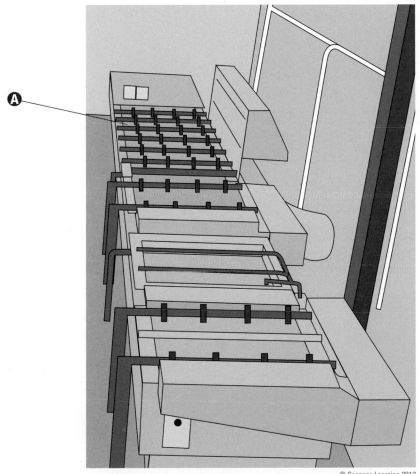

Industrial Machinery

A machine is considered to be an individual unit and, therefore, shall have a disconnecting means. Branch circuits protected by either fuses or circuit breakers can be the disconnecting means. The disconnecting means is not required to incorporate overcurrent protection ≫ *670.4(B)* ≪.

Where overcurrent protection (single circuit breaker or set of fuses) is furnished as part of the machine, 670.3 required markings shall be used and the supply conductors shall be considered either as feeders (or taps), as covered by 240.21 ≫ *670.4(C)* ≪.

If the machine has no branch-circuit short-circuit and ground-fault protective device, the overcurrent protective device's rating (or setting) shall be based on 430.52 and 430.53, as applicable ≫ *670.4(C)* ≪.

Where overcurrent protection is provided, per 670.4(B), the machine shall be marked "overcurrent protection provided at machine supply terminals" ≫ *670.3(B)* ≪.

A Article 670 covers the definition of, the nameplate data for, and the size and overcurrent protection of supply conductors to industrial machinery ≫ *670.1* ≪.

B The rating or setting of the overcurrent protective device for the circuit supplying the machine shall not be greater than the sum of the largest rating or setting of the branch-circuit short-circuit and ground-fault protective device provided with the machine, plus 125% of the full-load current rating of all resistance heating loads, plus the sum of the full-load currents of all other motors and apparatus that could be in operation at the same time ≫ *670.4(C)* ≪.

C Article 110, Part II provisions apply to working space around electrical equipment operating at 600 volts, nominal, or less to ground.

D The overcurrent protective device's rating (or setting) for the circuit supplying the machine must comply with 670.4(C).

E The selected supply conductor size shall have an ampacity not less than 125% of the FLC rating for all resistance heating loads, plus 125% of the highest rated motor's FLC rating, plus the sum of the FLC ratings of all other connected motors and apparatus, based on their duty cycle, that may be simultaneously operated ≫ *670.4(A)* ≪.

F A permanent nameplate listing supply voltage, number of phases, frequency, FLC, maximum short-circuit and ground-fault protective device ampere rating, largest motor or load ampere rating, short-circuit interrupting rating of the machine overcurrent-protective device, and diagram number(s) shall be attached to the control equipment enclosure/machine, or the number of the index to the electrical drawings, remaining plainly visible after installation ≫ *670.3(A)* ≪.

G Industrial machinery (machine) is defined as a power-driven machine (or a group of machines working together in a coordinated manner), not portable by hand during operation, used to process material by cutting; forming; pressure; electrical, thermal, or optical techniques; lamination; or a combination of these processes. It can include associated equipment used to transfer material or tooling, including fixtures; assemble/disassemble; inspect; test; or package. (Associated electrical equipment, including logic controller[s] and associated software [or logic] together with the machine actuators and sensors, are all part of the industrial machine.) ≫ *670.2* ≪

> **N O T E**
>
> For information on the workspace requirements for equipment containing supply conductor terminals, see 110.26. For information on the workspace requirements for machine power and control equipment, see NFPA 79-2012, *Electrical Standard for Industrial Machinery* ≫ *670.1 Informational Note No. 2* ≪.

> **WARNING**
>
> Industrial machinery shall not be installed where the available fault current exceeds its short-circuit current rating as marked in accordance with 670.3(A)(4) ≫ *670.5* ≪.

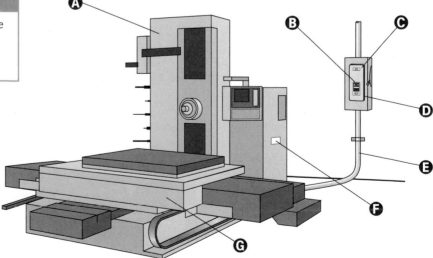

Capacitors

Surge capacitors, or capacitors included as a component part of other apparatus and conforming with such apparatus' requirements, are excluded from Article 460 requirements ≫ *460.1* ≪.

Article 460 also covers capacitor installation in hazardous (classified) locations as modified by Articles 501 through 503 ≫ *460.1* ≪.

Capacitors containing more than 3 gallons (11 L) of flammable liquid shall be enclosed in vaults or outdoor fenced enclosures to comply with Article 110, Part III. This limit applies to any single unit in capacitor installations ≫ *460.2(A)* ≪.

If a motor installation includes a capacitor connected on the load side of a motor overload device, the motor overload device's rating/setting shall be based on the improved power factor of the motor circuit. Disregard the capacitor effect in determining the motor-circuit conductor rating per 430.22 ≫ *460.9* ≪.

A capacitor connected on the load side of a motor overload protective device does not require a separate overcurrent device ≫ *460.8(B) Exception* ≪.

A separate disconnecting means is not required where a capacitor is connected on a motor controller's load side ≫ *460.8(C) Exception* ≪.

Ⓐ Each capacitor shall have a nameplate stating the manufacturer's name, rated voltage, frequency, kilovar or amperes, number of phases, and, if filled with a combustible liquid, the volume of liquid. Any nonflammable liquid filling shall be indicated on the nameplate. The nameplate shall also identify a capacitor having a discharge device inside the case ≫ *460.12* ≪.

Ⓑ Capacitors shall have a means of discharging stored energy ≫ *460.6* ≪.

Ⓒ Article 460 covers the installation of capacitors on electric circuits ≫ *460.1* ≪.

Ⓓ Capacitor cases shall be connected to an equipment grounding conductor ≫ *460.10* ≪. Capacitor cases shall not be connected to an

equipment grounding conductor where capacitor units are supported on a structure designed to operate at other than ground potential ≫ *460.10 Exception* ≪.

Ⓔ Capacitor residual voltage shall be reduced to 50 volts, nominal, or less within 1 minute after the capacitor disconnects from the supply source ≫ *460.6(A)* ≪. The discharge circuit shall either (1) be permanently connected to the capacitor/capacitor-bank terminals or (2) have automatic means of connecting it to the capacitor-bank terminals on line voltage removal. Manual means of switching or connecting the discharge circuit shall not be used ≫ *460.6(B)* ≪.

Ⓕ Each capacitor bank's ungrounded conductor shall have a disconnecting means. The disconnecting means shall simultaneously open all ungrounded conductors. It can disconnect the capacitor from the line as a regular operating procedure. The rating of the disconnecting means shall be at least 135% of the capacitor's rated current ≫ *460.8(C)* ≪.

Ⓖ The capacitor circuit conductor's ampacity shall be at least 135% of the capacitor's rated current. The ampacity of the conductors that connect a capacitor to a motor's terminals, or to motor circuit conductors, shall not be less than one-third of the motor circuit conductor's ampacity and in no case less than 135% of the capacitor's rated current ≫ *460.8(A)* ≪.

Ⓗ Each capacitor bank's ungrounded conductor shall have an overcurrent device. The overcurrent device's rating (or setting) shall be as low as practicable ≫ *460.8(B)* ≪.

> **NOTE**
>
> Requirements for capacitors over 1000 volts, nominal, are in Part II (460.24 through 460.28).

CAUTION

Capacitors, accessible to unauthorized/unqualified persons, shall be enclosed, located, or guarded so that neither persons nor conducting materials can come into accidental contact with exposed energized parts, terminals, or associated buses. However, no additional guarding is required for enclosures accessible only to authorized, qualified persons ≫ *460.2(B)* ≪.

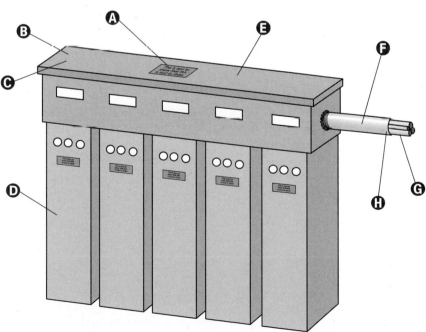

Summary

- A service may have up to six disconnecting means in a single enclosure, in a group of separate enclosures, or in (or on) a switchboard, or in switchgear.
- A service-entrance conductor's raceway shall not contain nonservice conductors.
- Conductors can be tapped, without overcurrent protection at the tap, to a feeder as specified in 240.21(B)(1) through (5).
- Conductors can connect to a transformer secondary, without overcurrent protection at the secondary, as specified in 240.21(C)(1) through (6).
- The provisions of 240.4(B) shall not be permitted for tap conductors.
- Article 450 covers not only transformers, but also includes transformer vault provisions.
- Taps can connect the grounding electrode conductors in a service consisting of more than a single enclosure.
- Cable tray provisions are located in Article 392.
- Article 430 covers motors, motor branch-circuit and feeder conductors (and their protection), motor overload protection, motor-control circuits, motor controllers, and motor-control centers.
- Article 610 covers electrical equipment and wiring installations used in connection with cranes, hoists, monorail hoists, and all runways.
- Electric arc welding, resistance welding apparatus, and similar welding equipment provisions are located in Article 630.
- Article 669 requirements apply to the installation of the electrical components and accessory equipment supplying electroplating, anodizing, electropolishing, and electrostripping power and controls.
- Article 670 covers the definition of, the nameplate data for, and the size and overcurrent protection of supply conductors to industrial machinery.
- Provisions pertaining to the installation of capacitors on electric circuits are contained in Article 460.

Unit 17 Competency Test

NEC Reference	Answer	
_____	_____	1. Terminals of motors and controllers shall be suitably marked or colored where necessary to indicate _____.
_____	_____	2. Capacitors containing more than _____ gallons of flammable liquid shall be enclosed in vaults or outdoor fenced enclosures complying with Article 110, Part III.
_____	_____	3. Cable tray systems shall not be used in _____ or where subject to severe physical damage.

      I. hoistways

     II. hazardous locations

    III. corrosive areas

        a) I only   b) I and II only   c) I and III only   d) I, II, and III

_____	_____	4. Wires that are used as crane (or hoist) contact conductors shall be secured at the ends by means of approved strain insulators and shall be mounted on approved insulators so that the extreme limit of displacement of the wire does not bring the latter within less than _____ in. from the surface wired over.
_____	_____	5. Multispeed motors shall be marked with the code letter designating the locked-rotor kilovolt-ampere (kVA) per horsepower (hp) for the highest speed at which the motor _____.
_____	_____	6. The walls and roofs of transformer vaults shall be constructed of materials that have approved structural strength for the conditions with a minimum fire resistance of _____.

NEC Reference	Answer
_____	_____

7. Where cable trays support individual conductors and where the conductors pass from one cable tray to another, from a cable tray to raceway(s), or from cable tray to equipment where the conductors are terminated, the distance between cable trays or between the cable tray and the raceway(s) or the equipment shall not exceed ____ ft.

8. The residual voltage of a 480-volt capacitor shall be reduced to _____, nominal, or less, within _____ after the capacitor is disconnected from the source of supply.

9. The disconnecting means for motor circuits rated 1000 volts, nominal, or less, shall have an ampacity rating not less than _____ of the full-load current rating of the motor.

10. Steel cable trays shall not be used as equipment grounding conductors for circuits with ground-fault protection above _____.

11. Where no branch-circuit short-circuit and ground-fault protective device is provided with the industrial machine, the rating or setting of the overcurrent protective device shall be based on _____, as applicable.

12. Branch-circuit conductors supplying one or more units of electroplating equipment shall have an ampacity of not less than _____ of the total connected load.

13. Vaults containing more than _____ transformer capacity shall be provided with a drain or other means that will carry off any accumulation of oil or water in the vault unless local conditions make this impracticable.

14. Conductors that supply one or more arc welders shall be protected by an overcurrent device rated or set at not more than _____ of the conductor ampacity.

15. The full-load current of a 25-hp synchronous, 460-volt, 3-phase motor at unity power factor is _____.

16. The ampacity of capacitor circuit conductors shall not be less than _____ of the rated current of the capacitor.

17. Hoist and monorail hoist and their trolleys that are not used as part of an overhead traveling crane shall not require individual motor overload protection, provided the largest motor does not exceed _____ and all motors are under manual control of the operator.

18. A door sill or curb that is of an approved height that will confine the oil from the largest transformer within the vault shall be provided, and in no case shall the height be less than _____ in.

19. Branch-circuit conductors that supply a single motor used in a continuous duty application shall have an ampacity of not less than _____ of the motor full-load current rating as determined by 430.6(A)(1), or not less than specified in 430.22(A) through (G).

20. A(n) _____ is a power-driven machine (or a group of machines working together in a coordinated manner), not portable by hand while working, that is used to process material by cutting; forming; pressure; electrical, thermal, or optical techniques; lamination; or a combination of these processes. It can include associated equipment used to transfer material or tooling, including fixtures, to assemble/disassemble, to inspect or test, or to package.

21. Cable trays containing welding cables shall have a permanent sign attached to the cable tray at intervals not greater than ____ ft.

22. For raceway terminating at the tray, a listed _____ or adapter shall be used to securely fasten the raceway to the cable tray system.

NEC Reference   Answer

_____ _____ 23. Where the motor-control circuit transformer rated primary current is less than 2 amperes, an overcurrent device rated or set at not more than _____ of the rated primary current shall be permitted in the primary circuit.

_____ _____ 24. For cranes and hoists, the dimension of the working space in the direction of access to live parts that are likely to require examination, adjustment, servicing, or maintenance while energized shall be a minimum of ____ ft.

_____ _____ 25. All transformer vault ventilation openings to the indoors shall be provided with automatic closing fire dampers that operate in response to a vault fire. Such dampers shall possess a standard fire rating of not less than _____.

_____ _____ 26. The ampacity of the supply conductors for a resistance welder that may be operated at different times, at different values of primary current, or duty cycle shall not be less than _____ of the rated primary current for manually operated nonautomatic welders.

_____ _____ 27. Cable trays shall be permitted to extend _____ through partitions and walls, or _____ through platforms and floors in wet or dry locations where the installations, complete with installed cables, are made in accordance with the requirements of 300.21.

_____ _____ 28. Conductors for a motor used in a short-time, intermittent, periodic, or varying duty application shall have an ampacity of not less than the percentages of the motor nameplate current ratings shown in _____, unless the authority having jurisdiction grants special permission for conductors of lower ampacity.

_____ _____ 29. Each doorway leading into a transformer vault from the building interior shall be provided with a tight-fitting door that has a minimum fire rating of _____.

_____ _____ 30. _____ shall be placed in all service-entrance motor-control centers to isolate service busbars and terminals from the remainder of the motor control center.

_____ _____ 31. Crane, hoist, and monorail hoist motor branch circuits shall be protected by fuses or inverse-time circuit breakers that have a rating in accordance with _____.

_____ _____ 32. For a nonautomatically started motor, the overload protection shall be permitted to be shunted or cut out of the circuit during the starting period of the motor if the device by which the overload protection is shunted or cut out shall not be left in the starting position and if fuses or inverse-time circuit breakers rated or set at not over _____ of the full-load current of the motor are located in the circuit so as to be operative during the starting period of the motor.

_____ _____ 33. The discharge circuit of a 480-volt capacitor or 480-volt capacitor bank shall not be connected in which way(s)?

    I. Permanently connected to the terminals

    II. Automatic means of connecting it to the terminals

    III. Manual means of connecting the discharge circuit

      a) I or II only b) I or III only c) III only d) I, II, or III

_____ _____ 34. Each controller shall be capable of starting and stopping the motor it controls and shall be capable of interrupting the _____ of the motor.

UNIT 18

Special Occupancies

Objectives

After studying this unit, the student should:

▶ know that portions of occupancies intended for the assembly of more than 99 persons are referred to as *places of assembly*.

▶ be familiar with definitions pertinent to motion picture (and television studio) audience areas, theaters, and similar locations.

▶ have a basic understanding not only of motion picture (and television studio) audience area, theater, and similar location general requirements, but also stage switchboards, stage equipment, and theater dressing room provisions.

▶ know the location of provisions pertaining to carnivals, circuses, exhibitions, fairs, traveling attractions, and similar functions.

▶ have a solid grasp of requirements for motion picture/television studios (and similar locations), including cellulose nitrate film storage vaults.

▶ be well acquainted with motion picture projector provisions.

▶ thoroughly understand manufactured building requirements.

▶ be familiar with agricultural building provisions including what constitutes a bonding and equipotential plane.

▶ be apprised of requirements for mobile homes, manufactured homes, and mobile home parks.

▶ know the location of provisions relating to recreational vehicles and recreational vehicle parks.

▶ have a good understanding of floating building requirements.

▶ be familiar with the provisions for marinas and boatyards.

Introduction

Unit 18—Special Occupancies covers a variety of mostly unrelated occupancies. This unit addresses most of the occupancy types found in Article 518 through Article 555. Unlike other Chapter 5 articles in the *NEC*, "Places of Assembly" (Article 518) is not a specific occupancy type. Places of assembly are buildings, or portions of buildings, designed (or intended) for the assembly of 100 or more persons. "Theaters, Audience Areas of Motion Picture and Television Studios, Performance Areas, and Similar Locations" (Article 520) covers all buildings, or that part of a building or structure, indoor or outdoor, designed for presentation, dramatic, musical, motion picture projection, or similar purposes; and specific audience seating areas within motion picture or television studios. "Carnivals, Circuses, Fairs, and Similar Events" (Article 525) addresses portable wiring and equipment installations for carnivals, circuses, exhibitions, fairs, traveling attractions, and similar functions, including wiring in and on all structures. Article 530, "Motion Picture and Television Studios and Similar Locations," requirements apply to television studios and motion picture studios using either film or electronic cameras, except as provided in 520.1, and exchanges, factories, laboratories, stages, or a segment of the building in which film or tape (more than 7/8 inch [22 mm] in width) is exposed, developed, printed, cut, edited, rewound, repaired, or stored. Article 540, "Motion Picture Projection Rooms," provisions apply to motion picture projection rooms, motion picture projectors, and associated equipment (both professional and non-professional types) using incandescent, carbon arc, xenon, or other light source equipment that is known to develop hazardous gases, dust, or radiation. "Manufactured Buildings" (Article 545) details the requirements for manufactured buildings and building components. Article 547, "Agricultural Buildings," provisions apply to certain agricultural buildings, parts of buildings, and adjacent areas. Article 550, "Mobile Homes, Manufactured Homes, and Mobile Home Parks," requirements apply to the location types identified by the article's title. Article 551, "Recreational Vehicles and Recreational

Vehicle Parks," provisions apply to the electrical conductors and equipment installed within (or on) recreational vehicles, the conductors connecting recreational vehicles to an electrical supply, as well as the installation of equipment and devices related to electrical systems within a recreational vehicle park. "Floating Buildings" (Article 553) covers wiring, services, feeders, and grounding for, of course, floating buildings. Finally, Article 555, "Marinas and Boatyards," describes wiring and equipment installations in locations comprising fixed or floating piers, wharfs, docks, and other areas in marinas, boatyards, boat basins, boathouses, and similar occupancies that are, or can be, used for the purpose of repair, berthing, launching, storage, or fueling of small craft and the moorage of floating buildings.

MOTION PICTURE (AND TELEVISION STUDIO) AUDIENCE AREAS, PERFORMANCE AREAS, THEATERS, AND SIMILAR LOCATIONS

General Requirements

The fixed wiring methods employed shall be metal raceway, nonmetallic raceway encased in 2 in. (50 mm) or more of concrete, Type MI cable, Type MC cable, or AC cable containing an insulated equipment grounding conductor sized per Table 250.122 》520.5(A)《.

Fixed wiring method stipulations in Article 640 apply to audio signal processing, amplification, and reproduction equipment; Article 800 to communication circuits; Article 725 to Class 2 and Class 3 remote control and signaling circuits; and Article 760 to fire alarm circuits 》520.5(A) Exception《.

Portable switchboards, stage set lighting, stage effects, and other wiring without a fixed location is permitted with approved flexible cords and cables as provided elsewhere in Article 520. Such wiring shall not be fastened by uninsulated staples or nailing 》520.5(B)《.

Nonmetallic-sheathed cable, Type AC cable, ENT, and PVC, can be installed where applicable building code does not require the building (or portions thereof) to be of fire-rated construction 》520.5(C)《.

The number of conductors permitted in any metal conduit, PVC, or EMT for circuits, or for remote-control conductors, shall not exceed the percentage fill shown in Table 1 of Chapter 9. Where contained within an auxiliary gutter or wireway, the sum of the cross-sectional area of all internal conductors at any point shall not exceed 20% of the interior cross-sectional area of the auxiliary gutter or wireway. The 30-conductor limitation of 366.6 and 376.22 shall not apply 》520.6《.

Live parts shall be enclosed or guarded, preventing accidental contact by persons and objects. All switches shall be of the externally operable type. Dimmers, including rheostats, shall be installed in cases (or cabinets) that enclose all live parts 》520.7《.

Portable stage/studio lighting equipment and portable power distribution equipment are permitted for temporary outdoor use, provided the equipment is (1) supervised by qualified personnel while energized and (2) barriered from the general public 》520.10《.

A Article 520 covers all buildings/structures, or parts thereof, indoor or outdoor, intended for presentation, dramatic, musical, motion picture projection, or similar purposes as well as specific audience seating areas within motion picture or television studios 》520.1《.

CAUTION

Any size branch circuit supplying receptacle(s) can also supply stage set lighting. The receptacle's voltage rating shall not be less than the circuit voltage. Receptacle ampere rating and branch-circuit conductor ampacity shall at least equal the branch-circuit overcurrent device ampere rating. Table 210.21(B)(2) shall not apply 》520.9《.

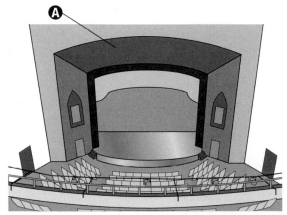

Definitions

A **Border light:** A permanently installed overhead strip light ≫520.2≪.

B Breakout assemblies can contain listed, hard usage (junior hard service) cords where all of the following conditions are met:

1. The cords connect a single multipole connector (containing two or more branch circuits) and multiple two-pole, 3-wire connectors.

2. No cord in the breakout assembly exceeds 20 ft (6.0 m) in length.

3. The breakout assembly's entire length is protected from physical damage by attachment to a pipe, truss, tower, scaffold, or other substantial support structure.

4. All branch circuits feeding the breakout assembly are protected by overcurrent devices rated at no more than 20 amperes ≫520.68(A)(5)≪.

C **Breakout assembly:** An adapter used to join a multipole connector (containing two or more branch circuits) to multiple individual branch-circuit connectors ≫520.2≪.

D **Portable equipment:** Equipment intended for movement from place to place and fed with portable cords or cables ≫520.2≪.

E Flexible conductors, including cable extensions, supplying portable stage equipment shall be listed extra-hard usage cords or cables ≫520.68(A)(1)≪.

F **Connector strip:** A metal wireway containing pendant or flush receptacles ≫520.2≪.

G **Footlight:** A border light installed on (or in) the stage ≫520.2≪.

H **Portable power distribution unit:** A power distribution box containing receptacles and overcurrent devices ≫520.2≪.

I Portable power distribution units shall comply with (a) through (e):

(a) The construction shall prevent the exposure of current-carrying parts.

(b) Receptacles shall comply with 520.45 and shall have branch-circuit overcurrent protection within the box. Fuses and circuit breakers shall be protected from physical damage. Flexible cords or cables supplying pendant receptacles or cord connectors shall be listed for extra-hard usage.

(c) Busbars shall have an ampacity equal to the sum of the ampere ratings of all the connected circuits. Lugs shall be provided for the master cable's connection.

(d) Power accepting flanged surface inlets (recessed plugs) shall be rated in amperes.

(e) Cable arrangement shall alleviate tension at the termination points and be adequately protected when passing through enclosures ≫520.62≪.

J **Proscenium:** The wall and arch separating the stage from the auditorium (house) ≫520.2≪.

K **Stand lamp (work light):** A portable stand containing a general-purpose luminaire or lampholder with guard for stage or auditorium general illumination ≫520.2≪.

L **Strip light:** A luminaire with a row of multiple lamps ≫520.2≪.

M **Two-fer:** An adapter cable containing one male plug and two female cord connectors used to join two loads to one branch circuit ≫520.2≪.

N Adapters, two-fers, and other single/multiple circuit outlet devices shall comply with all of the following:

(a) Each receptacle and corresponding cable shall have the same current and voltage rating as its supply plug. Utilization in a stage circuit having a greater current rating is not allowed.

(b) All conductors shall be wired in accordance with 520.67.

(c) Adapter and two-fer conductors shall be listed extra-hard usage or listed hard usage (junior hard service) cord. The overall length of hard usage (junior hard service) cord is restricted to 3.3 ft (1.0 m) ≫520.69≪.

O Stand lamps can be supplied by listed, hard usage cord provided the cord is not subject to physical damage and is protected by an overcurrent device rated at not over 20 amperes ≫520.68(A)(2)≪.

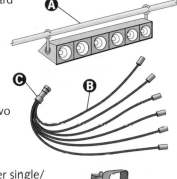
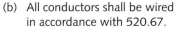
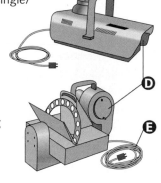

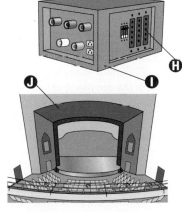
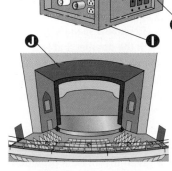

Fixed Stage Switchboards

Fixed stage switchboards shall comply with 520.21(1) through (4):

(1) Fixed stage switchboards shall be listed.

(2) Fixed stage switchboards shall be readily accessible but shall not be required to be located on or adjacent to the stage. Multiple fixed stage switchboards shall be permitted at different locations.

(3) A fixed stage switchboard shall contain overcurrent protective devices for all branch circuits supplied by that switchboard.

(4) A fixed stage switchboard shall be permitted to supply both stage and non-stage equipment ⟫ *520.21* ⟪.

Dimmers installed in ungrounded conductors shall have overcurrent protection no greater than 125% of the dimmer rating and shall be disconnected from all ungrounded conductors when the dimmer's master or individual switch (or circuit breaker) is in the open position ⟫ *520.25(A)* ⟪.

A stage switchboard shall be one (or a combination) of the following types:

(A) Dimmers and switches operated by handles mechanically linked to control devices.

(B) Devices operated electrically from a pilot-type control console or panel. Pilot control panels can be part of the switchboard or can be located elsewhere.

(C) A stage switchboard having interconnected circuits is a secondary switchboard (patch panel) or panelboard remote to the primary stage switchboard and shall contain overcurrent protection. If the dimmer panel provides the required branch-circuit overcurrent protection, it can be omitted from the intermediate switchboard.

(D) A stage switchboard containing only overcurrent protective devices and no control elements ⟫ *520.26* ⟪.

Stage switchboard supply feeders shall be one of the following:

1. A single feeder disconnected by a single disconnect device.

2. Unlimited feeder quantities are permitted, provided all feeders are part of a single system. Where combined, neutral conductors in a given raceway shall have sufficient ampacity to carry the maximum unbalanced current supplied by multiple feeder conductors within the same raceway. It is not necessary that the ampacity be greater than that of the neutral conductor supplying the primary stage switchboard. Parallel neutral conductors shall comply with 310.10(H).

3. Separate feeders of a single primary stage switchboard shall each have a disconnecting means. The primary stage switchboard shall be permanently and conspicuously labeled with the number and location of disconnecting means. If the disconnecting means are located in multiple distribution switchboards, the primary stage switchboard shall have barriers corresponding with these multiple locations ⟫ *520.27(A)* ⟪.

When calculating supply capacity to switchboards, it is permissible to consider the maximum intended switchboard load in a given installation provided that (1) all switchboard supply feeders are protected by an overcurrent device with a rating no greater than the feeder's ampacity and (2) opening the overcurrent device does not affect proper operation of egress or emergency lighting systems ⟫ *520.27(C)* ⟪.

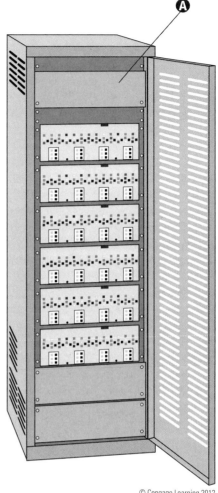

Ⓐ

© Cengage Learning 2012

Ⓐ A stage switchboard is a switchboard, panelboard, or rack containing dimmers or relays with associated overcurrent protective devices, or overcurrent protective devices alone, used primarily to feed stage equipment ⟫ *520.2* ⟪.

> ### N O T E
>
> For the purpose of ampacity adjustment,
> the following shall apply:
>
> (1) The neutral conductor of feeders supplying solid-state, phase-control 3-phase, 4-wire dimming systems shall be considered a current-carrying conductor.
>
> (2) The neutral conductor of feeders supplying solid-state, sine wave 3-phase, 4-wire dimming systems shall not be considered a current-carrying conductor.
>
> (3) The neutral conductor of feeders supplying systems that use or may use both phase-control and sine wave dimmers shall be considered as current-carrying ⟫ *520.27(B)* ⟪.

Portable Switchboards on Stage

Although portable switchboards shall contain overcurrent protection for branch circuits, the requirements of 210.23 shall not apply ≫*520.52*≪.

With the exception of busbars, all conductors within the switchboard shall be stranded ≫*520.53(F)(1)*≪.

Only listed extra-hard usage cords or cables shall be permitted to supply a portable switchboard. The supply cords (or cable) shall terminate within the switchboard enclosure, in an externally operable fused master switch, circuit breaker, or an identified connector assembly. The supply cords or cable (and connector assembly) shall have current ratings not less than the total load connected to the switchboard and shall be protected by overcurrent devices ≫*520.53(H)(1)*≪.

Single-pole portable cable connectors, where used, shall be listed and of the locking type ≫*520.53(K)*≪.

Ⓐ The enclosure shall contain a pilot light connected to the board's supply circuit so that opening the master switch does not cut off the lamp's supply. This lamp shall be on an individual branch circuit having overcurrent protection rated or set at no more than 15 amperes ≫*520.53(G)*≪.

Ⓑ Portable switchboards and feeders for on-stage use shall comply with 520.53(A) through (P) ≫*520.53*≪.

Ⓒ Only externally operable, enclosed type switches and circuit breakers shall be used ≫*520.53(C)*≪.

Ⓓ Portable switchboards shall be housed within substantially constructed enclosures, that remain open during operation. Wooden enclosures shall be fully lined with sheet metal no less than 0.020 in. (0.51 mm) thick. The sheet metal shall either be properly galvanized, enameled, or otherwise coated to prevent corrosion, or be of a corrosion-resistant material ≫*520.53(A)*≪.

Ⓔ Road show connection panel (a type of patch panel): A panel designed to allow for road show connection of portable stage switchboards to fixed lighting outlets via permanently installed supplementary circuits. The panel, supplementary circuits, and outlets shall comply with all of the following:

(a) Circuits shall originate from grounding-type polarized inlets whose current and voltage rating match the fixed-load receptacles.

(b) Circuits transferred between fixed and portable switchboards shall transfer all circuit conductors simultaneously.

(c) Supplementary circuit supply devices shall have branch-circuit overcurrent protection. Each supplementary circuit, within the road show connection panel and theater, shall be protected by branch-circuit overcurrent protective devices installed within the road show connection panel.

(d) Panel construction shall be in accordance with Article 408 ≫*520.50*≪.

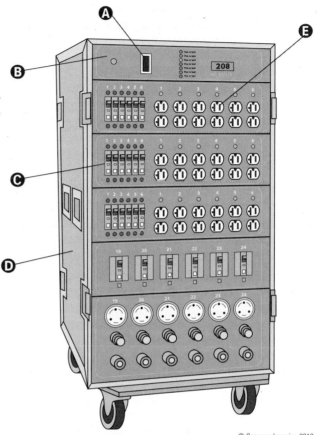

© Cengage Learning 2012

CAUTION

Unless an exception is met, only qualified personnel can route portable supply conductors, make/break supply connectors (and other supply connections), and energize/de-energize supply services. The portable switchboard shall, by means of permanent and conspicuous markings, reflect this requirement ≫*520.53(P)*≪.

NOTE

Supply conductors can have a maximum of three interconnections (mated connector pairs) where the total length from supply to switchboard does not exceed 100 ft (30 m). Should the total length from supply to switchboard exceed 100 ft (30 m), one additional supply conductor interconnection is permitted for each additional 100 ft (30 m) ≫*520.53(J)*≪.

NOTE

Warning sign(s) and label(s) installed on portable switchboards that are used on stage shall comply with 110.21(B) ≫*520.53(K)*≪.

Stage Equipment Other than Switchboards

Arrange footlights, border lights, and proscenium sidelights so that no branch circuit supplying such equipment carries more than a 20-ampere load ≫ *520.41(A)* ≪ .

Article 210 provisions can be applied to heavy-duty lampholder circuits ≫ *520.41(B)* ≪ .

The conductors for foot, border, proscenium, or portable strip lights and connector strips shall have insulation suitable for the conductor's operating temperature, but in no case less than 125°C (257°F). The ampacity of the 125°C (257°F) conductors shall be that of 60°C (140°F) conductors ≫ *520.42* ≪ .

All drops from connector strips shall be 90°C (194°F) wire sized to the ampacity of 60°C (140°F) cords and cables, with no more than 6 in. (150 mm) of conductor extending into the connector strip. Section 310.15(B)(3)(a) shall not apply ≫ *520.42* ≪ .

The ampacity of listed multiconductor extra-hard usage type cords and cables not in direct contact with equipment containing heat-producing elements can be determined by Table 520.44. If so determined, the maximum load current in any conductor shall not exceed the values in Table 520.44 ≫ *520.44(C)(2)* ≪ .

Receptacles for on-stage electrical equipment shall be rated in amperes. Conductors supplying receptacles shall comply with Articles 310 and 400 ≫ *520.45* ≪ .

Connector strip, drop box, floor pocket, and similar receptacles used to connect portable stage-lighting equipment shall be pendant, or mounted in suitable pockets or enclosures, and shall satisfy 520.45 requirements. Connector strip and drop box supply cables shall meet 520.44(C) specifications ≫ *520.46* ≪ .

Bare-bulb lamps in backstage and ancillary areas, where contact with scenery is possible, shall be located and guarded to remain free from physical damage. A minimum air space of at least 2 in. (50 mm) between such lamps and any combustible material shall be provided ≫ *520.47* ≪ . For the purpose of this section, decorative lamps installed in scenery shall not be considered to be backstage lamps ≫ *520.47 Exception* ≪ .

The operating circuit of an electrical device releasing stage smoke ventilators shall be (1) normally closed and (2) controlled by at least two externally operable switches with one such switch being readily accessible on stage and the other positioned as dictated by the AHJ. The device shall be designed for the connected circuit's full voltage, with no resistance inserted. The device shall be located in the loft above the scenery and have been enclosed in a suitable metal box with a tight, self-closing door ≫ *520.49* ≪ .

Ⓐ Borders and proscenium sidelights shall be (1) constructed per 520.43 specifications, (2) suitably stayed and supported, and (3) designed so that the reflector flanges (or other adequate guards) protect the lamps from mechanical damage and from accidental contact with any combustible material ≫ *520.44(A)* ≪ .

Ⓑ Cords and cables for the supply to border lights, drop boxes, and connector strips shall be listed for extra-hard usage and shall be suitably supported. Such cords and cables shall be employed only where flexible conductors are necessary. Conductor ampacity shall satisfy 400.5 provisions ≫ *520.44(C)(1)* ≪ .

Ⓒ Where the metal trough construction specified in 520.43(A) is not used, footlights shall consist of individual outlets whose lampholders are wired with RMC, IMC, FMC, Type MC cable, or mineral-insulated metal-sheathed cable. The circuit conductors shall be soldered to the lampholder terminals ≫ *520.43(B)* ≪ .

Ⓓ The current supply to disappearing footlights shall automatically disconnect when the footlights are fully retracted ≫ *520.43(C)* ≪ .

Ⓔ Metal trough footlights containing circuit conductors shall be made of oxidation-resistant sheet metal no lighter than 0.032 in. (0.81 mm). Lampholder terminals shall be spaced at least ½ in. (13 mm) from the metal trough. The circuit conductors shall be soldered to the lampholder terminals ≫ *520.43(A)* ≪ .

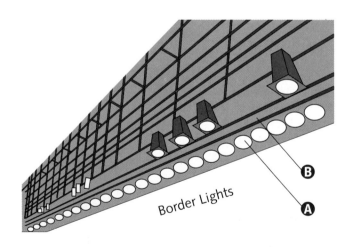

Border Lights

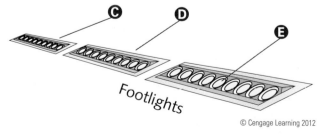

Footlights

Theater Dressing Rooms

A All dressing room lights and receptacles adjacent to the mirror(s) and above the dressing table counter(s) shall be controlled by wall switches within the dressing room(s) ≫520.73≪.

B Each switch controlling receptacles adjacent to the mirror(s) and above the dressing table counter(s) shall have a pilot light located outside the dressing room, adjacent to the door, indicating energized receptacles ≫520.73≪. Note: Although the pilot light must be installed outside the dressing room, the switch must be located inside the dressing room.

C In theater dressing rooms, all exposed incandescent lamps less than 8 ft (2.5 m) from the floor shall be equipped with open-end guards riveted to the outlet box cover or otherwise sealed and locked in place ≫520.72≪.

D Receptacles away from dressing table counter(s) and mirror(s) do not require switches ≫520.73≪.

> **NOTE**
>
> Fixed wiring in motion picture and television studio dressing rooms shall be installed according to Chapter 3 methods ≫**530.31**≪. (Sections 520.71 through 520.73 provisions apply only to theater dressing rooms, and not to motion picture and television studio dressing rooms.)

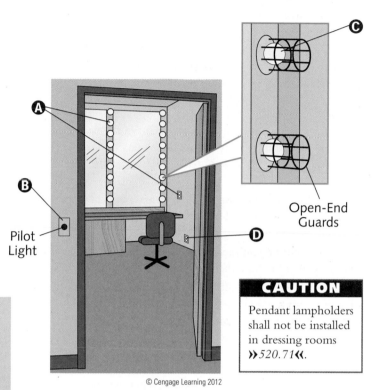

Open-End Guards

Pilot Light

> **CAUTION**
>
> Pendant lampholders shall not be installed in dressing rooms ≫520.71≪.

© Cengage Learning 2012

CARNIVALS, CIRCUSES, FAIRS, AND SIMILAR EVENTS

Ground-Fault Circuit-Interrupter (GFCI) Protection

Egress lighting shall not be protected by a GFCI ≫**525.23(C)**≪.

A GFCI protection for personnel shall be provided for the following: (1) all 125-volt, single-phase, 15- and 20-ampere non-locking type receptacles used for disassembly/reassembly or readily accessible to the general public, and (2) equipment that is readily accessible to the general public and supplied from a 125-volt, single-phase, 15- or 20-ampere branch circuit. The GFCI can be an integral part of the attachment plug or located in the power-supply cord within 12 in. (300 mm) of the attachment plug. Listed cord sets incorporating GFCI for personnel shall be permitted ≫525.23(A)≪.

> **NOTE**
>
> Receptacles that are not accessible from grade level and only facilitate quick disconnecting/reconnecting of electrical equipment do not require GFCI protection. These receptacles shall be of the lockable type ≫**525.23(B)**≪.

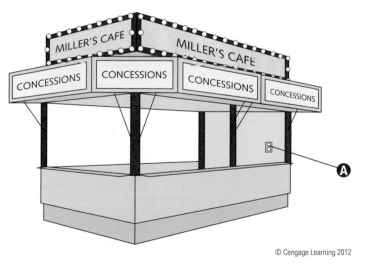

© Cengage Learning 2012

General Requirements

Permanent structures shall comply with Articles 518 and 520 》》*525.3(B)* 《《.

The following equipment connected to the same source shall be bonded: (1) metal raceways and metal-sheathed cable; (2) metal enclosures of electrical equipment; and (3) metal frames and metal parts of portable structures, trailers, trucks, or other equipment that contain or support electrical equipment. The equipment grounding conductor of the circuit supplying the equipment in items (1), (2), or (3) that is likely to energize the metal frame or part shall be permitted to serve as the bonding means 》》*525.30* 《《.

Nonconductive mats are permitted to cover publically accessible flexible cords or cable. Arrangement of cables and mats shall not present a tripping hazard 》》*525.20(G)* 《《.

No amusement ride, attraction, tent, or similar structure shall support wiring for any other ride or structure unless specifically designed for that purpose 》》*525.20(F)* 《《.

All equipment to be grounded shall be connected to an equipment grounding conductor of a type recognized by 250.118 and installed in accordance with Parts VI and VII of Article 250. The equipment grounding conductor shall be connected to the system grounded conductor at the service disconnecting means or, in the case of separately derived system such as a generator, at the generator or first disconnecting means supplied by the generator. The grounded circuit conductor shall not be connected to the equipment grounding conductor on the load side of the service disconnecting means or on the load side of a separately derived system disconnecting means 》》*525.31* 《《.

Ⓐ Conductors shall have a vertical clearance to ground per 225.18. These clearances apply only to exterior wiring of tents and concessions (not within) 》》*525.5(A)* 《《.

Ⓑ Tent and concession interior electrical wiring for temporary lighting shall be securely installed and mechanically protected where subject to physical damage. Protect all temporary general illumination lamps from accidental breakage by installing a suitable luminaire or lampholder with a guard 》》*525.21(B)* 《《.

Ⓒ Article 525 covers the installation of portable wiring and equipment for carnivals, circuses, fairs, and similar functions, including wiring in or on all structures 》》*525.1* 《《.

Ⓓ Portable structures shall be maintained not less than 15 ft (4.5 m) in any direction from overhead conductors operating at 600 volts or less, except for the conductors supplying the portable structure. Portable structures included in 525.3(D) shall comply with Table 680.8(A) 》》*525.5(B)(1)* 《《.

Ⓔ Electrical equipment and wiring methods in or on portable structures shall be provided with mechanical protection where such equipment or wiring methods are subject to physical damage 》》*525.6* 《《.

Ⓕ A means to disconnect each portable structure from all ungrounded conductors shall be provided. The disconnecting means shall be located within sight of and within 6 ft (1.8 m) of the operator's station. The disconnecting means shall be readily accessible to the operator, including when the ride is in operation. Where accessible to unqualified persons, the disconnecting means shall be lockable. A shunt trip device that opens the fused disconnect or circuit breaker when a switch located in the ride operator's console is closed shall be a permissible method of opening the circuit 》》*525.21(A)* 《《.

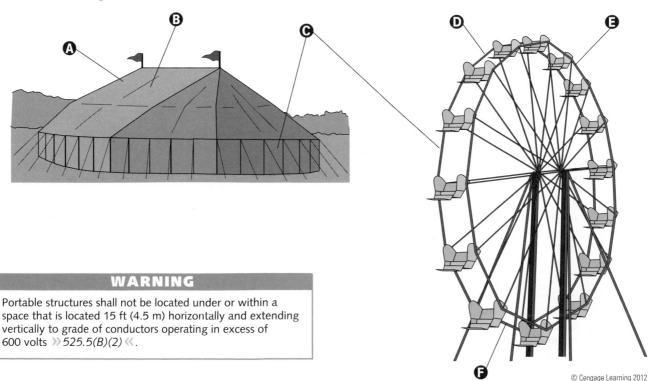

WARNING

Portable structures shall not be located under or within a space that is located 15 ft (4.5 m) horizontally and extending vertically to grade of conductors operating in excess of 600 volts 》》*525.5(B)(2)* 《《.

© Cengage Learning 2012

PLACES OF ASSEMBLY

Comprehensive Provisions

The wiring of any such building or area containing a projection booth, stage platform, or area for the presentation of theatrical or musical productions (fixed or portable) shall comply with Article 520. These requirements also apply to associated audience seating and all equipment used in the production, including portable equipment and associated wiring that is not connected to permanent wiring **》518.2(C)《**.

The fixed wiring methods shall be metal raceways, flexible metal raceways, nonmetallic raceways encased in not less than 2 in. (50 mm) of concrete, Type MI, MC, or AC cable. The wiring method shall itself qualify as an equipment grounding conductor according to 250.118 or shall contain an insulated equipment grounding conductor sized in accordance with Table 250.122 **》518.4(A)《**.

ENT and PVC are not approved for use in other spaces used for environmental air in accordance with 300.22(C) **》518.4(C)《**.

Portable switchboards and portable power distribution equipment shall be supplied only from listed power outlets of sufficient voltage and ampere rating. Such power outlets shall be protected by overcurrent devices. Such overcurrent devices and power outlets shall not be accessible to the general public. Provisions for connection of an equipment grounding conductor shall be provided. The neutral conductor of feeders supplying solid-state phase control, 3-phase, 4-wire dimmer systems shall be considered a current-carrying conductor for purposes of ampacity adjustment. The neutral conductor of feeders supplying solid-state sine wave, 3-phase, 4-wire dimming systems shall not be considered a current-carrying conductor for purposes of ampacity adjustment **》518.5《**.

Ⓐ Temporary wiring for display booths within exhibition halls (as in trade shows) shall be permitted to be installed in accordance with Article 590. Flexible cables and cords, approved for hard or extra-hard usage, laid on floors shall be protected from contact by the general public. The GFCI requirements of 590.6 shall not apply. All other GFCI requirements of the *NEC* shall apply.

Where GFCI protection for personnel is supplied by plug and cord connection to the branch circuit or to the feeder, the GFCI protection shall be listed as portable GFCI protection or provide a level of protection equivalent to a portable GFCI, whether assembled in the field or at the factory **》518.3(B)《**.

Ⓑ Where an assembly occupancy forms a portion of a building containing other occupancies, Article 518 applies only to that portion of the building considered an assembly occupancy. Occupancy of any room or space for assembly purposes by less than 100 persons in a building of other occupancy, and incidental to such other occupancy, shall be classified as part of the other occupancy and subject to the provisions applicable thereto **》518.2(B)《**.

Ⓒ In addition to the wiring methods of 518.4(A), nonmetallic-sheathed cable, Type AC cable, ENT, and PVC can be installed in those buildings (or portions thereof) for which the applicable building code does not require fire-rated construction **》518.4(B)《**.

Ⓓ ENT and PVC can be installed within club rooms, conference and meeting rooms in hotels (or motels), courtrooms, dining facilities, restaurants, mortuary chapels, museums, libraries, and places of religious worship where:

1. The ENT or PVC is concealed within walls, floors, and ceilings where the enclosing structure provides a thermal material barrier having at least a 15-minute finish rating as identified in fire-rated assembly listings.

2. Such tubing or conduit is installed above suspended ceilings where the ceilings act as a thermal material barrier having at least a 15-minute finish rating according to fire-rated assembly listings **》518.4(C)《**.

Ⓔ Except for the assembly occupancies explicitly covered by 520.1, this article covers all buildings or portions of buildings or structures designed or intended for the gathering together of 100 or more persons for such purposes as deliberation, worship, entertainment, eating, drinking, amusement, awaiting transportation, or similar purposes **》518.1《**.

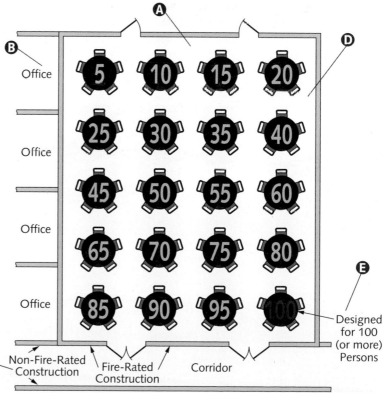

Office

Office

Office

Office

Office

Non-Fire-Rated Construction

Fire-Rated Construction

Corridor

Designed for 100 (or more) Persons

MOTION PICTURE (AND TELEVISION) STUDIOS AND SIMILAR LOCATIONS

Comprehensive Requirements

Portable stage or studio lighting equipment and portable power distribution equipment can be for temporary use outdoors if (1) supervised by qualified personnel while energized and (2) barriered from the general public ≫530.6≪.

Stage or set permanent wiring shall be Type MC cable, Type AC cable containing an insulated equipment grounding conductor sized per Table 250.122, Type MI cable, or approved raceways ≫530.11≪.

The wiring for stage set lighting and other supply wiring not fixed as to location shall be achieved with listed hard usage flexible cords/cables. Where subject to physical damage, such flexible cords/cables shall be of the listed extra-hard usage type. Cable splices or taps are allowed, provided the total connected load does not exceed the cable's maximum ampacity ≫530.12(A)≪.

Stage effects and electrical equipment used as stage properties can be wired with single- or multiconductor listed flexible cords or cables where protected from physical damage and secured to the scenery by approved cable ties or insulated staples. Splices (or taps) can only be made with listed devices in a circuit protected at no more than 20 amperes ≫530.12(B)≪.

Switches used for studio stage set lighting and effects (on the stages and lots and on location) shall be of the externally operable type. Where contactors are used as the disconnecting means for fuses, an individual externally operable switch, suitably rated, for the control of each contactor shall be located at a distance of not more than

6 ft (1.8 m) from the contactor, in addition to remote-control switches. A single externally operable switch shall be permitted to simultaneously disconnect all the contactors on any one location board, where located at a distance of not more than 6 ft (1.8 m) from the location board ≫530.13≪. (Location board is defined in 530.2 definitions.)

Portable luminaires and work lights shall have flexible cords, composition or metal-sheathed porcelain sockets, and substantial guards ≫530.16≪. Portable luminaires used as properties in a motion picture or television stage set, on a studio stage or lot, or on location are, for this purpose, not considered portable luminaires ≫530.16 Exception≪.

Automatic overcurrent protective devices (circuit breakers or fuses) for motion picture studio stage set lighting and associated cables must comply with 530.18(A) through (G).

Ⓐ Article 530 requirements apply to motion picture/television studios using either film or electronic cameras (except as provided in 520.1) as well as exchanges, factories, laboratories, stages, or a building segment in which film or tape wider than $7/_8$ in. (22 mm) is exposed, developed, printed, cut, edited, rewound, repaired, or stored ≫530.1≪.

Ⓑ Stage set: A specific area set up with temporary scenery and properties planned for a particular motion picture or television production scene ≫530.2≪.

Ⓒ Television studio or motion picture stage (sound stage): All or part of a building, usually insulated from outside noise and natural light, used by the entertainment industry for motion picture, television, or commercial production purposes ≫530.2≪.

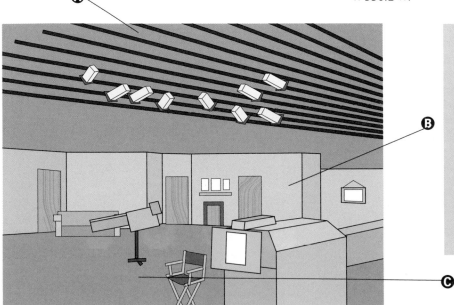

> **NOTE**
>
> Type MC cable, Type MI cable, Type AC cable containing an insulated equipment grounding conductor, metal raceways, and all non-current-carrying metal parts of appliances, devices, and equipment shall be connected to an equipment grounding conductor. This shall not apply to pendant and portable lamps, to portable stage lighting and stage sound equipment, or to other portable and special stage equipment operating at not over 150 volts dc to ground ≫530.20≪.

Comprehensive Requirements *(continued)*

The maximum ampacity allowed on a given conductor, cable, or cord size is dictated by the applicable tables of Articles 310 and 400 》*530.18*《.

It shall be permissible to apply Table 530.19(A) demand factors to that portion of the maximum possible connected load for studio or stage set lighting for all permanently installed feeders between substations and stages and to all permanently installed feeders between the main stage switchboard and stage distribution centers (or location boards) 》*530.19(A)*《.

Portable feeders can have a demand of 50% of maximum possible connected load 》*530.19(B)*《.

Plugs and receptacles, including cord connectors and flanged surface devices, shall be rated in amperes. The voltage rating of plugs and receptacles shall not be less than the nominal circuit voltage. All ac circuit plug and receptacle ampere ratings shall not be less than the feeder (or branch-circuit) overcurrent device ampere rating. Table 210.21(B)(2) shall not apply 》*530.21(A)*《.

Any ac single-pole portable cable connectors used shall be listed and of the locking type. Sections 400.10, 406.7, and 406.8 shall not apply to listed single-pole separable connections, or to single-conductor cable assemblies incorporating listed single-pole separable connectors. Paralleled sets of current-carrying single-pole separable connectors acting as input devices shall carry prominent warning labels indicating the presence of internal parallel connections. Single-pole separable connectors shall comply with at least one provision of 530.22(A)(1) through (3) 》*530.22(A)*《.

Cellulose Nitrate Film Storage Vaults

A Lamps in cellulose nitrate film storage vaults shall be installed in rigid luminaires of the glass-enclosed and gasketed type 》*530.51*《.

B Lamps shall be controlled outside the vault by a switch having a pole in each ungrounded conductor and provided with a pilot light indicating "on" or "off." This switch shall disconnect every ungrounded conductor terminating in any outlet within the vault from any supply source 》*530.51*《.

> ### CAUTION
> Unless otherwise permitted in 530.51, no receptacles, outlets, heaters, portable lights, or other portable electric equipment are allowed inside cellulose nitrate film storage vaults 》*530.52*《.

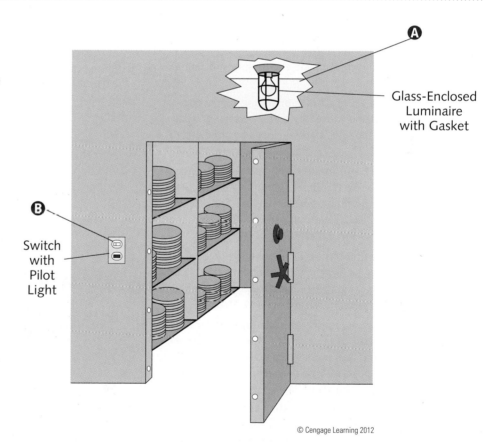

A — Glass-Enclosed Luminaire with Gasket

B — Switch with Pilot Light

© Cengage Learning 2012

MOTION PICTURE PROJECTION ROOMS

General Provisions

Nonprofessional projectors: Any type other than described in 540.2 》》*540.2* 《《. These projectors, including miniature types, employing cellulose acetate (safety) film may be operated without a projection room 》》*540.31* 《《.

Motor-generator sets, transformers, rectifiers, rheostats, and similar equipment for the supply (or control) of current to projection (or spotlight equipment) using nitrate film shall be located in a separate room. If inside the projection room, they shall be located (or guarded) so that arcs/sparks cannot contact film, and the commutator end(s) of motor generator sets shall comply with *one* of the following conditions:

1. Be of the totally enclosed, enclosed fan-cooled, or enclosed pipe-ventilated type.

2. Be enclosed in separate rooms or housings built of noncombustible material constructed so as to exclude flyings or lint with approved ventilation from a source of clean air.

3. Have the motor-generator brush or sliding-contact end enclosed by solid metal covers.

4. Have brushes or sliding contacts enclosed in substantial, tight metal housings.

5. Have the upper half of the brush or sliding-contact end of the motor-generator enclosed by a wire screen (or perforated metal), and the lower half enclosed by solid metal covers.

6. Have wire screens or perforated metal placed at the commutator of brush ends. No dimension of any opening in the wire screen or perforated metal shall exceed 0.05 in. (1.27 mm), regardless of the shape of the opening and of the material used 》》*540.11(A)* 《《.

Extraneous equipment (switches, overcurrent devices, etc.), not normally a part of projection, sound reproduction, flood or other special effect lamps, shall not be placed in projection rooms unless an exception is met 》》*540.11(B)* 《《.

Conductors supplying outlets for arc and xenon projectors (professional-type) shall have an ampacity not less than the projector current rating, but in no case smaller than 8 AWG. Conductors for incandescent-type projectors shall conform to the normal wiring standards of 210.24 》》*540.13* 《《.

Insulated conductors having an operating temperature rating no less than 200°C (392°F) shall be used on all lamps (or other equipment) where the ambient temperature at the installed conductors will exceed 50°C (122°F) 》》*540.14* 《《.

Ⓐ Article 540 provisions apply to motion picture projection rooms, motion picture projectors, and associated equipment, both professional and nonprofessional types, that use incandescent, carbon arc, xenon, or other light source equipment capable of producing hazardous gases, dust, or radiation 》》*540.1* 《《.

Ⓑ Every professional-type projector shall be located within a permanently constructed projection room approved for the type of building in which it is located. All projection ports, spotlight ports, viewing ports, and similar openings shall be completely enclosed by use of glass or other approved material. Such rooms do not qualify as hazardous (classified) locations as defined in article 500 》》*540.10* 《《.

Ⓒ A minimum 30-in. (750-mm) wide working space shall be provided on each side and at the rear of each motion picture projector, floodlight, spotlight, or similar equipment 》》*540.12* 《《. Adjacent pieces of equipment can be served by one such space 》》*540.12 Exception* 《《.

Ⓓ **Professional projector:** Projectors using either (1) 35- or 70-mm film (a minimum of 1⅜ in. [35 mm] wide with 5.4 edge perforations per inch [212 perforations per meter]) or (2) a type using carbon arc, xenon, or other light source equipment producing hazardous gases, dust, or radiation 》》*540.2* 《《.

Ⓔ Projectors and enclosures for arc, xenon, and incandescent lamps and rectifiers, transformers, rheostats, and similar equipment shall be listed 》》*540.20* 《《.

Ⓕ Projectors and other equipment shall be marked with the manufacturer's name (or trademark) as well as the voltage and current for which they are designed, per 110.21 》》*540.21* 《《.

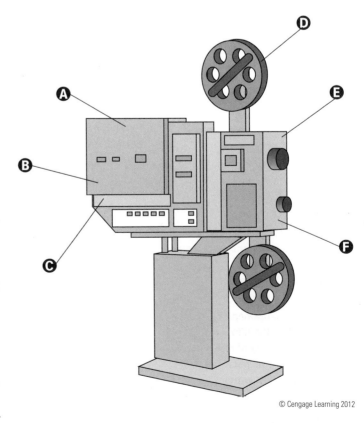

MANUFACTURED BUILDINGS

Manufactured Building Requirements

Fittings and connectors to be concealed at the time of on-site assembly (where tested, identified, and listed to applicable standards) are permitted for on-site module/component interconnection. Such fittings and connectors shall equal the employed wiring method in insulation, temperature rise, and fault-current withstand. They shall also be able to endure the vibration and minor relative motions occurring in the manufactured building components »545.13«.

In closed construction, cables can be secured only at cabinets, boxes, or fittings where using 10 AWG or smaller conductors and protection against physical damage is provided »545.4(B)«.

Provisions shall be made to route the service-entrance conductors, underground service conductors, service-lateral, feeder, or branch-circuit supply to the service or building disconnecting means conductors »545.5«.

Install service-entrance conductors after the building is site-erected, unless the point of attachment location is known prior to manufacture »545.6«.

Exposed conductors and equipment shall be protected during manufacturing, packaging, transit, and setup at the building site »545.8«.

Boxes of dimensions other than those required in Table 314.16(A) can be installed where tested, identified, and listed to applicable standards »545.9(A)«.

Any box no larger than 100 in.³ (1650 cm³) and intended for closed-construction mounting shall be affixed with anchors (or clamps) to provide a rigid and secure installation »545.9(B)«.

A receptacle (or switch) with integral enclosure and mounting means can be installed where such is tested, identified, and listed to applicable standards »545.10«.

Prewired panels and building components shall provide for the bonding, or bonding and grounding, of all exposed metals likely to be energized, per Article 250, Parts V, VI, and VII »545.11«.

Provisions shall be made to route a grounding electrode conductor from the service, feeder, or branch-circuit supply to the point of attachment to the grounding electrode »545.12«.

Closed construction: Any building, building component, assembly, or system manufactured in such a manner that concealed parts of manufacture processes cannot be inspected after building site installation without disassembly, damage, or destruction »545.2».

Building component: Any subsystem, subassembly, or other system designed for use in, integral with, or as part of a structure. A building component can be structural, electrical, mechanical, plumbing, fire protection, and other health and safety systems »545.2«.

Building system: Plans, specifications, and documentation for a manufactured building system, or for a type of system building component, such as structural, electrical, mechanical, plumbing, fire protection, and other systems affecting health and safety, including variations thereof specifically permitted by regulation, provided the variations are submitted as part of the building system or as an amendment thereto »545.2«.

A Article 545 covers manufactured building and building components requirements »545.1«.

B **Manufactured building:** Any closed construction building made (or assembled) in manufacturing facilities, whether on or off the building site, for installation, or assembly and installation, on the building site, excluding manufactured homes, mobile homes, park trailers, or recreational vehicles »545.2«.

C All raceways and cable wiring methods included in the *NEC* and other wiring systems specifically intended and listed for use in manufactured buildings shall be permitted with listed fittings and with fittings listed and identified for manufactured buildings »545.4(A)«.

> ### N O T E
> Service equipment shall be installed in accordance with 230.70 »545.7«.

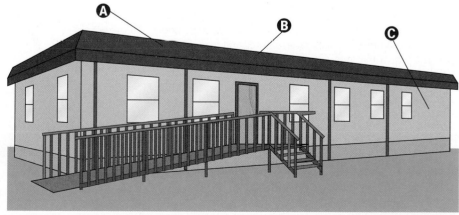

AGRICULTURAL BUILDINGS

Agricultural Building Provisions

Motors and other rotating electrical machinery shall either be totally enclosed or designed to minimize the entrance of dust, moisture, or corrosive particles **»547.7«**.

Equipotential planes shall be connected to the electrical grounding system. The bonding conductor shall be solid copper, insulated, covered or bare, and not smaller than 8 AWG. The means of bonding to wire mesh or conductive elements shall be by pressure connectors or clamps of brass, copper, copper alloy, or an equally substantial approved means. Slatted floors that are supported by structures that are a part of an equipotential plane shall not require bonding **»547.10(B)«**.

For the purpose of 547.10, the term *livestock* does not include poultry **»547.10«**.

A This article's provisions apply to agricultural buildings, part of buildings, or adjacent areas of similar nature as specified in (A) or (B).

(A) Agricultural buildings where excessive dust, and dust in combination with water, may accumulate. Included are all areas of poultry, livestock, and fish confinement systems, where litter dust or feed dust (including mineral feed particles) may accumulate.

(B) Agricultural buildings with a corrosive atmosphere. Such buildings include areas where:
(1) poultry and animal excrement can cause corrosive vapors,
(2) corrosive particles may mix with water,
(3) periodic washing for cleaning and sanitizing with water and cleansing agents creates a damp, even wet, environment,
(4) similar conditions exist **»547.1«**.

B The wiring method(s) employed shall be types UF, NMC, copper SE cables, jacketed Type MC cable, PVC, LFNC, or other location-suitable cables/raceways having approved termination fittings. Article 502 wiring methods are permitted for areas described in 547.1(A) **»547.5(A)«**.

C An **equipotential plane** is an area where a wire mesh (or other conductive element) is (1) embedded in or placed under concrete; (2) bonded to all metal structures and fixed nonelectrical equipment that may become energized; and (3) connected to the electrical grounding system to minimize voltage potentials within the plane and between the planes, the grounded equipment, and the earth **»547.2«**.

D All cables shall be secured within 8 in. (200 mm) of each cabinet, box, or fitting. Nonmetallic boxes, fittings, conduit,

and cables shall be permitted to be mounted directly to any building surface covered by this article without maintaining the ¼-in. (6-mm) airspace in accordance with 300.6(D) **»547.5(B)«**.

E Luminaires shall comply with all of the following:
A. By installation, the entrance of dust, foreign matter, moisture, and corrosive material shall be minimized.
B. If subject to physical damage, shall have a suitably protective guard.
C. If exposed to moisture, whether from condensation, cleansing water, or solution, shall be listed as suitable for use in wet locations **»547.8«**.

F Indoors: Equipotential planes shall be installed in confinement areas with concrete floors where metallic equipment is located that may become energized and is accessible to livestock **»547.10(A)(1)«**. Outdoors: Equipotential planes shall be installed in concrete slabs where metallic equipment is located that may become energized and is accessible to livestock. The equipotential plane shall encompass the area where the livestock stands while accessing metallic equipment that may become energized **»547.10(A)(2)«**.

G Where an equipment grounding conductor is installed underground within a location falling under the scope of Article 547, it shall be insulated or covered **»547.5(F)«**.

> ### N O T E
> All 125-volt, single-phase, 15- and 20-ampere general-purpose receptacles installed in areas having an equipotential plane, outdoors, damp or wet locations, and dirt confinement areas for livestock shall have GFCI protection **»547.5(G)«**.

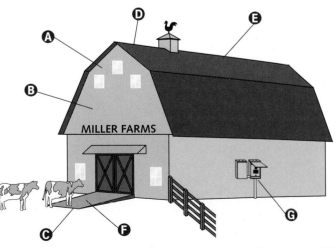

© Cengage Learning 2012

> ### CAUTION
> Protect all electrical wiring and equipment subject to physical damage **»547.5(E)«**.

MOBILE HOMES, MANUFACTURED HOMES, AND MOBILE HOME PARKS

General Requirements

Mobile home: A factory-assembled structure transportable in one or more sections, built on a permanent chassis, designed for use as a dwelling having no permanent foundation where connected to the required utilities, and including the plumbing, heating, air-conditioning, and electric systems contained therein ≫ *550.2* ≪.

Manufactured home: A structure, transportable in one or more sections, that in traveling mode, is 8 body-ft (2.4 m) or more in width, or 40 body-ft (12.2 m) or more in length, or, when erected on site, is 320 ft² (29.7 m²) or more, built on a permanent chassis and designed for dwelling use (with or without a permanent foundation) when connected therein ≫ *550.2* ≪.

At least one receptacle outlet shall be installed outdoors accessible at grade level, and not more than 6½ ft (2.0 m) above grade. A receptacle outlet located in an outside accessible compartment qualifies as an outdoor receptacle. These receptacle outlets shall have GFCI protection for personnel ≫ *550.13(B) and (D)(8)* ≪.

If a pipe heating cable outlet is installed, the outlet shall be:

1. Located within 2 ft (600 mm) of the cold water inlet.

2. Connected to an interior branch circuit, other than a small-appliance branch circuit. A bathroom receptacle circuit can be utilized for this purpose.

3. On a circuit where all outlets are on the load side of the ground-fault circuit-interrupter protection for personnel.

4. Mounted on the mobile home's underside, although this is not considered to be the outdoor receptacle outlet required in 550.13 (D)(8) ≫ *550.13(E)* ≪.

Use only approved and listed fixed-type wiring methods to join portions of a circuit that must be electrically joined and are located in adjacent mobile home sections after installation on its support foundation. The circuit's junction shall be accessible for disassembly when preparing the home for relocation ≫ *550.19(A)* ≪.

A mobile home not intended as a dwelling unit is not required to meet this article's provisions pertaining to the number or capacity of circuits required. Included are those equipped for sleeping purposes only, contractor's on-site offices, construction job dormitories, mobile studio dressing rooms, banks, clinics, or mobile stores, or those intended for the display or demonstration of merchandise or machinery. Such nondwelling units shall, however, meet all other applicable requirements of this article if provided with an electrical installation to be energized from a 120-volt or 120/240-volt ac power supply system. If design or available power supply systems dictate a different voltage, make adjustments in accordance with the appropriate articles/sections for the voltage used ≫ *550.4(A)* ≪.

Ⓐ Article 550 provisions cover the electrical conductors and equipment in or on mobile and manufactured homes, mobile and manufactured home electrical supply conductors, and the installation of electrical wiring, luminaires, equipment, and appurtenances related to electrical installations within a mobile home park up to the mobile home service-entrance conductors or, if none, the service equipment ≫ *550.1* ≪.

Ⓑ For *Code* purposes (unless otherwise indicated), the term *mobile home* includes manufactured homes ≫ *550.2 Mobile Home* ≪.

Ⓒ This article's provisions apply to mobile homes designed to connect to a wiring system rated 120/240 volts, nominal, 3-wire ac, with grounded neutral conductor ≫ *550.4(C)* ≪.

Ⓓ Outdoor or under-chassis line-voltage (120 volts, nominal, or higher) wiring exposed to moisture or physical damage shall be protected by a conduit or raceway approved for use in wet locations or where subject to physical damage. The conductors shall be listed for use in wet locations ≫ *550.15(H)* ≪.

CAUTION

All electrical materials, devices, appliances, fittings, and other equipment shall be (1) listed or labeled by a qualified testing agency and (2) connected in an approved manner when installed ≫ *550.4(D)* ≪.

NOTE

Mobile homes not installed in mobile home parks shall comply with the provisions of this article ≫ *550.4(B)* ≪.

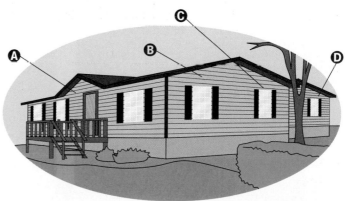

Power Supply

Cords with adapters and pigtail ends, extension cords, and similar items shall not be attached to or shipped with a mobile home 》*550.10(B)* 《.

A cord passing through walls or floors shall be protected by means of conduits and bushings (or equivalent). The cord can be installed within the walls, provided a continuous raceway (maximum 1¼ in. [32 mm] in size) is installed from the branch-circuit panelboard to the underside of the mobile home floor 》*550.10(G)* 《.

The power-supply cord's attachment plug cap and any connector cord assembly or receptacle shall be permanently protected against corrosion and mechanical damage, if such devices are externally located while the mobile home is in transit 》*550.10(H)* 《.

If the calculated load exceeds 50 amperes, or if a permanent feeder is used, the supply shall be by one of the following means:

1. **One mast weatherhead installation, installed per Article 230, containing four continuous, insulated, color-coded feeder conductors, one of which shall be an equipment grounding conductor**

2. **A metal raceway or PVC from the internal disconnecting means to the mobile home's underside, with provisions for attachment to the raceway on the mobile home's underside via a suitable junction box (or fitting) (with or without conductors as in 550.10[I][1])** 》*550.10(I)* 《.

Ⓐ The attachment plug cap shall not only be a 3-pole, 4-wire, grounding type, rated 50 amperes, 125/250 volts with a configuration as shown in Figure 550.10(C) but also shall be intended for use with a 50-ampere, 125/250-volt receptacle configuration as shown in Figure 550.10(C). It shall be listed, individually or as part of a power-supply cord assembly, for such purpose, and molded to (or installed on) the flexible cord at the point where the cord enters the attachment plug cap 》*550.10(C)* 《.

Ⓑ In a right-angle cap configuration, the grounding member shall be the farthest from the cord 》*550.10(C)* 《.

Ⓒ The mobile home power supply shall be a feeder assembly consisting of a single listed 50-ampere mobile home power-supply cord or a permanently installed feeder 》*550.10(A)* 《.

Ⓓ The power-supply cord shall bear the marking: "FOR USE WITH MOBILE HOMES—40 AMPERES" or "FOR USE WITH MOBILE HOMES—50 AMPERES" 》*550.10(E)* 《.

Ⓔ The cord shall be a four-conductor listed type. One conductor shall be identified by a continuous green color, or a continuous green color with one or more yellow stripes, for use as the grounding conductor 》*550.10(B)* 《.

Ⓕ A suitable clamp (or equivalent) shall be provided at the panelboard knockout to afford cord strain relief and effectively prevent strain transmission to the terminals when the power-supply cord is handled as intended 》*550.10(B)* 《.

Ⓖ A mobile home power-supply cord, if present, shall be permanently attached to the panelboard either directly or via a junction box also permanently connected thereto, with the free end terminating in an attachment plug cap 》*550.10(B)* 《.

Ⓗ From the end of the power-supply cord (including bared leads) to the face of the attachment plug cap, the cord shall not be less than 21 ft (6.4 m) or more than 36½ ft (11 m) in length. From the face of the attachment plug cap to the mobile home entry point, the cord shall be at least 20 ft (6.0 m) long 》*550.10(D)* 《.

N O T E

The feeder assembly shall enter the mobile home through an exterior wall, the floor, or roof 》*550.10(F)*《 .

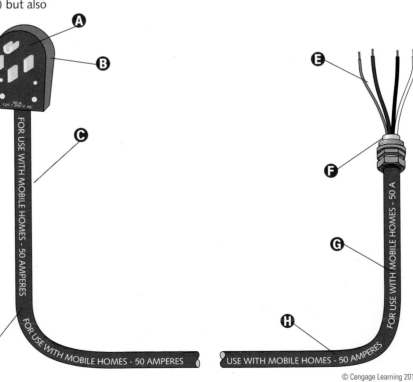

Panelboards and Branch Circuits

A panelboard shall be rated no less than 50 amperes and employ a two-pole circuit breaker rated 40 amperes for a 40-ampere supply cord, or 50 amperes for a 50-ampere supply cord. A panelboard employing a disconnect switch and fuses shall be rated 60 amperes and employ a single two-pole, 60-ampere fuseholder with 40- or 50-ampere main fuses for 40- or 50-ampere supply cords, respectively. The fuse size shall be plainly marked on the outside of the panelboard »*550.11(A)*«.

Each mobile home shall have branch-circuit distribution equipment that Includes circuit breaker or fuse overcurrent protection for each branch circuit. The branch-circuit overcurrent devices shall be rated as follows: (1) not more than the circuit conductors and (2) not more than 150% of a single appliance rated 13.3 amperes (or more) supplied by an individual branch circuit, but (3) not more than the overcurrent protection size and type marked on the air conditioner (or other motor-operated appliance)»*550.11(B)*«.

A metal nameplate on the outside adjacent to the feeder assembly entrance shall read:

THIS CONNECTION FOR 120/240-VOLT, 3-POLE, 4-WIRE, 60-HERTZ, _____-AMPERE SUPPLY

Fill in the blank with the correct ampere rating »*550.11(D)*«.

Multiply 3 volt-amperes/ft^2 (33 volt-amperes/m^2) by the outside dimensions of the mobile home (coupler excluded) and divide by 120 volts to determine the number of 15- or 20-ampere lighting circuits »*550.12(A)*«.

There shall be one or more adequately rated circuits in accordance with the following:

1. Ampere rating of fixed appliances not over 50% of circuit rating if lighting outlets (receptacles, other than kitchen, dining area, and laundry, considered as lighting outlets) exist on the same circuit.

2. For fixed appliances on a circuit without lighting outlets, the sum of rated amperes shall not exceed the branch-circuit rating. Motor loads or continuous loads shall not exceed 80% of the branch-circuit rating.

3. The rating of a single cord- and plug-connected appliance on a circuit having no other outlets shall not exceed 80% of the circuit rating.

4. The rating of a range branch circuit shall be based on the range demand as specified in 550.18(B)(5) »*550.12(D)*«.

A Each mobile home shall have a single disconnecting means consisting of a circuit breaker, or a switch and fuses, and its accessories installed in a readily accessible location near the supply cord (or conductor's) entry point into the mobile home. The main circuit breakers (or fuses) shall be plainly marked "Main." This equipment shall contain a solderless-type grounding connector (or bar) with sufficient terminals for all grounding conductors. The terminations of the grounded circuit conductors shall be insulated in accordance with 550.16(A) »*550.11(A)*«.

B While the panelboard shall be located in an accessible location, it shall not be located in a bathroom or clothes closet. A clear working space at least 30 in. (750 mm) wide and 30 in. (750 mm) in front of the panelboard shall be provided. This space shall extend from the floor to the top of the panelboard »*550.11(A)*«.

C The bottom of the distribution equipment, whether circuit breaker or fused type, shall be at least 24 In. (600 mm) above the mobile home's floor level »*550.11(A)*«.

D The distribution equipment shall have a rating not less than the calculated load »*550.11(A)*«.

E Where circuit breakers provide branch-circuit protection, 240-volt circuits shall be protected by a 2-pole common or companion trip, or circuit breakers with identified handle ties »*550.11(C)*«.

F Determine the required number of branch circuits in accordance with 550.12(A) through (E) »*550.12*«.

G The branch-circuit equipment can be combined with the disconnecting means as a single assembly. Such a combination can be designated as a panelboard. If a fused panelboard is used, the maximum fuse size for the mains shall be plainly marked with ¼-in. (6-mm) or taller lettering visible during fuse replacements. If plug fuses and fuseholders are used, they shall be tamper-resistant Type S, enclosed in dead-front fuse panelboards. Electrical panelboards containing circuit breakers shall also be dead-front type »*550.11*«.

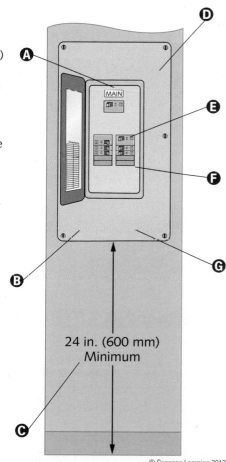

24 in. (600 mm) Minimum

Grounding

Cord-connected appliances (such as washing machines, clothes dryers, refrigerators, and the electrical system of gas ranges) shall be grounded via a cord having an equipment grounding conductor and grounding-type attachment plug ≫ *550.16(B)(3)* ≪ .

Metallic pipes (gas, water, and waste) as well as metallic air-circulating ducts are considered bonded if connected to the chassis terminal by clamps, solderless connectors, or appropriate grounding-type straps ≫ *550.16(C)(3)* ≪ .

Any metallic exterior covering is considered bonded if (1) the metal panels overlap one another and are securely attached to the wood or metal frame parts by metallic fasteners and (2) the lower panel of the metallic exterior covering is secured at opposite ends by metallic fasteners at a chassis cross member by two metal straps per mobile home unit (or section). The bonding strap shall be at least 4 in. (100 mm) wide and of the same material as the skin, or of a material of equal or better electrical conductivity. The straps shall be fastened with paint-penetrating fittings such as screws and starwashers or equivalent ≫ *550.16(C)(4)* ≪ .

In the electrical system, all exposed metal parts, enclosures, frames, luminaire canopies, etc. shall be effectively bonded to the grounding terminal or enclosure of the panelboard ≫ *550.16(B)(2)* ≪ .

A Grounding terminals shall be of the solderless type, listed as pressure-terminal connectors, and recognized for the wire size employed ≫ *550.16(C)(2)* ≪ .

B The supply cord or permanent feeder's green-colored insulated grounding wire shall be connected to the grounding bus in the panelboard or disconnecting means ≫ *550.16(B)(1)* ≪ .

C The bonding conductor shall be solid or stranded, insulated or bare, but shall be 8 AWG copper minimum, or the equivalent. The bonding conductor routing shall prevent physical damage exposure ≫ *550.16(C)(2)* ≪ .

D All potentially energizable exposed, non-current-carrying metal parts shall be effectively bonded to either the grounding terminal or the panelboard's enclosure. A bonding conductor shall be connected between the panelboard and an accessible terminal on the chassis ≫ *550.16(C)(1)* ≪ .

E Remove and discard bonding screws, straps, or buses in the panelboard or in appliances ≫ *550.16(A)(1)* ≪ .

F The grounded circuit conductor (neutral) shall be insulated from the grounding conductors, from equipment enclosures, and other grounded parts. The grounded circuit terminals in panelboards, ranges, clothes dryers, counter-mounted cooking units, and wall-mounted ovens shall be insulated from the equipment enclosure ≫ *550.16(A)(1)* ≪ .

G In a mobile home, grounding of metal parts (both electrical and nonelectrical) shall be accomplished by connection to a panelboard's grounding bus and shall be connected through the green-colored insulated conductor in the supply cord (or the feeder wiring) to the grounding bus in the service-entrance equipment, located adjacent to the mobile home ≫ *550.16* ≪ .

N O T E

Connections of ranges and clothes dryers with 120/240-volt, 3-wire ratings shall be made with 4-conductor cord and three-pole, 4-wire, grounding-type plugs or by Type AC cable, Type MC cable, or conductors enclosed in flexible metal conduit ≫ *550.16(A)(2)* ≪ .

CAUTION

Neither the mobile home's frame nor any appliance frame shall be connected to the grounded circuit conductor (neutral) in the mobile home ≫ *550.16* ≪ .

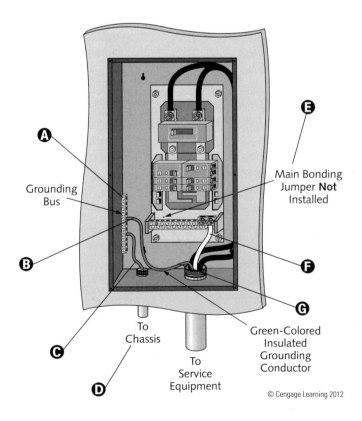

Grounding Bus

Main Bonding Jumper **Not** Installed

To Chassis

To Service Equipment

Green-Colored Insulated Grounding Conductor

© Cengage Learning 2012

Services and Feeders

The mobile home park secondary electrical distribution system supplying mobile home lots shall be single-phase, 120/240 volts, nominal. For Part III purposes, where the park service exceeds 240 volts, nominal, transformers and secondary panelboards shall be treated as services ≫ *550.30* ≪.

The **manufactured home** service equipment can be installed in or on a manufactured home, provided all of the conditions in 550.32(B) are met ≫ *550.32(B)* ≪.

Park electrical wiring systems shall be calculated (at 120/240 volts) on the larger of (1) 16,000 volt-amperes for each mobile home lot or (2) the load calculated per 550.18 for the largest typical mobile home that each lot accepts. The feeder or service load can be calculated in accordance with Table 550.31. No demand factor is allowed for any other load, except as provided in this *Code* ≫ *550.31* ≪.

Feeder conductors shall consist of either a listed cord, factory installed in accordance with 550.10(B), or a permanently installed feeder consisting of four insulated, color-coded conductors that shall be identified by the factory or field marking of the conductors in compliance with 310.110. Equipment grounding conductors shall not be identified by stripping the insulation ≫ *550.33(A)(1)* ≪. Feeder conductors shall be installed in compliance with 250.32(B) ≫ *550.33(A)(2)* ≪.

For an existing feeder that is installed between the service equipment and a disconnecting means, as covered in 550.32(A), the equipment grounding conductor can be omitted where the grounded circuit conductor is grounded at the disconnecting means in accordance with 250.32(B) Exception ≫ *550.33(A) Exception* ≪.

Ⓐ Mobile home service equipment shall be rated no less than 100 amperes at 120/240 volts. A mobile home feeder assembly shall be connected by a permanent wiring method. Power outlets used as mobile home service equipment can also contain receptacles rated up to 50 amperes with appropriate overcurrent protection. Fifty-ampere receptacles shall conform to Figure 550.10(C)'s configuration ≫ *550.32(C)* ≪.

Ⓑ Additional receptacles are allowed for connection of electrical equipment located outside the mobile home. All such 125-volt, single-phase, 15- and 20-ampere receptacles shall be protected by a listed GFCI ≫ *550.32(E)* ≪.

Ⓒ Any mobile home service equipment utilizing a 125/250-volt receptacle shall be marked as follows:

> TURN DISCONNECTING SWITCH
> OR CIRCUIT BREAKER OFF BEFORE INSERTING
> OR REMOVING PLUG. PLUG MUST
> BE FULLY INSERTED OR REMOVED.

The marking shall be located adjacent to the service equipment's receptacle outlet ≫ *550.32(G)* ≪.

Ⓓ The mobile home service equipment shall be located adjacent to, not mounted in or on, the mobile home. It shall be located in sight from and within 30 ft (9.0 m) from the

exterior wall of the mobile home it serves. The service equipment can be located elsewhere on the premises, provided a disconnecting means suitable for service equipment is located as stated previously for service equipment and is rated not less than that required for service equipment per 550.32(C). Grounding at the disconnecting means shall be compliant with 250.32 ≫ *550.32(A)* ≪.

Ⓔ Mobile home service equipment or the local external disconnecting means permitted in 550.32(A) shall provide a means for connecting (by a fixed wiring method) an accessory building, structure, or additional electrical equipment located outside the mobile home ≫ *550.32(D)* ≪.

Ⓕ Outdoor mobile home disconnecting means shall be installed so that the bottom of the disconnecting means enclosure is at least 2 ft (600 mm) above finished grade (or working platform). Installation shall ensure that the center of the operating handle's grip, in the highest position, is no more than 6 ft 7 in. (2.0 m) above the finished grade (or working platform) ≫ *550.32(F)* ≪.

CAUTION

Mobile home and manufactured home lot feeder circuit conductors shall: (1) have a capacity not less than the loads supplied, (2) be rated at not less than 100 amperes, and (3) be permitted to be sized in accordance with 310.15(B)(7) ≫ *550.33(B)* ≪.

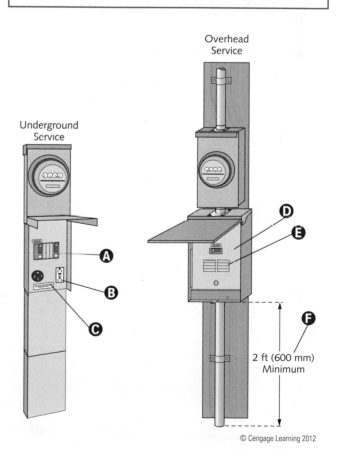

Underground Service

Overhead Service

2 ft (600 mm) Minimum

RECREATIONAL VEHICLES AND RECREATIONAL VEHICLE PARKS

Recreational Vehicles

Recreational vehicle park: A plot of land on which multiple recreational vehicle sites are located, established, or maintained for general public recreational vehicle occupancy as temporary living quarters for recreation or vacation purposes 》》*551.2*《《.

Recreational vehicle site: A plot of ground within a recreational vehicle park set aside for the accommodation of a recreational vehicle on a temporary basis or used as a camping unit site 》》*551.2*《《.

Recreational vehicle stand: The area of a recreational vehicle site intended for the placement of a recreational vehicle 》》*551.2*《《.

Generator installations must comply with 551.30(A) through (E).

Recreational vehicle main power-supply assemblies must meet 551.44(A) through (D) requirements.

Recreational vehicle panelboard provisions are found in 551.45(A) through (C).

The power-supply assembly (or assemblies) shall be factory supplied/installed and be of a type specified in 551.46(A)(1) or (2) 》》*551.46(A)*《《.

Recreational vehicle power-supply attachment plugs must meet 551.46(C) requirements.

Recreational vehicle wiring methods must comply with 551.47(A) through (S).

Ⓐ Article 551 provisions cover the electrical conductors and equipment installed within (or on) recreational vehicles, the conductors that connect recreational vehicles to an electrical supply, and the installation of electrical-related equipment and devices within a recreational vehicle park 》》*551.1*《《.

Ⓑ Recreational vehicle electrical equipment and material, intended for connection to a wiring system rated 120 volts, nominal, 2-wire with ground, or a wiring system rated 120/240 volts, nominal, 3-wire with an equipment grounding conductor, shall be listed and installed in accordance with the requirements of Parts I, II, III, IV, and V of Article 551. Electrical equipment connected line-to-line shall have a voltage rating of 208–230 volts 》》*551.40(A)*《《.

Ⓒ **Recreational vehicle:** A vehicular-type unit primarily designed as temporary living quarters for recreational, camping, or travel use, which either has its own motive power or is mounted on (or drawn) by another vehicle. Basically this includes travel trailers, camping trailers, truck campers, and motor homes 》》*551.2*《《. (See 551.2 for other word and phrase definitions relating to recreational vehicles and recreational vehicle parks.)

Ⓓ Every appliance shall be accessible for inspection, service, repair, and replacement without removal of permanent construction. Means shall be provided to securely fasten appliances in place during travel 》》*551.57*《《.

Ⓔ **Motor home:** A vehicular unit designed to provide temporary living quarters for recreational, camping, or travel use built on, or permanently attached to, a self-propelled motor vehicle chassis, or on a chassis cab/van that is an integral part of the total vehicle 》》*551.2*《《.

> **N O T E**
>
> A recreational vehicle not used for the purpose defined in 551.2 is not required to comply with Part IV provisions relating to the number (or capacity) of circuits. The recreational vehicle shall, however, meet all other applicable requirements of this article if it has an electrical installation that can be energized from a 120-volt, a 208Y/120-volt, or 120/240-volt, nominal, ac power-supply system 》》*551.4(A)*《《.

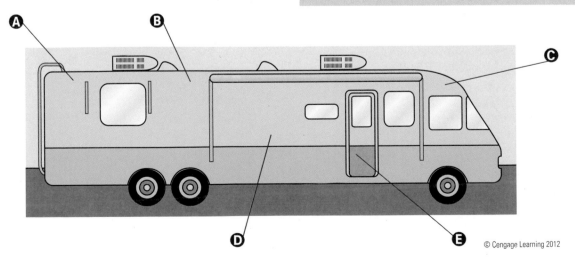

© Cengage Learning 2012

Recreational Vehicle Parks

Where park service exceeds 240 volts, transformers and secondary panelboards shall be treated as services ⟩⟩*551.73(B)*⟨⟨.

Recreational vehicle site feeder-circuit conductors shall have an ampacity not less than the loads supplied and shall be rated not less than 30 amperes. The neutral conductors shall have an ampacity not less than the ungrounded conductors ⟩⟩*551.73(D)*⟨⟨.

No grounding electrode connection can be made to the grounded conductor on the load side of the service disconnecting means except as covered in 250.30(A) for separately derived systems and 250.32(B) Exception for separate buildings ⟩⟩*551.76(D)*⟨⟨.

Where provided on back-in sites, the recreational vehicle site electrical equipment shall be located on the parked vehicle's left (road) side, on a line that is 5 to 7 ft (1.5 to 2.1 m) from the left edge (driver's side of the parked RV) and from 15 ft (4.5 m) forward of the rear of the stand ⟩⟩*551.77(A)*⟨⟨.

A disconnecting switch (or circuit breaker) shall be provided in the site supply equipment for disconnecting the recreational vehicle power supply ⟩⟩*551.77(B)*⟨⟨.

All site supply equipment shall be accessible by an unobstructed entrance or passageway at least 2 ft (600 mm) wide and 6½ ft (2.0 m) high ⟩⟩*551.77(C)*⟨⟨.

Sufficient space shall be constantly available around all electrical equipment for ready and safe operation, in accordance with 110.26 ⟩⟩*551.77(E)*⟨⟨.

Ⓐ At least 70% of all electrically supplied recreational vehicle sites shall be equipped with a 30-ampere, 125-volt receptacle conforming to Figure 551.46(C)(1). Additional receptacle configurations conforming to 551.81 can be included. Dedicated tent sites having a 15- or 20-ampere electrical supply can be excluded when determining 30- or 50-ampere receptacles ⟩⟩*551.71*⟨⟨.

Ⓑ The recreational vehicle park secondary electrical distribution system to 50-ampere recreational vehicle sites shall be supplied from a branch circuit of the voltage class and rating of the receptacle. Other recreational vehicle sites with 125-volt, 20- and 30-ampere receptacles can be derived from any grounded distribution system supplying 120-volt, single-phase power. The neutral conductors can be reduced in size below the minimum required ungrounded conductor size for 240-volt, line-to-line, permanently connected loads only ⟩⟩*551.72*⟨⟨.

Ⓒ Every electrical equipped recreational vehicle site shall have at least one 20-ampere, 125-volt receptacle ⟩⟩*551.71*⟨⟨.

Ⓓ Twenty percent (or more) of all electrically supplied recreational vehicle sites shall be equipped with a 50-ampere, 125/250-volt receptacle conforming to the configuration identified in Figure 551.46(C)(1). Every recreational

vehicle site equipped with a 50-ampere receptacle shall also be equipped with a 30-ampere, 125-volt receptacle conforming to Figure 551.46(C)(1). These electrical supplies can include additional receptacles configured per 551.81 ⟩⟩*551.71*⟨⟨.

Ⓔ Where the site supply equipment contains a 125/250-volt receptacle, the equipment shall be marked as follows: "Turn disconnecting switch or circuit breaker off before inserting or removing plug. Plug must be fully inserted or removed." The marking shall be located on the equipment adjacent to the receptacle outlet ⟩⟩*551.77(F)*⟨⟨.

Ⓕ Electrical service and feeders shall be calculated on the basis of no less than 9600 volt-amperes per site equipped with 50-ampere, 208Y/120-volt, or 120/240-volt supply facilities; 3600 volt-amperes per site equipped with both 20-ampere and 30-ampere supply facilities; 2400 volt-amperes per site equipped with only 20-ampere supply facilities; and 600 volt-amperes per dedicated tent site. Table 551.73 demand factors are the minimum allowable demand factors that can be used in calculating load for service and feeders. If the recreational vehicle site's electrical supply has more than one receptacle, the calculated load is computed for the highest rated receptacle only ⟩⟩*551.73(A)*⟨⟨.

Ⓖ Site supply equipment shall be located no less than 2 ft (600 mm) and no more than 6½ ft (2.0 m) above the ground ⟩⟩*551.77(D)*⟨⟨.

Ⓗ Underground service, feeder, branch-circuit, and recreational vehicle site feeder-circuit conductors must be installed according to 551.80(A) and (B).

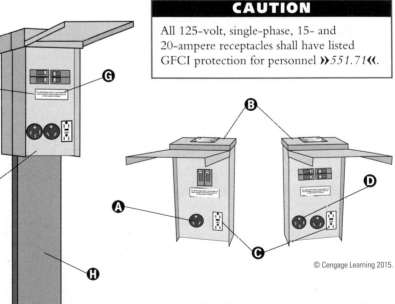

> ### CAUTION
> All 125-volt, single-phase, 15- and 20-ampere receptacles shall have listed GFCI protection for personnel ⟩⟩*551.71*⟨⟨.

© Cengage Learning 2015.

> ### NOTE
> In all areas subject to vehicular movement, open conductors of not over 1000 volts, nominal, shall have a vertical clearance of at least 18 ft (5.5 m) and a minimum 3-ft (900-mm) horizontal clearance. In all other areas, clearances shall conform to 225.18 and 225.19 ⟩⟩*551.79*⟨⟨.

FLOATING BUILDINGS

Floating Building Provisions

The service equipment for a floating building shall be located adjacent to, but not in (or on), the building or any floating structure. The main overcurrent protective device that feeds the floating structure shall have ground-fault protection not exceeding 100 milliamperes (mA). Ground fault protection of each individual branch or feeder circuit shall be permitted as a suitable alternative »553.4«.

One set of service conductors can serve multiple service equipment sets »553.5«.

Each floating building shall be supplied by a single set of feeder conductors from its service equipment »553.6«. Where the floating building has multiple occupancy, each occupant can be supplied by a single set of feeder conductors extended from the occupant's service equipment to the corresponding panelboard »553.6 Exception«.

Flexibility of the wiring system shall be maintained between the floating buildings and the supply conductors. All wiring shall be installed so that water surface motion and changes in water level do not result in unsafe conditions »553.7(A)«.

LFMC, or LFNC with approved fittings, is allowed for feeders and wherever flexible connections are required. Extra-hard usage portable power cable listed for both wet locations and sunlight resistance is permitted for a feeder to a floating building where flexibility is required. Other raceways suitable for the location can be installed where flexibility is not required »553.7(B)«.

In accordance with 553.8, grounding at floating buildings shall comply with the following:

(A) Grounding of both electrical and nonelectrical parts in a floating building shall be through connection to a grounding bus in the building panelboard.

(B) The equipment grounding conductor shall be installed with the feeder conductors and connected to a grounding terminal in the service equipment.

(C) The equipment grounding conductor shall be an insulated copper conductor with a continuous outer finish that is either green or green with one or more yellow stripes. For conductors larger than 6 AWG, or where multiconductor cables are used, reidentification of conductors as allowed in 250.119(A)(2)(b) and (A)(2)(c) or 250.119(B)(2) and (B)(3) shall be permitted.

(D) The grounding terminal in the service equipment shall be grounded by connection through an insulated grounding electrode conductor to a grounding electrode on shore.

The grounded circuit conductor (neutral) shall be an insulated conductor identified in conformance with 200.6. The neutral conductor shall be connected to the equipment grounding terminal in the service equipment, and, except for that connection, it shall be insulated from the equipment grounding conductors, equipment enclosures, and all other grounded parts. The neutral conductor terminals in the panelboard and in ranges, clothes dryers, counter-mounted cooking units, and the like shall be insulated from the enclosures »553.9«.

Where cord-connected appliances require grounding, it shall be accomplished by means of an equipment grounding conductor in the cord and a grounding-type attachment plug »553.10(B)«.

A Article 553 covers wiring, services, feeders, and grounding for floating buildings »553.1«.

B **Floating building:** A building unit as defined in Article 100 that floats on water, is moored in a permanent location, and has a premises wiring system served through connection by permanent wiring to an electrical supply system not located on the premises »553.2«.

NOTE

All electrical system enclosures and exposed metal parts shall be connected to the grounding bus »553.10(A)«.

CAUTION

All metal parts in contact with the water, all metal piping, and all non-current-carrying metal parts that are likely to become energized shall be connected to the grounding bus in the panelboard »553.11«.

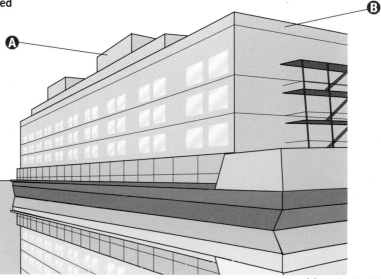

MARINAS AND BOATYARDS

Marina and Boatyard Requirements

The following items shall be connected to an equipment grounding conductor run with the circuit conductors in the same raceway, cable, or trench:

1. Metal boxes, metal cabinets, and all other metal enclosures

2. Metal frames of utilization equipment

3. Grounding terminals of grounding-type receptacles ≫ *555.15(A)* ≪

Branch-circuit insulated equipment grounding conductors shall terminate either at a grounding terminal in a remote panelboard or at the main service-equipment grounding terminal ≫ *555.15(D)* ≪ .

A feeder supplying a remote panelboard shall have an insulated equipment grounding conductor extending from a service-equipment grounding terminal to a grounding terminal in the remote panelboard ≫ *555.15(E)* ≪ .

Floating dock or marina service equipment shall be located adjacent to, rather than on or in, the floating structure ≫ *555.7* ≪ .

The load calculation specifications for each feeder or service circuit supplying receptacles providing shore power for boats are located in 555.12.

A Article 555 covers wiring and equipment installation in the areas comprising fixed (or floating) piers, wharfs, docks, and other areas in marinas, boatyards, boat basins, boat-houses, and similar occupancies designed (or planned) for the purpose of repair, berthing, launching, storage, or fueling of small craft and the moorage of floating buildings. This article does not cover private, noncommercial docking facilities constructed or occupied for the use of owner or residents or the associated single-family dwelling ≫ *555.1* ≪ .

B Electrical equipment enclosures installed on piers above deck level shall be securely and substantially supported by structural members, independent of any conduit connected to them. If enclosures are not attached to mounting surfaces by means of external ears or lugs, the internal screw heads shall be sealed to prevent seepage of water through mounting holes ≫ *555.10(A)* ≪ .

C Receptacles intended to supply shore power to boats shall be housed in marina power outlets listed for wet locations, listed enclosures protected from the weather, or listed weatherproof enclosures. The weatherproof integrity of the assembly shall not be affected when the receptacles are in use with any of properly configured booted or nonbooted attachment plug/cap inserted ≫ *555.19(A)(1)* ≪ .

D Disconnecting means shall be provided to isolate each boat from its supply connection(s) ≫ *555.17* ≪ . The disconnecting means shall consist of a circuit breaker, switch, or both, and shall be properly identified as to which receptacle it controls ≫ *555.17(A)* ≪ .

E Receptacles providing shore power for boats shall be rated at least 30 amperes and shall be single type ≫ *555.19(A)(4)* ≪ .

F Electrical equipment enclosures on piers shall be located so as not to interfere with mooring lines ≫ *555.10(B)* ≪ .

G Receptacles rated 30 and 50 amperes shall be of the locking and grounding type ≫ *555.19(A)(4)(a)* ≪ . Receptacles rated 60 and 100 amperes shall be of the pin and sleeve type ≫ *555.19(A)(4)(b)* ≪ .

N O T E

Fifteen– and 20-ampere, single-phase, 125-volt receptacles installed outdoors, in boathouses, in buildings or structures used for storage, maintenance, or repair where portable electrical hand tools, electrical diagnostic equipment, or portable lighting equipment are to be used shall be provided with GFCI protection for personnel. Receptacles in other locations shall be protected in accordance with 210.8(B) ≫ *555.19(B)(1)* ≪ .

CAUTION

Each single receptacle supplying shore power to boats shall be supplied from a marina power outlet or panelboard by an individual branch circuit of the voltage class and rating corresponding to the receptacle's rating ≫ *555.19(A)(3)* ≪ .

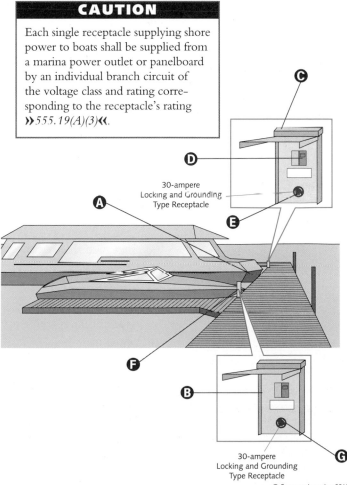

30-ampere
Locking and Grounding
Type Receptacle

30-ampere
Locking and Grounding
Type Receptacle

© Cengage Learning 2012

Summary

- Article 518 covers all buildings, portions of buildings, or structures intentionally designed for the assembly of 100 or more persons.
- Article 520 provisions apply to motion picture (and television) studio audience areas, performance areas, theaters, and similar locations.
- While Article 520, Part II applies to fixed stage switchboards, Part IV applies to portable on-stage switchboards.
- Stage equipment, other than switchboards (fixed and portable), are covered in Article 520, Parts III and V, respectively.
- Theater dressing room provisions are located in Article 520, Part VI (520.71 through 520.73).
- Article 525 covers the installation of portable wiring and equipment for carnivals, circuses, exhibitions, fairs, traveling attractions, and similar functions, including wiring in (or on) all structures.
- Motion picture/television studios and similar locations are located in Article 530.
- The provisions of Article 540 apply to motion picture projector rooms, motion picture projectors, and associated equipment.

- Article 545 covers requirements for manufactured buildings and their components.
- The requirements of Article 547 apply to agricultural buildings (all or in part), or adjacent areas of like nature as specified in 547.1(A) and (B).
- Article 550 (mobile homes, manufactured homes, and mobile home parks) is divided into three parts: Part I (general), Part II, (mobile and manufactured homes), and Part III (services and feeders).
- The provisions of Article 551 cover the electrical conductors and equipment installed within (or on) recreational vehicles, the conductors connecting recreational vehicles to an electrical supply, and the installation of electrical equipment and devices within a recreational vehicle park.
- Article 553 covers floating building wiring, services, feeders, and grounding.
- Article 555 covers wiring and equipment installation in the areas comprising fixed (or floating) piers, wharfs, docks, and other areas in marinas, boatyards, boat basins, boathouses, and similar occupancies designed (or planned) for repair, berthing, launching, storage, or fueling of small craft and the moorage of floating buildings.

Unit 18 Competency Test

NEC Reference   Answer

_____ _____ 1. A(n) _____ is any building that is of closed construction and is made or assembled in manufacturing facilities on or off the building site for installation, or assembly and installation on the building site, other than manufactured homes, mobile homes, park trailers, or recreational vehicles.

_____ _____ 2. Every recreational vehicle site with an electrical supply shall be equipped with at least _____ 20-ampere, 125-volt receptacle. A minimum of _____% of all recreational vehicle sites with an electrical supply shall each be equipped with a 30-ampere, 125-volt receptacle.

_____ _____ 3. Outdoor mobile home disconnecting means shall be installed so the bottom of the enclosure containing the disconnecting means is not less than _____ ft above finished grade or the working platform.

_____ _____ 4. A(n) _____ is an adapter cable containing one male and two female cord connectors used to connect two loads to one branch circuit.

_____ _____ 5. Agricultural building luminaires shall be installed to minimize the entrance of _____.

_____ _____ 6. All exposed incandescent lamps in theater dressing rooms, where less than _____ ft from the floor, shall be equipped with open-end guards riveted to the outlet box cover or otherwise sealed or locked in place.

_____ _____ 7. In exhibition halls used for display booths, as in trade shows, the temporary wiring shall be permitted to be installed in accordance with Article _____.

NEC Reference	Answer

8. Recreational vehicle site feeder-circuit conductors shall have an ampacity not less than the loads supplied and shall be rated at not less than _____.

9. A(n) _____ is a factory-assembled structure or structures transportable in one or more sections that is built on a permanent chassis and designed to be used as a dwelling without a permanent foundation where connected to the required utilities, and includes the plumbing, heating, air-conditioning, and electric systems contained therein.

10. The service equipment for floating docks or marinas shall be located _____ the floating structure.

11. Stage cables for motion picture studio stage set lighting shall be protected by means of overcurrent devices set at not more than _____% of the ampacity given in the applicable tables of Articles 310 and 400.

12. Fixed stage switchboards shall be _____ but shall not be required to be located on or adjacent to the stage.

13. A means to disconnect each portable structure from all ungrounded conductors shall be provided for each carnival ride. The disconnecting means shall be located within sight and within _____ ft of the operator's station.

14. Each motion picture projector, floodlight, spotlight, or similar equipment shall have clear working space not less than _____ in. wide on each side and at the rear thereof.

15. The overall length of a mobile home power-supply cord, measured from the end of the cord, including bared leads, to the face of the attachment plug cap shall not be less than _____ ft and shall not exceed _____ ft.

16. A travel trailer is a vehicular unit, mounted on wheels, designed to provide temporary living quarters for recreational, camping, or travel use, of such size or weight as not to require special highway movement permits when towed by a motorized vehicle, and of gross trailer area less than _____ ft².

17. A 15-ampere duplex receptacle installed in a television audience seating area can be supplied by a _____-ampere branch-circuit.

    I. 15

   II. 20

  III. 30

     a) II only   b) I only   c) either I or II   d) I, II, or III

18. Single receptacles that provide shore power for boats shall be rated not less than _____ amperes.

19. Occupancy of any room or space for assembly purposes by less than _____ persons in a building of other occupancy, and incidental to such other occupancy, shall be classified as part of the other occupancy and subject to the provisions applicable thereto.

20. Open conductors of not over 1000 volts, nominal, shall have a horizontal clearance of not less than _____ ft and a vertical clearance of not less than _____ ft in all areas subject to recreational vehicle movement.

21. Manufactured building service-entrance conductors shall be installed after erection at the building site unless the _____ is known prior to manufacture.

22. In a fixed stage switchboard in a theater, the circuit supplying a solid-state dimmer shall not exceed _____ volts between conductors unless the dimmer is listed specifically for higher voltage operation.

NEC Reference   Answer

_____ _____ 23. The service equipment shall be located in sight from and not more than _____ ft from the exterior wall of the mobile home it serves.

_____ _____ 24. Conductors supplying outlets for arc and xenon projectors of the professional type shall not be smaller than _____ and shall have an ampacity not less than the projector current rating.

_____ _____ 25. Each recreational vehicle site service shall be calculated on the basis of not less than _____ volt-amperes per site equipped with 50-ampere, 208Y/120-volt, or 120/240-volt supply facilities; and _____ volt-amperes per site equipped with only 20-ampere supply facilities that are not dedicated to tent sites.

_____ _____ 26. Every recreational vehicle site equipped with a _____ shall also be equipped with a 30-ampere, 125-volt receptacle.

_____ _____ 27. A _____ is a building unit, as defined in Article 100, that floats on water, is moored in a permanent location, and has a premises wiring system served through connection by permanent wiring to an electricity supply system not located on the premises.

_____ _____ 28. The power supply to the mobile home shall be a feeder assembly consisting of not more than one listed _____-ampere mobile power-supply cord or a permanently installed feeder.

_____ _____ 29. Portable structures at carnivals shall be maintained not less than _____ ft in any direction from overhead conductors operating at 1000 volts or less, except for the conductors supplying the portable structure.

_____ _____ 30. Lamps in cellulose nitrate film storage vaults shall be installed in rigid luminaires of the _____ type.

_____ _____ 31. A professional projector is a type of projector using _____ film that has a minimum width of _____ in. and has on each edge _____ perforations per inch, or a type using carbon arc, xenon, or other light source equipment that develops hazardous gases, dust, or radiation.

_____ _____ 32. In closed construction of manufactured buildings, cables shall be permitted to be secured only at cabinets, boxes, or fittings where _____ or smaller conductors are used and protection against physical damage is provided.

_____ _____ 33. All cables installed in agricultural buildings shall be secured within _____ in. of each cabinet, box, or fitting.

_____ _____ 34. A minimum of _____% of all recreational vehicle sites with electrical supply shall each be equipped with a 50-ampere, 125/250-volt receptacle.

_____ _____ 35. Footlights, border lights, and proscenium side lights shall be arranged so that no branch circuit supplying such equipment will carry a load exceeding _____.

Specific Equipment

Objectives

After studying this unit, the student should:

▶ have a good understanding of electric sign and outline lighting system requirements, including neon and skeleton tubing.

▶ understand manufactured wiring system provisions.

▶ be able to properly install office partition wiring (power, communication, etc.).

▶ be familiar with electric vehicle charging system stipulations.

▶ thoroughly understand information technology equipment requirements.

▶ be able to locate provisions relating to fire pumps.

▶ have a basic understanding of elevator, dumbwaiter, escalator, moving walk, platform lift, and stairway chairlift requirements.

▶ be familiar with definitions pertinent to swimming pools, fountains, and similar installations.

▶ understand what constitutes an equipotential bonding grid (for permanently installed pools) and what items must be bonded to the grid.

▶ have a solid grasp of equipment, panelboard, receptacle, and lighting (including underwater lighting) provisions pertaining to permanently installed swimming pools.

▶ be able to accurately define the term storable swimming pool and know the applicable requirements.

▶ be familiar with spa and hot tub provisions.

▶ know where to find requirements for therapeutic pools and tubs.

▶ understand the fundamental hydromassage bathtub provisions.

Introduction

Bringing this text to a close, this unit addresses a wide variety of electrical equipment requirements. *NEC* Chapter 6 is the basis for numerous equipment illustrations throughout this unit. While most of the articles of Chapter 6 are covered here, four additional articles are described under Unit 17—Industrial Locations. Those articles are 610 ("Cranes and Hoists"), 630 ("Electric Welders"), 669 ("Electroplating"), and 670 ("Industrial Machinery"). Some of the less commonly used articles are not referenced.

Beginning with "Pipe Organs" (650), this unit recounts some of the many Chapter 6 articles. Subsequently described in illustrative detail (with corresponding article numbers in parentheses) are "Electric Signs and Outline Lighting" (600); "Manufactured Wiring Systems" (604); "Office Furnishings" (605); "Electric Vehicle Charging Systems" (625); "Audio Signal Processing, Amplification, and Reproduction Equipment" (640); "Information Technology Equipment" (645); "Sensitive Electronic Equipment" (647); and "Fire Pumps" (695). Next, this unit devotes several illustrations to Article 620's topics: "Elevators, Dumbwaiters, Escalators, Moving Walks, Platform Lifts, and Stairway Chairlifts." Finally, the expansive subject of "Swimming Pools, Fountains, and Similar Installations" (Article 680) is introduced. Because Article 680 covers a broad range of equipment and installation applications, a significant number of illustrations have been dedicated to these requirements. It is imperative that the reader have a thorough understanding of the words and phrases found in 680.2 relating specifically to swimming pools, fountains, and similar locations. One particularly notable definition is that of **storable swimming or wading pool.** A misconception held by many is that a storable pool is, in simple terms, any aboveground swimming pool. Observe, however, that a storable pool has a maximum depth of only 42 in. (1.0 m). While some of the smaller aboveground pools fall within this parameter, many, if not most, have a depth of at least 48 in (1.2 m). It must be concluded, therefore, that any aboveground pool having a maximum depth greater than 42 in. (1.0 m) qualifies as a permanently installed pool and must be compliant with Article 680, Parts I and II. Storable/portable spas and hot tubs are also included in the definition of storable swimming pools. While space constraints prohibit an all-encompassing discussion of these provisions, the more predominant features are illustrated and discussed. Because of the potentially fatal nature of mistakes, study this unit carefully together with Article 680 when installing or wiring a pool, spa, hydromassage bathtub, or similar item.

As throughout this text, state or local jurisdictions may alter electrical requirements. Obtain a copy of any additional rules and regulations for your specific area.

EQUIPMENT

Pipe Organs

Installations of digital/analog-sampled sound production technology and associated audio signal processing, amplification, reproduction equipment, and wiring installed as part of a pipe organ shall comply with the appropriate Article 640 provisions 》*650.3(A)*《.

Installations of optical fiber cables shall be in accordance with Parts I and V of Article 770 》*650.3(B)*《.

The power source shall be a transformer-type rectifier having a direct current (dc) potential of 30 volts or less 》*650.4*《.

The rectifier shall be bonded to the equipment grounding conductor according to the provisions in Article 250, Parts V, VI, VII, and VIII 》*650.5*《.

Use a minimum of 28 AWG conductors for electronic signal circuits, and at least 26 AWG for electromagnetic valve supply and the like. A main common-return conductor in the electromagnetic supply shall be 14 AWG or larger 》*650.6(A)*《.

Conductors shall have thermoplastic or thermosetting insulation 》*650.6(B)*《.

Except for the common-return conductor and conductors inside the organ proper, the organ sections and the console conductors shall be cabled. An additional covering can enclose both cable and common-return conductors. Alternately, a separate common-return conductor is allowed to contact the cable 》*650.6(C)*《.

Each cable shall have an outer covering, either overall or on individual subassemblies of grouped conductors. Tape is an acceptable covering. If not housed in metal raceway, the covering shall be flame-retardant or the cable/cable

subassembly shall be wrapped by closely wound listed fireproof tape 》*650.6(D)*《.

Securely fastened cables can attach directly to the organ structure without insulating supports. Cables shall not be placed in contact with other conductors. Abandoned cables that are not terminated at equipment shall be identified with a tag 》*650.7*《.

Arrange circuits so that 26 and 28 AWG conductors are protected from overcurrent by an overcurrent device rated at no more than 6 amperes. Other conductor sizes shall be protected in accordance with their ampacity. A common-return conductor does not require overcurrent protection 》*650.8*《.

Ⓐ Article 650 covers electrically operated pipe organ circuits and components employed for sounding apparatus and keyboard control 》*650.1*《.

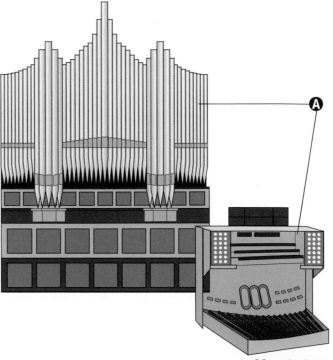

Electric Signs and Outline Lighting

Other than lamps and neon tubing, live parts shall be enclosed. An additional enclosure is not required for transformers and power supplies provided with an integral enclosure, including a primary and secondary circuit splice enclosure ≫ *600.8* ≪.

Signs and outline lighting systems with lampholders for incandescent lamps shall be marked to indicate the maximum allowable lamp wattage per lampholder. The markings shall be permanently installed, in letters at least ¼ in. (6 mm) high and shall be located where visible during relamping ≫ *600.4(B)* ≪.

Other than branch circuits that supply neon tubing installations, branch circuits that supply signs and outline lighting systems shall not exceed a 20-ampere rating ≫ *600.5(B)(2)* ≪.

The wiring method supplying signs and outline lighting systems shall terminate within a sign, an outline lighting system enclosure, a suitable box, or a conduit body ≫ *600.5(C)(1)* ≪.

Metal or nonmetallic poles used to support signs can enclose supply conductors, provided the poles and conductors are installed in accordance with 410.30(B) ≫ *600.5(C)(3)* ≪.

If operated by external electronic or electromechanical controllers, sign or outline lighting systems shall have a disconnecting means located within sight of the controller or within the controller enclosure. The disconnecting means shall disconnect both the sign (or outline lighting system) and the controller from all ungrounded supply conductors. The disconnecting means shall be designed so that no pole can be operated independently and shall be lockable in accordance with 110.25 ≫ *600.6(A)(3)* ≪.

Ⓐ Each sign and outline lighting system, feeder circuit, or branch circuit supplying a sign, outline lighting system, or skeleton tubing shall be controlled by an externally operable switch (or circuit breaker) that opens all ungrounded conductors. The switch or circuit breaker shall open all ungrounded conductors simultaneously on multiwire branch circuits in accordance with 210.4(B) ≫ *600.6* ≪. The disconnecting means shall be within sight of the sign or outline lighting system that it controls. If out of the line of sight from any energizable section, the disconnecting means shall be lockable in accordance with 110.25 ≫ *600.6(A)(2)* ≪.

Ⓑ Article 600 covers the installation of conductors, equipment, and field wiring for electric signs and outline lighting, regardless of voltage. All installations and equipment using neon tubing, such as signs, decorative elements, skeleton tubing, or art forms, are covered by Article 600 ≫ *600.1* ≪.

Ⓒ Signs and outline lighting systems shall be marked with the manufacturer's name, trademark or other identification, input voltage, and current rating ≫ *600.4(A)* ≪. The markings required in 600.4(A) and listing labels shall not be required to be visible after installation but shall be permanently applied in a location visible during servicing ≫ *600.4(C)* ≪.

Ⓓ Signs and outline lighting system equipment in wet locations, unless of the listed watertight type, shall be weatherproof and, if necessary, have drain holes in accordance with 600.9(D)(1) through (3) ≫ *600.9(D)* ≪.

Ⓔ Metal equipment of signs, outline lighting, and skeleton tubing systems shall be grounded by connection to the equipment grounding conductor of the supply branch circuit(s) or feeder using the types of equipment grounding conductors specified in 250.118 ≫ *600.7(A)(1)* ≪. (See additional grounding and bonding provisions in 600.7[A] and [B].)

Ⓕ Unless protected from physical damage, sign or outline lighting system equipment shall be at least 14 ft (4.3 m) above areas accessible to vehicles ≫ *600.9(A)* ≪.

Ⓖ Marking labels shall be permanent, durable, and, when in wet locations, shall be weatherproof ≫ *600.4(D)* ≪.

> **NOTE**
>
> Each commercial building and occupancy open to pedestrians shall have at least one accessible outlet at every tenant space entrance for sign or outline lighting system use. The outlet(s) shall be supplied by an exclusively dedicated branch circuit rated 20 amperes (or more). Service halls or corridors are not considered pedestrian accessible ≫ *600.5(A)* ≪.

> **NOTE**
>
> The disconnect shall be located at the point the feeder circuit or branch circuit(s) supplying a sign or outline lighting system enters a sign enclosure or a pole in accordance with 600.5(C)(3) and shall disconnect all wiring where it enters the enclosure of the sign or pole ≫ *600.6(A)(1)* ≪.

> **CAUTION**
>
> Fixed, mobile, or portable electric signs, section signs, outline lighting, and retrofit kits, regardless of voltage, shall be listed, provided with installation instructions, and installed in conformance with that listing, unless otherwise approved by special permission ≫ *600.3* ≪. As a system, outline lighting does not have to be listed when it consists of listed luminaires wired in accordance with Chapter 3 ≫ *600.3(B)* ≪.

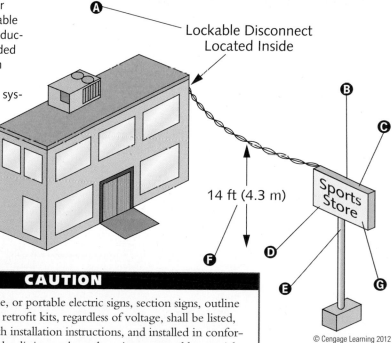

Lockable Disconnect Located Inside

14 ft (4.3 m)

Sports Store

© Cengage Learning 2012

Neon and Skeleton Tubing

Article 600, Part II provisions apply to field-installed skeleton tubing, field-installed secondary circuits, and outline lighting. These requirements are in addition to Part I requirements *》600.30《*.

Neon tubing: Electric-discharge luminous tubing, including cold cathode luminous tubing, that is manufactured into shapes to illuminate signs, forming letters (or parts thereof), skeleton tubing, outline lighting, and other decorative elements or art forms, and filled with various inert gases *》600.2《*.

Skeleton tubing: Neon tubing that is, itself, the sign or outline lighting and is not attached to an enclosure or sign body *》600.2《*.

The supply branch circuits of neon tubing installations shall not be rated in excess of 30 amperes *》600.5(B)(1)《*.

Neon tubing, other than listed, dry-location, portable signs that are readily accessible to pedestrians shall be protected from physical damage *》600.9(B)《*.

(A) Sharp bends in insulated conductors shall be avoided *》600.32(D)《*.

(B) Only one conductor shall be installed per run of conduit or tubing *》600.32(A)(2)《*.

(C) Conductors shall be installed in rigid metal conduit, intermediate metal conduit, liquidtight flexible nonmetallic conduit, flexible metal conduit, liquidtight flexible metal conduit, electrical metallic tubing, metal enclosures, on insulators in metal raceways, or other equipment listed for use with neon secondary circuits over 1000 volts *》600.32(A)(1)《*.

(D) Metal equipment of signs, outline lighting, and skeleton tubing systems shall be grounded by connection to the equipment grounding conductor of the supply branch circuit(s) or feeder using the types of equipment grounding conductors specified in 250.118 *》600.7(A)(1)《*.

(E) Secondary conductors shall be separated from each other and from all objects other than insulators or neon tubing by at least 1½ in. (38 mm) of space. GTO cable, installed in metal conduit or tubing, requires no spacing between the cable insulation and the conduit or tubing *》600.32(E)《*.

(F) All conductors' insulation shall extend no less than 2½ in. (65 mm) beyond the metal conduit or tubing *》600.32(G)《*.

(G) Neon tubing's length and design shall not cause a continuous overcurrent beyond the capacity of the transformer or electronic power supply *》600.41(A)《*.

(H) Neon tubing secondary conductor(s) shall be supported within 6 in. (150 mm) of the electrode connection *》600.42(D)《*.

(I) Connections shall be made by twisting the wires together, by use of a connection device, or by use of an electrode receptacle. Connections shall be electrically and mechanically secure and shall be within an enclosure listed for the purpose *》600.42(C)《*.

(J) Electrode terminals shall not be accessible to unqualified persons *》600.42(B)《*.

> **NOTE**
>
> Conductor provisions pertaining to this illustration apply to neon secondary circuit conductors over 1000 volts, nominal (600.32). Section 600.31 covers neon secondary circuit conductors of 1000 volts or less, nominal.

> **NOTE**
>
> Ballasts, transformers, and electronic power supplies must meet the requirements in 600.21 through 600.23.

> **CAUTION**
>
> Equipment having an open circuit voltage exceeding 1000 volts shall not be installed in (or on) dwelling occupancies *》600.32(I)《*.

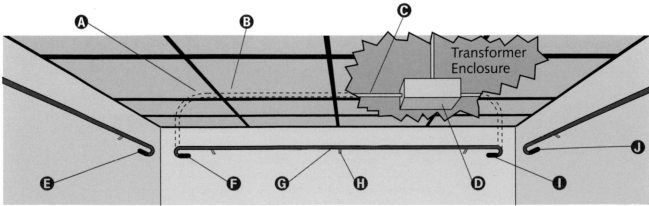

Transformer Enclosure

© Cengage Learning 2012

Manufactured Wiring Systems

Where listed for the purpose, manufactured wiring system assemblies can be used in outdoor locations »*604.4 Exception No. 2*«.

Conduit shall be listed FMC, or listed liquidtight flexible conduit, containing nominal 600-volt 8 to 12 AWG copper-insulated conductors, having a copper equipment grounding conductor (bare or insulated) equal in size to the ungrounded conductor »*604.6(A)(2)*«. (See Exceptions No. 1 through 3.)

Hard-usage flexible cord (minimum 12 AWG conductors) can be part of a listed factory-made assembly, not exceeding 6 ft (1.8 m) in length, as a transition method between manufactured wiring system components and utilization equipment, not permanently secured to the building structure. The cord must be visible for the entire length, shall not be subject to physical damage, and shall be provided with identified strain relief »*604.6(A)(3)*«. (See exception for listed electric-discharge luminaires.)

A Article 604 provisions apply to field-installed wiring using off-site manufactured subassemblies for branch circuits, remote control circuits, signaling circuits, and communications circuits in accessible areas »*604.1*«.

B The type of cable, flexible cord, or conduit shall be clearly marked on each section »*604.6(B)*«.

C Receptacles and connectors shall be (1) of the locking type, uniquely polarized and identified for the purpose, and (2) part of a listed assembly appropriate for the system »*604.6(C)*«.

D Manufactured wiring systems shall be secured and supported in accordance with the applicable cable or conduit article for the cable or conduit type employed »*604.7*«.

E Ceiling grid-supported wires shall not be used to support cables (and raceways). Secure support shall be provided via wires and associated fittings *and* shall be installed in addition to the ceiling grid support wires »*300.11(A)*«. (This illustration does not show cable supports.)

F A **manufactured wiring system** is a system containing component parts that are assembled in the process of manufacture and cannot be inspected at the building site without damage or destruction to the assembly and used for the connection of luminaires, utilization equipment, continuous plug-in type busways, and other devices »*604.2*«.

G Cable shall be one of the following:

(1) Listed Type AC cable containing nominal 600-volt, 8- to 12-AWG insulated copper conductors with a bare or insulated copper equipment grounding conductor equivalent in size to the ungrounded conductor.

(2) Listed Type MC cable containing nominal 600-volt, 8- to 12-AWG insulated copper conductors with a bare or insulated copper equipment grounding conductor equivalent in size to the ungrounded conductor.

(3) Listed Type MC cable containing nominal 600-volt, 8- to 12-AWG insulated copper conductors with a grounding conductor and armor assembly listed and identified for grounding in accordance with 250.118(10). The combined metallic sheath and grounding conductor shall have a current-carrying capacity equivalent to the ungrounded copper conductor »*604.6(A)(1)*«.

H Manufactured wiring systems are permitted in accessible and dry locations, and in ducts, plenums, and other air-handling spaces where listed for this application and installed per 300.22 »*604.4*«. In concealed spaces, one end of tapped cable can extend into hollow walls for direct switch/outlet termination points »*604.4 Exception No. 1*«.

> **N O T E**
>
> Manufactured wiring systems shall not be used where limited by the applicable article in Chapter 3 for the wiring method used in its construction »**604.5**«.

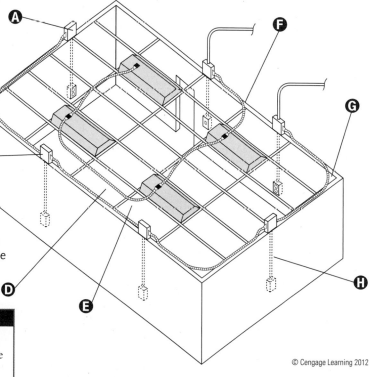

© Cengage Learning 2012

> **CAUTION**
>
> All connector openings shall be designed to prevent inadvertent contact with live parts or capped to effectively close the connector openings »*604.6(C)*«.

Office Furnishings

Wiring systems shall be identified as suitable for providing power for lighting accessories and utilization equipment used within office furnishings. A wired partition shall not extend from floor to ceiling ≫ *605.3* ≪ .

Office furnishings of the freestanding type (not fixed) shall be permitted to be connected to the building electrical system by one of the wiring methods of Chapter 3 ≫ *605.7* ≪ .

Individual office furnishings of the freestanding type, or groups of individual office furnishings that are electrically connected, are mechanically contiguous, and do not exceed 30 ft (9.0 m) when assembled, shall be permitted to be connected to the building's electrical system by means of a single flexible cord and plug, provided *all* of the following conditions are met:

(A) The flexible power-supply cord shall be of the extra-hard usage type, with 12 AWG or larger conductors, with an insulated equipment grounding conductor, and not more than 2 ft (600 mm) in length.

(B) The receptacle(s) supplying power shall be on a separate circuit serving only the office furnishing and no other loads and shall be located not more than 12 in. (300 mm) from the office furnishing to which it is connected.

(C) An individual office furnishing or groups of interconnected individual office furnishings shall not contain more than thirteen 15-ampere, 125-volt receptacle outlets.

(D) An individual office furnishing or groups of interconnected office furnishings shall not contain multiwire circuits ≫ *605.9* ≪ .

Ⓐ Article 605 covers electrical equipment, lighting accessories, and wiring systems that connect to, are contained within, or are installed on office furnishings ≫ *605.1* ≪ .

Ⓑ Office furnishing are defined as cubicle panels, partitions, study carrels, workstations, desks, shelving systems, and storage units that may be mechanically and electrically interconnected to form an office furnishing system ≫ *605.2* ≪ .

> **CAUTION**
>
> Each multiwire branch circuit shall be provided with a means that will simultaneously disconnect all ungrounded conductors at the point where the branch circuit originates ≫ *210.4(B)* ≪ .

Ⓒ The electrical connection between office furnishings shall be either a flexible assembly identified for use with office furnishings or shall be permitted to be installed using flexible cord, provided *all* of the following conditions are met: (1) the cord is extra-hard usage type with 12 AWG or larger conductors and shall have an insulated equipment grounding conductor; (2) the office furnishings are mechanically contiguous; (3) the cord is not longer than necessary for maximum positioning of the office furnishing but is in no case to exceed 2 ft (600 mm); and (4) the cord terminates at an attachment plug-and-cord connector with strain relief ≫ *605.5* ≪ .

Ⓓ Lighting equipment shall be listed and identified for use with office furnishings and shall comply with 605.6(A), (B), and (C) ≫ *605.6* ≪ .

Ⓔ All conductors and connections shall be contained within wiring channels (metal or other material) identified as suitable for the conditions of use. These channels shall be free of projections or other conditions that may damage conductor insulation ≫ *605.4* ≪ .

Ⓕ Chapters 5, 6, and 7 supplement or modify the general rules in Chapters 1 through 4. Chapters 1 through 4 apply except as amended by Chapters 5, 6, and 7 for the particular conditions ≫ *90.3* ≪ .

> **NOTE**
>
> Office furnishings that are fixed (secured to building surfaces) shall be permanently connected to the building electrical system by one of Chapter 3's wiring methods ≫ *605.6* ≪ .

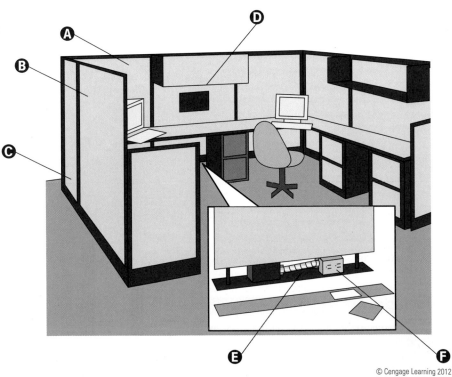

Electric Vehicle Charging System

Electric vehicle storage battery: A battery, comprised of one or more rechargeable electrochemical cells, that has no provision for the release of excessive gas pressure during normal charging and operation, or for the addition of water or electrolyte for external measurements of electrolyte-specific gravity ⟩⟩*625.2* ⟨⟨.

Output cable to the electric vehicle: An assembly consisting of a length of flexible EV cable and an electric vehicle connector (supplying power to the electric vehicle) ⟩⟩*625.2* ⟨⟨.

Power-supply cord: An assembly consisting of an attachment plug and length of flexible cord that connects the electric vehicle supply equipment (EVSE) to a receptacle ⟩⟩*625.2* ⟨⟨.

Plug-in hybrid electric vehicle (PHEV): A type of electric vehicle intended for on-road use with the ability to store and use off-vehicle electrical energy in the rechargeable energy storage system, and having a second source of motive power ⟩⟩*625.2* ⟨⟨.

Rechargeable energy storage system: Any power source that has the capability to be charged and discharged ⟩⟩*625.2* ⟨⟨. Batteries, capacitors, and electromechanical flywheels are examples of rechargeable energy storage systems ⟩⟩*625.2 Informational Note* ⟨⟨.

Overcurrent protection for feeders and branch circuits supplying electric vehicle supply equipment shall be sized for continuous duty and shall have a rating of not less than 125% of the maximum load of the electric vehicle supply equipment. Where noncontinuous loads are supplied from the same feeder or branch circuit, the overcurrent device shall have a rating of not less than the sum of the noncontinuous loads plus 125% of the continuous loads ⟩⟩*625.40* ⟨⟨.

Mechanical ventilation shall not be required where electric vehicle storage batteries are used, or where the electric vehicle supply equipment is listed for charging electric vehicles indoors without ventilation and marked in accordance with 625.15(B) ⟩⟩*625.52(A)* ⟨⟨.

Ⓐ Unless specifically listed and marked for the location, the electric vehicle supply equipment's coupling means shall be

> ### NOTE
> All electrical materials, devices, fittings, and associated equipment shall be listed ⟩⟩*625.5* ⟨⟨.

> ### WARNING
> Mechanical ventilation (such as a fan) shall be provided where the electric vehicle supply equipment is listed for charging electric vehicles that require ventilation for indoor charging, and is marked in accordance with 625.15(C). The ventilation shall include both supply and exhaust equipment and shall be permanently installed and located to intake from, and vent directly to, the outdoors. Positive-pressure ventilation systems shall be permitted only in vehicle charging buildings or areas that have been specifically designed and approved for that application. Mechanical ventilation requirements shall be determined using one of the methods found in 625.52(B)(1) through (4) ⟩⟩*625.52(B)* ⟨⟨.

stored or located at a height of not less than 18 in. (450 mm) above the floor level for indoor locations and 24 in. (600 mm) above the grade level for outdoor locations ⟩⟩*625.50* ⟨⟨.

Ⓑ Article 625 provisions cover the electrical conductors and equipment external to an electric vehicle that connect it to an electrical supply by conductive or inductive means, as well as the installation of equipment and devices related to electric vehicle charging ⟩⟩*625.1* ⟨⟨.

Ⓒ Electric vehicle: On-road-worthy automotive-type vehicle (such as a passenger automobile, bus, truck, van, neighborhood electric vehicles, electric motorcycles, and the like) primarily powered by an electric motor drawing current from a rechargeable storage battery, fuel cell, photovoltaic array, or other electric current source. Plug-in hybrid electric vehicles (PHEV) are considered electric vehicles. Article 625 excludes off-road self-propelled electric vehicles (such as industrial trucks, hoists, lifts, transports, golf carts, airline ground support equipment, tractors, boats, etc.) ⟩⟩*625.2* ⟨⟨.

Ⓓ Electric vehicle connector: A device that, when electrically coupled (conductive or inductive) to an electric vehicle inlet, establishes an electrical connection to the electric vehicle for the purpose of power transfer and information exchange. This device is part of the electric vehicle coupler ⟩⟩*625.2* ⟨⟨.

Ⓔ Electric vehicle coupler: A mating electric vehicle inlet and electric vehicle connector set ⟩⟩*625.2* ⟨⟨.

Ⓕ Electric vehicle inlet: The device on the electric vehicle into which the electric vehicle connector is electrically coupled (conductive or inductive) for power transfer and information exchange. This device is part of the electric vehicle coupler. For *NEC* purposes, the electrical vehicle inlet is considered part of the vehicle rather than part of the electric vehicle supply equipment ⟩⟩*625.2* ⟨⟨.

Ⓖ Electric vehicle supply equipment: Includes the conductors (ungrounded, grounded, and equipment grounding conductors), electric vehicle connectors, attachment plugs, and all other fittings, devices, power outlets, or apparatus purposely installed to transfer energy between the premises wiring and the electric vehicle ⟩⟩*625.2* ⟨⟨.

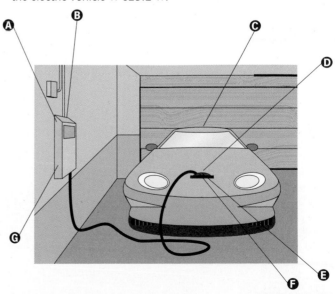

Audio Signal Processing, Amplification, and Reproduction Equipment

Grounding of separately derived systems with *60 volts to ground* shall meet 647.6 requirements ≫ *640.7(B)* ≪.

Equipment access shall not be denied by accumulated wires and cables that prevent panel removal, including suspended ceiling panels ≫ *640.5* ≪.

Wireways and auxiliary gutters shall be connected to an equipment grounding conductor(s), to an equipment bonding jumper, or to the grounded conductor where permitted or required by 250.92(B)(1) or 250.142. Where the wireway or auxiliary gutter does not contain power-supply wires, the equipment grounding conductor shall not be required to be larger than 14 AWG copper or its equivalent. Where the wireway or auxiliary gutter contains power-supply wires, the equipment grounding conductor shall not be smaller than specified in 250.122 ≫ *640.7(A)* ≪.

Isolated grounding-type receptacles are permitted as described in 250.146(D), as well as for the implementation of other Article 250 compliant technical power systems. For separately derived systems with 60 volts to ground, the branch-circuit equipment grounding conductor shall be terminated per 647.6(B) ≫ *640.7(C)* ≪.

Portable equipment: Equipment fed with portable cords or cables and intended for movement from one place to another ≫ *640.2* ≪.

Temporary equipment: Portable wiring and equipment used for events of a transient or temporary nature where equipment removal is presumed at the event's conclusion ≫ *640.2* ≪.

Technical power system (often referred to as *tech power*): An electrical distribution system with grounding in accordance with 250.146(D), where the equipment grounding conductor is isolated from the premises grounded conductor except at a single grounded termination point within a branch-circuit panelboard, at the originating (main breaker) branch-circuit panelboard, or at the premises grounding electrode ≫ *640.2* ≪.

Grouped (or bundled) insulated conductors of different systems, in close physical contact within the same raceway (or other enclosure), or in portable cords (or cables), shall comply with 300.3(C)(1) ≫ *640.8* ≪.

Ⓐ Article 640 covers equipment and wiring for audio signal generation, recording, processing, amplification and reproduction; distribution of sound; public address; speech input systems; temporary audio system installations; and electronic musical instruments (including organs). This encompasses audio systems subject to Article 517, Part VI; and Articles 518, 520, 525, and 530 ≫ *640.1* ≪.

Ⓑ **Loudspeaker:** Equipment that converts an alternating-current (ac) electric signal into an acoustic signal. The term *speaker* is commonly used to mean loudspeaker ≫ *640.2* ≪.

Ⓒ **Mixer:** Equipment used to combine and level-match a multiplicity of electronic signals, as from microphones, electronic instruments, and recorded audio ≫ *640.2* ≪.

Ⓓ **Audio signal processing equipment:** Electrically operated equipment that produces or processes electronic signals that, when appropriately amplified and reproduced via loudspeaker, output an acoustic signal within the range of normal human hearing (typically 20 Hz to 20 kHz). Within Article 640, the terms *equipment* and *audio equipment* are equivalent to audio signal processing equipment ≫ *640.2* ≪.

Ⓔ **Audio system:** The totality of equipment and interconnected wiring used to fabricate a fully functional audio signal processing, amplification, and reproduction system ≫ *640.2* ≪.

Ⓕ **Equipment rack:** A framework for the equipment support or enclosure, either portable or stationary ≫ *640.2* ≪.

> **NOTE**
>
> Audio signal processing, amplification, and reproduction equipment, cables, and circuits shall be installed in a neat, workmanlike manner ≫ *640.6(A)* ≪.

> **CAUTION**
>
> Cables installed exposed on the surface of ceilings and sidewalls shall be supported in such a manner that the audio distribution cables will not be damaged by normal building use. Such cables shall be secured by straps, staples, cable ties, hangers, or similar fittings designed and installed so as not to damage the cable. The installation shall conform to 300.4 and 300.11(A) ≫ *640.6(B)* ≪.

© Cengage Learning 2012

Information Technology Equipment

(A) An approved means shall be provided to disconnect power to all electronic equipment in the information technology equipment room or in designated zones within the room. There shall also be a similar approved means to disconnect the power to all dedicated HVAC systems serving the room or designated zones and shall cause all required fire/smoke dampers to close. Unless the disconnecting means is for a critical operations data system as covered in 645.10(B), the disconnecting means shall comply with the following:

(1) Remote disconnect means controls shall be located at approved locations readily accessible in case of fire to authorized personnel and emergency responders.

(2) The remote disconnect means controls for the control of electronic equipment power and HVAC systems shall be grouped and identified. A single means to control both systems shall be permitted.

(3) Where multiple zones are created, each zone shall have an approved means to confine fire or products of combustion to within the zone.

(4) Additional means to prevent unintentional operation of remote disconnect controls shall be permitted ›› *645.10 and 645.10(A)* ‹‹ .

(B) Article 645 covers equipment, power-supply wiring, and equipment interconnected wiring, as well as grounding of information technology equipment and systems in an information technology equipment room ›› *645.1* ‹‹ .

(C) Article 645 shall be permitted to provide alternate wiring methods to the provisions of Chapter 3 and Article 708 for power wiring, Parts I and III of Article 725 for signaling wiring, and Parts I and V of Article 770 for optical fiber cabling, provided all of the conditions listed in 645.4(1) through (6) are met ›› *645.4* ‹‹ .

(D) Information technology equipment shall be permitted to be connected to a branch circuit by a power-supply cord:

(1) Power-supply cords shall not exceed 15 ft (4.5 m).

(2) Power cords shall be listed and a type permitted for use on listed information technology equipment or shall be constructed of listed flexible cord and listed attachment plugs and cord connectors of a type permitted for information technology equipment ›› *645.5(B)* ‹‹ .

(E) The branch-circuit conductors supplying one or more units of information technology equipment shall have an ampacity not less than 125% of the total connected load ›› *645.5(A)* ‹‹ .

(F) Cables (power, communications, connecting, and interconnecting), cord and plug connections, and receptacles associated with the information technology equipment are permitted under a raised floor, provided 645.5(E)(1) through (6) requirements are met ›› *645.5(E)* ‹‹ .

(G) Uninterruptible power supply (UPS) systems installed within the information technology equipment room, and their supply and output circuits, shall comply with 645.10 unless: (1) the installation qualifies under the provisions of Article 685, or (2) the power source is limited to 750 volt-amperes or less derived either from UPS equipment or from battery circuits

integral to electronic equipment. The disconnecting means shall also disconnect the battery from its load ›› *645.11* ‹‹ .

(H) One of the conditions required in order to use the alternate wiring methods in Article 645 pertains to the HVAC system. A heating/ventilating/air-conditioning (HVAC) system shall be provided in one of the methods identified in 645.4(2) a or b.

a. A separate HVAC system that is dedicated for information technology equipment use and is separated from other areas of occupancy; or

b. An HVAC system that serves other occupancies and meets all of the following: (1) also serves the information technology equipment room, (2) provides fire/smoke dampers at the point of penetration of the room boundary, and (3) activates the damper operation upon initiation by smoke detector alarms, by operation of the disconnecting means required by 645.10, or by both ›› *645.4(2)* ‹‹ .

(I) All exposed non-current-carrying metal parts of an information technology system shall either be bonded to the equipment grounding conductor in accordance with Parts I, V, VI, VII, and VIII of Article 250 or be double-insulated. Signal reference structures, where installed, shall be bonded to the equipment grounding conductor provided for the information technology equipment. Any auxiliary grounding electrode(s) installed for information technology equipment shall be installed in accordance with 250.54 ›› *645.15* ‹‹ .

(J) It is not necessary to secure in place cables (power, communications, connecting, and interconnecting) and associated boxes, connectors, plugs, and receptacles that are listed as part of, or for, information technology equipment ›› *645.5(F)* ‹‹ .

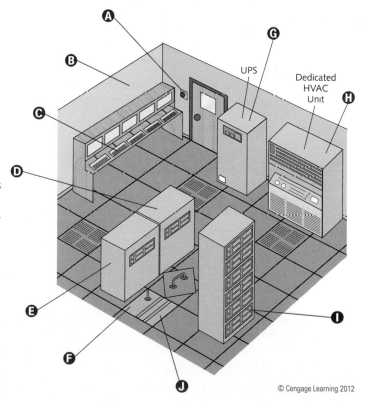

UPS

Dedicated HVAC Unit

Information Technology Equipment *(continued)*

> **N O T E**
>
> Separately derived power systems shall be installed in accordance with the provisions of Parts I and II of Article 250. Power systems derived within listed information technology equipment that supply information technology systems through receptacles or cable assemblies supplied as part of this equipment shall not be considered separately derived for the purpose of applying 250.30 **》645.14《**.

> **WARNING**
>
> Where exposed to physical damage, supply circuits and interconnecting cables shall be protected 》 *645.5(D)* 《.

SENSITIVE ELECTRONIC EQUIPMENT

Technical Power-System Receptacles

Ⓐ Article 647 covers the installation and wiring of sensitive electronic equipment that is connected to a separately derived system operating at 120 volts line-to-line and 60 volts to ground 》 *647.1* 《.

> **N O T E**
>
> While isolated ground receptacles are permitted as described in 250.146(D), the branch–circuit equipment grounding conductor shall be terminated per 647.6(B) **》647.7(B)《**.

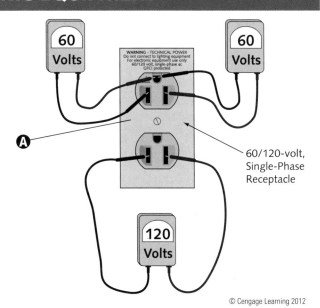

60/120-volt, Single-Phase Receptacle

© Cengage Learning 2012

Receptacle Installation

Where receptacles are used as a means of connecting equipment, all of the conditions in 647.7(A)(1) through (4) shall be met 》 *647.7(A)* 《.

Ⓐ All receptacle outlet strips, adapters, receptacle covers, and faceplates shall be marked as required by 647.7(A)(2):

WARNING—TECHNICAL POWER

Do not connect to

lighting equipment.

For electronic equipment use only.

60/120 V 1ø ac

GFCI protected.

Ⓑ All 125-volt receptacles used for 60/120-volt technical power shall be uniquely configured and identified for use with this class of system 》 *647.7(A)(4)* 《.

> **CAUTION**
>
> Clear markings on all junction box covers shall indicate the distribution panel and the system voltage **》647.4(B)《**.

Ⓒ All 15- and 20-ampere receptacle outlets shall be GFCI protected 》 *647.7(A)(1)* 《.

Ⓓ A 125-volt, single-phase, 15- or 20-ampere-rated receptacle having one of its current-carrying poles connected to a grounded circuit conductor shall be located within 6 ft (1.8 m) of all permanently installed 15- or 20-ampere-rated 60/120-volt technical power-system receptacles 》 *647.7(A)(3)* 《.

> **N O T E**
>
> The warning sign(s) or label(s) shall comply with 110.21(B) **》647.7(A)(2)《**.

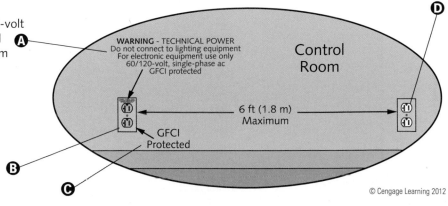

© Cengage Learning 2012

Separately Derived Systems with 60 Volts to Ground

A Standard single-phase panelboards and distribution equipment with higher voltage ratings can be used ≫ *647.4(A)* ≪.

B The transformer secondary center tap of the 60/120-volt, 3-wire system shall be grounded as provided in 250.30 ≫ *647.6(A)* ≪.

C A separately derived 120-volt, single-phase, 3-wire system with 60 volts on each of two ungrounded conductors to a grounded neutral conductor can be used to reduce objectionable noise in sensitive electronic equipment locations provided that (1) the system is installed only in commercial or industrial occupancies, (2) the system's use is restricted to areas under close supervision by qualified personnel, and (3) all requirements in 647.4 through 647.8 are met ≫ *647.3* ≪.

D Common-trip, two-pole circuit breakers or a combination two-pole fused disconnecting means that are identified for use at the system voltage shall be provided for both ungrounded conductors in all feeders and branch circuits ≫ *647.4(A)* ≪.

E Branch circuits and feeders shall be provided with a means to simultaneously disconnect all ungrounded conductors ≫ *647.4(A)* ≪.

F The system shall be clearly marked on the panel's face or inside the panel's door ≫ *647.4(A)* ≪.

G Permanently wired utilization equipment and receptacles shall be grounded via an equipment grounding conductor, run with the circuit conductors, to an equipment grounding bus (prominently marked "Technical Equipment Ground") in the originating branch-circuit panelboard. The grounding bus shall be connected to the grounded conductor on the line side of the separately derived system's disconnecting means. The grounding shall be sized as specified in Table 250.122 and be run with the feeder conductors. The technical equipment grounding bus need not be bonded to the panelboard enclosure ≫ *647.6(B)* ≪.

> **N O T E**
>
> All feeders and branch-circuit conductors installed under 647.4 shall be identified as to system at all splices and terminations by color, marking, tagging, or equally effective means. The means of identification shall be posted at each branch-circuit panelboard and at the building's disconnecting means ≫*647.4(C)*≪.

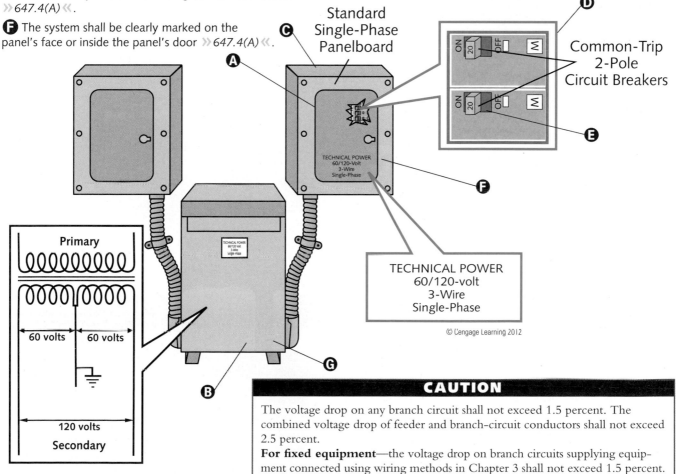

Standard Single-Phase Panelboard

Common-Trip 2-Pole Circuit Breakers

TECHNICAL POWER 60/120-Volt 3-Wire Single-Phase

Primary

60 volts 60 volts

120 volts

Secondary

TECHNICAL POWER 60/120-volt 3-Wire Single-Phase

© Cengage Learning 2012

> **CAUTION**
>
> The voltage drop on any branch circuit shall not exceed 1.5 percent. The combined voltage drop of feeder and branch-circuit conductors shall not exceed 2.5 percent.
>
> **For fixed equipment**—the voltage drop on branch circuits supplying equipment connected using wiring methods in Chapter 3 shall not exceed 1.5 percent. The combined voltage drop of feeder and branch-circuit conductors shall not exceed 2.5 percent.
>
> **For cord-connected equipment**—the voltage drop on branch circuits supplying receptacles shall not exceed 1 percent. For the purposes of making this calculation, the load connected to the receptacle outlet shall be considered to be 50 percent of the branch-circuit rating. The combined voltage drop of feeder and branch-circuit conductors shall not exceed 2.0 percent ≫*647.4(D)*≪.

Fire Pumps

Electric motor-driven fire pumps shall have a reliable power source ⟩⟩ *695.3* ⟨⟨ . See 695.3(A) for individual source requirements and 695.3(B) for multiple source requirements.

A fire pump can be supplied by a separate service ⟩⟩ *230.2(A)* and *695.3(A)(1)* ⟨⟨ .

If service or system voltage differs from the fire pump motor's utilization voltage, one or more transformers (protected by disconnecting means and overcurrent protective devices) can be installed between the system supply and the fire pump controller in accordance with 695.5(A) and (B), or with (C). Only transformers covered in (C) can supply loads unrelated to the fire pump system ⟩⟩ *695.5* ⟨⟨ .

Power circuits and wiring methods shall comply with the requirements in 695.6(A) through (J), and, as permitted in 230.90(A), Exception No. 4; 230.94, Exception No. 4; 240.13; 230.208; 240.4(A); and 430.31 ⟩⟩ *695.6* ⟨⟨ .

The location of electric motor-driven fire pump controllers and power transfer switches shall be within sight from and as close as practicable to the motors they control ⟩⟩ *695.12(A)* ⟨⟨ . Engine-driven fire pump controllers shall be located as close as practical to and within sight from the engines they control ⟩⟩ *695.12(B)* ⟨⟨ .

Fire pump controllers and power transfer switches, by location or protection, shall remain undamaged by water escaping from pumps or pump connections ⟩⟩ *695.12(E)* ⟨⟨ .

The disconnecting means shall be marked "Fire Pump Disconnecting Means." The minimum 1-in. (25-mm)-high letters shall be visible without opening enclosure doors or covers ⟩⟩ *695.4(B)(3)(c)* ⟨⟨ .

Fire pump supply conductors on the load side of the final disconnecting means and overcurrent device(s) permitted by 695.4(B), or conductors that connect directly to an on-site standby generator, shall be kept entirely independent of all other wiring. The conductors shall supply only loads that are directly associated with the fire pump system, and they shall be protected from potential damage by fire, structural failure, or operational accident. Where routed through a building, the conductors shall (1) be encased in at least 2 in. (50 mm) of concrete or (2) be protected by a fire-rated assembly listed to achieve a minimum fire rating of 2 hours and dedicated to the fire pump circuit(s) or (3) be a listed electrical circuit protective system with a minimum 2-hour fire rating ⟩⟩ *695.6(A)(2)* ⟨⟨ .

Automatic protection against overloads is not allowed on power circuits. Except for protection of transformer primaries provided in 695.5(C)(2), branch-circuit and feeder conductors shall be protected against short circuits only ⟩⟩ *695.6(C)* ⟨⟨ .

All engine controller/battery wiring shall be protected against physical damage, being installed in accordance with the controller and engine manufacturer's instructions ⟩⟩ *695.6(F)* ⟨⟨ .

A All wiring from the controllers to the pump motors shall be in RMC, IMC, EMT, LFMC, or LFNC Type LFNC-B, listed Type MC cable with an impervious covering, or Type MI cable. Electrical connections at motor terminal boxes shall be made with a listed means of connection. Twist-on, insulation-piercing-type, and soldered wire connectors shall not be permitted to be used for this purpose ⟩⟩ *695.6(D)* ⟨⟨ .

B Conductors supplying fire pump motor(s), pressure maintenance pumps, and associated accessory equipment shall have a minimum rating of 125% of the sum of the fire pump motor(s) and pressure maintenance motor(s) full-load current(s), plus 100% of the associated fire pump accessory equipment ⟩⟩ *695.6(B)(1)* ⟨⟨ . Conductors supplying only fire pump motor(s) shall have a minimum rating in accordance with 430.22 and shall comply with the voltage drop requirements in 695.7 ⟩⟩ *695.6(B)(2)* ⟨⟨ .

C Article 695 covers the installation of (1) electric power sources and interconnecting circuits and (2) switching and control equipment dedicated to fire pump drivers ⟩⟩ *695.1(A)* ⟨⟨ .

> ### N O T E
>
> Article 695 does not cover (1) the performance, maintenance, and acceptance testing of the fire pump system and the system components' internal wiring, (2) the installation of pressure maintenance (jockey or makeup) pumps, and (3) transfer equipment upstream of the fire pump transfer switch(es) ⟩⟩ *695.1(B)* ⟨⟨ .
>
> Taps used to supply fire pump equipment, if provided with service equipment and installed per service-entrance conductor requirements, can be connected to the supply side of the service disconnecting means ⟩⟩ *230.82(5)* ⟨⟨ .

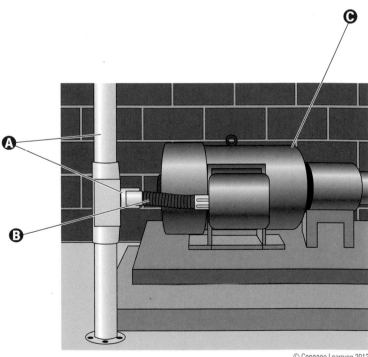

© Cengage Learning 2012

ELEVATORS, DUMBWAITERS, ESCALATORS, MOVING WALKS, PLATFORM LIFTS, AND STAIRWAY CHAIRLIFTS

General

Control system: Overall system governing starting, stopping, direction of motion, acceleration, speed, and retardation of the moving member »*620.2*«.

Motion controller: Electrical device(s) for that portion of the control system governing acceleration, speed, retardation, and stoppage of the moving member »*620.2*«.

Motor controller: The control system's operative units, that is, the starter device(s) and power conversion equipment that drive an electric motor, or the pumping unit used to power hydraulic control equipment »*620.2*«.

Operation controller: Electric device(s) for that portion of the control system that initiates starting, stopping, and direction of motion in response to an operating device's signal »*620.2*«.

Door operator controller and door motor branch circuits, as well as feeders to motor controllers, driving machine motors, machine brakes, and motor-generator sets shall have a circuit voltage of 1000 volts or less »*620.3(A)*«.

Operating device: Devices used to activate the operation controller, such as the car switch, push buttons, key, or toggle switch(es) »*620.2*«.

Conductors and optical fibers within hoistways, escalator and moving walk wellways, platform lifts, stairway chairlift runways, machinery spaces, control spaces, in/on cars, and machine/control rooms (exclusive of traveling cables connecting the car or counterweight and hoistway wiring) shall be installed in RMC, IMC, EMT, PVC, or wireways; or shall be Type MC, MI, or AC cable unless otherwise permitted in 620.21(A) through (C) »*620.21*«.

Ⓐ Working space shall be provided around controllers, disconnecting means, and other electrical equipment in accordance with 110.26(A). Where

maintenance and supervision conditions ensure that only qualified persons will examine, adjust, service, and maintain the equipment, 110.26(A) clearance requirements shall not be required where any of the conditions in 620.5(A) through (D) are met »*620.5*«.

Ⓑ Circuit voltage of heating/air-conditioning equipment branch circuits located on the elevator car shall not exceed 1000 volts »*620.3(C)*«.

Ⓒ Lighting circuits shall comply with Article 410 requirements »*620.3(B)*«.

Ⓓ The minimum conductor size for traveling cables that feed lighting circuits is 14 AWG copper. Provided the ampacity equals or surpasses that of 14 AWG copper, 20 AWG (or larger) copper conductors are permitted in parallel. For traveling cable circuits, other than lighting, 20 AWG copper is the minimum size »*620.12(A)*«.

Ⓔ The motor controller rating shall comply with 430.83. When the controller is marked as power limited, thereby limiting the available power to the motor, the rating of the controller can be less than the nominal rating of the elevator motor »*620.15*«.

Ⓕ Article 620 covers the installation of electric equipment and wiring relating to elevators, dumbwaiters, escalators, moving walks, platform lifts, and stairway chairlifts »*620.1*«.

Ⓖ All live parts of electrical apparatus in (and around) elevator and dumbwaiter hoistways, landings, and cars; escalator and moving walk wellways or landings; and machinery space runways of platform lifts and stairway chairlifts shall be enclosed to guard against accidental contact »*620.4*«.

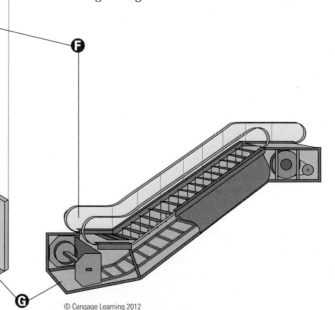

© Cengage Learning 2012

Elevator Wiring

Flexible cords and cables that are components of listed equipment used in circuits operating at 30 volts rms (or less) or 42 volts dc (or less) are permitted in lengths not exceeding 6 ft (1.8 m) provided the cords and cables are (1) supported and protected from physical damage and (2) of a jacketed and flame-retardant type »*620.21(A)(2)(c)* «.

Motor generators, machine motors, or pumping unit motors and valves that are located adjacent to, or underneath, control equipment and provided with extra-length terminal leads and are no longer than 6 ft (1.8 m) can be connected directly to controller terminal studs without regard to the carrying-capacity requirements of Articles 430 and 445. Auxiliary gutters are permitted in machine/control rooms between controllers, starters, and similar apparatus »*620.21(A)(3)(b)* «.

Existing or listed equipment conductors can be grouped together (taped or corded) without being installed in a raceway. Such cable groups should by location be protected from physical damage, being supported at intervals of 3 ft (900 mm) or less »*620.21(A)(3)(d)* «.

Ⓐ FMC, LFMC, or LFNC of ⅜ in. nominal trade size or larger and no longer than 6 ft (1.8 m) can be installed between control panels and machine motors, machine brakes, motor-generator sets, disconnecting means, and pumping unit motors and valves »*620.21(A)(3)(a)* «. LFNC, as defined in 356.2(2), can be installed in lengths exceeding 6 ft (1.8 m) »*620.21(A)(3)(a) Exception* «.

Ⓑ FMC, LFMC, or LFNC of ⅜ in. nominal trade size (or larger) and not exceeding 6 ft (1.8 m) in length, allowed for installation on cars, shall be securely fastened in an oil-free location »*620.21(A)(2)(a)* «. LFNC of ⅜ in. nominal trade size or larger, as defined by 356.2(2), is permitted in lengths greater than 6 ft (1.8 m) »*620.21(A)(2)(a) Exception* «.

Ⓒ Hard-service cords and junior hard-service cords conforming to Article 400 (Table 400.4) requirements are permitted as flexible connections between the car's fixed wiring and the car door/gate devices. Only hard-service cords are permitted as flexible connections for top-of-car operating devices or car-top work lights. Devices or luminaires shall be grounded via an equipment grounding conductor run with the circuit conductors. Cables having smaller conductors as well as other insulation (jacket) types and thicknesses can serve as flexible connections between the car's fixed wiring and the car door/gate devices, if specifically listed for this use »*620.21(A)(2)(b)* «.

Ⓓ Cords and cables of listed cord- and plug-connected equipment shall not be required to be installed in a raceway »*620.21 Exception* «.

Ⓔ Class 2 power-limited circuit cables installed in hoistways between risers, signal equipment, and operating devices shall be of a jacketed flame-retardant type, securely supported and protected from physical damage »*620.21(A)(1)(a)* «.

Ⓕ The following wiring methods shall be permitted on the counterweight assembly and on the car assembly in lengths not to exceed 6 ft (1.8 m): (1) flexible metal conduit; (2) liquidtight flexible metal conduit; (3) liquidtight flexible nonmetallic conduit; and (4) flexible cords and cables, or conductors grouped together and taped or corded, shall be permitted to be installed without a raceway. They shall be located to be protected from physical damage and shall be of a flame-retardant type and shall be part of the following:

a. Listed equipment

b. A driving machine, or

c. A driving machine brake »*620.21(A)(2)(d)* and *620.21(A)(4)* «.

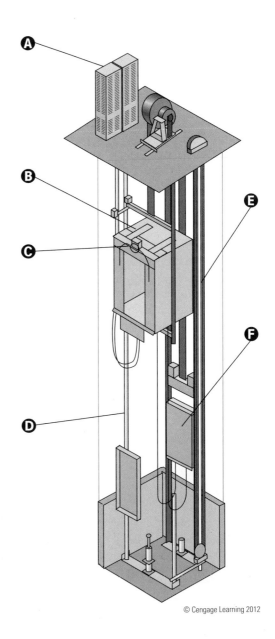

© Cengage Learning 2012

Escalator and Moving Walks Wiring

A FMC, LFMC, or LFNC are acceptable in escalator and moving walk wellways. FMC or liquidtight flexible conduit of ⅜ inch nominal trade size is permitted in lengths of 6 ft (1.8 m) or less ≫ *620.21(B)(1)* ≪. LFNC of ⅜ in. nominal trade size or larger, as defined in 356.2(2), can be installed in lengths greater than 6 ft (1.8 m) ≫ *620.21(B)(1) Exception* ≪.

B Class 2 power-limited circuit cables can be installed within escalators and moving walkways provided the cables are (1) supported and protected from physical damage and (2) of a jacketed and flame-retardant type ≫ *620.21(B)(2)* ≪.

> ### N O T E
> Platform lift and stairway chairlift raceways shall be installed according to 620.21(C) provisions. Escalators, moving walks, platform lifts, and stairway chairlifts shall comply with Article 250 ≫***620.84***≪.

C Hard-service cords conforming to Article 400 (Table 400.4) requirements are permitted as flexible connections on escalator and moving walk control panels and disconnecting means where the entire control panel and disconnecting means can be removed from machine spaces per 620.5 ≫ *620.21(B)(3)* ≪.

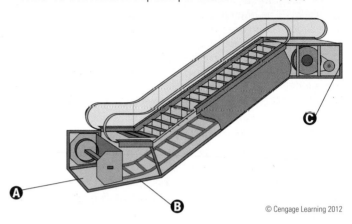

© Cengage Learning 2012

Conductor Installation

Auxiliary gutters are not subject to the length restrictions of 366.12(2), or to the number of conductors per 366.22 ≫ ***620.35*** ≪.

A Only electric wiring, raceways, and cables used in direct connection with the elevator or dumbwaiter are permitted inside the hoistway, machine rooms, control rooms, machinery spaces, and control spaces. Wiring for signals; for communication with the car; for lighting, heating, air conditioning, and ventilating the elevator car; for fire-detecting systems; for pit sump pumps; and for lighting and ventilating the hoistway are allowed ≫ *620.37(A)* ≪.

B Cables or raceway supports in a hoistway, escalator/moving walk wellway, or platform lift and stairway chairlift runway shall be securely fastened to the guide rail, escalator/moving walk truss, or the hoistway, wellway, or runway construction ≫ *620.34* ≪.

C Optical fiber cables and conductors for operating devices, operation and motion control, power, signaling, fire, alarm, lighting, heating, and air-conditioning circuits of 1000 volts (or less) can run in the same traveling cable (or raceway) system if (1) all conductors are insulated for the maximum voltage applied to any conductor within the cable (or raceway) system and (2) all live parts of the equipment are insulated from ground for the same maximum voltage. Such a traveling cable (or raceway) can also include shielded conductors or coaxial cable(s), if insulated for the maximum voltage applied to any conductor within the cable (or raceway) system. Conductors can be suitably shielded for telephone, audio, video, or higher frequency communications circuits ≫ *620.36* ≪.

D Bonding of elevator rails (car or counterweight) to a lightning protection system grounding-down conductor(s) is permitted. The lightning protection system shall be located outside the hoistway. Elevator rails or other hoistway equipment shall not be used as the grounding-down conductor for lightning protection systems ≫ *620.37(B)* ≪.

E Main feeders supplying elevator and dumbwaiter power shall be installed outside the hoistway unless (1) by special permission, elevator feeders are permitted within an existing hoistway containing no spliced conductors or (2) feeders are permitted inside the hoistway for elevators with driving machine motors located in the hoistway, on the car, or on the counterweight ≫ *620.37(C)* ≪.

F The cross-sectional area's sum of the individual conductors in a wireway shall be no more than 50% of the wireway's interior cross-sectional area. Vertically run wireways shall be securely supported at intervals of 15 ft (4.5 m) or less and shall have a maximum of one joint between supports. Adjoining wireway sections shall be securely fastened together to form a rigid joint ≫ *620.32* ≪.

> ### CAUTION
> Electrical equipment and wiring used for elevators, dumbwaiters, escalators, moving walks, platform lifts, and stairway lifts in garages shall comply with Article 511 requirements ≫*620.38*≪.

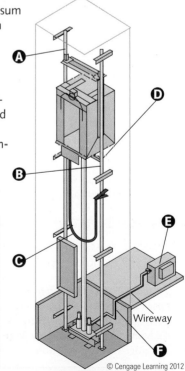

Wireway

© Cengage Learning 2012

Branch Circuits and Traveling Cables

Ⓐ Each machine room or control room and machinery space or control space requires at least one 125-volt, single-phase, 15- or 20-ampere duplex receptacle ≫ 620.23(C) ≪.

Ⓑ A separate branch circuit shall supply each elevator car's lights, receptacle(s), auxiliary lighting power source, and ventilation. The branch circuit's overcurrent protective device shall be located in the elevator machine room or control room/machinery space or control space. Required lighting shall not be connected to the load side of a ground-fault circuit interrupter ≫ 620.22(A) ≪. A separate branch circuit is required for each elevator car's air-conditioning/heating unit. Locate this branch circuit's overcurrent protective device in the elevator machine room or control room/machinery space or control space ≫ 620.22(B) ≪.

Ⓒ Traveling cables shall be suspended at both the car and hoistway ends (or counterweight end if applicable), thereby reducing the strain on individual copper conductors to a minimum. Support traveling cables by any one of the following means:

1. Steel supporting member(s)

2. Looping the cables around supports for unsupported lengths less than 100 ft (30 m)

3. Suspending from the supports by a means that automatically tightens around the cable with increased tension, for unsupported lengths up to 200 ft (60 m) ≫ 620.41 ≪

Ⓓ Careful location of traveling cable supports can reduce to a minimum the possibility of damage due to cables contacting hoistway construction/equipment. Provide suitable guards where necessary to protect the cables against damage ≫ 620.43 ≪.

Ⓔ Each hoistway pit shall have at least one 125-volt, single-phase, 15- or 20-ampere duplex receptacle ≫ 620.24(C) ≪.

Ⓕ A separate branch circuit shall supply the machine room or control room/machinery space or control space lighting and receptacle(s). Required lighting shall not be connected to a GFCI's load side ≫ 620.23(A) ≪. Locate the lighting switch at the machine room or control room/machinery space or control space point of entry ≫ 620.23(B) ≪.

Ⓖ Metal raceways and cables of Types MC, MI, or AC attached to an elevator car shall be bonded to the metal parts of the car that are bonded to the equipment grounding conductor ≫ 620.81 ≪.

Ⓗ Traveling cables that are suitably supported and protected from physical damage shall be permitted to be run without the use of a raceway in either or both of the following:

(a) When used inside the hoistway, on the elevator car, hoistway wall, counterweight, or controllers and machinery that are located inside the hoistway, provided the cables are in the original sheath.

(b) From inside the hoistway, to elevator controller enclosures and to elevator car and machine room, control room, machinery space, and control space connections that are located outside the hoistway for a distance not exceeding 6 ft (1.8 m) in length as measured from the first point of support on the elevator car or hoistway wall, or counterweight where applicable, provided the conductors are grouped together and taped or corded, or in the original sheath. These traveling cables shall be permitted to be continued to this equipment ≫ 620.44 ≪.

Ⓘ Hoistway pit lighting and receptacle(s) shall be supplied by a separate branch circuit. In no case should required lighting be connected to the load side of a GFCI ≫ 620.24(A) ≪. The lighting switch's location shall be readily accessible from the pit access door ≫ 620.24(B) ≪.

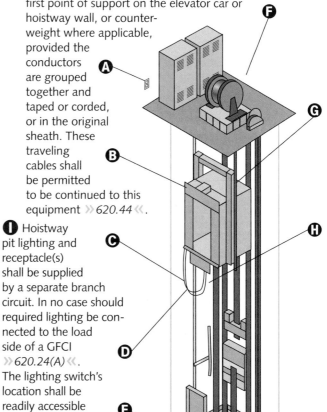

© Cengage Learning 2012

CAUTION

Each 125-volt, single-phase, 15- and 20-ampere receptacle installed in pits or hoistways, on elevator car tops, and in escalator and moving walk wellways shall be of the GFCI type. All 125-volt, single-phase, 15- and 20-ampere receptacles installed in machine rooms and machinery spaces shall have GFCI protection for personnel. A single receptacle supplying a permanently installed sump pump does not require GFCI protection ≫ 620.85 ≪.

NOTE

For electric elevators, all motor frames, elevator machines, controllers, and metal electrical-equipment enclosures in (or on) the car or in the hoistway shall be bonded per Article 250, Parts V and VII ≫ 620.82 ≪.

Protection and Control

Each unit shall have a single means for disconnecting all ungrounded main power supply conductors that is designed so that no pole operates independently. Where multiple driving machines are connected to a single elevator, escalator, moving walk, or pumping unit, a single means shall be provided to disconnect the motor(s) and control valve operating magnets 》*620.51*《.

The location of the disconnecting means shall be readily accessible to qualified persons 》*620.51(C)*《. Depending on the application, the location must comply with 1, 2, 3, or 4 in 620.51(C).

Where there are multiple driving machines in a machine room, the disconnecting means shall be numbered to correspond with the identifying number of the driving machine being controlled. The disconnecting means shall have a sign identifying the location of the supply side overcurrent protective device 》*620.51(D)*《.

Each elevator car shall have a single means for disconnecting all ungrounded car light, receptacle(s), and ventilation power-supply conductors. The disconnecting means shall be an enclosed externally operable fused motor-circuit switch or circuit breaker that is lockable open in accordance with 110.25 and shall be located in that elevator car's machine/control room. Where there is no machine room or control room, the disconnecting means shall be located in a machinery space or control space outside the hoistway that is readily accessible to only qualified persons. Each disconnecting means shall be numbered to correspond with the identifying number of the car whose light source it controls. The disconnecting means shall have a sign identifying the location of the supply side overcurrent protective device 》*620.53*《.

Operating devices and control/signaling circuits shall be protected against overcurrent according to 725.43 and 725.45. Class 2 power-limited circuits shall be protected against overcurrent in accordance with the requirements of Chapter 9, Notes to Tables 11(A) and 11(B) 》*620.61(A)*《.

Duty on elevator and dumbwaiter driving machine motors and driving motors of motor generators (used with generator field control) is rated as *intermittent*. Such motors shall be permitted to be protected against overload per 430.33 》*620.61(B)(1)*《.

Duty on escalator and moving walk driving machine motors is rated as *continuous*. Protect such motors against overload in accordance with 430.32 》*620.61(B)(2)*《. Escalator and moving walk driving machine motors, and driving motors of motor-generator sets, shall be protected against running overload as provided in Table 430.37 》*620.61(B)(3)*《.

Platform lift and stairway chairlift driving machine motor duty is rated as *intermittent*. Such motors can be protected against overload according to 430.33 》*620.61(B)(4)*《.

A Elevator, dumbwaiter, escalator, and moving walk driving machines; motor-generator sets; motor controllers; and disconnecting means shall be installed in a room or space dedicated for that purpose, unless otherwise permitted in 620.71(A) or (B). The room or enclosure shall be secured against unauthorized access 》*620.71*《.

> **NOTE**
>
> Elevator emergency and standby power systems must comply with 620.91.

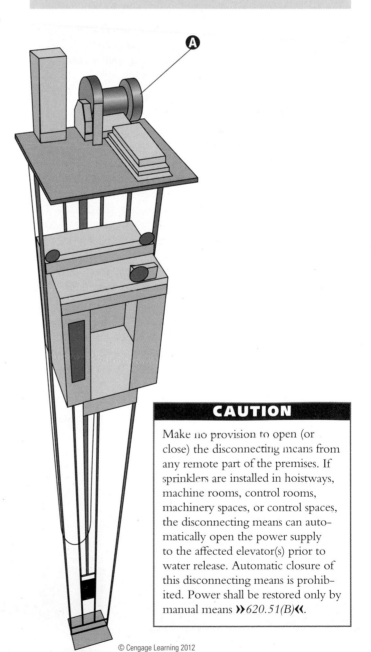

A

> **CAUTION**
>
> Make no provision to open (or close) the disconnecting means from any remote part of the premises. If sprinklers are installed in hoistways, machine rooms, control rooms, machinery spaces, or control spaces, the disconnecting means can automatically open the power supply to the affected elevator(s) prior to water release. Automatic closure of this disconnecting means is prohibited. Power shall be restored only by manual means 》*620.51(B)*《.

© Cengage Learning 2012

SWIMMING POOLS, FOUNTAINS, AND SIMILAR INSTALLATIONS

General

All electrical equipment installed in the water, walls, or decks of pools, fountains, and similar applications shall comply with the provisions of Article 680 》》*680.4* 《《.

Except as modified by this article, wiring and equipment in or adjacent to pools and fountains shall comply with other applicable requirements of the *Code* including provisions identified in Table 680.3 》》*680.3* 《《.

Provisions for storable pools, storable spas, and storable hot tubs are located in Article 680, Part III (680.30 through 34).

The provisions of Article 680, Part I and Part VI shall apply to therapeutic pools and tubs in health care facilities, gymnasiums, athletic training rooms, and similar areas 》》*680.60* 《《.

Hydromassage bathtub requirements are located in Article 680, Part VII.

A Article 680 provisions apply to the electrical wiring construction (and installation) and for equipment in or adjacent to all swimming, wading, therapeutic, and decorative pools, fountains, hot tubs, spas, and hydromassage bathtubs, whether permanently installed or storable, as well as to similar metallic auxiliary equipment (such as pumps, filters, etc.). The term *body of water* used throughout Part I applies to all bodies of water covered in this scope unless otherwise amended 》》*680.1* 《《.

B **Fountain:** Includes fountains, ornamental pools, display pools, and reflection pools, but drinking fountains are not included 》》*680.2* 《《.

C Part I and Part V provisions apply to all permanently installed fountains as defined in 680.2. Fountains that have water common to a pool shall additionally comply with the requirements in Part II of Article 680. Part V does not cover self-contained, portable fountains. Portable fountains shall comply with Parts II and III of Article 422 》》*680.50* 《《.

D **Permanently installed decorative fountains and reflection pools:** Those that are constructed in or on the ground, or in a building so that the fountain cannot be readily disassembled, with or without electrical circuits of any nature. These units are constructed primarily for their aesthetic value and are not intended for swimming or wading 》》*680.2* 《《.

E **Spa or hot tub:** A hydromassage pool or tub for recreational or therapeutic use, not located in health care facilities, designed for immersion of users, and usually having a filter, heater, and motor-driven blower. It may be installed either indoors or out and on or in the ground or supporting structure. Generally, a spa or hot tub is not designed to be drained after each use 》》*680.2* 《《.

F Spa and hot tub requirements can be found in Article 680, Part IV (680.40 through 44).

G Equipotential bonding provisions for permanently installed pools are located in 680.26.

H **Pool:** Manufactured or field-constructed equipment designed to contain water on a permanent or semipermanent basis, and used for swimming, wading, immersion, or therapeutic purposes 》》*680.2* 《《.

I Permanently installed pool underwater luminaire requirements are located in 680.23.

J **Permanently installed swimming, wading, immersion, and therapeutic pools:** Those that are constructed totally or partially in the ground, all others with depth capacities greater than 42 in. (1.0 m), and all pools installed inside of a building, regardless of water depth, whether or not served by electrical circuits of any nature 》》*680.2* 《《.

K Permanently installed pools must comply with Article 680, Part II.

> **NOTE**
>
> Swimming pool overhead conductor clearance requirements are located in 680.8 and Table 680.8(A).

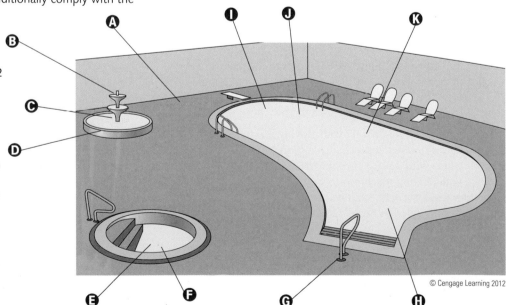

© Cengage Learning 2012

Definitions

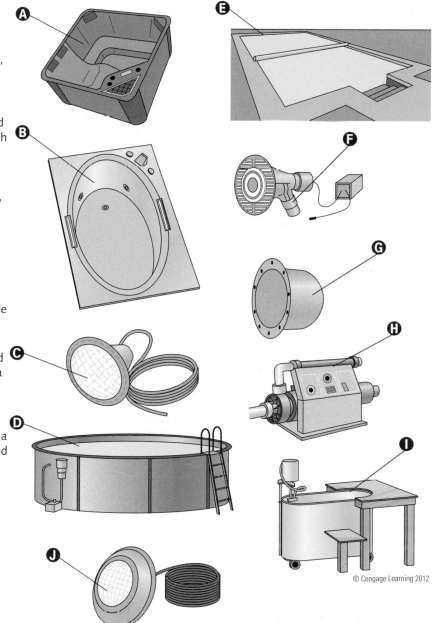

A **Self-contained spa or hot tub:** Factory-fabricated unit consisting of a spa or hot tub vessel having integrated water-circulating, heating, and control equipment. Equipment may include pumps, air blowers, heaters, lights, controls, sanitizer generators, etc. ≫ *680.2* ≪.

B **Hydromassage bathtub:** A permanently installed bathtub with recirculation piping, pump, and associated equipment. It is designed to accept, circulate, and discharge water at each use ≫ *680.2* ≪.

C **Wet-niche luminaire:** A luminaire intended for installation in a pool or fountain structure's forming shell, where completely surrounded by water ≫ *680.2* ≪.

D **Storable swimming, wading, or immersion pools; or storable/portable spas and hot tubs:** Those constructed on or above the ground having a maximum water depth capacity of 42 in. (1.0 m), or a pool, spa, or hot tub with nonmetallic, molded polymeric walls (or inflatable fabric walls) regardless of dimension ≫ *680.2* ≪.

E **Pool cover, electrically operated:** Motor-driven equipment designed to cover and uncover the pool's water surface by means of a flexible sheet or rigid frame ≫ *680.2* ≪.

F **Cord- and plug-connected lighting assembly:** A lighting assembly consisting of a luminaire intended for installation in the wall of a spa, hot tub, or storable pool, having a cord- and plug-connected transformer ≫ *680.2* ≪.

G **Forming shell:** A support structure designed for a wet-niche luminaire assembly and intended for pool or fountain structure mounting ≫ *680.2* ≪.

H **Packaged spa or hot tub equipment assembly:** A factory-fabricated unit consisting of water-circulating, heating, and control equipment mounted on a common base and that operates as a spa or hot tub. Equipment may include pumps, air blowers, heaters, lights, controls, sanitizer generators, etc. ≫ *680.2* ≪.

I **Self-contained therapeutic tubs or hydrotherapeutic tanks:** A factory-fabricated unit consisting of a therapeutic tub or hydrotherapeutic tank with integrated water-circulating, heating, and control equipment. Equipment may include pumps, air blowers, heaters, lights, controls, sanitizer generators, etc. ≫ *680.2* ≪.

J **No-niche luminaire:** A luminaire intended for above or below water installation without a niche ≫ *680.2* ≪.
Note: Not to be mistaken for a dry-niche luminaire.

Dry-niche luminaire: A luminaire intended for installation in the floor or wall of a pool, spa, or fountain in a niche that is sealed against water entry ≫ *680.2* ≪.

© Cengage Learning 2012

Conductive Pool Shells and Perimeter Surfaces

The equipotential bonding required by 680.26 shall be installed to reduce voltage gradients in the pool area ≫ *680.26(A)* ≪ .

Ⓐ The parts specified in 680.26(B)(1) through (B)(7) shall be bonded together using solid copper conductors, insulated covered, or bare, not smaller than 8 AWG or with rigid metal conduit of brass or other identified corrosion-resistant metal. Connections to bonded parts shall be made in accordance with 250.8. An 8 AWG or larger solid copper bonding conductor provided to reduce voltage gradients in the pool area shall not be required to be extended or attached to remote panelboards, service equipment, or electrodes ≫ *680.26(B)* ≪ .

Ⓑ Bonding to conductive pool shells shall be provided as specified in 680.26(B)(1)(a) or 680.26(B)(1)(b). Poured concrete, pneumatically applied or sprayed concrete, and concrete block with painted or plastered coatings shall all be considered conductive materials due to water permeability and porosity. Vinyl liners and fiberglass composite shells shall be considered to be nonconductive materials ≫ *680.26(B)(1)* ≪ .

Ⓒ Unencapsulated structural reinforcing steel shall be bonded together by steel tie wires or the equivalent. Where structural reinforcing steel is encapsulated in a nonconductive compound, a copper conductor grid shall be installed in accordance with 680.26(B)(1)(b) ≫ *680.26(B)(1)(a)* ≪ . A copper conductor grid shall be provided and shall: (1) be constructed of minimum 8 AWG bare solid copper conductors bonded to each other at all points of crossing (the bonding shall be in accordance with 250.8 or other approved means), (2) conform to the contour of the pool, (3) be arranged in a 12-in. × 12-in. (300-mm × 300-mm) network of conductors in a uniformly spaced perpendicular grid pattern with a tolerance of 4 in. (100 mm), and (4) be secured within or under the pool no more than 6 in. (150 mm) from the outer contour of the pool shell ≫ *680.26(B)(1)(b)* ≪ .

Ⓓ The perimeter surface shall extend for 3 ft (1 m) horizontally beyond the inside walls of the pool and shall include unpaved surfaces as well as poured concrete surfaces and other types of paving. Perimeter surfaces less than 3 ft (1 m) separated by a permanent wall or building 5 ft (1.5 m) in height or more shall require equipotential bonding on the pool side of the permanent wall or building. Bonding to perimeter surfaces shall be provided as specified in 680.26(B)(2)(a) or 680.26(B)(2)(b) and shall be attached to the pool reinforcing steel or copper conductor grid at a minimum of four (4) points uniformly spaced around the perimeter of the pool. For nonconductive pool shells, bonding at four points shall not be required ≫ *680.26(B)(2)* ≪ .

(a) Structural reinforcing steel shall be bonded in accordance with 680.26(B)(1)(a).

(b) Where structural reinforcing steel is not available or is encapsulated in a nonconductive compound, a copper conductor(s) shall be utilized where the following requirements are met:

(1) At least one minimum 8 AWG bare solid copper conductor shall be provided.

(2) The conductors shall follow the contour of the perimeter surface.

(3) Only listed splices shall be permitted.

(4) The required conductor shall be 18 to 24 in. (450 to 600 mm) from the inside walls of the pool.

(5) The required conductor shall be secured within or under the perimeter surface 4 to 6 in. (100 to 150 mm) below the subgrade ≫ *680.26(B)(2)(a) and (b)* ≪ .

> **NOTE**
>
> All metallic parts of the pool structure, including reinforcing metal not addressed in 680.26(1)(a), shall be bonded. Where reinforcing steel is encapsulated with a nonconductive compound, the reinforcing steel shall not be required to be bonded ≫ *680.26(B)(3)* ≪ .

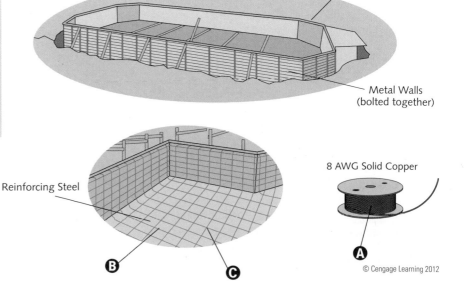

Metal Walls (bolted together)

8 AWG Solid Copper

Reinforcing Steel

© Cengage Learning 2012

Receptacles

Ⓐ Receptacles that provide power for water-pump motors or for other loads directly related to the circulation and sanitation system shall be located at least 10 ft (3.0 m) from the inside walls of the pool or not less than 6 ft (1.83 m) from the inside walls of the pool if they meet all of the following conditions: (1) consist of single receptacles; (2) are of the grounding type; and (3) have GFCI protection ≫ *680.22(A)(2)* ≪.

Ⓑ Where a permanently installed pool is installed, no fewer than one 125-volt 15- or 20-ampere receptacle on a general-purpose branch circuit shall be located a minimum of 6 ft (1.83 m), but no more than 20 ft (6.0 m), from the pool's inside wall. This receptacle shall not be more than 6 ft, 6 in. (2.0 m) above the floor, platform, or grade level serving the pool ≫ *680.22(A)(1)* ≪.

Ⓒ Where receptacles providing power for water-pump motors or for other loads directly related to the circulation and sanitation system do not meet all the conditions in 680.22(A)(2)(1) through (3), they must be located at least 10 ft (3.0 m) from the inside walls of the pool.

CAUTION

Fifteen- and 20-ampere, 125- and 250-volt receptacles installed in a wet location shall have an enclosure that is weatherproof whether or not the attachment plug cap is inserted. An outlet box hood installed for this purpose shall be listed, and shall be identified as "extra-duty." All 15- and 20-ampere, 125- and 250-volt nonlocking-type receptacles shall be listed weather-resistant type ≫ *406.9(B)(1)* ≪.

NOTE

All 15- and 20-ampere, single-phase, 125-volt receptacles located within 20 ft (6.0 m) of the pool's inside walls shall be protected by a GFCI ≫ *680.22(A)(4)* ≪. This distance is determined by measuring the shortest path that the appliance supply cord (connected to the receptacle) would follow without piercing a floor, wall, ceiling, hinged or sliding panel doorway, window opening, or other effective permanent barrier ≫ *680.22(A)(5)* ≪.

NOTE

Other receptacles shall be not less than 6 ft (1.83 m) from the inside walls of a pool ≫ *680.22(A)(3)* ≪.

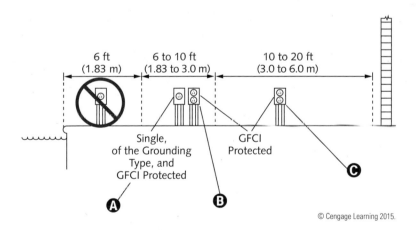

6 ft (1.83 m)

6 to 10 ft (1.83 to 3.0 m)

10 to 20 ft (3.0 to 6.0 m)

Single, of the Grounding Type, and GFCI Protected **Ⓐ**

GFCI Protected

Ⓑ

Ⓒ

© Cengage Learning 2015.

Equipotential Bonding

The parts specified in 680.26(B)(1) through (B)(7) shall be bonded together using solid copper conductors, insulated, covered, or bare, not smaller than 8 AWG or with rigid metal conduit of brass or other identified corrosion-resistant metal. Connections to bonded parts shall be made in accordance with 250.8. An 8 AWG or larger solid copper bonding conductor provided to reduce voltage gradients in the pool area shall not be required to be extended or attached to remote panelboards, service equipment, or electrodes ≫ *680.26(B)* ≪.

All metallic parts of the pool structure, including reinforcing metal not addressed in 680.26(1)(a), shall be bonded ≫ *680.26(B)(3)* ≪.

Equipotential Bonding *(continued)*

A All fixed metal parts shall be bonded including, but not limited to, metal-sheathed cables and raceways, metal piping, metal awnings, metal fences, and metal door and window frames *»680.26(B)(7)«*. Fixed metal parts that are not required to be bonded include: (1) those separated from the pool by a permanent barrier that prevents contact by a person, (2) those greater than 5 ft (1.5 m) horizontally from the inside walls of the pool, and (3) those greater than 12 ft (3.7 m) measured vertically above the pool's maximum water level, or as measured vertically above any observation stands, towers, or platforms, or any diving structures *»680.26(B)(7) Exception No. 1, 2 and 3«*.

B Bond all metal forming shells and mounting brackets of no-niche luminaires, unless a listed low-voltage lighting system is used that does not require bonding *»680.26(B)(4)«*.

C Bond any diving structure, observation stands, towers, or platforms unless one of the exceptions in 680.26(B)(7) is met *»680.26(B)(7)«*.

D Bond all metal fittings within or attached to the pool structure *»680.26(B)(5)«*.

E Isolated parts not more than 4 in. (100 mm) in any dimension that do not penetrate the pool's structure more than 1 in. (25 mm) do not require bonding *»680.26(B)(5)«*.

F Bonding to conductive pool shells shall be provided as specified in 680.26(B)(1)(a) or 680.26(B)(1)(b) *»680.26(B)(1)«*. (See section in this text titled "Conductive Pool Shells and Perimeter Surfaces.")

G Bond all metal fittings within or attached to the pool structure *»680.26(B)(5)«*. Note: Each ladder/handrail anchor must be bonded unless not over 4 in. (100 mm) in any dimension.

H Bond all metal parts of equipment associated with pool covers, including electric motors *»680.26(B)(6)«*.

I Bond metal parts of electrical equipment associated with the pool water circulating system, including pump motors. (Metal parts of listed equipment incorporating an approved system of double insulation and providing a means for

grounding internal inaccessible, non-current-carrying metal parts shall not be bonded) *»680.26(B)(6)«*.
Note: Pool equipment must be bonded to the equipotential bonding grid regardless of the intervening distance or location to the pool.

J Where reinforcing steel is encapsulated with a non-conductive compound, the reinforcing steel shall not be required to be bonded *»680.26(B)(3)«*.

K Unencapsulated structural reinforcing steel shall be bonded together by steel tie wires or the equivalent *»680.26(B)(1)(a)«*.

L Bonding to perimeter surfaces shall be provided as specified in 680.26(B)(2)(a) or 680.26(B)(2)(b) and shall be attached to the pool reinforcing steel or copper conductor grid at a minimum of four (4) points uniformly spaced around the perimeter of the pool. The perimeter surface shall extend for 3 ft (1 m) horizontally beyond the inside walls of the pool and shall include unpaved surfaces as well as poured concrete surfaces and other types of paving. Perimeter surfaces less than 3 ft (1 m) separated by a permanent wall or building 5 ft (1.5 m) in height or more shall require equipotential bonding on the pool side of the permanent wall or building *»680.26(B)(2)«*.

M For pool water heaters rated at more than 50 amperes and having specific instructions regarding bonding and grounding, only those parts designated to be bonded shall be bonded and only those parts designated to be grounded shall be grounded *»680.26(B)(6)(b)«*.

WARNING

Where none of the bonded parts is in direct connection with the pool water, the pool water shall be in direct contact with an approved corrosion-resistant conductive surface that exposes not less than 9 in.² (5800 mm²) of surface area to the pool water at all times. The conductive surface shall be located where it is not exposed to physical damage or dislodgement during usual pool activities, and it shall be bonded in accordance with 680.26(B) *»680.26(C)«*.

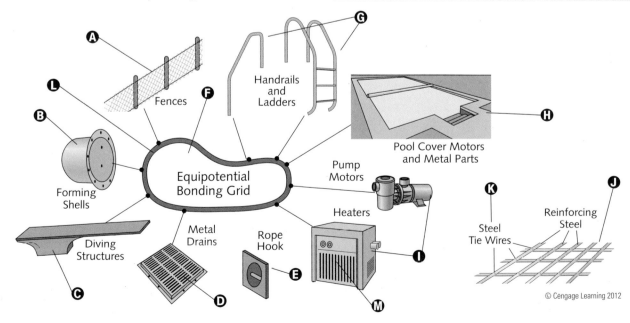

Fences

Handrails and Ladders

Pool Cover Motors and Metal Parts

Pump Motors

Heaters

Forming Shells

Equipotential Bonding Grid

Diving Structures

Metal Drains

Rope Hook

Steel Tie Wires

Reinforcing Steel

© Cengage Learning 2012

Luminaires, Lighting Outlets, and Ceiling-Suspended (Paddle) Fans

For installations in indoor pool areas, the clearances shall be the same as for outdoor areas unless modified as provided in this paragraph. If the branch circuit supplying the equipment is protected by a GFCI, the following equipment shall be permitted at a height not less than 7 ft 6 in. (2.3 m) above the maximum pool water level: (1) totally enclosed luminaires; and (2) ceiling-suspended (paddle) fans identified for use beneath ceiling structures such as provided on porches or patios ≫ *680.22(B)(2)* ≪.

Cord- and plug-connected luminaires shall meet the same 680.7 specifications, where installed within 16 ft (4.9 m) of any point along the water's surface, measured radially ≫ *680.22(B)(5)* ≪.

The provisions of 680.27(C) shall apply to all pool deck areas, including a covered pool, where electrically operated comfort heating units are installed within 20 ft (6.0 m) of the inside wall of the pool.

(1) **Unit heaters:** Unit heaters shall be rigidly mounted to the structure and shall be of the totally enclosed or guarded type. Unit heaters shall not be mounted over the pool or within the area extending 5 ft (1.5 m) horizontally from the inside walls of a pool.

(2) **Permanently wired radiant heaters:** Radiant electric heaters shall be suitably guarded and securely fastened to their mounting device(s). Heaters shall not be installed over a pool or within the area extending 5 ft (1.5 m) horizontally from the inside walls of the pool and shall be mounted at least 12 ft (3.7 m) vertically above the pool deck unless otherwise approved.

(3) **Radiant heating cables not permitted:** Radiant heating cables embedded in or below the deck shall not be permitted ≫ *680.27(C)* ≪.

Ⓐ For new installations in outdoor pool areas, luminaires, lighting outlets, and ceiling-suspended (paddle) fans shall not be installed over the pool, or over the area extending 5 ft (1.5 m) horizontally from the pool's inside walls, unless located not less than 12 ft (3.7 m) above the maximum water level ≫ *680.22(B)(1)* ≪.

Ⓑ Existing luminaires and lighting outlets located less than 5 ft (1.5 m), measured horizontally from the pool's inside walls, shall be (1) at least 5 ft (1.5 m) above the maximum water level surface; (2) rigidly attached to the existing structure; and (3) protected by a GFCI ≫ *680.22(B)(3)* ≪.

Ⓒ Luminaires and lighting outlets, and ceiling-suspended (paddle) fans installed in the area extending between 5 ft (1.5 m) and 10 ft (3.0 m) horizontally from a pool's inside walls shall be protected by a GFCI, unless 5 ft (1.5 m) above the maximum water level and rigidly attached to the pool's adjacent (or enclosing) structure ≫ *680.22(B)(4)* ≪.

Ⓓ Listed low-voltage luminaires not requiring grounding, not exceeding the low-voltage contact limit, and supplied by listed transformers or power supplies that comply with 680.23(A)(2) shall be permitted to be located less than 5 ft (1.5 m) from the inside walls of the pool ≫ *680.22(B)(6)* ≪.

> **NOTE**
>
> Locate all switching devices on the property at least 5 ft (1.5 m) horizontally from a pool's inside walls, unless separated from the pool by a solid fence, wall, or other permanent barrier. A switch listed as being acceptable for use within 5 ft (1.5 m) is also permitted ≫ *680.22 (C)* ≪.

> **CAUTION**
>
> Other outlets shall be not less than 10 ft (3.0 m) from the inside walls of the pool. Measurements shall be determined in accordance with 680.22(A)(5) ≫ *680.22(D)* ≪. Other outlets may include, but are not limited to, remote-control, signaling, fire alarm, and communications circuits ≫ *680.22(D) Informational Note* ≪.

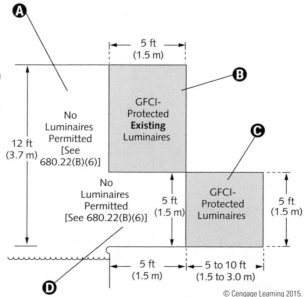

No Luminaires Permitted [See 680.22(B)(6)]

12 ft (3.7 m)

No Luminaires Permitted [See 680.22(B)(6)]

5 ft (1.5 m)

GFCI-Protected **Existing** Luminaires

GFCI-Protected Luminaires

5 ft (1.5 m)

5 ft (1.5 m)

5 ft (1.5 m)

5 to 10 ft (1.5 to 3.0 m)

© Cengage Learning 2015.

Underwater Luminaires (Permanently Installed Pools)

Requirements for luminaires installed below the pool's normal water level are covered in 680.23(A) through (F).

Wet-niche luminaire requirements are covered in 680.23(B)(1) through (6).

The equipment grounding conductor terminals of a junction box or transformer/other enclosure in the supply circuit to a wet-niche or no-niche luminaire and the field-wiring chamber of a dry-niche luminaire shall be connected to the panelboard's equipment grounding terminal. This terminal shall be directly connected to the panelboard enclosure ›› *680.24 (F)* ‹‹.

Ⓐ Luminaires mounted in walls shall be installed so that the top of the luminaire lens is at least 18 in. (450 mm) below the pool's normal water level, unless the luminaire is listed and identified for use at a depth of not less than 4 in. (100 mm) below the pool's normal water level ›› *680.23(A)(5)* ‹‹.

Ⓑ The 8 AWG bonding conductor termination in the forming shell shall be covered or encapsulated in a listed potting compound to protect it from the possibly deteriorating effect of pool water ›› *680.23(B)(2)(b)* ‹‹.

Ⓒ The luminaire shall be bonded and secured to the forming shell by a positive locking device that ensures a low-resistance contact, and which can only be removed from the forming shell by use of a tool. Luminaires listed for the application and having no non-current-carrying metal parts do not require bonding ›› *680.23(B)(5)* ‹‹.

Ⓓ Other than listed low-voltage luminaires not requiring grounding, all through-wall lighting assemblies, wet-niche, dry-niche, or no-niche luminaires shall be connected to an insulated 12 AWG or larger equipment grounding conductor sized per Table 250.122 ›› *680.23(F)(2)* ‹‹.

> ### NOTE
>
> Conductors on the load side of a GFCI or transformer, used to comply with 680.23(A)(8) provisions, shall not occupy raceways, boxes, or enclosures containing other conductors, unless the other conductors are either protected by GFCIs or are grounding conductors. Feed-through type GFCI supply conductors can occupy the same enclosure. GFCIs are permitted in a panelboard containing circuits protected by means other than GFCIs ›› *680.23(F)(3)* ‹‹.

> ### CAUTION
>
> All wet-niche luminaires shall be removable from the water for inspection, relamping, or other maintenance. The forming shell location and length of cord in the forming shell shall permit personnel to place the removed luminaire on the deck or other dry location for such maintenance. The luminaire maintenance location shall be accessible without entering or going in the pool water ›› *680.23(B)(6)* ‹‹.

Ⓔ All metal forming shells shall be bonded to an equipotential bonding grid with a solid copper conductor (insulated, covered, or bare) no smaller than 8 AWG ›› *680.26(B)(4)* ‹‹.

Ⓕ A junction box connected to a conduit extending directly to a forming shell or mounting bracket of a no-niche luminaire must comply with 680.24(A)(1) and (2).

Ⓖ The equipment grounding conductor shall be (1) an insulated copper conductor and (2) installed along with the circuit conductors ›› *680.23(F)(2)* ‹‹.

Ⓗ A GFCI shall be installed in the branch circuit supplying luminaires operating at more than the low voltage contact limit such that there is no shock hazard during relamping. The installation of the GFCI shall be such that there is no shock hazard with any likely fault-condition combination that involves a person in a conductive path from any ungrounded part of the branch circuit or the luminaire to ground ›› *680.23(A)(3)* ‹‹. Low voltage contact limit is defined as a voltage not exceeding the following values: (1) 15 volts (RMS) for sinusoidal ac, (2) 21.2 volts peak for nonsinusoidal ac, (3) 30 volts for continuous dc (4) 12.4 volts peak for dc that is interrupted at a rate of 10 to 200 Hz ›› *680.2* ‹‹.

Ⓘ Conduit shall extend from the forming shell to a junction box (or other enclosure) conforming to the requirements as provided in 680.24. Conduit must be RMC, IMC, LFNC, or PVC. Metal conduit shall be approved and shall be made of brass or other approved corrosion-resistant metal ›› *680.23(B)(2) and (B)(2)(a)* ‹‹.

Ⓙ When using nonmetallic conduit, an 8 AWG insulated copper bonding jumper (solid or stranded) shall be installed in the conduit with provisions for terminating in the forming shell, junction box/transformer enclosure, or GFCI enclosure, unless using a listed low-voltage lighting system that does not require grounding ›› *680.23(B)(2)(b)* ‹‹.

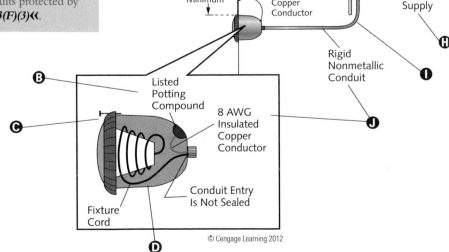

18 in. (450 mm) Minimum

8 AWG Solid Copper Conductor

To GFCI-Protected Supply

Rigid Nonmetallic Conduit

Listed Potting Compound

8 AWG Insulated Copper Conductor

Conduit Entry Is Not Sealed

Fixture Cord

© Cengage Learning 2012

Junction Boxes for Underwater Lighting

A junction box connected to a conduit that extends directly to a forming shell or mounting bracket of a no-niche luminaire shall meet the requirements of 680.24(A) through (F).

An enclosure for a transformer, GFCI, or a similar device connected to a conduit extending directly to a forming shell or mounting bracket of a no-niche luminaire must comply with 680.24(B) requirements.

Junction boxes, transformer and power-supply enclosures, and GFCI enclosures connected to a conduit extending directly to a forming shell or mounting bracket of a no-niche luminaire shall have a number of grounding terminals that exceeds the number of conduit entries by at least one »*680.24(D)*«.

Strain relief shall be provided for the termination of an underwater luminaire's flexible cord within a junction box, transformer enclosure, GFCI, or other enclosure »*680.24(E)*«.

A Locate the junction box at least 4 ft (1.2 m) from the pool's inside wall, unless separated from the pool by a solid fence, wall, or other permanent barrier »*680.24(A)(2)(b)*«.

B A junction box connected to a conduit extending directly to a forming shell or mounting bracket of a no-niche luminaire must comply with 680.24(A)(1)(1) through (3).

C The junction box shall be listed as a swimming pool junction box »*680.24(A)(1)*«.

D The junction box shall be comprised of copper, brass, suitable plastic, or other approved corrosion-resistant material »*680.24(A)(1)(2)*«.

E The junction box shall have electrical continuity (between every connected metal conduit and the grounding terminals) by means of copper, brass, or other approved corrosion-resistant metal that is integral with the box »*680.24(A)(1)(3)*«.

F Measured from the inside of the bottom of the box, the junction box shall be located at least 8 in. (200 mm) above the maximum pool water level »*680.24(A)(2)(a)*«. (Note: The bottom of the junction box must not be less than 4 in. (100 mm) above the ground level or pool deck.)

G The junction box shall be equipped with threaded entries/hubs or a nonmetallic hub »*680.24(A)(1)(1)*«.

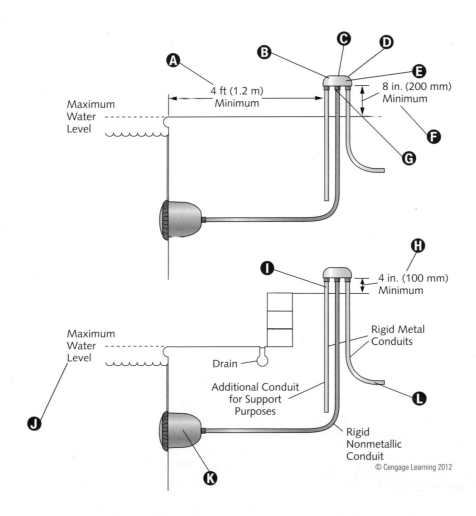

© Cengage Learning 2012

Junction Boxes for Underwater Lighting *(continued)*

H Measured from the inside of the bottom of the box, the junction box shall be located no less than 4 in. (100 mm) above the ground level, or pool deck ≫ *680.24(A)(2)(a)* ≪. (Note: The bottom of the junction box must not be less than 8 in. [200 mm] above the maximum pool water level.)

I A swimming pool junction box must comply with 314.23(E) support provisions. It must be supported by *two* or more conduits (rigid or intermediate metal conduit only) threaded wrenchtight into the enclosure or hubs. Secure each conduit within 18 in. (450 mm) of the enclosure since all entries are on the same side. Note: PVC is not permitted for swimming pool junction box support.

J Maximum water level is defined as the highest level that water can reach before it spills out ≫ *680.2* ≪.

> ### N O T E
> Branch-circuit wiring on the supply side of enclosures and junction boxes connected to conduits run to wet-niche and no-niche luminaires, and the field wiring compartments of dry-niche luminaires, shall be installed using RMC, IMC, LFNC, PVC conduit, or RTRC. Where installed on buildings, EMT shall be permitted, and where installed within buildings, ENT, Type MC cable, EMT, or Type AC cable shall be permitted. In all cases, an insulated equipment grounding conductor sized in accordance with Table 250.122 but not less than 12 AWG shall be required ≫***680.23(F)(1)***≪.
> (See exception for connecting to transformers.)

K Forming shells shall be installed for the mounting of all wet-niche underwater luminaires and shall be equipped with provisions for conduit entries. Metal parts of the luminaire and forming shell in contact with the pool water shall be of brass or other approved corrosion-resistant metal. All forming shells used with nonmetallic conduit systems, other than those that are part of a listed low-voltage lighting system not requiring grounding, shall include provisions for terminating an 8 AWG copper conductor ≫ *680.23(B)(1)* ≪.

L The equipment grounding conductor terminals of a junction box, transformer enclosure, or other enclosure in the supply circuit to a wet-niche or no-niche luminaire and the field-wiring chamber of a dry-niche luminaire shall be connected to the equipment grounding terminal of the panelboard. This terminal shall be directly connected to the panelboard enclosure ≫ *680.24(F)* ≪.

> **CAUTION**
>
> Junction boxes and enclosures mounted above the grade of the finished walkway around the pool shall not be located in the walkway unless afforded additional protection, such as by location under diving boards, adjacent to fixed structures, and the like ≫*680.24(C)*≪.

Equipment and Feeders

The following equipment shall be grounded in accordance with Parts V, VI, an VII of Article 250 and connected by wiring methods of Chapter 3, except as modified by Article 680:

1. Through-wall lighting assemblies and underwater luminaires, other than those low-voltage lighting products listed for the application without a grounding conductor
2. All electrical equipment located within 5 ft (1.5 m) of the inside wall of the specified body of water
3. All electrical equipment associated with the recirculating system of the specified body of water
4. Junction boxes
5. Transformer and power-supply enclosures
6. GFCIs
7. Panelboards (not part of the service equipment) that supply any electrical equipment associated with the specified body of water ≫ *680.6* ≪

A separate building's panelboard can supply swimming pool equipment if the feeder meets 250.32(B) grounding requirements. If an equipment grounding conductor is installed, it shall be an insulated conductor ≫ *680.25(B)(2)* ≪.

Branch circuits for pool-associated motors in the interior of dwelling units, or in the interior of accessory buildings associated with a dwelling unit, can be installed in any of Chapter 3's wiring methods where meeting the requirements of this section. Where run in a cable assembly, the equipment grounding conductor can be uninsulated, but it shall be enclosed within the cable assembly's outer sheath ≫ *680.21(A)(4)* ≪.

Electric equipment shall not be installed in rooms (or pits) that do not have drainage that prevents water accumulation during normal operation or filter maintenance ≫ *680.11* ≪.

A permanently installed pool can have listed cord- and plug-connected pool pumps that incorporate an approved double insulation system and provide a grounding means for the pump's internal and inaccessible, non-current-carrying metal parts. This listed cord- and plug-connected pool pump shall be connected to any wiring method recognized in Chapter 3

Equipment and Feeders *(continued)*

that is suitable for the location. **Where the bonding grid is connected to the equipment grounding conductor of the motor circuit in accordance with the second sentence of 680.26(B)(6)(a), the branch-circuit wiring shall comply with 680.21(A)** ≫ *680.21(B)* ≪ .

Ⓐ The branch circuits for pool-associated motors shall be installed in RMC, IMC, PVC, or Type MC cable listed for the location. Other wiring methods and materials are permitted in specific locations or applications as covered in this section. Any wiring method employed shall contain an insulated copper equipment grounding conductor sized per 250.122 but not smaller than 12 AWG ≫ *680.21(A)(1)* ≪ . Where installed on or within buildings, EMT is permitted ≫ *680.21(A)(2)* ≪ .

Ⓑ One or more means to simultaneously disconnect all ungrounded conductors shall be provided for all utilization equipment other than lighting. Each means shall be readily accessible and within sight from its equipment and shall be located at least 5 ft (1.5 m) horizontally from the inside walls of a pool, spa, fountain, or hot tub unless separated from the open water by a permanently installed barrier that provides a 5-ft (1.5-m) reach path or greater. This horizontal distance is

to be measured from the water's edge along the shortest path required to reach the disconnect ≫ *680.12* ≪ .

Ⓒ Section 680.25 provisions apply to any feeder on the supply side of panelboards supplying branch circuits for pool equipment covered in Part II and on the load side of the service equipment or the source of a separately derived system ≫ *680.25* ≪ . Feeders shall be installed in RMC or IMC. If not subject to physical damage, the following wiring methods shall be permitted: (1) LFNC, (2) PVC, (3) RTRC, (4) EMT where installed on or within a building, (5) ENT where installed within a building, and (6) Type MC cable where installed within a building and if not subject to corrosive environment ≫ *680.25(A)* ≪ .

Ⓓ Where connections must be flexible at (or adjacent to) the motor, LFMC or LFNC with approved fittings are permitted ≫ *680.21(A)(3)* ≪ .

WARNING

Outlets supplying pool pump motors connected to single-phase, 120-volt through 240-volt branch circuits, whether by receptacle or direct connection, shall be provided with GFCI protection for personnel ≫ *680.21(C)* ≪ .

N O T E

The only underground wiring permitted under the pool or within the area extending 5 ft (1.5 m) horizontally from the pool's inside wall is the wiring necessary to supply pool equipment (permitted by this article). Where space limitations prevent wiring outside the 5-ft (1.5-m) restricted area, such wiring is permitted if installed in complete raceway systems of RMC, IMC, or a nonmetallic raceway system. All metal conduit shall be corrosion resistant and suitable for the location ≫ **680.10** ≪ . Minimum burial depths are provided in Table 680.10.

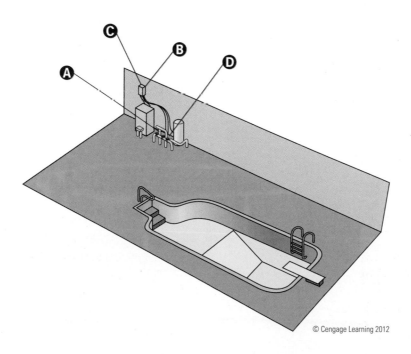

Storable Pools, Storable Spas, and Storable Hot Tubs

A luminaire installed in or on a storable pool's, storable spa's, or storable hot tub's wall shall be part of a cord- and plug-connected lighting assembly. This assembly shall be listed as an assembly for the purpose, and have the following construction features:

1. No exposed metal parts

2. A luminaire lamp that is suitable for use at the supplied voltage

3. An impact-resistant polymeric lens, luminaire body, and transformer enclosure

4. A transformer or power supply meeting 680.23(A)(2) requirements with a primary rating not over 150 volts ≫ *680.33(A)* ≪

A lighting assembly (without a transformer or power supply) whose luminaire lamp(s) operate at 150 volts or less can be cord- and plug-connected where the assembly is listed as an assembly for the purpose. The installation shall comply with 680.23(A)(5), and the assembly shall have the following construction features:

1. No exposed metal parts

2. An impact-resistant polymeric lens and luminaire body

3. A GFCI with open neutral conductor protection as an integral part of the assembly

4. A permanent connection to the GFCI with open-neutral protection

5. Compliance with 680.23(A) requirements ≫ *680.33(B)* ≪

A Storable swimming, wading, or immersion pool; or storable/portable spas and hot tubs: Those that are constructed on or above the ground and are capable of holding water to a maximum depth of 42 in. (1.0 m), or a pool, spa, or hot tub with nonmetallic, molded polymeric walls or inflatable fabric walls regardless of dimension ≫ *680.2* ≪.

B All 15- and 20-ampere, 125- and 250-volt receptacles installed in a wet location shall have an enclosure that is weatherproof whether or not the attachment plug cap is inserted. An outlet box hood installed for this purpose shall be listed, and shall be identified as "extra-duty." All 15- and 20-ampere, 125- and 250-volt nonlocking-type receptacles shall be listed weather-resistant type ≫ *406.9(B)(1)* ≪.

C A cord-connected pool filter pump shall incorporate an approved system of double insulation or its equivalent and shall be provided with means for grounding only the internal and nonaccessible non-current-carrying metal parts of the appliance. The means for grounding shall be an equipment grounding conductor run with the power-supply conductors in the flexible cord that is properly terminated in a grounding-type attachment plug having a fixed grounding contact member. Cord-connected pool filter pumps shall be provided with a GFCI that is an integral part of the attachment plug or located in the power-supply cord within 12 in. (300 mm) of the attachment plug ≫ *680.31* ≪.

D Receptacles shall not be located less than 6 ft (1.83 m) from the inside walls of a storable pool, storable spa, or storable hot tub. In determining these dimensions, the distance to be measured shall be the shortest path the supply cord of an appliance connected to the receptacle would follow without piercing a floor, wall, ceiling, doorway with hinged or sliding door, window opening, or other effective permanent barrier ≫ *680.34* ≪.

CAUTION

All electrical equipment, including power-supply cords, used with storable pools shall be protected by ground-fault circuit interrupters. All 125-volt, 15- and 20-ampere receptacles located within 20 ft (6.0 m) of the inside walls of a storable pool, storable spa, or storable hot tub shall be protected by a GFCI. In determining these dimensions, the distance to be measured shall be the shortest path the supply cord of an appliance connected to the receptacle would follow without piercing a floor, wall, ceiling, doorway with hinged or sliding door, window opening, or other effective permanent barrier ≫ *680.32* ≪. If flexible cords are used, see 400.4 ≫ *680.32 Informational Note* ≪.

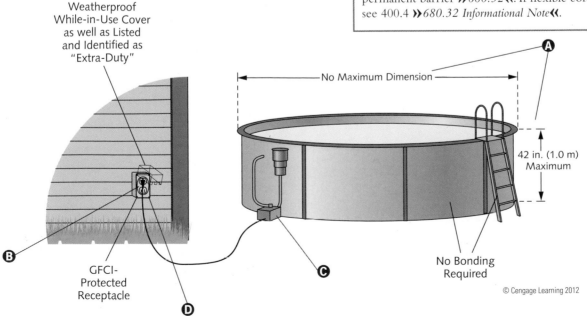

Weatherproof While-in-Use Cover as well as Listed and Identified as "Extra-Duty"

No Maximum Dimension

42 in. (1.0 m) Maximum

No Bonding Required

B

GFCI-Protected Receptacle

C

D

A

© Cengage Learning 2012

Spas and Hot Tubs

Receptacles providing spa or hot tub power shall be GFCI protected »*680.43(A)(3)*«.

An indoor spa or hot tub shall comply with Article 680 provisions (Parts I and II) unless modified by 680.43 and shall be connected by a Chapter 3–approved wiring method. Listed spa and hot tub packaged units rated 20 amperes or less can be cord- and plug-connected to facilitate the removal or disconnection of the unit for maintenance and repair »*680.43 and Exception No. 1*«.

The equipotential bonding requirements for perimeter surfaces in 680.26(B)(2) shall not apply to a listed self-contained spa or hot tub installed above a finished floor. For a dwelling unit(s) only, where a listed spa or hot tub is installed indoors, the wiring method requirements of 680.42(C) shall also apply »*680.43 Exception No. 2 and No. 3*«.

All of the following shall be grounded: (1) electric equipment located within 5 ft (1.5 m) of the spa or hot tub's inside wall and (2) electric equipment associated with the spa or hot tub's circulating system »*680.43(F)*«.

Ⓐ Luminaires, lighting outlets, and ceiling-suspended (paddle) fans located over the spa/hot tub less than 5 ft (1.5 m) from the spa or hot tub's inside walls shall be a minimum of 7 ft 6 in. (2.3 m) above the maximum water level and shall be GFCI protected unless the mounting height is at least 12 ft (3.7 m) »*680.43(B)(1)(b)*«.

Ⓑ Luminaires meeting either (1) or (2) of the following requirements *and* protected by a GFCI can be installed less than 7 ft 6 in. (2.3 m) over a spa or hot tub:

1. Recessed luminaires having a glass or plastic lens and nonmetallic or electrically isolated metal trim, suitable for damp location use

2. Surface-mounted luminaires, with a glass or plastic globe, and a nonmetallic body or a metallic body isolated from contact, and suitable for use in damp locations »*680.43(B)(1)(c)*«.

Ⓒ Luminaires, lighting outlets, and ceiling-suspended (paddle) fans located 12 ft (3.7 m) or more above the maximum water level do not require GFCI protection »*680.43(B)(1)(a)*«.

Ⓓ Switches shall be located at least 5 ft (1.5 m) measured horizontally from the spa or hot tub's inside walls »*680.43(C)*«.

Ⓔ A readily accessible, clearly labeled emergency shutoff or control switch for stopping the motor(s) that provide recirculating and jet system power shall be installed at least 5 ft (1.5 m) away, adjacent to, and within sight from the spa or hot tub. Note: This requirement does not apply to single-family dwellings »*680.41*«.

Ⓕ Receptacles of 125 volts and 30 amperes or less and within 10 ft (3.0 m) of the spa or hot tub's inside walls shall be protected by a GFCI »*680.43(A)(2)*«. Determine this distance by measuring the shortest path the supply cord of an appliance connected to the receptacle would follow without piercing a building's floor, wall, ceiling, doorway with hinged or sliding door, window opening, or other effective permanent barrier »*680.43(A)(4)*«.

Ⓖ At least one 125-volt, 15- or 20-ampere general-purpose branch-circuit receptacle shall be located a minimum of 6 ft (1.83 m), and no more than 10 ft (3.0 m), from the inside wall of an indoor spa or hot tub »*680.43(A)*«.

Ⓗ Locate property receptacles at least 6 ft (1.83 m) measured horizontally from the spa or hot tub's inside walls »*680.43(A)(1)*«.

> **N O T E**
>
> A spa or hot tub installed **outdoors** shall comply with Article 680, Part I and II provisions unless otherwise permitted in the following: (A) listed packaged units utilizing a factory-installed assembly control panel or panelboard can be connected with 6 ft (1.8 m) or less of liquidtight flexible conduit external to the spa or hot tub enclosure in addition to the length needed within the enclosure to make the electrical connection, or can be cord- and plug-connected with a cord no longer than 15 ft (4.6 m), if protected by a GFCI »*680.42(A)*«.

GFCI Not Required

12 ft (3.7 m)

GFCI Required

Nonmetallic Luminaire GFCI Protected

7¼ ft (2.3 m)

Emergency Shutoff

6 ft (1.83 m)

GFCI Protected

Maximum Water Level

© Cengage Learning 2012

Spa or Hot Tub Installed Indoors

Spas and Hot Tubs *(continued)*

NOTE

Bonding by metal-to-metal mounting on a common frame or base shall be permitted. The metal bands or hoops used to secure wooden staves shall not be required to be bonded as required in 680.26. Equipotential bonding of perimeter surfaces in accordance with 680.26(B)(2) shall not be required to be provided for spas and hot tubs where all of the following conditions apply:

(1)   The spa or hot tub shall be listed as a self-contained spa for aboveground use.
(2)   The spa or hot tub shall not be identified as suitable only for indoor use.
(3)   The installation shall be in accordance with the manufacturer's instructions and shall be located on or above grade.
(4)   The top rim of the spa or hot tub shall be at least 28 in. (710 mm) above all perimeter surfaces that are within 30 in. (760 mm), measured horizontally from the spa or hot tub. The height of nonconductive external steps for entry to or exit from the self-contained spa shall not be used to reduce or increase this rim height measurement ≫*680.42(B)*≪.

Pools and Tubs for Therapeutic Use

Outlet(s) supplying the following shall be protected by a GFCI: (1) self-contained therapeutic tub or hydrotherapeutic tank; (2) packaged therapeutic tub or hydrotherapeutic tank; or (3) field-assembled therapeutic tub or hydrotherapeutic tank ≫*680.62(A)*≪.

A listed self-contained unit or listed packaged equipment assembly marked to indicate that integral GFCI protection is provided for all electrical parts within the unit or assembly (pumps, air-blowers, heaters, lights, controls, sanitizer generators, wiring, etc.), does not require that the outlet supply be GFCI protected ≫*680.62(A)(1)*≪.

A therapeutic tub or hydrotherapeutic tank rated greater than 250 volts or rated 3-phase or with a heater load exceeding 50 amperes does not require supply protection by means of a GFCI ≫*680.62(A)(2)*≪.

All receptacles within 6 ft (1.83 m) of a therapeutic tub shall be protected by a GFCI ≫*680.62(E)*≪.

Ⓐ Article 680, Part I and Part VI provisions apply to therapeutically used pools and tubs in health care facilities, gymnasiums, athletic training rooms, and similar areas. Portable therapeutic appliances shall comply with Parts II and III of Article 422 ≫*680.60*≪.

Ⓑ All tub-associated metal parts shall be bonded by any of the following methods: (1) interconnection of threaded metal piping and fittings; (2) metal-to-metal mounting on a common frame or base; (3) suitable metal clamp connections; or (4) the provision of a solid copper bonding jumper (insulated, covered, or bare) no smaller than 8 AWG ≫*680.62(C)*≪.

Ⓒ Tubs used for the therapeutic submersion and treatment of patients, which during normal use are not easily moved from place to place, or which are fastened or secured at a given location (including associated piping systems), shall conform to Part VI of Article 680 ≫*680.62*≪.

Ⓓ All of the following shall be connected to an equipment grounding conductor: (1) electrical equipment located within 5 ft (1.5 m) of the tub's inside wall and (2) electrical equipment associated with the tub's circulating system ≫*680.62(D)(1)*≪.

Ⓔ The following parts shall be bonded together:

1.   All metal fittings within, or attached to, the tub structure.

2.   Metal parts of electrical equipment associated with the tub's water circulating system, including pump motors.

3.   Metal-sheathed cables and raceways and metal piping within 5 ft (1.5 m) of the tub's inside walls and not separated from the tub by a permanent barrier.

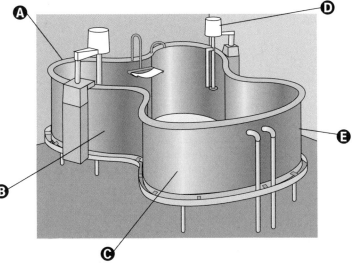

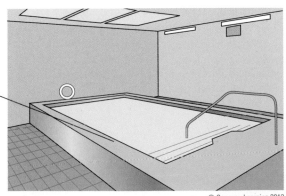

© Cengage Learning 2012

Pools and Tubs for Therapeutic Use *(continued)*

4. All metal surfaces within 5 ft (1.5 m) of the tub's inside walls not separated by a permanent barrier.

5. Electrical devices and controls unrelated to the therapeutic tubs shall either be located at least 5 ft (1.5 m) away from such units, or be bonded to the therapeutic tub system 》 *680.62(B)* 《.

F Therapeutic pools constructed in or on the ground or within a building in such a manner that the pool cannot be

> **CAUTION**
>
> All luminaires used in therapeutic tub areas shall be of the totally enclosed type 》*680.62(F)*《.

Hydromassage Bathtubs

Hydromassage bathtubs as defined in 680.2 shall comply with Part VII of this article. They shall not be required to comply with other parts of this article 》 *680.70* 《**.**

A All bathroom receptacles shall have GFCI protection for personnel 》 *210.8(A)(1) and (B)(1)* 《.

B All 125-volt, single-phase receptacles not exceeding 30 amperes and located within 6 ft (1.83 m) measured horizontally of the inside walls of a hydromassage tub shall be protected by a ground-fault circuit interrupter 》 *680.71* 《.

C Hydromassage bathtubs and their associated electrical components shall be on an individual branch circuit(s) and protected by a readily accessible ground-fault circuit interrupter 》 *680.71* 《.

D Both metal piping systems and grounded metal parts in contact with the circulating water shall be bonded together using a solid copper bonding jumper, insulated, covered, or bare, not smaller than 8 AWG. The bonding jumper shall be connected to the terminal on the circulating pump motor that is intended for this purpose. The bonding jumper shall not be required to be connected to a double-insulated circulating pump motor. The 8 AWG or larger solid copper bonding jumper shall be required for equipotential bonding in the area of the hydromassage bathtub and shall not be required to be extended or attached to any remote panelboard, service equipment, or any electrode. The 8 AWG or larger solid copper bonding jumper shall be long enough to terminate on a replacement non-double-insulated pump motor and shall be terminated to the equipment grounding

> **NOTE**
>
> Luminaires, switches, receptacles, and other electrical equipment located in the same room, and not directly associated with a hydromassage bathtub, shall be installed in accordance with the requirements of Chapters 1 through 4 in the *NEC* covering the installation of that equipment in bathrooms 》*680.72*《.

readily disassembled shall comply with Article 680, Parts I and II 》 *680.61* 《. The limitations of 680.22(B)(1) through (C)(4) do not apply if all luminaires are of the totally enclosed type 》 *680.61 Exception* 《.

> **NOTE**
>
> Small conductive surfaces not likely to become energized, such as air and water jets and drain fittings not connected to metallic piping, and towel bars, mirror frames, and similar nonelectrical equipment not connected to metal framing, shall not be required to be bonded 》*680.62(B) Exception*《.

conductor of the branch circuit of the motor when a double-insulated circulating pump motor is used 》 *680.74* 《.

E Hydromassage bathtub electrical equipment shall be accessible without damage to the building structure or finish. Where the hydromassage bathtub is cord- and plug-connected with the supply receptacle accessible only through a service access opening, the receptacle shall be installed so that its face is within direct view and not more than 1 ft (300 mm) of the opening 》 *680.73* 《.

> **CAUTION**
>
> A receptacle shall not be installed within a bathtub or shower space 》*406.9(C)*《.

> **WARNING**
>
> The GFCI protecting the hydromassage bathtub branch circuit shall be readily accessible 》 *680.71* 《.

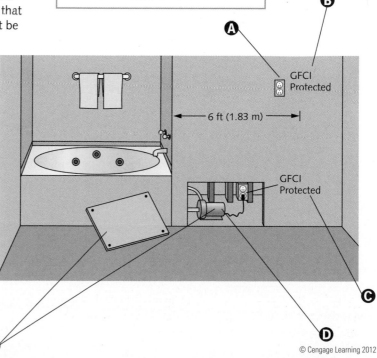

GFCI Protected

6 ft (1.83 m)

GFCI Protected

© Cengage Learning 2012

Summary

- Article 600 covers the conductor and equipment installation for electric signs and outline lighting (including neon and skeleton tubing).
- Manufactured wiring system provisions are located in Article 604.
- Article 605 ("Office Furnishings") covers electrical equipment, lighting accessories, and wiring systems used to connect, or contained within, or installed on office furnishings.
- Electric vehicle charging system provisions are located in Article 625.
- Audio signal processing, amplification, and reproduction equipment requirements can be found in Article 640.
- Article 645 covers information technology equipment.
- Electrically operated pipe organ circuits and components employed for sounding apparatus and keyboard control are located in Article 650.
- Article 695 ("Fire Pumps") covers the installation of electric power sources and interconnecting circuits as well as switching and control equipment dedicated to fire pump drivers.
- Elevator, dumbwaiter, escalator, moving walk, platform lift, and stairway chairlift provisions are contained in Article 620.

- Article 680 provisions apply to the construction and installation of electrical wiring and equipment in or near all swimming, wading, therapeutic, and decorative pools, fountains, hot tubs, spas, and hydromassage bathtubs, whether permanently installed or storable, as well as to metallic auxiliary equipment (such as pumps, filters, etc.).
- The parts specified in 680.26(B)(1) through (B)(7) shall be bonded together using solid copper conductors, insulated, covered, or bare, not smaller than 8 AWG or with rigid metal conduit of brass or other identified corrosion-resistant metal.
- Very specific and detailed receptacle and lighting (including underwater) requirements are stipulated in Article 680.
- While Article 680, Part II covers permanently installed pools, requirements for storable pools, storable spas, and storable hot tubs are found in Part III.
- Indoor spa and hot tub installations are covered in 680.43(A) through (G).
- Article 680, Part VI provisions apply to pools and tubs for therapeutic use in health care facilities, gymnasiums, athletic training rooms, and similar areas.
- Hydromassage bathtubs (and associated electrical components) are covered in Article 680, Part VII.

Unit 19 Competency Test

NEC Reference	Answer	
_____	_____	1. Storable swimming pools are capable of holding water to a maximum depth of _____ in.
_____	_____	2. Overcurrent protection for feeders and branch circuits supplying electric vehicle supply equipment shall be sized for _____ duty and shall have a rating of not less than _____% of the maximum load of the electric vehicle supply equipment.
_____	_____	3. A clearly labeled emergency shutoff or control switch for the purpose of stopping the motor(s) that provide power to the recirculation system and jet system shall be installed at a point readily accessible to the users and at least _____ ft away, adjacent to, and within sight from the spa or hot tub.
_____	_____	4. Transformers and power supplies used for the supply of underwater luminaires shall incorporate either a transformer of the isolated winding type with an ungrounded secondary that has a _____ between the primary and secondary windings.
_____	_____	5. Duty on elevator driving machine motors shall be rated as _____.
_____	_____	6. Not more than _____ ft of high-voltage cable shall be permitted in nonmetallic conduit from a high-voltage terminal of a neon transformer (over 1000 volts, nominal) supply to the first neon tube.
_____	_____	7. All fixed metal parts that are within _____ ft horizontally of the inside walls of a permanently installed swimming pool and within _____ ft above the pool's maximum water level shall be bonded.

8. Electrically operated pipe organ circuits shall be arranged so that 26 and 28 AWG conductors shall be protected by an overcurrent device rated at not more than _____.

9. All live parts of electrical apparatus in the hoistways, at the landings, in or on the cars of elevators and dumbwaiters, in the wellways or the landings of escalators or moving walkways, or in the runways and machinery spaces of platform lifts and stairway chairlifts shall be _____.

10. Unless specifically listed and marked for the location, the coupling means of the electric vehicle supply equipment shall be stored or located at a height of not less than _____ in. above the floor level for indoor locations and _____ in. above the grade level for outdoor locations.

11. A(n) _____ is a structure designed to support a wet-niche luminaire assembly and intended for mounting in a pool or fountain structure.

12. Switches, flashers, and similar devices controlling transformers and electronic power supplies shall be rated for controlling inductive loads or have a current rating not less than _____ of the current rating of the transformer.

    a) 80%   b) 100%   c) 125%   d) 200%

13. Audio system equipment supplied by branch-circuit power shall not be placed horizontally within _____ ft of the inside wall of a pool, spa, hot tub, or fountain, nor within _____ ft of the prevailing or tidal high water mark.

14. Where a permanently installed pool is installed, at least one 125-volt, 15- or 20-ampere receptacle on a general-purpose branch circuit shall be located a minimum of _____ ft from and not more than _____ ft from the inside wall of the pool.

15. The voltage at a fire pump controller's line terminals shall not drop more than _____ below normal (controller-rated voltage) under motor starting conditions.

16. Luminaires located over a spa or within 5 ft (1.5 m) from the inside walls of the spa shall be permitted to be not less than _____ ft above the maximum water level where protected by a GFCI.

17. Traveling cables that are suitably supported and protected from physical damage shall be permitted to be run without the use of a raceway from inside the hoistway, to elevator controller enclosures and to elevator car and machine room, control room, machinery space, and control space connections that are located outside the hoistway for a distance not exceeding _____ ft in length as measured from the first point of support on the elevator car or hoistway wall, or counterweight where applicable, provided the conductors are grouped together and taped or corded, or in the original sheath.

18. Underground wiring shall not be permitted under the pool or within the area extending _____ ft horizontally from the inside wall of the pool unless this wiring is necessary to supply pool equipment permitted by Article 680.

19. Hydromassage bathtub electrical equipment shall be _____ without damaging the building structure or building finish.

20. All 15- and 20-ampere, single-phase 125-volt receptacles located within _____ ft of the inside wall of the pool shall be protected by a GFCI.

21. Manufactured wiring system cable shall be listed Type AC cable or listed Type MC cable containing nominal 600-volt _____ AWG copper-insulated conductors with a bare or insulated copper equipment grounding conductor equivalent in size to the ungrounded conductor.

NEC Reference   Answer

_____   _____ 22. Luminaires mounted in permanently installed swimming pool walls shall be installed with the top of the luminaire lens at least _____ in. below the normal water level of the pool.

_____   _____ 23. Sign or outline lighting system equipment shall be at least _____ ft above areas accessible to vehicles unless protected from physical damage.

_____   _____ 24. Existing luminaires and lighting outlets located less than _____ ft measured horizontally from the inside walls of a pool shall be at least _____ ft above the surface of the maximum water level, shall be rigidly attached to the existing structure, and shall be protected by a GFCI.

_____   _____ 25. A fixed or stationary electric sign installed inside a fountain shall be at least _____ ft inside the fountain measured from the outside edges of the fountain.

_____   _____ 26. The conductors to the hoistway door interlocks from the hoistway riser shall be flame retardant and suitable for a temperature of not less than _____.

_____   _____ 27. In new outdoor pool areas, ceiling-suspended (paddle) fans shall not be installed above the pool or over the area extending _____ ft horizontally from the inside walls of the pool unless no part of the fan is less than _____ ft above the maximum water level.

_____   _____ 28. The branch-circuit conductors supplying one or more units of information technology equipment shall have an ampacity not less than _____ of the total connected load.

_____   _____ 29. A junction box connected to a conduit that extends directly to a forming shell shall be provided with a number of grounding terminals that shall be _____.

_____   _____ 30. Electric vehicle supply equipment shall be provided with a(n) _____ that de-energizes the electric vehicle connector whenever the electric connector is uncoupled from the electric vehicle.

_____   _____ 31. Branch circuits that supply neon tubing installations shall not be rated in excess of _____.

_____   _____ 32. The termination of the 8 AWG bonding jumper in a permanently installed swimming pool forming shell shall be covered with or encapsulated in a(n) _____ to protect the connection from the possible deteriorating effect of pool water.

_____   _____ 33. Conductors supplying a fire pump motor(s), pressure maintenance pumps, and associated fire pump accessory equipment shall have a rating not less than _____ of the sum of the fire pump motor(s) and pressure maintenance motor(s) full-load current(s), and _____ of the associated fire pump accessory equipment.

_____   _____ 34. Switching devices shall be located at least _____ ft horizontally from the inside walls of a pool unless separated from the pool by a solid fence, wall, or other permanent barrier.

_____   _____ 35. Article _____ covers electrical equipment, lighting accessories, and wiring systems used to connect, contained within, or installed on office furnishings.

_____   _____ 36. A junction box connected to a conduit that extends directly to a forming shell shall be located not less than _____ in., measured from the inside of the bottom of the box, above the ground level, or pool deck, or not less than _____ in. above the maximum pool water level, whichever provides the greater elevation.

Index